CHARLES PROTEUS
STEINMETZ

Lectures on Electrical Engineering

EDITED BY PHILIP L. ALGER
CONSULTING PROFESSOR OF ELECTRICAL ENGINEERING
RENSSELAER POLYTECHNIC INSTITUTE

IN
THREE VOLUMES

I

Elements of Electrical Engineering

DOVER PUBLICATIONS, INC.
NEW YORK

Published in Canada by General Publishing Company, Ltd., 30 Lesmill Road, Don Mills, Toronto, Ontario.

Published in the United Kingdom by Constable and Company, Ltd., 10 Orange Street, London WC 2.

This Dover edition, first published in 1971, is a reprinting, in three volumes, of selected portions of the Steinmetz Electrical Engineering Library, originally published between 1915 and 1920 in nine volumes by the McGraw-Hill Book Company, Inc. The correspondence between the present edition and the original set is indicated on page vi of the General Preface.

International Standard Book Number: 0-486-62514-1
Library of Congress Catalog Card Number: 70-137004

Manufactured in the United States of America
Dover Publications, Inc.
180 Varick Street
New York, N. Y. 10014

GENERAL PREFACE

Steinmetz is well described by his adopted middle name, Proteus, for the range of his interests and accomplishments was vast. To his friends he was known simply as "the Doctor," but by others he was referred to variously as the wizard of Schenectady, forger of thunderbolts, and latter-day Vulcan. Harvard University called him "the foremost expert in applied electricity of this country," and he was a prolific writer on both technical and general subjects.

But it is as a teacher that he is best known, and with good reason. Steinmetz had a remarkable ability to explain things. He liked to talk with children, telling them why the stars twinkle, what makes the wind blow, and what makes flowers bloom. He was able to explain engineering problems in the same way, clearly and simply; he would resolve them into distinct, readily understood elements, and then solve them step by step, using approximations freely in order to emphasize method rather than detailed calculation.

In Steinmetz's day alternating-current phenomena were considered very mysterious, and there were very few textbooks available to explain them. Thus, a whole generation of engineers eagerly read the nine volumes of his Union College lectures and monographs published by McGraw-Hill, some of which went into five editions. Steinmetz realized that many engineers did not have the time or inclination for a rigorous mathematical presentation, and therefore endeavored in this series "to give a descriptive exposition of the physical nature and meaning, the origin and effects, of these phenomena, with the use of very little and only the simplest form of mathematics."

In view of the remarkable clarity of Steinmetz's explanations, and the breadth of his treatment of fundamental concepts, his writings of fifty years ago still offer one of the best ways of getting a basic electrical engineering education. They are also of historical interest: Steinmetz's use of the complex number $a + jb$ for alternating-current circuit calculations was just one of many unique contributions to electrical engineering recorded in these books.

This Dover edition of the lectures, in three volumes, includes all the material still of use to present-day students of engineering, with the exception of Volume 2 of the McGraw-Hill set, which is already available in both hardcover and paperback in a new edition revised by the present writer (*Mathematics for Science and Engineering*, by Philip L. Alger, McGraw-Hill, Second Edition, 1969). The relationship between the new Dover edition and the 9-volume McGraw-Hill edition of the lectures is:

Dover Edition	McGraw-Hill Edition
Volume I: *Elements of Electrical Engineering*	Volume 3, pp. 1–120; Volume 4, pp. 1–206; Volume 5, pp. 1–228
Volume II: *Electric Waves and Impulses*	Volume 4, pp. 341–465; Volume 7, complete; Volume 8, pp. 3–219; Volume 4, pp. 466–473 (Appendix)
Volume III: *Transient Electric Phenomena*	Volume 8, pp. 223–683

Omitted from the Dover edition are Volume 1, which consists of general lectures too diffuse to be of present-day interest, Volumes 6 and 9, and parts of the other volumes which deal with rotating machines, with more complex theories, or with apparatus now obsolete.

These lectures present concepts usually studied by electrical engineers in the first three years of college. They should be readily understood by any intelligent technician or engineering apprentice, so that by his own efforts he can gain a working knowledge of electrical engineering.

Some necessary changes have been made in the original text: a number of misprints have been silently emended, and cross references have been altered so as to be useful to the reader of these three new volumes. Where footnotes have been added they are differentiated from Steinmetz's own by the use of brackets.

Steinmetz occasionally refers the reader to his book *Engineering Mathematics* for a more detailed mathematical explanation than that given in the text. As mentioned above, this book is now available in a revised edition.

The reader should take note of the few changes in terminology that have occurred since Steinmetz's day:

1. Condensers are now called *capacitors*, condensance is

called *capacitance*, and condensive reactance is known as *capacitive reactance.*

2. A vector which indicates the phase position of a sinusoidally varying quantity is now known as a *phasor*; Steinmetz refers to these simply as vectors.

3. Wave screens are now called *wave filters.*

Philip L. Alger

Schenectady, New York
March, 1969

The books listed below will be useful as supplements to this work.

Alger, Philip L.: *Mathematics for Science and Engineering* (Based on *Engineering Mathematics* by C. P. Steinmetz), Second Edition; McGraw-Hill, 1969.

———: *The Nature of Induction Machines*; Gordon and Breach Science Publishers, 1965.

Bewley, L. V.: Alternating Current Machinery; The Macmillan Co., 1949.

Clement, P. R. and W. C. Johnson: *Electrical Engineering Science*; McGraw-Hill, 1960.

Concordia, C.: *Synchronous Machines*; John Wiley, 1951.

Dow, W. G.: *Fundamentals of Engineering Electronics*; John Wiley, 1952.

Fano, Chu, and Adler: *Electromagnetic Fields, Energy and Forces*; John Wiley, 1960.

Fitzgerald, A. E. and C. Kingsley, Jr.: *Electric Machinery*, Second Edition; McGraw-Hill, 1961.

Guillemin, E. A.: *Introductory Circuit Theory*; John Wiley, 1953.

Ham, J. M. and G. R. Slemon: *Scientific Basis of Electrical Engineering*; John Wiley, 1961.

Ku, Y. H.: *Electric Energy Conversion*; Ronald Press, 1959.

Lewis, W. W. and C. F. Goodheart: *Basic Electric Circuit Theory*; Ronald Press, 1958.

Majmudar, Harit: *Electromechanical Energy Converters*; Allyn and Bacon, 1965.

Messerle, H. K.: *Dynamic Circuit Theory*; Pergamon Press, 1965.

Scott, W. T.: *The Physics of Electricity and Magnetism*; John Wiley, 1959.

Slemon, G. R.: *Magnetoelectric Devices*; John Wiley, 1966.

Veinott, C. G.: *Fractional Horsepower Electric Motors*, Second Edition; McGraw-Hill, 1948.

Zimmerman, H. J. and S. J. Mason: *Electric Circuit Theory*; John Wiley, 1959.

The following books provide more information about the life and thought of C. P. Steinmetz.

Alger, Philip L. and E. Caldecott: *Steinmetz—The Philosopher*; Mohawk Development Service, Inc., P.O. Box 7, Bellevue Station, Schenectady, New York 12306.

Miller, Floyd: *The Electrical Genius of Liberty Hall*; McGraw-Hill, 1962.

Miller, J. A.: *Modern Jupiter*; The American Society of Mechanical Engineers, 1958.

PREFACE TO VOLUME I

In this volume are included substantially all of Steinmetz's writings on magnetism, dielectrics, electric circuits, transformers, and the materials used in electrical engineering. Several of these topics, particularly magnetism and transmission lines, are first considered in an elementary way, and then reconsidered in greater length and depth in later chapters. The sequence of chapters is in accord with Steinmetz's notion that the best way to learn a subject is first to make a general survey, to see the scope of the subject and its relation to other things already known, and then to make a more detailed study, filling in the fine points by going over the ground repeatedly.

The first part of the book is devoted to general theory at an elementary level, covering the whole range of magnetism, simple d-c and a-c circuits, transmission lines, fields of force, and transformers, with numerous examples worked out to illustrate each principle. The second part covers much of the same ground in more detail, with special emphasis on the use of *general numbers* (complex numbers, $a + jb$), in the solution of a-c circuits. The phenomena of hysteresis, eddy currents (skin effect), dielectric losses and corona, and the distributed L and C constants of transmission lines are clearly explained. The third part is devoted to transformers—their nature, circuit calculations, performance, and reactances. In the fourth part, the phenomena of electric conduction in solids, liquids, and gases are considered, with particular emphasis on the instability of arc circuits (singing arcs). Additional aspects of magnetism and hysteresis are treated, and simple formulas are given for magnetic forces, particularly those due to the high short-circuit currents in transformers.

Philip L. Alger

Schenectady, New York
March, 1969

CONTENTS FOR VOLUME I

PART I
GENERAL THEORY

PART II
ALTERNATING-CURRENT PHENOMENA

Chapter II. Instantaneous Values and Integral Values

Chapter III. Law of Electromagnetic Induction

Chapter IV. Vector Representation

Chapter V. Symbolic Method

Chapter VI. Topographic Method

CHAPTER VII. POLAR COORDINATES AND POLAR DIAGRAMS

Circuits

CHAPTER VIII. ADMITTANCE, CONDUCTANCE, SUSCEPTANCE

CHAPTER IX. CIRCUITS CONTAINING RESISTANCE, INDUCTANCE, AND CAPACITY

CHAPTER X. RESISTANCE AND REACTANCE OF TRANSMISSION LINES

Chapter XVI. Power, and Double-Frequency Quantities in General

Induction Apparatus

Chapter XVII. The Alternating-Current Transformer

PART III
THEORY OF ELECTRIC CIRCUITS

Chapter I. Electric Conduction; Solid and Liquid Conductors

PART I

GENERAL THEORY

1. MAGNETISM AND ELECTRIC CURRENT

1. A magnet pole attracting (or repelling) another magnet pole of equal strength at unit distance with unit force[1] is called a *unit magnet pole.*

The space surrounding a magnet pole is called a *magnetic field of force*, or *magnetic field.*

The magnetic field at unit distance from a unit magnet pole is called a *unit magnetic field*, and is represented by one line of magnetic force (or shortly "one line") per square centimeter, and from a unit magnet pole thus issue a total of 4π lines of magnetic force.

The total number of lines of force issuing from a magnet pole is called its *magnetic flux.*

The magnetic flux Φ of a magnet pole of strength m is,

$$\Phi = 4\pi m.$$

At the distance l from a magnet pole of strength m, and therefore of flux $\Phi = 4\pi m$, assuming a uniform distribution in all directions, the magnetic field has the intensity,

$$H = \frac{\Phi}{4\pi l^2} = \frac{m}{l^2}.$$

since the Φ lines issuing from the pole distribute over the area of a sphere of radius l, that is, the area $4\pi l^2$.

A magnetic field of intensity H exerts upon a magnet pole of strength m the force,

$$mH.$$

Thus two magnet poles of strengths m_1 and m_2, and distance l from each other, exert upon each other the force,

$$\frac{m_1 m_2}{l^2}.$$

[1] That is, at 1 cm. distance with such force as to give to the mass of 1 gram the acceleration of 1 cm. per second.

2. Electric currents produce magnetic fields also; that is, the space surrounding the conductor carrying an electric current is a magnetic field, which appears and disappears and varies with the current producing it, and is indeed an essential part of the phenomenon called an electric current.

Thus an electric current represents a *magnetomotive force* (m.m.f.).

The magnetic field of a straight conductor, whose return conductor is so far distant as not to affect the field, consists of lines of force surrounding the conductor in concentric circles. The intensity of this magnetic field is directly proportional to the current strength and inversely proportional to the distance from the conductor.

Since the lines of force of the magnetic field produced by an electric current return into themselves, the magnetic field is a *magnetic circuit.* Since an electric current, at least a steady current, can exist only in a closed circuit, electricity flows in an *electric circuit.* The magnetic circuit produced by an electric current surrounds the electric circuit through which the electricity flows, and inversely. That is, the electric circuit and the magnetic circuit are *interlinked* with each other.

Unit current in an electric circuit is the current which produces in a magnetic circuit of unit length the field intensity 4π, that is, produces as many lines of force per square centimeter as issue from a unit magnet pole.

At unit distance from an electric conductor carrying unit current, that is, in a magnetic circuit of length 2π, the field intensity is $\frac{4\pi}{2\pi} = 2$, and at the distance 2 the field intensity is unity; that is, unit current is the current which, in a straight conductor, whose return conductor is so far distant as not to affect its magnetic field, produces field intensity 2 at unit distance from the conductor.

One-tenth of unit current is the practical unit, called *one ampere.*

3. One ampere in an electric circuit or turn, that is, one ampere-turn, thus produces in a magnetic circuit of unit length the field intensity 0.4π, and in a magnetic circuit of length l the field intensity $\frac{0.4\pi}{l}$, and F ampere-turns produce in a magnetic circuit of length l the field intensity:

$$H = \frac{0.4\pi F}{l} \text{ lines of force per sq. cm.}$$

regardless whether the F ampere-turns are due to F amperes in a single turn, or 1 amp. in F turns, or $\frac{F}{n}$ amperes in n turns.

F, that is, the product of amperes and turns, is called *magnetomotive force* (m.m.f.).

The m.m.f. per unit length of magnetic circuit, or ratio,

$$f = \frac{\text{m.m.f.}}{\text{length of magnetic circuit}}$$

is called the *magnetizing force*, or *magnetic gradient.*

Hence, m.m.f. is expressed in *ampere-turns;* magnetizing force in *ampere-turns per centimeter* (or in practice frequently ampere-turns per inch), field intensity in lines of magnetic force per square centimeter.

At the distance l from the conductor of a loop or circuit of F ampere-turns, whose return conductor is so far distant as not to affect the field, assuming the m.m.f. $= F$, since the length of the magnetic circuit $= 2\,\pi l$, we obtain as the magnetizing force,

$$f = \frac{F}{2\,\pi l},$$

and as the field intensity,

$$H = 0.4\,\pi f = \frac{0.2\,F}{l}.$$

4. The magnetic field of an electric circuit consisting of two parallel conductors (or any number of conductors, in a polyphase system), as the two wires of a transmission line, can be considered as the superposition of the separate fields of the conductors (consisting of concentric circles). Thus, if there are I amperes in a circuit consisting of two parallel conductors (conductor and return conductor), at the distance l_1 from the first and l_2 from the second conductor, the respective field intensities are,

$$H_1 = \frac{0.2\,I}{l_1},$$

and

$$H_2 = \frac{0.2\,I}{l_2},$$

and the resultant field intensity, if $\tau =$ angle between the directions of the two fields,

$$H = \sqrt{H_1^2 + H_2^2 + 2\,H_1 H_2 \cos\tau},$$

$$= \frac{0.2\,I}{l_1 l_2}\sqrt{l_1^2 + l_2^2 + 2\,l_1 l_2 \cos\tau}.$$

In the plane of the conductors, where the two fields are in the same or opposite direction, the resultant field intensity is,

$$H = \frac{0.2\,I\,(l_1 \pm l_2)}{l_1 l_2},$$

where the plus sign applies to the space between, the minus sign the space outside of the conductors.

The resultant field of a circuit of parallel conductors consists of excentric circles, interlinked with the conductors, and crowded together in the space between the conductors as shown in Fig. 1 by drawn lines.

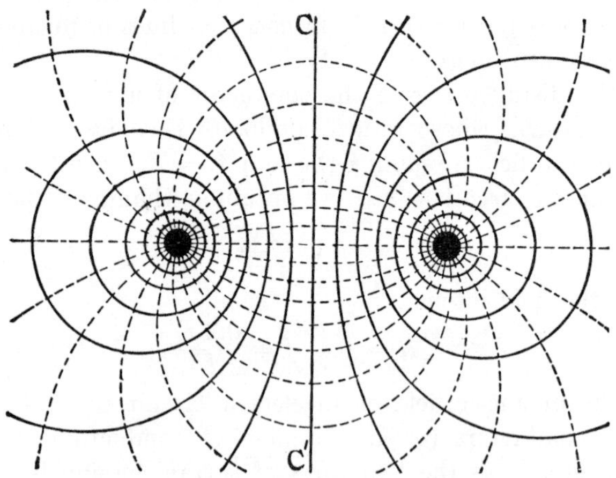

FIG. 1.—Magnetic field of parallel conductors.

The magnetic field in the interior of a spiral (solenoid, helix, coil) carrying an electric current consists of straight lines.

5. If a conductor is coiled in a spiral of l centimeter axial length of spiral, and N turns, thus $n = \frac{N}{l}$ turns per centimeter length of spiral, and I = current, in amperes, in the conductor, the m.m.f. of the spiral is

$$F = IN,$$

and the magnetizing force in the middle of the spiral, assuming the latter of very great length,

$$f = nI = \frac{N}{l}\,I,$$

thus the field intensity in the middle of the spiral or solenoid,

$$\begin{aligned} H &= 0.4\,\pi f \\ &= 0.4\,\pi n I. \end{aligned}$$

Strictly this is true only in the middle part of a spiral of such length that the m.m.f. consumed by the external or magnetic return circuit of the spiral is negligible compared with the m.m.f. consumed by the magnetic circuit in the interior of the spiral, or in an endless spiral, that is, a spiral whose axis curves back into itself, as a spiral whose axis is curved in a circle.

Magnetomotive force F applies to the total magnetic circuit, or part of the magnetic circuit. It is measured in ampere-turns.

Magnetizing force f is the m.m.f. per unit length of magnetic circuit. It is measured in ampere-turns per centimeter.

Field intensity H is the number of lines of force per square centimeter.

If l = length of the magnetic circuit or a part of the magnetic circuit,

$$F = lf, \qquad f = \frac{F}{l},$$

$$H = 0.4\,\pi f \qquad f = \frac{H}{0.4\,\pi},$$

$$= 1.257\,f \qquad f = 0.796\,H.$$

6. The preceding applies only to magnetic fields in air or other unmagnetic materials.

If the medium in which the magnetic field is established is a "magnetic material," the number of lines of force per square centimeter is different and usually many times greater. (Slightly less in diamagnetic materials.)

The ratio of the number of lines of force in a medium, to the number of lines of force which the same magnetizing force would produce in air (or rather in a vacuum), is called the *permeability* or magnetic conductivity μ of the medium.

The number of lines of force per square centimeter in a magnetic medium is called the *magnetic induction B*. The number of lines of force produced by the same magnetizing force in air, or rather, in the vacuum, is called the *field intensity H*.

In air, magnetic induction B and field intensity H are equal.

As a rule, the magnetizing force in a magnetic circuit is changed by the introduction of the magnetic material, due to the change of distribution of the magnetic flux.

The permeability of air = 1 and is constant.

The permeability of iron and other magnetic materials varies with the magnetizing force between a little above 1 and values beyond 10,000 in soft iron.

The magnetizing force f in a medium of permeability μ produces the field intensity $H = 0.4\,\pi f$ and the magnetic induction $B = 0.4\,\pi\mu f$.

EXAMPLES

7. (1) A pull of 2 grams at 4 cm. radius is required to hold a horizontal bar magnet 12 cm. in length, pivoted at its center, in a position at right angles to the magnetic meridian. What is the intensity of the poles of the magnet, and the number of lines of magnetic force issuing from each pole, if the horizontal intensity of the terrestrial magnetic field $H = 0.2$, and the acceleration of gravity = 980?

The distance between the poles of the bar magnet may be assumed as five-sixths of its length.

Let m = intensity of magnet poles. $l = 5$ is the radius on which the terrestrial magnetism acts.

Thus $2\,mHl = 2\,m$ = torque exerted by the terrestrial magnetism.

2 grams weight = $2 \times 980 = 1960$ units of force. These at 4 cm. radius give the torque $4 \times 1960 = 7840\,g$ cm.

Hence $2\,m = 7840$.

$m = 3920$ is the strength of each magnet pole and

$\Phi = 4\,\pi m = 49{,}000$, the number of lines of force issuing from each pole.

8. (2) A conductor carrying 100 amp. runs in the direction of the magnetic meridian. What position will a compass needle assume, when held below the conductor at a distance of 50 cm., if the intensity of the terrestrial magnetic field is 0.2?

The intensity of the magnetic field of 100 amp., 50 cm. from the conductor, is $H = \frac{0.2\,I}{l} = 0.2 \times \frac{100}{50} = 0.4$, the direction is at right angles to the conductor, that is, at right angles to the terrestrial magnetic field.

If τ = angle between compass needle and the north pole of the magnetic meridian, l_0 = length of needle, m = intensity of its magnet pole, the torque of the terrestrial magnetism is $Hml_0 \sin\tau = 0.2\,ml_0 \sin\tau$, the torque of the current is

$$Hml_0 \cos\tau = \frac{0.2\,Iml_0 \cos\tau}{l} = 0.4\,ml_0 \cos\tau.$$

In equilibrium, $0.2\,ml_0 \sin \tau = 0.4\,ml_0 \cos \tau$, or $\tan \tau = 2$, $\tau = 63.4°$.

9. (3) What is the total magnetic flux per $l = 1000$ m. length, passing between the conductors of a long distance transmission line carrying I amperes of current, if $l_d = 0.82$ cm. is the diameter of the conductors (No. 0 B. & S.), $l_s = 45$ cm. the spacing or distance between them?

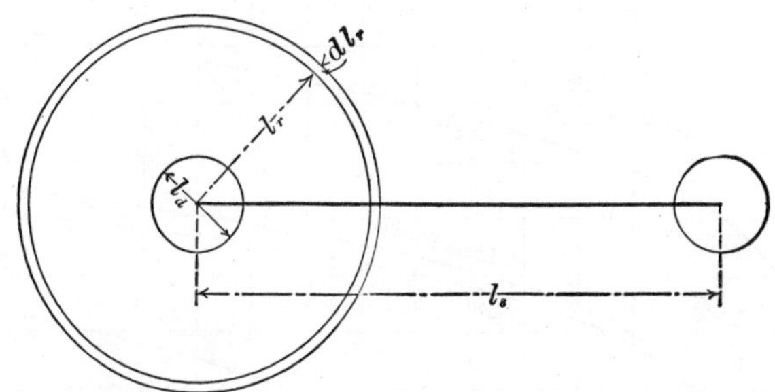

FIG. 2.—Diagram of transmission line for inductance calculation.

At distance l_r from the center of one of the conductors (Fig. 2), the length of the magnetic circuit surrounding this conductor is $2\,\pi l_r$, the m.m.f., I ampere-turns; thus the magnetizing force $f = \frac{I}{2\,\pi l_r}$, and the field intensity $H = 0.4\,\pi f = \frac{0.2\,I}{l_r}$, and the flux in the zone dl_r is $d\Phi = \frac{0.2\,Ildl_r}{l_r}$, and the total flux from the surface of the conductor to the next conductor is,

$$\Phi = \int_{\frac{l_d}{2}}^{l_s} \frac{0.2\,Ildl_r}{l_r} =$$

$$0.2Il\Big[\log_\epsilon l_r\Big]_{\frac{l_d}{2}}^{l_s} = 0.2\,Il \log_\epsilon \frac{2l_s}{l_d}.$$

The same flux is produced by the return conductor in the same direction, thus the total flux passing between the transmission wires is,

$$2\,\Phi = 0.4\,Il \log_\epsilon \frac{2l}{l_d}$$

or per 1000 m. $= 10^5$ cm. length,

$$2\,\Phi = 0.4 \times 10^5\,I \log_\epsilon \frac{90}{0.82} = 0.4 \times 10^5 \times 4.70\,I = 0.188 \times 10^6\,I,$$

or 0.188 I megalines or millions of lines per line of 1000 m. of which 0.094 I megalines surround each of the two conductors.

10. (4) In an alternator each pole has to carry 6.4 millions of lines, or 6.4 megalines magnetic flux. How many ampere-turns per pole are required to produce this flux, if the magnetic

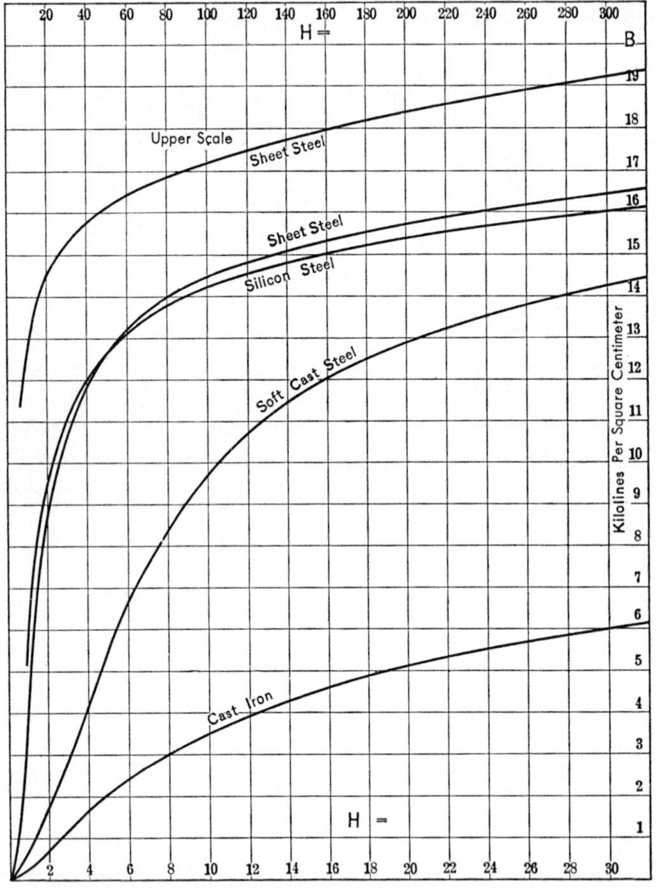

FIG. 3.—Magnetization curves of various irons.

circuit in the armature of laminated iron has the cross section of 930 sq. cm. and the length of 15 cm., the air-gap between stationary field poles and revolving armature is 0.95 cm. in length and 1200 sq. cm. in section, the field pole is 26.3 cm. in length and 1075 sq. cm. in section, and is of laminated iron, and the outside return circuit or yoke has a length per pole of 20 cm. and 2250 sq. cm. section, and is of cast iron?

The magnetic densities are: $B_1 = 6880$ in the armature, $B_2 = 5340$ in the air-gap, $B_3 = 5950$ in the field pole, and $B_4 = 2850$ in the yoke. The permeability of sheet iron is $\mu_1 = 2550$ at $B_1 = 6880$, $\mu_3 = 2380$ at $B_3 = 5950$. The permeability of cast iron is $\mu_4 = 280$ at $B_4 = 2850$. Thus the field intensity $\left(H = \frac{B}{\mu}\right)$ is: $H_1 = 2.7$, $H_2 = 5340$, $H_3 = 2.5$, $H_4 = 10.2$.

The magnetizing force $\left(f = \frac{H}{0.4\,\pi}\right)$ is, $f_1 = 2.15$, $f_2 = 4250$, $f_3 = 1.99$, $f_4 = 8.13$ ampere-turns per centimeter. Thus the m.m.f. $(F = fl)$ is: $F_1 = 32, F_2 = 4040$, $F_3 = 52$, $F_4 = 163$, or the total m.m.f. per pole is

$$F = F_1 + F_2 + F_3 + F_4 = 4290 \text{ ampere-turns.}$$

The permeability μ of magnetic materials varies with the density B, thus tables or curves have to be used for these quantities. Such curves are usually made out for density B and magnetizing force f, so that the magnetizing force f corresponding to the density B can be derived directly from the curve. Such a set of curves is given in Fig. 3.

2. MAGNETISM AND E.M.F.

11. In an electric conductor moving relatively to a magnetic field, an e.m.f. is generated proportional to the rate of cutting of the lines of magnetic force by the conductor.

Unit e.m.f. is the e.m.f. generated in a conductor cutting one line of magnetic force per second.

10^8 times unit e.m.f. is the practical unit, called the *volt.*

Coiling the conductor n fold increases the e.m.f. n fold, by cutting each line of magnetic force n times.

In a closed electric circuit the e.m.f. produces an *electric current.*

The ratio of e.m.f. to electric current produced thereby is called the *resistance* of the electric circuit.

Unit resistance is the resistance of a circuit in which unit e.m.f. produces unit current.

10^9 times unit resistance is the practical unit, called the *ohm.*

The ohm is the resistance of a circuit, in which 1 volt produces 1 amp.

The resistance per unit length and unit section of a conductor is called its resistivity, ρ.

The resistivity ρ is a constant of the material, varying with the temperature.

The resistance r of a conductor of length l, area or section A, and resistivity ρ is $r = \frac{l\rho}{A}$.

12. If the current in the electric circuit changes, starts, or stops, the corresponding change of the magnetic field of the current generates an e.m.f in the conductor carrying the current, which is called the *e.m.f. of self-induction.*

If the e.m.f. in an electric circuit moving relatively to a magnetic field produces a current in the circuit, the magnetic field produced by this current is called its *magnetic reaction.*

The fundamental law of self-induction and magnetic reaction is that these effects take place in such a direction as to oppose their cause (Lentz's law).

Thus the e.m.f. of self-induction during an increase of current is in the opposite direction, during a decrease of current in the same direction as the e.m.f. producing the current.

The magnetic reaction of the current produced in a circuit moving out of a magnetic field is in the same direction, in a circuit moving into a magnetic field in opposite direction to the magnetic field.

Essentially, this law is nothing but a conclusion from the law of conservation of energy.

EXAMPLES

13. (1) An electromagnet is placed so that one pole surrounds the other pole cylindrically as shown in section in Fig. 4, and a copper cylinder revolves between these poles at 3000 rev. per min. What is the e.m.f. generated between the ends of this cylinder, if the magnetic flux of the electromagnet is $\Phi = 25$ megalines?

During each revolution the copper cylinder cuts 25 megalines. It makes 50 rev. per sec. Thus it cuts $50 \times 25 \times 10^6 = 12.5 \times 10^8$ lines of magnetic flux per second. Hence the generated e.m.f. is $E = 12.5$ volts.

Such a machine is called a "unipolar," or more properly a "non-polar" or an "acyclic," generator.

14. (2) The field spools of the 20-pole alternator in Example 4, page 8, are wound each with 616 turns of wire No. 7 (B. & S.), 0.106 sq. cm. in cross section and 160 cm. mean length of turn. The 20 spools are connected in series. How many amperes and how many volts are required for the excitation of this alternator field, if the resistivity of copper is 1.8×10^{-6} ohms per cm.^{3},[1]

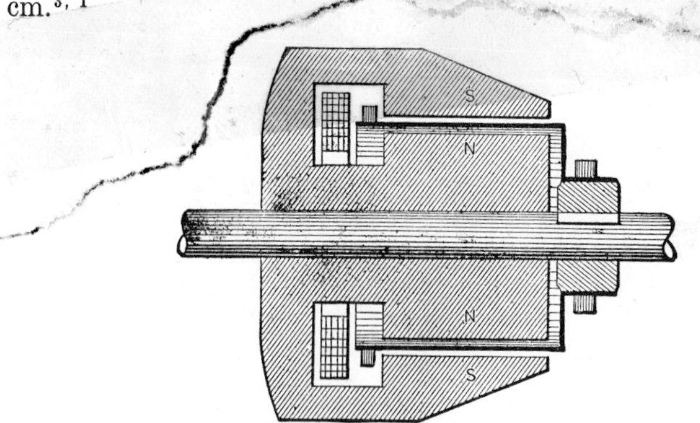

FIG. 4.—Unipolar generator.

Since 616 turns on each field spool are used, and 4290 ampere-turns required, the current is $\frac{4290}{616} = 6.95$ amp.

The resistance of 20 spools of 616 turns of 160 cm. length, 0.106 sq. cm. section, and 1.8×10^{-6} resistivity is,

$$\frac{20 \times 616 \times 160 \times 1.8 \times 10^{-6}}{0.106} = 33.2 \text{ ohms},$$

and the e.m.f. required, $6.95 \times 33.2 = 230$ volts.

3. GENERATION OF E.M.F.

15. A closed conductor, convolution or turn, revolving in a magnetic field, passes during each revolution through two positions of maximum inclosure of lines of magnetic force A in Fig. 5, and two positions of zero inclosure of lines of magnetic force B in Fig. 5.

[1] cm.^{3} refers to a cube whose side is 1 cm., and should not be confused with cu. cm.

Thus it cuts during each revolution four times the lines of force inclosed in the position of maximum inclosure.

If Φ = the maximum number of lines of force inclosed by the conductor, f = the frequency in revolutions per second or cycles, and n = number of convolutions or turns of the conductor, the lines of force cut per second by the conductor, and thus the *average generated e.m.f.* is,

$$E = 4\, fn\Phi \text{ absolute units,}$$
$$= 4\, fn\Phi\, 10^{-8} \text{ volts.}$$

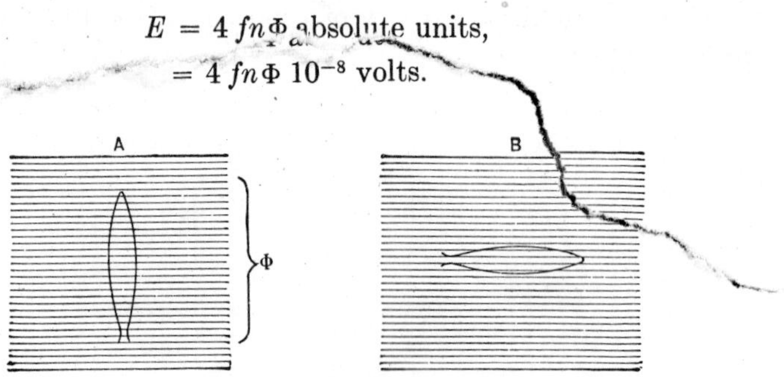

Fig. 5.—Generation of e.m.f.

If f is given in hundreds of cycles, Φ in megalines,

$$E = 4\, fn\Phi \text{ volts.}$$

If a coil revolves with uniform velocity through a uniform magnetic field, the magnetism inclosed by the coil at any instant is,

$$\Phi \cos \tau$$

where Φ = the maximum magnetism inclosed by the coil and τ = angle between coil and its position of maximum inclosure of magnetism.

The e.m.f. generated in the coil, which varies with the rate of cutting or change of $\Phi \cos \tau$, is thus,

$$e = E_0 \sin \tau,$$

where E_0 is the maximum value of e.m.f., which takes place for $\tau = 90°$, or at the position of zero inclosure of magnetic flux since in this position the rate of cutting is greatest.

Since avg. $(\sin \tau) = \frac{2}{\pi}$, the average generated e.m.f. is,

$$E = \frac{2}{\pi} E_0.$$

Since, however, we found above that

$$E = 4\, fn\Phi \text{ is the average generated e.m.f.,}$$

it follows that

$$E_0 = 2\, \pi fn\Phi \text{ is the } \textit{maximum}, \text{ and}$$
$$e = 2\, \pi fn\Phi \sin \tau \text{ the } \textit{instantaneous generated e.m.f.}$$

The interval between like poles forms 360 electrical-space degrees, and in the two-pole model these are identical with the mechanical-space degrees. With uniform rotation, Fig. 6, the space angle, τ, is proportional to time. Time angles are designated by θ, and with uniform rotation $\theta = \tau$, τ being measured in electrical-space degrees.

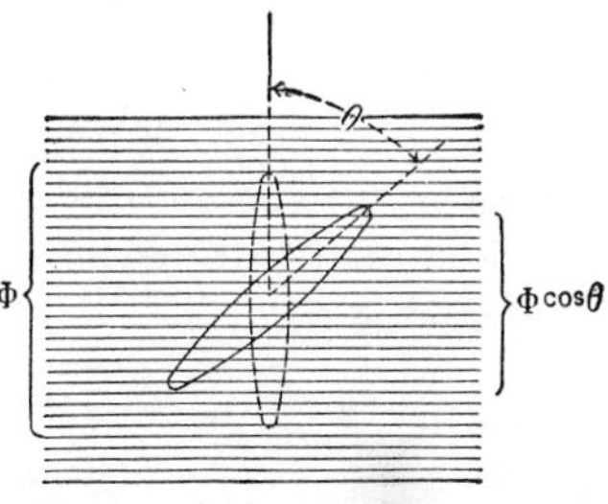

Fig. 6.—Generation of e.m.f. by rotation.

The period of a complete cycle is 360 time degrees, or $2\,\pi$ or $\frac{1}{f}$ seconds. In the two-pole model the period of a cycle is that of one complete revolution, and in a $2\, n_p$-pole machine, $\frac{1}{n_p}$ of that of one revolution.

Thus,
$$\theta = 2\, \pi ft$$
$$e = 2\, \pi fn\Phi \sin 2\, \pi ft.$$

If the time is not counted from the moment of maximum inclosure of magnetic flux, but t_1 = the time at this moment, we have

$$e = 2\, \pi fn\Phi \sin 2\, \pi f\,(t - t_1)$$

or,

$$e = 2\, \pi fn\Phi \sin\,(\theta - \theta_1),$$

where $\theta_1 = 2\, \pi ft_1$ is the angle at which the position of maximum inclosure of magnetic flux takes place, and is called its phase.

These e.m.fs. are alternating.

If at the moment of reversal of the e.m.f. the connections between the coil and the external circuit are reversed, the e.m.f. in the external circuit is pulsating between zero and E_0, but has the same average value E.

If a number of coils connected in series follow each other

successively in their rotation through the magnetic field, as the armature coils of a direct-current machine, and the connections of each coil with the external circuit are reversed at the moment of reversal of its e.m.f., their pulsating e.m.fs. superimposed in the external circuit make a more or less steady or continuous external e.m.f.

The average value of this e.m.f. is the sum of the average values of the e.m.fs. of the individual coils.

Thus in a direct-current machine, if Φ = maximum flux inclosed per turn, n = total number of turns in series from commutator brush to brush, and f = frequency of rotation through the magnetic field.

$$E = 4fn\Phi = \text{generated e.m.f. } (\Phi \text{ in megalines, } f \text{ in hundreds of cycles per second}).$$

This is the *formula of the direct-current generator.*

EXAMPLES

17. (1) A circular wire coil of 200 turns and 40 cm. mean diameter is revolved around a vertical axis. What is the horizontal intensity of the magnetic field of the earth, if at a speed of 900 rev. per min. the average e.m.f generated in the coil is 0.028 volt?

The mean area of the coil is $\frac{40^2\pi}{4} = 1255$ sq. cm., thus the terrestrial flux inclosed is $1255\,H$, and at 900 rev. per min. or 15 rev. per sec., this flux is cut $4 \times 15 = 60$ times per second by each turn, or $200 \times 60 = 12{,}000$ times by the coil. Thus the total number of lines of magnetic force cut by the conductor per second is $12{,}000 \times 1255\,H = 0.151 \times 10^8\,H$, and the average generated e.m.f. is $0.151\,H$ volts. Since this is = 0.028 volt, $H = 0.186$.

18. (2) In a 550-volt direct-current machine of 8 poles and drum armature, running at 500 rev. per min., the average voltage per commutator segment shall not exceed 11, each armature coil shall contain one turn only, and the number of commutator segments per pole shall be divisible by 3, so as to use the machine as three-phase converter. What is the magnetic flux per field pole?

550 volts at 11 volts per commutator segment gives 50, or as next integer divisible by 3, $n = 51$ segments or turns per pole.

8 poles give 4 cycles per revolution, 500 rev. per min. gives $^{500}/_{60} = 8.33$ rev. per sec. Thus the frequency is $f = 4 \times 8.33 = 33.3$ cycles per second.

The generated e.m.f. is $E = 550$ volts, thus by the formula of direct-current generator,

$$E = 4\,fn\Phi,$$

or,

$$550 = 4 \times 0.333 \times 51\,\Phi,$$
$$\Phi = 8.1 \text{ megalines per pole.}$$

19. (3) What is the e.m.f. generated in a single turn of a 20-pole alternator running at 200 rev. per min., through a magnetic field of 6.4 megalines per pole?

The frequency is $f = \dfrac{20 \times 200}{2 \times 60} = 33.3$ cycles.

$$e = E_0 \sin \tau,$$
$$E_0 = 2\,\pi fn\Phi,$$
$$\Phi = 6.4,$$
$$n = 1,$$
$$f = 0.333.$$

Thus, $E_0 = 2\,\pi \times 0.333 \times 6.4 = 13.4$ volts maximum, or $e = 13.4 \sin \theta$.

4. POWER AND EFFECTIVE VALUES

20. The power of the continuous e.m.f. E producing continuous current I is $P = EI$.

The e.m.f. consumed by resistance r is $E_1 = Ir$, thus the power consumed by resistance r is $P = I^2r$.

Either $E_1 = E$, then the total power in the circuit is consumed by the resistance, or $E_1 < E$, then only a part of the power is consumed by the resistance, the remainder by some counter e.m.f., $E - E_1$.

If an alternating current $i = I_0 \sin \theta$ passes through a resistance r, the power consumed by the resistance is,

$$i^2r = I_0^2r \sin^2 \theta = \frac{I_0^2r}{2}(1 - \cos 2\,\theta),$$

thus varies with twice the frequency of the current, between zero and I_0^2r.

The average power consumed by resistance r is,

$$\text{avg. } (i^2r) = \frac{I_0^2r}{2} = \left(\frac{I_0}{\sqrt{2}}\right)^2 r,$$

since avg. (cos) = 0.

Thus the alternating current $i = I_0 \sin \theta$ consumes in a resistance r the same power as a continuous current of intensity

$$I = \frac{I_0}{\sqrt{2}}.$$

The value $I = \frac{I_0}{\sqrt{2}}$ is called the *effective value* of the alternating current $i = I_0 \sin \theta$; since it gives the same effect.

Analogously $E = \frac{E_0}{\sqrt{2}}$ is the effective value of the alternating e.m.f., $e = E_0 \sin \theta$.

Since $E_0 = 2\pi f n \Phi$, it follows that

$$\begin{aligned} E &= \sqrt{2}\,\pi f n \Phi \\ &= 4.44\, f n \Phi \end{aligned}$$

is the *effective alternating e.m.f.* generated in a coil of turns n rotating at a frequency of f (in hundreds of cycles per second) through a magnetic field of Φ megalines of force.

This is the *formula of the alternating-current generator.*

21. The formula of the direct-current generator,

$$E = 4\, f n \Phi,$$

holds even if the e.m.fs. generated in the individual turns are not sine waves, since it is the average generated e.m.f.

The formula of the alternating-current generator,

$$E = \sqrt{2}\,\pi f n \Phi,$$

does not hold if the waves are not sine waves, since the ratios of average to maximum and of maximum to effective e.m.f. are changed.

If the variation of magnetic flux is not sinusoidal, the effective generated alternating e.m.f. is,

$$E = \gamma \sqrt{2}\,\pi f n \Phi.$$

γ is called the *form factor* of the wave, and depends upon its shape, that is, the distribution of the magnetic flux in the magnetic field.

Frequently *form factor* is defined as the ratio of the effective to the average value. This definition is undesirable since it gives for the sine wave, which is always considered the standard wave, a value differing from one.

EXAMPLES

22. (1) In a star-connected 20-pole three-phase machine, revolving at 33.3 cycles or 200 rev. per min., the magnetic flux per pole is 6.4 megalines. The armature contains one slot per pole and phase, and each slot contains 36 conductors. All these conductors are connected in series. What is the effective e.m.f. per circuit, and what the effective e.m.f. between the terminals of the machine?

Twenty slots of 36 conductors give 720 conductors, or 360 turns in series. Thus the effective e.m.f. is,

$$\begin{aligned} E_1 &= \sqrt{2}\,\pi f n \Phi \\ &= 4.44 \times 0.333 \times 360 \times 6.4 \\ &= 3400 \text{ volts per circuit.} \end{aligned}$$

The e.m.f. between the terminals of a star-connected three-phase machine is the resultant of the e.m.fs. of the two phases, which differ by 60 degrees, and is thus 2 sin 60° = $\sqrt{3}$ times that of one phase, thus,

$$\begin{aligned} E &= E_1 \sqrt{3} \\ &= 5900 \text{ volts effective.} \end{aligned}$$

23. (2) The conductor of the machine has a section of 0.22 sq. cm. and a mean length of 240 cm. per turn. At a resistivity (resistance per unit section and unit length) of copper of $\rho = 1.8 \times 10^{-6}$, what is the e.m.f. consumed in the machine by the resistance, and what the power consumed at 450 kw. output?

450 kw. output is 150,000 watts per phase or circuit, thus the current $I = \frac{150{,}000}{3400} = 44.2$ amperes effective.

The resistance of 360 turns of 240 cm. length, 0.22 sq. cm. section and 1.8×10^{-6} resistivity, is

$$r = \frac{360 \times 240 \times 1\,8 \times 10^{-6}}{0.22} = 0.71 \text{ ohms per circuit.}$$

44.2 amp. $\times$ 0.71 ohms gives 31.5 volts per circuit and $(44.2)^2 \times 0.71 = 1400$ watts per circuit, or a total of $3 \times 1400 = 4200$ watts loss.

24. (3) What is the self-inductance per wire of a three-phase line of 14 miles length consisting of three wires No. 0 ($l_d = 0.82$ cm.), 45 cm. apart, transmitting the output of this 450 kw. 5900-volt three-phase machine?

450 kw. at 5900 volts gives 44.2 amp. per line. 44.2 amp. effective gives $44.2\sqrt{2} = 62.5$ amp. maximum.

14 miles = 22,400 m. The magnetic flux produced by I amperes in 1000 m. of a transmission line of 2 wires 45 cm. apart and 0.82 cm. diameter was found in Example 3, page 7, as $2\,\Phi = 0.188 \times 10^6\, I$, or $\Phi = 0.094 \times 10^6\, I$ for each wire.

Thus at 22,300 m. and 62.5 amp. maximum, the flux per wire is

$$\Phi = 22.3 \times 62.5 \times 0.094 \times 10^6 = 131 \text{ megalines.}$$

Hence the generated e.m.f., effective value, at 33.3 cycles is,

$$\begin{aligned} E &= \sqrt{2}\,\pi f \Phi \\ &= 4.44 \times 0.333 \times 131 \\ &= 193 \text{ volts per line;} \end{aligned}$$

the maximum value is,

$$E_0 = E \times \sqrt{2} = 273 \text{ volts per line;}$$

and the instantaneous value,

$$e = E_0 \sin\,(\theta - \theta_1) = 273\ \sin\,(\theta - \theta_1);$$

or, since $\theta = 2\,\pi f t = 210\,t$ we have,

$$e = 273 \sin 210\,(t - t_1).$$

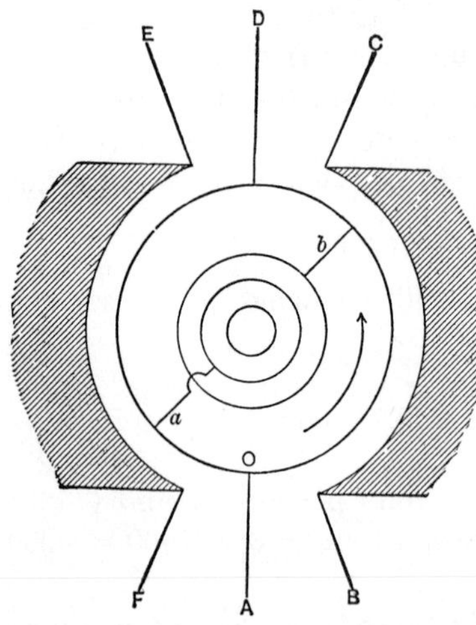

FIG. 7.—Diagram of bipolar generator.

25. (4) What is the form factor (a) of the e.m.f. generated in a single conductor of a direct-current machine having 80 per cent. pole arc and negligible spread of the magnetic flux at the pole corners, and (b) what is the form factor of the voltage between two collector rings connected to diametrical points of the armature of such a machine?

(a) In a conductor during the motion from position A, shown in Fig. 7, to position B, no e.m.f. is generated; from position B to C a constant e.m.f. e is generated, from C to E again no e.m.f., from E to F a constant e.m.f. $-\ e$, and from F to A again zero e.m.f. The e.m.f. wave thus is as shown in Fig. 8.

The average e.m.f. is

$$e_1 = 0.8\, e;$$

hence, with this average e.m.f., if it were a sine wave, the maximum e.m.f. would be

$$e_2 = \frac{\pi}{2} e_1 = 0.4\, \pi e,$$

and the effective e.m.f. would be

$$e_3 = \frac{e_2}{\sqrt{2}} = \frac{0.4\, \pi e}{\sqrt{2}}.$$

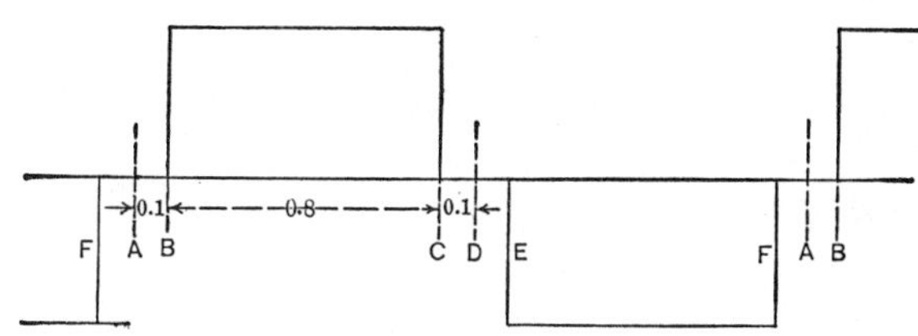

FIG. 8.—E.m.f. of a single conductor, direct-current machine 80 per cent. pole arc.

The actual square of the e.m.f. is e^2 for 80 per cent. and zero for 20 per cent. of the period, and the average or mean square thus is

$$0.8\, e^2,$$

and therefore the actual effective value,

$$e_4 = e\sqrt{0.8}.$$

The form factor γ, or the ratio of the actual effective value e_4 to the effective value e_3 of a sine wave of the same mean value and thus the same magnetic flux, then is

$$\gamma = \frac{e_4}{e_3} = \frac{\sqrt{10}}{\pi}$$
$$= 1.006;$$

that is, practically unity.

(*b*) While the collector leads a, b move from the position F, C, as shown in Fig. 6, to B, E, constant voltage E exists between them, the conductors which leave the field at C being replaced by the conductors entering the field at B. During the motion of the leads a, b from B, E to C, F, the voltage steadily decreases, reverses, and rises again, to $-E$, as the conductors entering the field at E have an e.m.f. opposite to that of the conductors leaving at C. Thus the voltage wave is, as shown by Fig. 9, triangular, with the top cut off for 20 per cent. of the half wave.

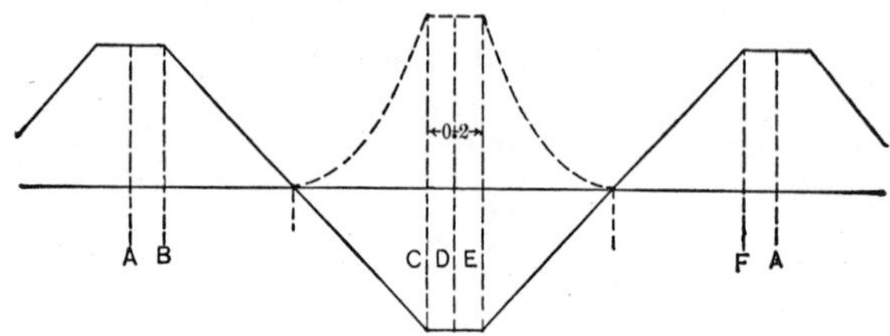

FIG. 9.—E.m.f. between two collector rings connected to diametrical points of the armature of a bipolar machine having 80 per cent. pole arc.

Then the average e.m.f. is

$$e_1 = 0.2\,E + 2 \times \frac{0.4\,E}{2} = 0.6\,E.$$

The maximum value of a sine wave of this average value is

$$e_2 = \frac{\pi}{2}\,e_1 = 0.3\,\pi E,$$

and the effective value corresponding thereto is

$$e_3 = \frac{e_2}{\sqrt{2}} = \frac{0.3\,\pi E}{\sqrt{2}}.$$

The actual voltage square is E^2 for 20 per cent. of the time, and rising on a parabolic curve from 0 to E^2 during 40 per cent. of the time, as shown in dotted lines in Fig. 9.

The area of a parabolic curve is width times one-third of height, or

$$\frac{0.4\,E^2}{3},$$

hence, the mean square of voltage is

$$0.2\,E^2 + 2 \times \frac{0.4\,E^2}{3} = \frac{1.4\,E^2}{3},$$

and the actual effective voltage is

$$e_4 = E\sqrt{\frac{1.4}{3}};$$

hence, the form factor is

$$\gamma = \frac{e_4}{e_3} = \frac{1}{\pi}\sqrt{\frac{280}{27}} = 1.025,$$

or, 2.5 per cent. higher than with a sine wave.

5. SELF-INDUCTANCE AND MUTUAL INDUCTANCE

26. The number of interlinkages of an electric circuit with the lines of magnetic force of the flux produced by unit current in the circuit is called the *inductance* of the circuit.

The number of interlinkages of an electric circuit with the lines of magnetic force of the flux produced by unit current in a second electric circuit is called the *mutual inductance* of the second upon the first circuit. It is equal to the mutual inductance of the first upon the second circuit, as will be seen, and thus is called the mutual inductance between the two circuits.

The number of interlinkages of an electric circuit with the lines of magnetic flux produced by unit current in this circuit and <u>not</u> interlinked with a second circuit is called the *self-inductance* of the circuit.

If i = current in a circuit of n turns, Φ = flux produced thereby and interlinked with the circuit, $n\Phi$ is the total number of interlinkages, and $L = \frac{n\Phi}{i}$ the inductance of the circuit.

If Φ is proportional to the current i and the number of turns n,

$$\Phi = \frac{ni}{\mathcal{R}}, \text{ and } L = \frac{n^2}{\mathcal{R}} \text{ the inductance.}$$

$\mathcal{R}$ is called the *reluctance* and ni the m.m.f. of the magnetic circuit.

In magnetic circuits the reluctance $\mathcal{R}$ has a position similar to that of resistance r in electric circuits.

The reluctance $\mathcal{R}$, and therefore the inductance, is not constant in circuits containing magnetic materials, such as iron, etc.

If $\mathcal{R}_1$ is the reluctance of a magnetic circuit interlinked with two electric circuits of n_1 and n_2 turns respectively, the flux produced by unit current in the first circuit and interlinked with the second circuit is $\frac{n_1}{\mathcal{R}_1}$ and the mutual inductance of the first upon the second circuit is $M = \frac{n_1 n_2}{\mathcal{R}_1}$, that is, equal to the mutual inductance of the second circuit upon the first circuit, as stated above.

If no flux leaks between the two circuits, that is, if all flux is interlinked with both circuits, and L_1 = inductance of the first, L_2 = inductance of the second circuit, and M = mutual inductance, then

$$M^2 = L_1 L_2.$$

If flux leaks between the two circuits, then $M^2 < L_1 L_2$.

In this case the total flux produced by the first circuit consists of a part interlinked with the second circuit also, the mutual inductance, and a part passing between the two circuits, that is, interlinked with the first circuit only, its self-inductance.

27. Thus, if L_1 and L_2 are the inductances of the two circuits, $\frac{L_1}{n_1}$ and $\frac{L_2}{n_2}$ is the total flux produced by unit current in the first and second circuit respectively.

Of the flux $\frac{L_1}{n_1}$ a part $\frac{S_1}{n_1}$ is interlinked with the first circuit only, S_1 being its self-inductance or leakage inductance, and a part $\frac{M}{n_2}$ interlinked with the second circuit also, M being the mutual inductance and $\frac{L_1}{n_1} = \frac{S_1}{n_1} + \frac{M}{n_2}$.

Thus, if

L_1 and L_2 = inductance,
S_1 and S_2 = self-inductance,
M = mutual inductance of two circuits of n and n_2 turns respectively, we have

$$\frac{L_1}{n_1} = \frac{S_1}{n_1} + \frac{M}{n_2} \qquad \frac{L_2}{n_2} = \frac{S_2}{n_2} + \frac{M}{n_1},$$

or

$$L_1 = S_1 + \frac{n_1}{n_2} M \qquad L_2 = S_2 + \frac{n_2}{n_1} M,$$

or

$$M^2 = (L_1 - S_1)(L_2 - S_2).$$

The practical unit of inductance is 10^9 times the absolute unit or 10^8 times the number of interlinkages per ampere (since 1 amp. = 0.1 unit current), and is called the *henry* (h); 0.001 of it is called the *milhenry* (mh.).

The number of interlinkages of i amperes in a circuit of

L henry inductance is $iL\ 10^8$ lines of force turns, and thus the e.m.f. generated by a change of current di in time dt is

$$e = -\frac{di}{dt} L\ 10^8 \text{ absolute units}$$

$$= -\frac{di}{dt} L \text{ volts.}$$

A change of current of 1 amp. per second in the circuit of 1 h. inductance generates 1 volt.

EXAMPLES

28. (1) What is the inductance of the field of a 20-pole alternator, if the 20 field spools are connected in series, each spool contains 616 turns, and 6.95 amp. produces 6.4 megalines per pole?

The total number of turns of all 20 spools is $20 \times 616 = 12{,}320$ Each is interlinked with 6.4×10^6 lines, thus the total number of interlinkages at 6.95 amp. is $12{,}320 \times 6.4 \times 10^6 = 78 \times 10^9$.

6.95 amp. = 0.695 absolute units, hence the number of interlinkages per unit current, or the inductance, is

$$\frac{78 \times 10^9}{0.695} = 112 \times 10^9 = 112 \text{ h.}$$

29. (2) What is the mutual inductance between an alternating transmission line and a telephone wire carried for 10 miles below and 1.20 m. distant from the one, 1.50 m. distant from the other conductor of the alternating line; and what is the e.m.f. generated in the telephone wire, if the alternating circuit carries 100 amp. at 60 cycles?

The mutual inductance between the telephone wire and the electric circuit is the magnetic flux produced by unit current in the telephone wire and interlinked with the alternating circuit, that is, that part of the magnetic flux produced by unit current in the telephone wire, which passes between the distances of 1.20 and 1.50 m.

At the distance l_x from the telephone wire the length of magnetic circuit is $2\pi l_x$. The magnetizing force $f = \frac{I}{2\pi l_x}$ if $I =$

current in telephone wire in amperes, and the field intensity $H = 0.4\,\pi f = \frac{0.2\,I}{l_x}$, and the flux in the zone dl_x is

$$d\Phi = \frac{0.2\,Il}{l_x}\,dl_x.$$

$$l = 10 \text{ miles} = 1610 \times 10^3 \text{ cm.}$$

thus,
$$\Phi = \int_{120}^{150} \frac{0.2\,Il}{l_x}\,dl_x$$

$$= 322 \times 10^3\,I \log\epsilon \frac{150}{120} = 72\,I\,10^3;$$

or, $72\,I\,10^3$ interlinkages, hence, for $I = 10$, or one absolute unit,

thus, $M = 72 \times 10^4$ absolute units $= 72 \times 10^{-5}$ h. $= 0.72$ mh.

100 amp. effective or 141.4 amp. maximum or 14.14 absolute units of current in the transmission line produces a maximum flux interlinked with the telephone line of $14.14 \times 0.72 \times 10^{-3} \times 10^9 = 10.2$ megalines. Thus the e.m.f. generated at 60 cycles is

$$E = 4.44 \times 0.6 \times 10.2 = 27.3 \text{ volts effective.}$$

6. SELF-INDUCTANCE OF CONTINUOUS-CURRENT CIRCUITS

30. Self-inductance makes itself felt in continuous-current circuits only in starting and stopping or, in general, when the current changes in value.

Starting of Current. If r = resistance, L = inductance of circuit, E = continuous e.m.f. impressed upon circuit, i = current in circuit at time t after impressing e.m.f. E, and di the increase of current during time moment dt, then the increase of magnetic interlinkages during time dt is

$$Ldi,$$

and the e.m.f. generated thereby is

$$e_1 = -L\frac{di}{dt}.$$

By Lentz's law it is negative, since it is opposite to the impressed e.m.f., its cause.

Thus the e.m.f. acting in this moment upon the circuit is

$$E + e_1 = E - L\frac{di}{dt},$$

and the current is

$$i = \frac{E + e_1}{r} = \frac{E - L\frac{di}{dt}}{r};$$

or, transposing,

$$-\frac{rdt}{L} = \frac{di}{i - \frac{E}{r}},$$

the integral of which is

$$-\frac{rt}{L} = \log_\epsilon\left(i - \frac{E}{r}\right) - \log_\epsilon c,$$

where $-\log_\epsilon c =$ integration constant.
This reduces to

$$i = \frac{E}{r} + c\epsilon^{-\frac{rt}{L}}$$

at $t = 0$, $i = 0$, and thus

$$-\frac{E}{r} = c.$$

Substituting this value, the current is

$$i = \frac{E}{r}\left(1 - \epsilon^{-\frac{rt}{L}}\right),$$

and the e.m.f. of inductance is

$$e_1 = ir - E = -E\epsilon^{-\frac{rt}{L}}.$$

At $t = \infty$,

$$i_0 = \frac{E}{r}, \qquad e_1 = 0.$$

Substituting these values,

$$i = i_0\left(1 - \epsilon^{-\frac{rt}{L}}\right)$$

and

$$e_1 = -ri_0\epsilon^{-\frac{rt}{L}}.$$

The expression $u = \frac{r}{L}$ is called the "*attenuation constant*," and its reciprocal, $\frac{L}{r}$, the "*time constant of the circuit*."[1]

[1] The name *time constant* dates back to the early days of telegraphy, where it was applied to the ratio: $\frac{L}{r}$, that is, the reciprocal of the attenuation constant. This quantity which had gradually come into disuse, again became of importance when investigating transient electric phenomena, and in this work it was found more convenient to denote the value $\frac{r}{L}$ as attenuation constant, since this value appears as one term of the more general constant of the electric circuit $\left(\frac{r}{L} + \frac{g}{C}\right)$.

Substituted in the foregoing equation this gives

$$i = \frac{E}{r}\left(1 - \epsilon^{-ut}\right)$$

and

$$e_1 = - E\epsilon^{-ut}.$$

At $t = \frac{1}{u}$

$$e_1 = - \frac{E}{r} = - 0.368\ E.$$

31. *Stopping of Current.* In a circuit of inductance L and resistance r, let a current $i_0 = \frac{E}{r}$ be produced by the impressed e.m.f. E, and this e.m.f. E be withdrawn and the circuit closed through a resistance r_1.

Let the current be i at the time t after withdrawal of the e.m.f. E and the change of current during time moment dt be di. di is negative, that is, the current decreases.

The decrease of magnetic interlinkages during moment dt is

$$Ldi.$$

Thus the e.m.f. generated thereby is

$$e_1 = - L\frac{di}{dt}.$$

It is negative since di is negative, and e_1 must be positive, that is, in the same direction as E, to maintain the current or oppose the decrease of current, its cause.

Then the current is

$$i = \frac{e_1}{r + r_1} = - \frac{L}{r + r_1}\frac{di}{dt};$$

or, transposing,

$$- \frac{r + r_1}{L}dt = \frac{di}{i},$$

the integral of which is

$$- \frac{r + r_1}{L}t = \log_\epsilon i - \log_\epsilon c,$$

where $- \log_\epsilon c =$ integration constant.

This reduces to
$$i = c\epsilon^{-\frac{r + r_1}{L}t},$$

for
$$t = 0, \qquad i_0 = \frac{E}{r} = c.$$

Substituting this value, the current is

$$i = \frac{E}{r} \epsilon^{-\frac{(r + r_1)t}{L}},$$

and the generated e.m.f. is

$$e_1 = i(r + r_1) = E\frac{r + r_1}{r}\epsilon^{-\frac{(r + r_1)t}{L}}.$$

Substituting $i_0 = \frac{E}{r}$, the current is

$$i = i_0 \epsilon^{-\frac{r + r_1}{L}t},$$

and the generated e.m.f. is

$$e_1 = i_0(r + r_1)\epsilon^{-\frac{r + r_1}{L}t}.$$

At $t = 0$,

$$e_1 = E\frac{r + r_1}{r};$$

that is, the generated e.m.f. is increased over the previously impressed e.m.f. in the same ratio as the resistance is increased.

When $r_1 = 0$, that is, when in withdrawing the impressed e.m.f. E the circuit is short circuited,

$$i = \frac{E}{r}\epsilon^{-\frac{rt}{L}} = i_0 \epsilon^{-\frac{rt}{L}} \text{ the current, and}$$

$$e_1 = E\epsilon^{-\frac{rt}{L}} = i_0 r \epsilon^{-\frac{rt}{L}} \text{ the generated e.m.f.}$$

In this case, at $t = 0$, $e_1 = E$, that is, the e.m.f. does not rise.

In the case $r_1 = \infty$, that is, if in withdrawing the e.m.f. E the circuit is broken, we have $t = 0$ and $e_1 = \infty$, that is, the e.m.f. rises infinitely.

The greater r_1, the higher is the generated e.m.f. e_1, the faster, however, do e_1 and i decrease.

If $r_1 = r$, we have at $t = 0$,

$$e_{11} = 2E, \qquad i = i_0,$$

and

$$e_{11} - i_0 r = E;$$

that is, if the external resistance r_1 equals the internal resistance r, at the moment of withdrawal of the e.m.f. E the terminal voltage is E.

The effect at the time t of the e.m.f. of inductance in stopping the current is

$$ie_1 = i_0^2 (r + r_1) \epsilon^{-2\frac{r + r_1}{L}t};$$

thus the total energy of the generated e.m.f.

$$W = \int_0^\infty ie_1 dt$$

$$= i_0^2 (r + r_1) \left[\epsilon^{-2\frac{r + r_1}{L}t}\right]_0^\infty \left(-\frac{L}{2(r + r_1)}\right) = \frac{i_0^2 L}{2};$$

that is, the energy stored as magnetism in a circuit of current i_0 and inductance L is

$$W = \frac{i_0^2 L}{2},$$

which is independent both of the resistance r of the circuit and the resistance r_1 inserted in breaking the circuit. This energy has to be expended in stopping the current.

EXAMPLES

32. (1) In the alternator field in Example 4, page 8, Example 2, page 11, and Example 1, page 23, how long a time after impressing the required e.m.f. $E = 230$ volts will it take for the field to reach (a) ½ strength, (b) $^9/_{10}$ strength?

(2) If 500 volts are impressed upon the field of this alternator, and a non-inductive resistance inserted in series so as to give the required exciting current of 6.95 amp., how long after impressing the e.m.f. $E = 500$ volts will it take for the field to reach (a) ½ strength, (b) $^9/_{10}$ strength, (c) and what is the resistance required in the rheostat?

(3) If 500 volts are impressed upon the field of this alternator without insertion of resistance, how long will it take for the field to reach full strength?

(4) With full field strength, what is the energy stored as magnetism?

(1) The resistance of the alternator field is 33.2 ohms (Example 2, page 11), the inductance 112 h. (Section 5, Example 1), the impressed e.m.f. is $E = 230$, the final value of current $i_0 = \frac{E}{r} = 6.95$ amp. Thus the current at time t is

$$i = i_0\left(1 - \epsilon^{-\frac{rt}{L}}\right)$$
$$= 6.95\,(1 - \epsilon^{-0.296\,t}).$$

(a) ½ strength: $i = \frac{i_0}{2}$, hence $(1 - \epsilon^{-0.296\,t}) = 0.5$.

$\epsilon^{-0.296\,t} = 0.5$, $-0.296\,t \log \epsilon = \log 0.5$, $t = \frac{-\log 0.5}{0.296 \log \epsilon}$, and $t = 2.34$ seconds.

(b) $\frac{9}{10}$ strength: $i = 0.9\,i_0$, hence $(1 - \epsilon^{-0.296\,t}) = 0.9$, and $t = 7.8$ seconds.

(2) To get $i_0 = 6.95$ amp., with $E = 500$ volts, a resistance $r = \frac{500}{6.95} = 72$ ohms, and thus a rheostat having a resistance of $72 - 33.2 = 38.8$ ohms, is required.

We then have

$$\begin{aligned} i &= i_0 \left(1 \quad \epsilon^{-\frac{rt}{L}}\right) \\ &= 6.95\,(1 - \epsilon^{-0.643\,t}). \end{aligned}$$

(a) $i = \frac{i_0}{2}$, after $t = 1.08$ seconds.

(b) $i = 0.9\,i_0$, after $t = 3.6$ seconds.

(3) Impressing $E = 500$ volts upon a circuit of $r = 33.2$, $L = 112$, gives

$$\begin{aligned} i &= \frac{E}{r}\left(1 - \epsilon^{-\frac{rt}{L}}\right) \\ &= 15.1\,(1 - \epsilon^{-0.296\,t}). \end{aligned}$$

$i = 6.95$, or full field strength, gives

$$6.95 = 15.1\,(1 - \epsilon^{-0.296\,t}).$$

$$1 - \epsilon^{-0.296\,t} = 0.46$$

and $t = 2.08$ seconds.

(4) The stored energy is

$$\begin{aligned} \frac{i_0^2 L}{2} &= \frac{6.95^2 \times 112}{2} = 2720 \text{ watt-seconds or joules} \\ &= 2000 \text{ foot-pounds.} \end{aligned}$$

(1 joule = 0.736 foot-pounds.)

Thus in case (3), where the field reaches full strength in 2.08 seconds, the average power input is $\frac{2000}{2.08} = 960$ foot pounds per second = 1.75 hp.

In breaking the field circuit of this alternator, 2000 foot-pounds of energy have to be dissipated in the spark, etc.

33. (5) A coil of resistance $r = 0.002$ ohm and inductance $L = 0.005$ mh., carrying current $I = 90$ amp., is short circuited.

(*a*) What is the equation of the current after short circuit?

(*b*) In what time has the current decreased to 0.1. its initial value?

(*a*) $i = I\epsilon^{-\frac{rt}{L}}$

$= 90\ \epsilon^{-400\,t}$.

(*b*) $i = 0.1\ I$, $\epsilon^{-400\,t} = 0.1$, after $t = 0.00576$ second.

(6) When short circuiting the coil in Example 5, an e.m.f. $E = 1$ volt is inserted in the circuit of this coil, in opposite direction to the current.

(*a*) What is equation of the current?

(*b*) After what time does the current become zero?

(*c*) After what time does the current reverse to its initial value in opposite direction?

(*d*) What impressed e.m.f. is required to make the current die out in 1/2000 second?

(*e*) What impressed e.m.f. E is required to reverse the current in 1/1000 second?

(*a*) If e.m.f. $-E$ is inserted, and at time t the current is denoted by i, we have

$$e_1 = -L\frac{di}{dt}, \text{ the generated e.m.f.;}$$

Thus,

$$-E + e_1 = -E - L\frac{di}{dt}, \text{ the total e.m.f.;}$$

and

$$i = \frac{-E + e_1}{r} = -\frac{E}{r} - \frac{L}{r}\frac{di}{dt}, \text{ the current;}$$

Transposing,

$$-\frac{r}{L}\,dt = \frac{di}{\frac{E}{r} + i},$$

and integrating,

$$-\frac{rt}{L} = \log_\epsilon\left(\frac{E}{r} + i\right) - \log_\epsilon c,$$

where $-\log_\epsilon c$ = integration constant.

At $t = 0$, $i = I$, thus $c = I + \frac{E}{r}$;

Substituting,

$$i = \left(I + \frac{E}{r}\right)\epsilon^{-\frac{rt}{L}} - \frac{E}{r},$$

$$i = 590\ \epsilon^{-400\,t} - 500.$$

(*b*) $i = 0,\ \epsilon^{-400\,t} = 0.85$, after $t = 0.000405$ second.

(*c*) $i = -I = -90,\ \epsilon^{-400\,t} = 0.694$, after $t = 0.00091$ second.

(*d*) If $i = 0$ at $t = 0.0005$, then

$$0 = (90 + 500\,E)\,\epsilon^{-0.2} - 500\,E,$$

$$E = \frac{0.18}{\epsilon^{0.2} - 1} = 0.81 \text{ volt.}$$

(*e*) If $i = -I = -90$ at $t = 0.001$, then

$$-90 = (90 + 500\,E)\,\epsilon^{-0.4} - 500\,E,$$

$$E = \frac{0.18\,(1 + \epsilon^{-0.4})}{1 - \epsilon^{-0.4}} = 0.91 \text{ volt.}$$

7. INDUCTANCE IN ALTERNATING-CURRENT CIRCUITS

34. An alternating current $i = I_0 \sin 2\pi ft$ or $i = I_0 \sin \theta$ can be represented graphically in rectangular coordinates by a curved line as shown in Fig. 10, with the instantaneous values

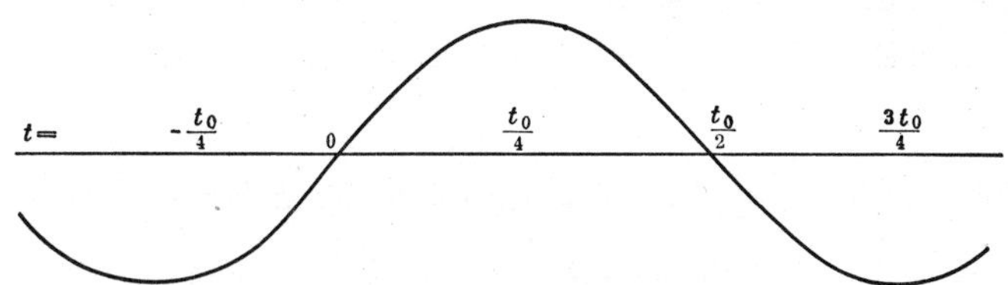

FIG. 10.—Alternating sine wave.

i as ordinates and the time t, or the arc of the angle corresponding to the time, $\theta = 2\pi ft$, as abscissas, counting the time from the zero value of the rising wave as zero point.

If the zero value of current is not chosen as zero point of time, the wave is represented by

$$i = I_0 \sin 2\pi f\,(t - t'),$$

or

$$i = I_0 \sin (\theta - \theta'),$$

where t' and θ' are respectively the time and the corresponding angle at which the current reaches its zero value in the ascendant.

If such a sine wave of alternating current $i = I_0 \sin 2\pi ft$ or $i = I_0 \sin \theta$ passes through a circuit of resistance r and inductance L, the magnetic flux produced by the current and thus its interlinkages with the current, $iL = I_0L \sin \theta$, vary in a wave line similar also to that of the current, as shown in Fig. 11 as Φ. The e.m.f. generated hereby is proportional to the change of iL, and is thus a maximum where iL changes most rapidly, or at its zero point, and zero where iL is a maximum, and according to Lentz's law it is positive during falling and negative during rising current. Thus this generated e.m.f. is a wave following the wave of current by the time $t = \frac{t_0}{4}$, where t_0 is time of one complete period, $= \frac{1}{f}$, or by the time angle $\theta = 90°$.

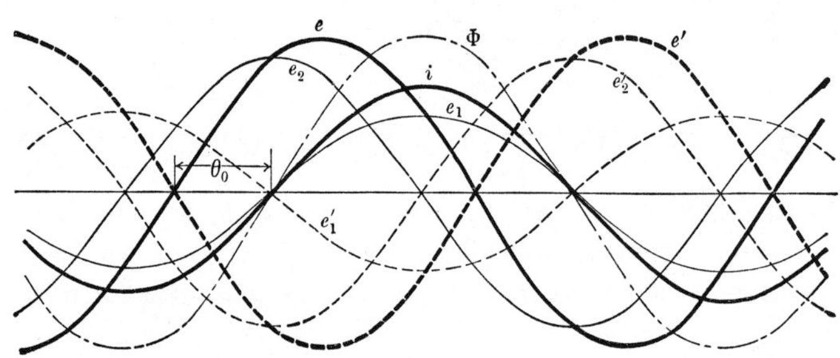

FIG. 11.—Self-induction effects produced by an alternating sine wave of current.

This e.m.f. is called the *counter e.m.f. of inductance.* It is

$$\begin{aligned} e'_2 &= -L\frac{di}{dt} \\ &= -2\pi f L I_0 \cos 2\pi f t. \end{aligned}$$

It is shown in dotted line in Fig. 11 as e'_2.

The quantity $2\pi f L$ is called the *inductive reactance* of the circuit, and denoted by x. It is of the nature of a resistance, and expressed in ohms. If L is given in 10^9 absolute units or henrys, x appears in ohms.

The counter e.m.f. of inductance of the current, $i = I_0 \sin 2\pi f t = I_0 \sin\theta$, of effective value

$$I = \frac{I_o}{\sqrt{2}}, \text{ is}$$

$$e'_2 = -xI_0 \cos 2\pi f t = -xI_0 \cos\theta,$$

having a maximum value of xI_0 and an effective value of

$$E_2 = \frac{xI_0}{\sqrt{2}} = xI;$$

that is, the effective value of the counter e.m.f. of inductance equals the reactance, x, times the effective value of the current, I, and lags 90 time degrees, or a quarter period, behind the current.

35. By the counter e.m.f. of inductance,

$$e'_2 = -xI_0 \cos \theta,$$

which is generated by the change in flux due to the passage of the current $i = I_0 \sin \theta$ through the circuit of reactance x, an equal but opposite e.m.f.

$$e_2 = xI_0 \cos \theta$$

is consumed, and thus has to be impressed upon the circuit. This e.m.f. is called the *e.m.f. consumed by inductance*. It is 90 time degrees, or a quarter period, ahead of the current, and shown in Fig. 11 as a drawn line e_2.

Thus we have to distinguish between counter e.m.f. of inductance 90 time degrees lagging, and e.m.f. consumed by inductance 90 time degrees leading.

These e.m.fs. stand in the same relation as action and reaction in mechanics. They are shown in Fig. 11 as e'_2 and as e_2.

The e.m.f. consumed by the resistance r of the circuit is proportional to the current,

$$e_1 = ri = rI_0 \sin \theta,$$

and in phase therewith, that is, reaches its maximum and its zero value at the same time as the current i, as shown by drawn line e_1 in Fig. 11.

Its effective value is $E_1 = rI$.

The resistance can also be represented by a (fictitious) counter e.m.f.,

$$e'_1 = -rI_0 \sin \theta,$$

opposite in phase to the current, shown as e'_1 in dotted line in Fig. 11.

The counter e.m.f. of resistance and the e.m.f. consumed by resistance have the same relation to each other as the counter e.m.f. of inductance and the e.m.f. consumed by inductance or inductive reactance.

36. If an alternating current $i = I_0 \sin \theta$ of effective value $I = \frac{I_0}{\sqrt{2}}$ exists in a circuit of resistance r and inductance L, that is, of reactance $x = 2\pi fL$, we have to distinguish:

E.m.f. consumed by resistance, $e_1 = rI_0 \sin \theta$, of effective value $E_1 = rI$, and in phase with the current.

Counter e.m.f. of resistance, $e'_1 = - rI_0 \sin \theta$, of effective value $E_1 = rI$, and in opposition or 180 time degrees displaced from the current.

E.m.f. consumed by reactance, $e_2 = xI_0 \cos \theta$, of effective value $E_2 = xI$, and leading the current by 90 time degrees or a quarter period.

Counter e.m.f. of reactance, $e'_2 = xI_0 \cos \theta$, of effective value $E'_2 = xI$, and lagging 90 time degrees or a quarter period behind the current.

The e.m.fs. consumed by resistance and by reactance are the e.m.fs. which have to be impressed upon the circuit to overcome the counter e.m.fs. of resistance and of reactance.

Thus, the total counter e.m.f. of the circuit is

$$e' = e'_1 + e'_2 = - I_0 (r \sin \theta + x \cos \theta),$$

and the total impressed e.m.f., or e.m.f. consumed by the circuit, is

$$e = e_1 + e_2 = I_0 (r \sin \theta + x \cos \theta).$$

Substituting

$$\frac{x}{r} = \tan \theta_0 \text{ and}$$

$$\sqrt{r^2 + r^2} = z,$$

it follows that

$$x = z \sin \theta_0, \qquad r = z \cos \theta_0,$$

and we have as the total impressed e.m.f.

$$e = zI_0 \sin (\theta + \theta_0),$$

shown by heavy drawn line e in Fig. 11, and total counter e.m.f.

$$e' = - zI_0 \sin (\theta + \theta_0),$$

shown by heavy dotted line e' in Fig. 11, both of effective value

$$e = zI.$$

For $\theta = - \theta_0$, $e = 0$, that is, the zero value of e is ahead of the zero value of current by the time angle θ_0, or the current lags behind the impressed e.m.f. by the angle θ_0.

θ_0 is called the *angle of lag* of the current, and $z = \sqrt{r^2 + x^2}$ the *impedance* of the circuit. e is called the e.m.f. consumed by impedance, e' the counter e.m.f. of impedance.

Since $E_1 = rI$ is the e.m.f. consumed by resistance,
$E_2 = xI$ is the e.m.f. consumed by reactance,
and $E = zI = \sqrt{r^2 + x^2}\, I$ is the e.m.f. consumed by impedance,
we have

$$E = \sqrt{E_1^2 + E_2^2}, \text{ the total e.m.f.}$$

and

$$E_1 = E \cos \theta_0,$$
$$E_2 = E \sin \theta_0, \text{ its components.}$$

The tangent of the angle of lag is

$$\tan \theta_0 = \frac{x}{r} = \frac{2\,\pi f L}{r},$$

and the time constant of the circuit is

$$\frac{L}{r} = \frac{\tan \theta_0}{2\,\pi f}.$$

The total e.m.f., e, impressed upon the circuit consists of two components, one, e_1, in phase with the current, the other one, e_2, in quadrature with the current.

Their effective values are

$$E,\ E \cos \theta_0,\ E \sin \theta_0.$$

EXAMPLES

37. (1) What is the reactance per wire of a transmission line of length l, if l_d = diameter of the wire, l_s = spacing of the wires, and f = frequency?

If I = current, in absolute units, in one wire of the transmission line, the m.m.f. is I; thus the magnetizing force in a zone dl_x at distance l_x from center of wire (Fig. 12) is $f = \frac{I}{2\,\pi l_x}$ and the field intensity in this zone is $H = 4\,\pi f = 2\frac{I}{l_x}$. Thus the magnetic flux in this zone is

$$d\Phi = H\, l dl_x = \frac{2\, I l dl_x}{l_x};$$

hence, the total magnetic flux between the wire and the return wire is

$$\Phi = \int_{\frac{l_d}{2}}^{l_s} d\Phi = 2\, Il \int_{\frac{l_d}{2}}^{l_s} \frac{dl_x}{l_x} = 2\, Il \log_\epsilon \frac{2\, l_s}{l_d},$$

neglecting the flux inside the transmission wire.

The inductance is

$$L = \frac{\Phi}{I} = 2\,l \log_\epsilon \frac{2\,l_s}{l_d} \text{ absolute units}$$
$$= 2\,l \log_\epsilon \frac{2\,l_s}{l_d}\, 10^{-9} \text{ h.,}$$

and the reactance $x = 2\,\pi f L = 4\,\pi f l \log_\epsilon \frac{2\,l_s}{l_d}$, in absolute units;

or $$x = 4\,\pi f l \log_\epsilon \frac{2\,l_s}{l_d}\, 10^{-9}, \text{ in ohms.}$$

38. (2) The voltage at the receiving end of a 33.3-cycle three-phase transmission line 14 miles in length shall be 5500

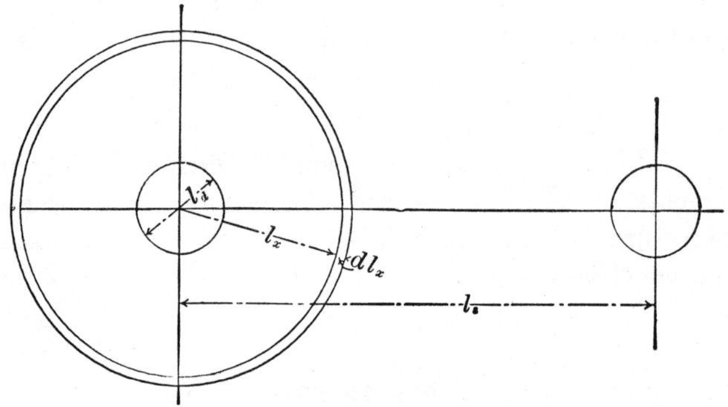

Fig. 12.—Diagram for calculation of inductance between two parallel conductors.

between the lines. The line consists of three wires, No. 0 B. & S. (l_d = 0.82 cm.), 18 in. (45 cm.) apart, of resistivity $\rho = 1.8 \times 10^{-6}$.

(*a*) What is the resistance, the reactance, and the impedance per line, and the voltage consumed thereby at 44 amp.?

(*b*) What is the generator voltage between lines at 44 amp. to a non-inductive load?

(*c*) What is the generator voltage between lines at 44 amp. to a load circuit of 45 degrees lag?

(*d*) What is the generator voltage between lines at 44 amp. to a load circuit of 45 degrees lead?

Here l = 14 miles = $14 \times 1.6 \times 10^5 = 2.23 \times 10^6$ cm.
l_d = 0.82 cm.

Hence the cross section, A = 0.528 sq. cm.

(*a*) Resistance per line, $r = \rho \frac{l}{A} = \frac{1.8 \times 10^{-6} \times 2.23 \times 10^{6}}{0.528}$ $= 7.60$ ohms.

Reactance per line, $x = 4\,\pi f l \log_\epsilon \frac{2\,l_s}{l_d} \times 10^{-9} = 4\pi \times 33.3 \times 2.23 \times 10^{6} \times \log_\epsilon 110 \times 10^{-9} = 4.35$ ohms.

The impedance per line, $z = \sqrt{r^2 + x^2} = 8.76$ ohms. Thus if $I = 44$ amp. per line,

the e.m.f. consumed by resistance is $E_1 = rI = 334$ volts,
the e.m.f. consumed by reactance is $E_2 = xI = 192$ volts,
and the e.m.f. consumed by impedance is $E_3 = zI = 385$ volts.

(*b*) 5500 volts between lines at receiving circuit give $\frac{5500}{\sqrt{3}} =$ 3170 volts between line and neutral or zero point (Fig. 13), or per line, corresponding to a maximum voltage of $3170\sqrt{2} = 4500$ volts. 44 amp. effective per line gives a maximum value of $44\sqrt{2} = 62$ amp.

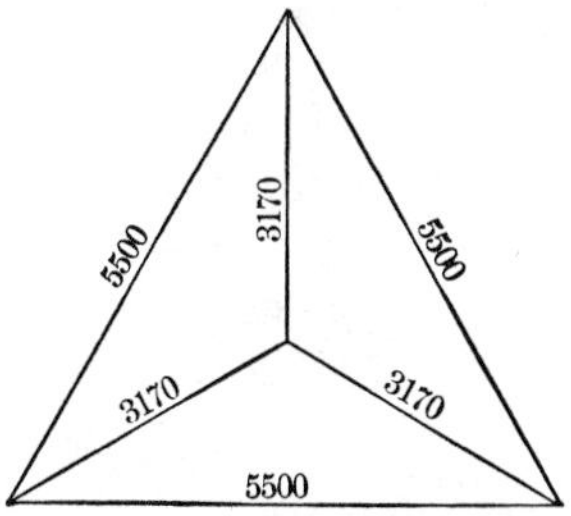

FIG. 13.—Voltage diagram for a three-phase circuit.

Denoting the current by $i = 62 \sin \theta$, the voltage per line at the receiving end with non-inductive load is $e = 4500 \sin \theta$.

The e.m.f. consumed by resistance, in phase with the current, of effective value 334, and maximum value $334\sqrt{2} = 472$, is

$$e_1 = 472 \sin \theta.$$

The e.m.f. consumed by reactance, 90 time degrees ahead of the current, of effective value 192, and maximum value $192\sqrt{2} = 272$, is

$$e_2 = 272 \cos \theta.$$

Thus the total voltage required per line at the generator end of the line is

$$e_0 = e + e_1 + e_2 = (4500 + 472) \sin \theta + 272 \cos \theta$$
$$= 4972 \sin \theta + 272 \cos \theta.$$

Denoting $\frac{272}{4972} = \tan \theta_0$, we have

$$\sin \theta_0 = \frac{\tan \theta_0}{\sqrt{1 + \tan^2 \theta_0}} = \frac{272}{4980}.$$

$$\cos \theta_0 = \frac{1}{\sqrt{1 + \tan^2 \theta_0}} = \frac{4972}{4980}.$$

Hence, $$e_0 = 4980 (\sin \theta \cos \theta_0 + \cos \theta \sin \theta_0) = 4980 \sin (\theta + \theta_0).$$

Thus θ_0 is the lag of the current behind the e.m.f. at the generator end of the line, = 3.2 time degrees, and 4980 the maximum voltage per line at the generator end; thus $E_0 = \frac{4980}{\sqrt{2}} = 3520$, the effective voltage per line, and $3520\sqrt{3} = 6100$, the effective voltage between the lines at the generator.

(*c*) If the current

$$i = 62 \sin \theta$$

lags in time 45 degrees behind the e.m.f. at the receiving end of the line, this e.m.f. is expressed by

$$e = 4500 \sin (\theta + 45) = 3170 (\sin \theta + \cos \theta);$$

that is, it leads the current by 45 time degrees, or is zero at $\theta = -45$ time degrees.

The e.m.f. consumed by resistance and by reactance being the same as in (*b*), the generator voltage per line is

$$e_0 = e + e_1 + e_2 = 3642 \sin \theta + 3442 \cos \theta.$$

Denoting $\frac{3442}{3642} = \tan \theta_0$, we have

$$e_0 = 5011 \sin (\theta + \theta_0).$$

Thus θ_0, the angle of lag of the current behind the generator e.m.f., is 43 degrees, and 5011 the maximum voltage; hence 3550 the effective voltage per line, and $3550\sqrt{3} = 6160$ the effective voltage between lines at the generator.

(*d*) If the current $i = 62 \sin \theta$ leads the e.m.f. by 45 degrees, the e.m.f. at the receiving end is

$$e = 4500 \sin (\theta - 45) = 3180 (\sin \theta - \cos \theta).$$

Thus at the generator end

$$e_0 = e + e_1 + e_2 = 3652 \sin \theta - 2908 \cos \theta.$$

Denoting $\frac{2908}{3652} = \tan \theta_0$, it is

$$e_0 = 4670 \sin (\theta - \theta_0).$$

Thus θ_0, the time angle of lead at the generator, is 39 degrees, and 4654 the maximum voltage; hence 3290 the effective voltage per line and 5710 the effective voltage between lines at the generator.

8. POWER IN ALTERNATING-CURRENT CIRCUITS

39. The power consumed by alternating current $i = I_0 \sin \theta$, of effective value $I = \frac{I_0}{\sqrt{2}}$, in a circuit of resistance r and reactance $x = 2 \pi f L$, is

$$p = ei,$$

where $e = zI_0 \sin (\theta + \theta_0)$ is the impressed e.m.f., consisting of the components

$$e_1 = rI_0 \sin \theta, \text{ the e.m.f. consumed by resistance}$$

and $$e_2 = xI_0 \cos \theta, \text{ the e.m.f. consumed by reactance.}$$

$z = \sqrt{r^2 + x^2}$ is the impedance and $\tan \theta_0 = \frac{x}{r}$ the phase angle of the circuit; thus the power is

$$\begin{aligned} p &= zI_0{}^2 \sin \theta \sin (\theta + \theta_0) \\ &= \frac{zI_0{}^2}{2} (\cos \theta_0 - \cos (2\,\theta + \theta_0)) \\ &= zI^2 (\cos \theta_0 - \cos (2\,\theta + \theta_0)). \end{aligned}$$

Since the average $\cos (2\,\theta + \theta_0) =$ zero, the average power is

$$\begin{aligned} P &= zI^2 \cos \theta_0 \\ &= rI^2 = E_1 I; \end{aligned}$$

that is, the power in the circuit is that consumed by the resistance, and independent of the reactance.

Reactance or self-inductance consumes no power, and the e.m.f. of self-inductance is a *wattless* or *reactive e.m.f.*, while the e.m.f. of resistance is a *power* or *active e.m.f.*

The wattless e.m.f. is in quadrature, the power e.m.f. in phase with the current.

In general, if $\theta =$ angle of time-phase displacement between the resultant e.m.f. and the resultant current of the circuit, $I =$ current, $E =$ impressed e.m.f., consisting of two components, one, $E_1 = E \cos \theta$, in phase with the current, the other, $E_2 = E \sin \theta$, in quadrature with the current, the power in the circuit is $IE_1 = IE \cos \theta$, and the e.m.f. in phase with the current $E_1 = E \cos \theta$ is a power e.m.f., the e.m.f. in quadrature with the current $E_2 = E \sin \theta$ a wattless or *reactive* e.m.f.

40. Thus we have to distinguish *power e.m.f.* and *wattless* or *reactive e.m.f.*, or power component of e.m.f., in phase with the current and wattless or reactive component of e.m.f., in quadrature with the current.

Any e.m.f. can be considered as consisting of two components, one, the power component, e_1, in phase with the current, and the other, the reactive component, e_2, in quadrature with the current. The sum of instantaneous values of the two components is the total e.m.f.

$$e = e_1 + e_2.$$

If E, E_1, E_2 are the respective effective values, we have

$$\begin{aligned} E &= \sqrt{E_1^2 + E_2^2}, \text{ since} \\ E_1 &= E \cos\theta, \\ E_2 &= E \sin\theta, \end{aligned}$$

where θ = phase angle between current and e.m.f.

Analogously, a current I due to an impressed e.m.f. E with a time-phase angle θ can be considered as consisting of two component currents,

$I_1 = I \cos\theta$, the *active* or *power* component of the current, and
$I_2 = I \sin\theta$, the *wattless* or *reactive* component of the current.

The sum of instantaneous values of the power and reactive components of the current equals the instantaneous value of the total current,

$$i_1 + i_2 = i,$$

while their effective values have the relation

$$I = \sqrt{I_1^2 + I_2^2}.$$

Thus an alternating current can be resolved in two components, the power component, in phase with the e.m.f., and the wattless or reactive component, in quadrature with the e.m.f.

An alternating e.m.f. can be resolved in two components: the power component, in phase with the current, and the wattless or reactive component, in quadrature with the current.

The power in the circuit is the current times the e.m.f. times the cosine of the time-phase angle, or is the power component of the current times the total e.m.f., or the power component of the e.m.f. times the total current.

EXAMPLES

41. (1) What is the power received over the transmission line in Example 2, page 36, the power lost in the line, the power put into the line, and the efficiency of transmission with non-inductive load, with 45-time-degree lagging load and 45-degree leading load?

The power received per line with non-inductive load is $P = EI = 3170 \times 44 = 139$ kw.

With a load of 45 degrees phase displacement, $P = EI \cos 45° = 98$ kw.

The power lost per line $P_1 = I^2R = 44^2 \times 7.6 = 14.7$ kw.

Thus the input into the line $P_0 = P + P_1 = 151.7$ kw. at non-inductive load,

and $\quad = 111.7$ kw. at load of 45 degrees phase displacement.

The efficiency with non-inductive load is

$$\frac{P}{P_0} = 1 - \frac{14.7}{151.7} = 90.3 \text{ per cent.}$$

and with a load of 45 degrees phase displacement is

$$\frac{P}{P_0} = 1 - \frac{14.7}{111.7} = 86.8 \text{ per cent.}$$

The total output is $3\,P = 411$ kw. and 291 kw., respectively.

The total input $3\,P_0 = 451.1$ kw. and 335.1 kw., respectively.

9. VECTOR DIAGRAMS

42. The best way of graphically representing alternating-current phenomena is by a vector diagram. The most frequently used vector diagram is the crank diagram. In this, sine waves of alternating currents, voltages, etc., are represented as projections of a revolving vector on the horizontal. That is, a vector equal in length to the maximum value of the alternating wave is assumed to revolve at uniform speed so as to make one complete revolution per period, and the projections of this revolving vector upon the horizontal then represent the instantaneous values of the wave.

Let, for instance, $\overline{OI}$ represent in length the maximum value of current $i = I \cos (\theta - \theta_0)$. Assume then a vector, $\overline{OI}$, to revolve, left-handed or in positive direction, so that it makes a complete revolution during each cycle or period. If then at a certain moment of time this vector stands in position $\overline{OI_1}$ (Fig. 14), the projection, $\overline{OA_1}$, of $\overline{OI_1}$ on $\overline{OA}$ represents the instantaneous value of the current at this moment. At a later moment $\overline{OI}$ has moved farther, to $\overline{OI_2}$, and the projection, $\overline{OA_2}$, of $\overline{OI_2}$ on $\overline{OA}$ is the instantaneous value. The diagram thus shows the instantaneous condition of the sine waves. Each sine wave

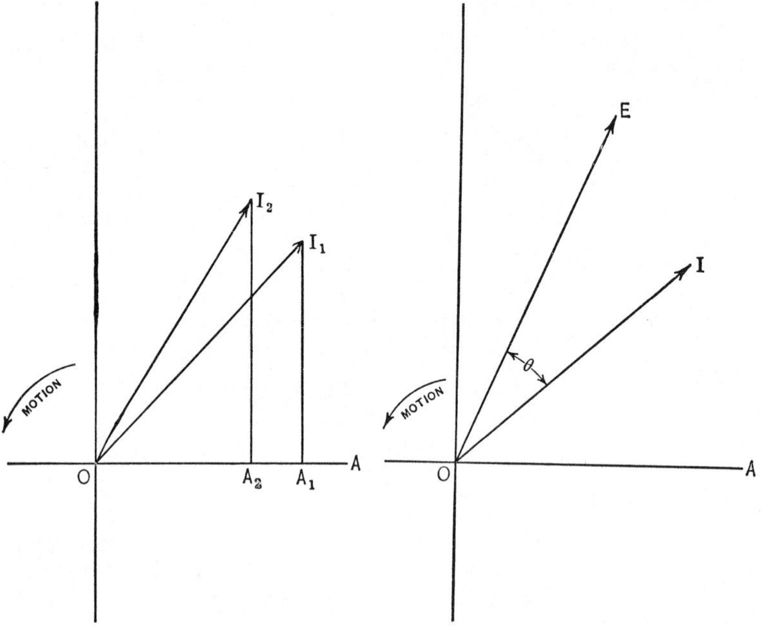

FIG. 14.—Crank diagram showing instantaneous values.

FIG. 15.—Crank diagram of an e.m.f. and current.

reaches the maximum at the moment when its revolving vector, $\overline{OI}$, passes the horizontal, and reaches zero when its revolving vector passes the vertical.

If Fig. 15 represents the crank diagram of a voltage $\overline{OE}$, and a current $\overline{OI}$, and if angle $AOE > AOI$, this means that the current $\overline{OI}$ is behind the voltage $\overline{OE}$, passes during the revolution the zero line or line of maximum intensity, $\overline{OA}$, later than the voltage; that is, the current lags behind the voltage.

In the vector diagram, the first quantity therefore can be put in any position. For instance, the current $\overline{OI}$, in Fig. 15, could be drawn in position $\overline{OI}$, Fig. 16. The voltage then being ahead of the current by angle $EOI = \theta$ would come into the position $\overline{OE}$, Fig. 16.

This vector diagram then shows graphically, by the projections of the vectors on the horizontal, the instantaneous values of the alternating waves at one moment of time. At any other moment

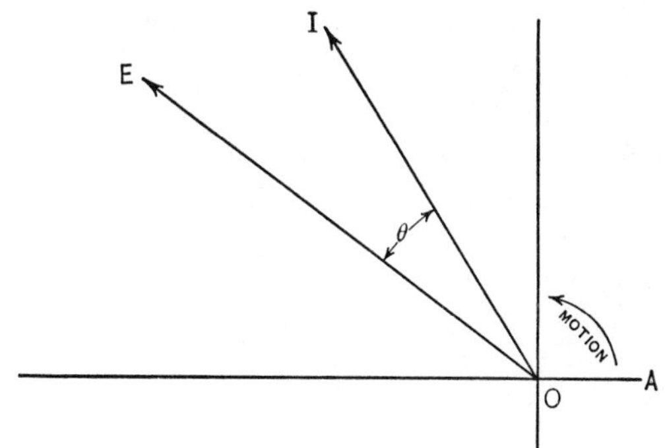

Fig. 16.—Crank diagram.

of time, the instantaneous values would be the projections of the vectors on another radius, corresponding to the other time. The angles between the vector representation are the phase differences between the vectors, and the angles each vector makes with the horizontal may be called its phase. The horizontal then

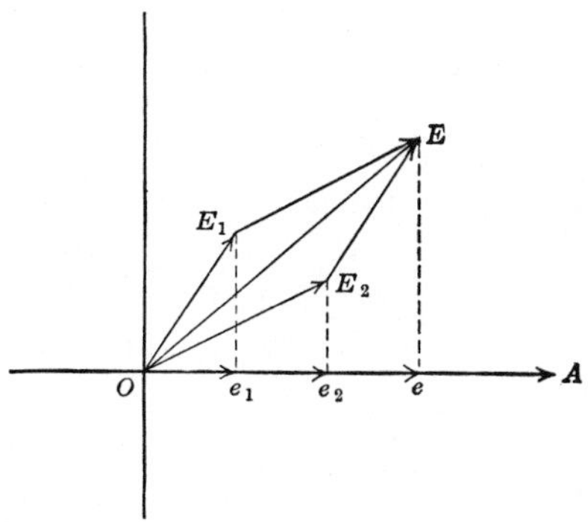

Fig. 17.—Vector diagram of two e.m.f.'s acting in the same circuit.

would be of phase zero. The phase of the first vector may be chosen at random; all other phases are determined thereby.

In this representation, the phase of an alternating wave is given by the time when its maximum value passes the horizontal.

Two voltages, e_1 and e_2, acting in the same circuit, give a resultant voltage e equal to the sum of their instantaneous values. Graphically, voltages e_1 and e_2 are represented in intensity and in phase by two revolving vectors, $\overline{OE_1}$ and $\overline{OE_2}$, Fig. 17. The instantaneous values are the projections $\overline{Oe_1}$, $\overline{Oe_2}$ of $\overline{OE_1}$ and $\overline{OE_2}$ upon the horizontal.

Since the sum of the projections of the sides of a parallelogram is equal to the projection of the diagonal, the sum of the projections $\overline{Oe_1}$ and $\overline{Oe_2}$ equals the projection $\overline{Oe}$ of $\overline{OE}$, the diagonal of the parallelogram with $\overline{OE_1}$ and $\overline{OE_2}$ as sides, and $\overline{OE}$ is thus the resultant e.m.f.; that is, graphically alternating sine waves of voltage, current, etc., are combined and resolved by the parallelogram or polygon of sine waves.

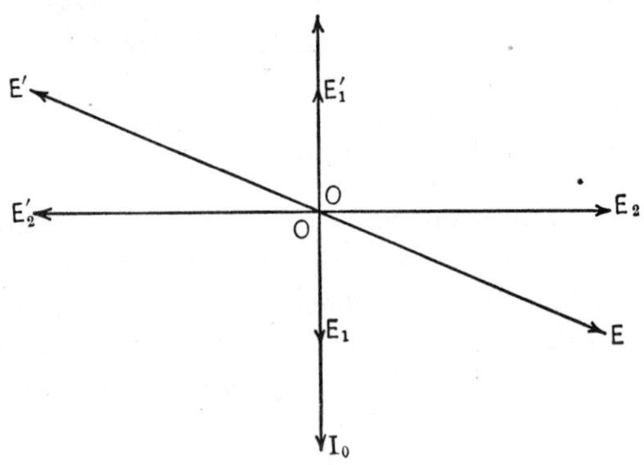

FIG. 18.—Vector diagram.

43. The sine wave of alternating current $i = I_0 \sin \theta$ is represented by a vector equal in length, $\overline{OI_0}$, to the maximum value I_0 of the wave, and located so that at time zero $\theta = 0$, its projection on the horizontal, is zero, and at times $\theta > 0$, but $< \pi$, the projection is positive. Thus this vector $\overline{OI_0}$ is the negative vertical, as shown in Fig. 18.

The voltage consumed by inductance, $e_2 = xI_0 \cos \theta$, is represented by a vector $\overline{OE_2}$ equal in length to xI_0, and located so that at $\theta = 0$, its projection on the horizontal is a maximum. That is, it is the zero vector $\overline{OE_2}$ in Fig. 18.

Analogously, the counter e.m.f. of self-inductance E'_2 is represented by vector $\overline{OE'_2}$ on the negative horizontal of Fig. 18; the voltage consumed by the resistance r, $e_1 = eI_0 \sin \theta$, is represented by vector $\overline{OE_1}$ equal to rI_0, and located on the negative vertical, and the counter e.m.f. of resistance by vector $\overline{OE'_1}$ on the positive vertical.

The counter e.m.f. of impedance:

$$\begin{aligned} e' &= -(rI_0 \sin\theta + xI_0 \cos\theta) \\ &= -zI_0 \sin(\theta + \theta_0) \end{aligned}$$

then is represented graphically as the resultant, by the parallelogram of sine waves of $\overline{OE'_1}$ and $\overline{OE'_2}$, that is, by a vector $\overline{OE'}$, equal in length to zI_0, and of phase $90 + \theta_0$.

The voltage consumed by impedance, or the impressed voltage, is represented by the vector $\overline{OE}$, equal and opposite in direction to the vector $\overline{OE'}$. This vector is the resultant of $\overline{OE_1}$ and $\overline{OE_2}$ and has the phase $\theta_0 - 90$, or $-(90 - \theta_0)$, as shown in Fig. 18.

An alternating wave is thus determined by the length and direction of its vector. The length is the maximum value, intensity or amplitude of the wave; the direction is the phase of its maximum value, usually called the phase of the wave.

44. As phase of the first quantity considered, as in the above instance the current, any direction can be chosen. The further quantities are determined thereby in direction or phase.

The zero vector $\overline{OA}$ is generally chosen for the most frequently used quantity or reference quantity, as for the current, if a number of e.m.fs. are considered in a circuit of the same current, or for the e.m.f., if a number of currents are produced by the same e.m.f., or for the generated e.m.f. in apparatus such as transformers and induction motors, synchronous apparatus, etc.

With the current as zero vector, all horizontal components of e.m.f. are power components, all vertical components are reactive components.

With the e.m.f. as zero vector, all horizontal components of current are power components, all vertical components of current are reactive components.

By measurement from the vector diagram numerical values can hardly ever be derived with sufficient accuracy, since the magnitudes of the different quantities used in the same diagram are usually by far too different, and the vector diagram is therefore useful only as basis for trigonometrical or other calculation, and to give an insight into the mutual relation of the different quantities, and even then great care has to be taken to distinguish between the two equal but opposite vectors, counter e.m.f. and e.m.f. consumed by the counter e.m.f., as explained before.

EXAMPLES

45. In a three-phase long-distance transmission line, the voltage between lines at the receiving end shall be 5000 at no load, 5500 at full load of 44 amp. power component, and proportional at intermediary values of the power component of the current; that is, the voltage at the receiving end shall increase proportional to the load. At three-quarters load the current shall be in phase with the e.m.f. at the receiving end. The generator excitation, however, and thus the (nominal) generated

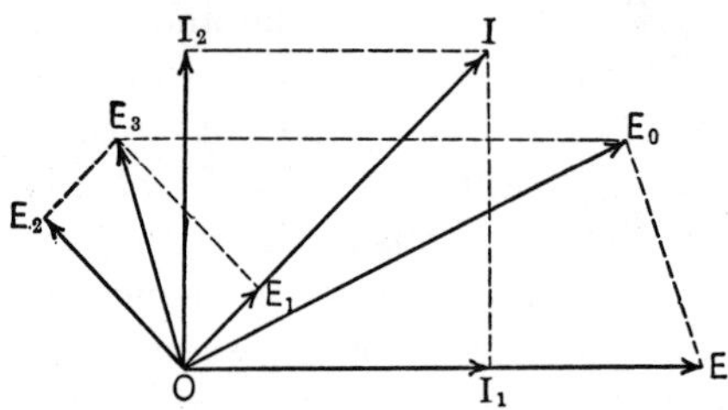

FIG. 19.—Vector diagram of e.m.f. and current in transmission line. Current leading.

e.m.f. of the generator shall be maintained constant at all loads, and the voltage regulation effected by producing lagging or leading currents with a synchronous motor in the receiving circuit. The line has a resistance $r_1 = 7.6$ ohms and a reactance $x_1 = 4.35$ ohms per wire, the generator is star connected, the resistance per circuit being $r_2 = 0.71$, and the (synchronous) reactance is $x_2 = 25$ ohms. What must be the wattless or reactive component of the current, and therefore the total current and its phase relation at no load, one-quarter load, one-half load, three-quarters load, and full load, and what will be the terminal voltage of the generator under these conditions?

The total resistance of the line and generator is $r = r_1 + r_2 = 8.31$ ohms; the total reactance, $x = x_1 + x_2 = 29.35$ ohms.

Let, in the polar diagram, Fig. 19 or 20, $\overline{OE} = E$ represent the voltage at the receiving end of the line, $\overline{OI}_1 = I_1$ the power component of the current corresponding to the load, in phase with $\overline{OE}$, and $\overline{OI}_2 = I_2$ the reactive component of the current in quadrature with $\overline{OE}$, shown leading in Fig. 19, lagging in Fig. 20.

We then have total current $I = \overline{OI}$.

Thus the e.m.f. consumed by resistance, $\overline{OE_1} = rI$, is in phase with I, the e.m.f. consumed by reactance, $\overline{OE_2} = xI$, is 90 degrees ahead of I, and their resultant is OE_3, the e.m.f. consumed by impedance.

$\overline{OE_3}$ combined with $\overline{OE}$, the receiver voltage, gives the generator voltage $\overline{OE_0}$.

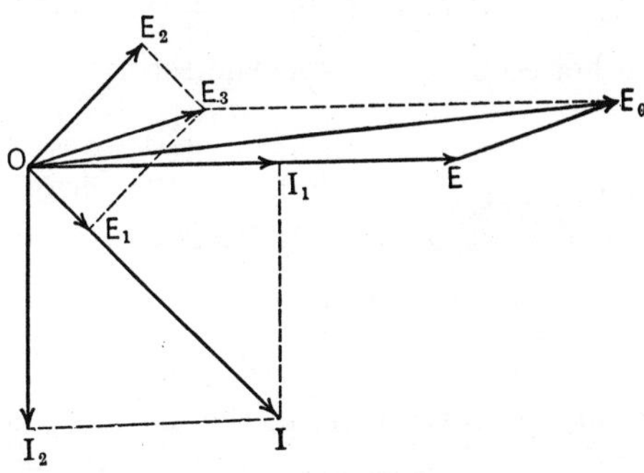

Fig. 20.—Vector diagram of e.m.f. and current in transmission line. Current lagging.

Resolving all e.m.fs. and currents into components in phase and in quadrature with the received voltage E, we have

	Phase component	Quadrature component
Current	I_1	$-I_2$
E.m.f. at receiving end of line, E =	E	0
E.m.f. consumed by resistance, E_1 =	rI_1	$-rI_2$
E.m.f. consumed by reactance, E_2 =	xI_2	$+xI_1$

Thus total e.m.f. or generator voltage,

$$E_0 = E + E_1 + E_2 = \quad E + rI_1 + xI_2 \quad xI_1 - rI_2$$

Herein the reactive lagging component of current is assumed as positive, the leading as negative.

The generator e.m.f. thus consists of two components, which give the resultant value

$$E_0 = \sqrt{(E + rI_1 + xI_2)^2 + (xI_1 - rI_2)^2};$$

substituting numerical values, this becomes

$$E_0 = \sqrt{(E + 8.31\,I_1 + 29.35\,I_2)^2 + (29.35\,I_1 - 8.31\,I_2)^2}.$$

At three-quarters load,

$$E = \frac{5375}{\sqrt{3}} = 3090 \text{ volts per circuit,}$$

$I_1 = 33, \qquad I_2 = 0$, thus

$E_0 = \sqrt{(3090 + 8.31 \times 33)^2 + (29.35 \times 33)^2} = 3520$ volts per line or $3520 \times \sqrt{3} = 6100$ volts between lines as (nominal) generated e.m.f. of generator.

Substituting these values, we have

$$3520 = \sqrt{(E + 8.31\,I_1 + 29.35\,I_2)^2 + (8.31\,I_2 - 29.35\,I_1)^2}.$$

The voltage between the lines at the receiving end shall be:

	No load	¼ load	½ load	¾ load	Full load
Voltage between lines,	5000	5125	5250	5375	5500
Thus, voltage per line $\div \sqrt{3}$, $E =$	2880	2950	3020	3090	3160
The power components of current per line, $I_1 =$	0	11	22	33	44

Herefrom we get by substituting in the above equation

	No load	¼ load	½ load	¾ load	Full load
Reactive component of current, $I_2 =$	−21.6	−16.2	−9.2	0	+9.7
hence, the total current, $I = \sqrt{I_1^2 + I_2^2} =$	21.6	19.6	23.9	33.0	45.05
and the power factor, $\frac{I_1}{I} = \cos\theta =$	0	56.0	92.0	100.0	97.7
the lag of the current, $\theta =$	90°	61°	23°	0°	−11.5°

the generator terminal voltage per line is

$$E' = \sqrt{(E + r_1 I_1 + x_1 I_2)^2 + (x_1 I_1 - r_1 I_2)^2}$$
$$= \sqrt{(E + 7.6\,I_1 + 4.35\,I_2)^2 + (4.35\,I_1 - 7.6\,I_2)^2}$$

thus:

	No load	¼ load	½ load	¾ load	Full load
Per line, $E' =$	2980	3106	3228	3344	3463
Between lines, $E'\sqrt{3} =$	5200	5400	5600	5800	6000

Therefore at constant excitation the generator voltage rises with the load, and is approximately proportional thereto.

10. HYSTERESIS AND EFFECTIVE RESISTANCE

46. If an alternating current $\overline{OI} = I$, in Fig. 21, exists in a circuit of reactance $x = 2\,\pi f L$ and of negligible resistance, the magnetic flux produced by the current, $\overline{O\Phi} = \Phi$, is in phase with the current, and the e.m.f. generated by this flux, or counter e.m.f. of self-inductance, $\overline{OE'''} = E''' = xI$, lags 90 degrees behind the current. The e.m.f. consumed by self-inductance or impressed e.m.f. $\overline{OE''} = E'' = xI$ is thus 90 degrees ahead of the current.

Inversely, if the e.m.f. $\overline{OE''} = E''$ is impressed upon a circuit of reactance $x = 2\pi fL$ and of negligible resistance, the current $\overline{OI} = I = \frac{E''}{x}$ lags 90 degrees behind the impressed e.m.f.

This current is called the *exciting* or magnetizing current of the magnetic circuit, and is wattless.

If the magnetic circuit contains iron or other magnetic material, energy is consumed in the magnetic circuit by a frictional resistance of the material against a change of magnetism, which is called *molecular magnetic friction.*

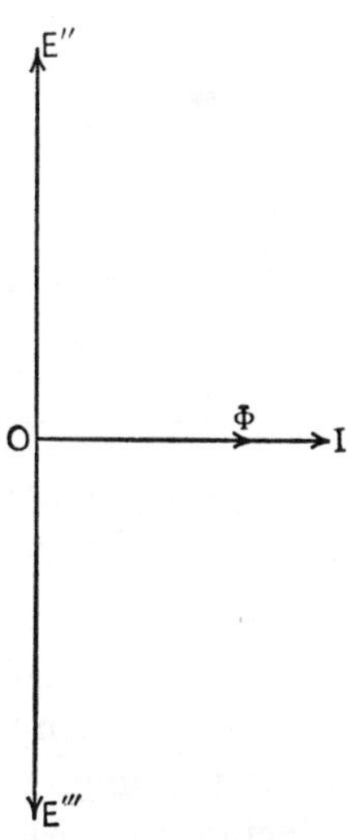

FIG. 21.—Phase relations of magnetizing current, flux and self-inductive e.m.f.

If the alternating current is the only available source of energy in the magnetic circuit, the expenditure of energy by molecular magnetic friction appears as a lag of the magnetism behind the m.m.f. of the current, that is, as *magnetic hysteresis*, and can be measured thereby.

Magnetic hysteresis is, however, a distinctly different phenomenon from molecular magnetic friction, and can be more or less eliminated, as for instance by mechanical vibration, or can be increased, without changing the molecular magnetic friction.

47. In consequence of magnetic hysteresis, if an alternating e.m.f. $\overline{OE''} = E''$ is impressed upon a circuit of negligible resistance, the exciting current, or current producing the magnetism, in this circuit is not a wattless current, or current of 90 degrees lag, as in Fig. 21, but lags less than 90 degrees, by an angle $90 - \alpha$, as shown by $\overline{OI} = I$ in Fig. 22.

Since the magnetism $\overline{O\Phi} = \Phi$ is in quadrature with the e.m.f. E'' due to it, angle α is the phase difference between the magnetism and the m.m.f., or the lead of the m.m.f., that is, the exciting current, before the magnetism. It is called the *angle of hysteretic lead*.

In this case the *exciting current* $\overline{OI} = I$ can be resolved in two components: the *magnetizing current* $\overline{OI_2} = I_2$, in phase with the magnetism $\overline{O\Phi} = \Phi$, that is, in quadrature with the e.m.f. $\overline{OE''} = E''$, and thus wattless, and the *magnetic power component of the current* or the *hysteresis current* $\overline{OI_1} = I_1$, in phase with the e.m.f. $OE'' = E''$, or in quadrature with the magnetism $\overline{O\Phi} = \Phi$.

Magnetizing current and hysteresis current are the two components of the exciting current.

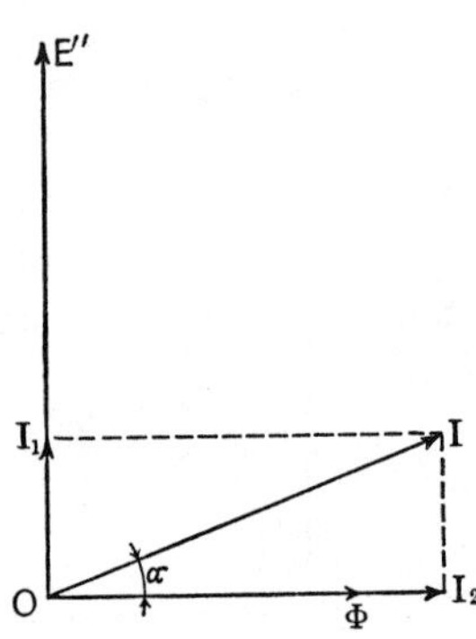

Fig. 22.—Angle of hysteretic lead.

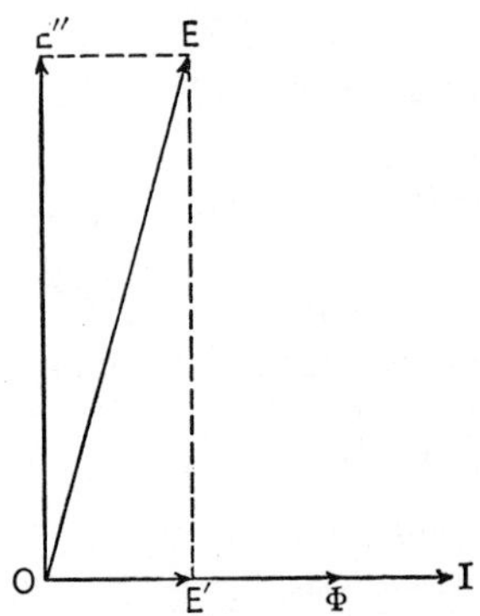

Fig. 23.—Effect of resistance on phase relation of impressed e.m.f. in a hysteresisless circuit.

If the circuit contains besides the reactance $x = 2\,\pi f L$, a resistance r, the e.m.f. $\overline{OE''} = E''$ in the preceding Figs. 21 and 22 is not the impressed e.m.f., but the e.m.f. consumed by self-inductance or reactance, and has to be combined, Figs. 23 and 24, with the e.m.f. consumed by the resistance, $\overline{OE'} = E' = Ir$, to get the impressed e.m.f. $\overline{OE} = E$.

Due to the hysteretic lead α, the lag of the current is less in Figs. 22 and 24, a circuit expending energy in molecular magnetic friction, than in Figs. 21 and 23, a hysteresisless circuit.

As seen in Fig. 24, in a circuit whose ohmic resistance is not negligible, the hysteresis current and the magnetizing current are not in phase and in quadrature respectively with the impressed e.m.f., but with the counter e.m.f. of inductance or e.m.f. consumed by inductance.

Obviously the magnetizing current is not quite wattless, since energy is consumed by this current in the ohmic resistance of the circuit.

Resolving, in Fig. 25, the impressed e.m.f. $\overline{OE} = E$ into two components, $\overline{OE_1} = E_1$ in phase, and $\overline{OE_2} = E_2$ in quadrature with the current $\overline{OI} = I$, the power component of the e.m.f., $\overline{OE_1} = E_1$, is greater than $E' = Ir$, and the reactive component $\overline{OE_2} = E_2$ is less than $E'' = Ix$.

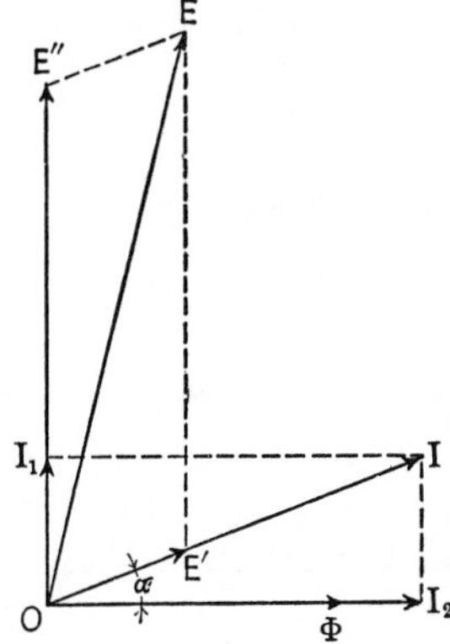

FIG. 24.—Effect of resistance on phase relation of impressed e.m.f. in a circuit having hysteresis.

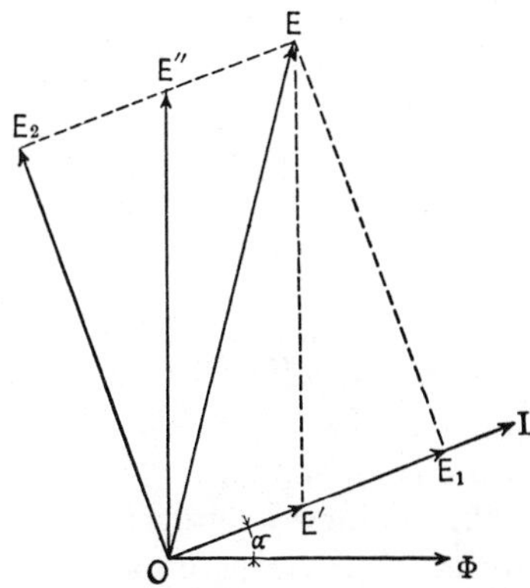

FIG. 25.—Impressed e.m.f. resolved into components in phase and in quadrature with the exciting current.

The value $r' = \frac{E_1}{I} = \frac{\text{power e.m.f.}}{\text{total current}}$ is called the *effective resistance*, and the value $x' = \frac{E_2}{I} = \frac{\text{wattless e.m.f.}}{\text{total current}}$ is called the *apparent* or *effective reactance* of the circuit.

48. Due to the loss of energy by hysteresis (eddy currents, etc.), the effective resistance differs from, and is greater than, the ohmic resistance, and the apparent reactance is less than the true or inductive reactance.

The loss of energy by molecular magnetic friction per cubic centimeter and cycle of magnetism is approximately

$$W = \eta B^{1.6},$$

where B = the magnetic flux density, in lines per sq. cm.

W = energy, in absolute units or ergs per cycle ($= 10^{-7}$ watt-seconds or joules), and η is called the coefficient of hysteresis.

In soft annealed sheet iron or sheet steel and in silicon steel, η varies from 0.60×10^{-3} to 2.5×10^{-3}, and can in average, for good material, be assumed as 1.5×10^{-3}.

The loss of power in the volume, V, at flux density B and frequency f, is thus

$$P = Vf\eta B^{1.6} \times 10^{-7}, \text{ in watts,}$$

and, if I = the exciting current, the hysteretic effective resistance is

$$r'' = \frac{P}{I^2} = Vf\eta\, 10^{-7} \frac{B^{1.6}}{I^2}.$$

If the flux density, B, is proportional to the current, I, substituting for B, and introducing the constant k, we have

$$r'' = \frac{kf}{I^{0.4}},$$

that is, the effective hysteretic resistance is inversely proportional to the 0.4 power of the current, and directly proportional to the frequency.

49. Besides hysteresis, eddy or Foucault currents contribute to the effective resistance.

Since at constant frequency the Foucault currents are proportional to the magnetism producing them, and thus approximately proportional to the current, the loss of power by Foucault currents is proportional to the square of the current, the same as the ohmic loss, that is, the effective resistance due to Foucault currents is approximately constant at constant frequency, while that of hysteresis decreases slowly with the current.

Since the Foucault currents are proportional to the frequency, their effective resistance varies with the square of the frequency, while that of hysteresis varies only proportionally to the frequency.

The total effective resistance of an alternating-current circuit increases with the frequency, but is approximately constant, within a limited range, at constant frequency, decreasing somewhat with the increase of magnetism.

EXAMPLES

50. A reactive coil shall give 100 volts e.m.f. of self-inductance at 10 amp. and 60 cycles. The electric circuit consists of 200 turns (No. 8 B. & S.) (= 0.013 sq. in.) of 16 in. mean length of turn. The magnetic circuit has a section of 6 sq. in. and a mean length of 18 in. of iron of hysteresis coefficient $\eta = 2.5 \times 10^{-3}$. An air gap is interposed in the magnetic circuit, of a section of 10 sq. in. (allowing for spread), to get the desired reactance.

How long must the air gap be, and what is the resistance, the reactance, the effective resistance, the effective impedance, and the power-factor of the reactive coil?

The coil contains 200 turns each 16 in. in length and 0.013 sq. in. in cross section. Taking the resistivity of copper as 1.8×10^{-6}, the resistance is

$$r_1 = \frac{1.8 \times 10^{-6} \times 200 \times 16}{0.013 \times 2.54} = 0.175 \text{ ohm},$$

where 2.54 is the factor for converting inches to centimeters. (1 inch = 2.54 cm.)

Writing $E = 100$ volts generated, $f = 60$ cycles per second, and $n = 200$ turns, the maximum magnetic flux is given by $E = 4.44\ fn\Phi$; or, $100 = 4.44 \times 0.6 \times 200\ \Phi$, and $\Phi = 0.188$ megaline.

This gives in an air gap of 10 sq. in. a maximum density $B = 18{,}800$ lines per sq. in., or 2920 lines per sq. cm.

Ten amperes in 200 turns give 2000 ampere-turns effective or $F = 2830$ ampere-turns maximum.

Neglecting the ampere-turns required by the iron part of the magnetic circuit as relatively very small, 2830 ampere-turns have to be consumed by the air gap of density $B = 2920$.

Since
$$B = \frac{4\,\pi F}{10\,l},$$

the length of the air gap has to be

$$l = \frac{4\,\pi F}{10\,B} = \frac{4\,\pi \times 2830}{10 \times 2920} = 1.22 \text{ cm., or } 0.48 \text{ in.}$$

With a cross section of 6 sq. in. and a mean length of 18 in., the volume of the iron is 108 cu. in., or 1770 cu. cm.

The density in the iron, $B_1 = \dfrac{188{,}000}{6} = 31{,}330$ lines per sq. in., or 4850 lines per sq. cm.

With an hysteresis coefficient $\eta = 2.5 \times 10^{-3}$, and density $B_1 = 4850$, the loss of energy per cycle per cubic centimeter is

$$\begin{aligned} W &= \eta B_1^{1.6} \\ &= 2.5 \times 10^{-3} \times 4850^{1.6} \\ &= 1980 \text{ ergs}, \end{aligned}$$

and the hysteresis loss at $f = 60$ cycles and the volume $V = 1770$ is thus

$$P = 60 \times 1770 \times 1980 \text{ ergs per sec.}$$
$$= 21.0 \text{ watts,}$$

which at 10 amp. represent an effective hysteretic resistance,

$$r_2 = \frac{21.0}{10^2} = 0.21 \text{ ohm.}$$

Hence the total effective resistance of the reactive coil is

$$r = r_1 + r_2 = 0.175 + 0.21 = 0.385 \text{ ohm;}$$

the effective reactance is

$$x = \frac{E}{I} = 10 \text{ ohms;}$$

the impedance is

$$z = 10.01 \text{ ohms;}$$

the power-factor is

$$\cos \theta = \frac{r}{z} = 3.8 \text{ per cent.;}$$

the total apparent power of the reactive coil is

$$I^2 z = 1001 \text{ volt-amperes,}$$

and the loss of power,

$$I^2 r = 38 \text{ watts.}$$

11. CAPACITY AND CONDENSERS

51. The charge of an electric condenser is proportional to the impressed voltage, that is, potential difference at its terminals, and to its capacity.

A condenser is said to have unit capacity if unit current existing for one second produces unit difference of potential at its terminals.

The practical unit of capacity is that of a condenser in which 1 amp. during one second produces 1 volt difference of potential.

The practical unit of capacity equals 10^{-9} absolute units. It is called a farad.

One farad is an extremely large capacity, and therefore one millionth of one farad, called microfarad, mf., is commonly used.

If an alternating e.m.f. is impressed upon a condenser, the charge of the condenser varies proportionally to the e.m.f., and thus there is current to the condenser during rising and from the condenser during decreasing e.m.f., as shown in Fig. 26.

That is, the current consumed by the condenser leads the impressed e.m.f. by 90 time degrees, or a quarter of a period.

Denoting f as frequency and E as effective alternating e.m.f. impressed upon a condenser of C mf. capacity, the condenser is charged and discharged twice during each cycle, and the time of one complete charge or discharge is therefore $\frac{1}{4f}$.

Since $E\sqrt{2}$ is the maximum voltage impressed upon the condenser, an average of $CE\sqrt{2}\,10^{-6}$ amp. would have to exist during one second to charge the condenser to this voltage, and

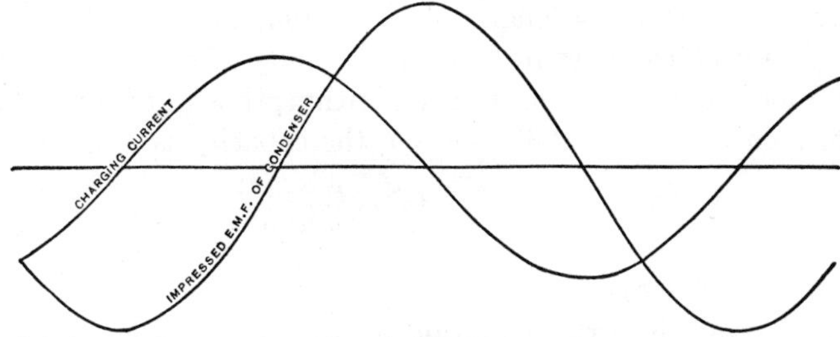

FIG. 26.—Charging current of a condenser across which an alternating e.m.f. is impressed.

to charge it in $\frac{1}{4f}$ seconds an average current of $4\,fCE\sqrt{2}\,10^{-6}$ amp. is required.

Since
$$\frac{\text{effective current}}{\text{average current}} = \frac{\pi}{2\sqrt{2}},$$
the effective current is $I = 2\pi fCE\,10^{-6}$; that is, at an impressed e.m.f. of E effective volts and frequency f, a condenser of C mf. capacity consumes a current of

$$I = 2\,\pi fCE\,10^{-6} \text{ amp. effective,}$$

which current leads the terminal voltage by 90 degrees or a quarter period.

Transposing, the e.m.f. of the condenser is

$$E = \frac{10^6\,I}{2\,\pi fC} = x_0 I.$$

The value $x_0 = \frac{10^6}{2\,\pi fC}$ is called the *condensive reactance* of the condenser.

Due to the energy loss in the condenser by dielectric hysteresis, the current leads the e.m.f. by somewhat less than 90 time degrees, and can be resolved into a *wattless charging current* and a *dielectric hysteresis current,* which latter, however, is generally so small as to be negligible, though in underground cables of poor quality, it may reach as high as 50 per cent. or more of the charging or wattless current of the condenser.

52. The capacity of one wire of a transmission line is

$$C = \frac{1.11 \times 10^{-6} \times l}{2 \log_\epsilon \frac{2\, l_s}{l_d}}, \text{ in mf.,}$$

where l_d = diameter of wire, cm.; l_s = distance of wire from return wire, cm.; l = length of wire, cm., and 1.11×10^{-6} = reduction coefficient from electrostatic units to mf.

The logarithm is the natural logarithm; thus in common logarithms, since $\log_\epsilon a = 2.303 \log_{10} a$, the capacity is

$$C = \frac{0.24 \times 10^{-6} \times l}{\log_{10} \frac{2\, l_s}{l_d}}, \text{ in mf.}$$

The derivation of this equation must be omitted here.

The charging current of a line wire is thus

$$I = 2\, \pi f C E\ 10^{-6},$$

where f = the frequency, in cycles per second, E = the difference of potential, effective, between the line and the neutral (E = $\frac{1}{2}$ line voltage in a single-phase, or four-wire quarter-phase system, $\frac{1}{\sqrt{3}}$ line voltage, or Y voltage, in a three-phase system).

EXAMPLES

53. In the transmission line discussed in the examples in 37, 38, 41 and 45, what is the charging current of the line at 6000 volts between lines, at 33.3 cycles? How many volt-amperes does it represent, and what percentage of the full-load current of 44 amp. is it?

The length of the line is, per wire, $l = 2.23 \times 10^6$ cm.
The distance between wires, $l_s = 45$ cm.
The diameter of transmission wire, $l_d = 0.82$ cm.

Thus the capacity, per wire, is

$$C = \frac{0.24 \times 10^{-6}\, l}{\log_{10} \frac{2\, l_s}{l_d}} = 0.26 \text{ mf.}$$

The frequency is $f = 33.3$,
The voltage between lines, 6000.

Thus per line, or between line and neutral point,

$$E = \frac{6000}{\sqrt{3}} = 3460 \text{ volts;}$$

hence, the charging current per line is

$$I_0 = 2\,\pi f C E\, 10^{-6}$$
$$= 0.19 \text{ amp.,}$$

or 0.43 per cent. of full-load current;

that is, negligible in its influence on the transmission voltage.

The volt-ampere input of the transmission is,

$$3\,I_0 E = 2000$$
$$= 2.0 \text{ kv-amp.}$$

12. IMPEDANCE OF TRANSMISSION LINES

54. Let r = resistance; $x = 2\,\pi f L$ = the reactance of a transmission line; E_0 = the alternating e.m.f. impressed upon the line; I = the line current; E = the e.m.f. at receiving end of the line, and θ = the angle of lag of current I behind e.m.f. E.

$\theta < 0$ thus denotes leading, $\theta > 0$ lagging current, and $\theta = 0$ a non-inductive receiver circuit.

The capacity of the transmission line shall be considered as negligible.

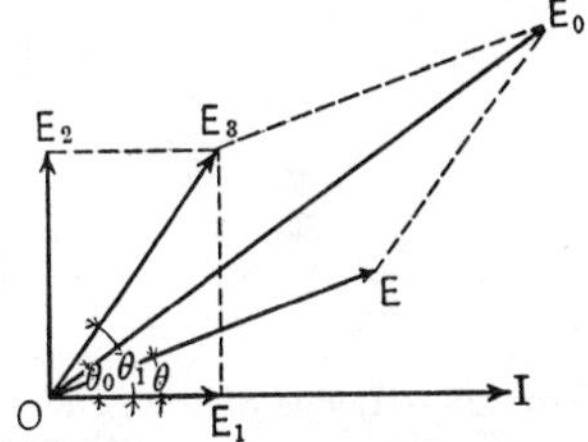

FIG. 27.—Vector diagram of current and e.m.fs. in a transmission line assuming zero capacity.

Assuming the phase of the current $\overline{OI} = I$ as zero in the polar diagram, Fig. 27, the e.m.f. E is represented by vector $\overline{OE}$, ahead of $\overline{OI}$ by angle θ. The e.m.f. consumed by resistance r is $\overline{OE_1} = E_1 = Ir$ in phase with the current, and the e.m.f. consumed by reactance x is $\overline{OE_2} = E_2 = Ix$, 90 time degrees ahead of the current; thus the total e.m.f. consumed by the line, or e.m.f. consumed by impedance, is the resultant $\overline{OE_3}$ of $\overline{OE_1}$ and $\overline{OE_2}$, and is $E_3 = Iz$.

Combining $\overline{OE_3}$ and $\overline{OE}$ gives $\overline{OE_0}$, the e.m.f. impressed upon the line.

Denoting $\tan \theta_1 = \frac{x}{r}$ the time angle of lag of the line impedance, it is, trigonometrically,

$$\overline{OE_0}^2 = \overline{OE}^2 + \overline{EE_0}^2 - 2\,\overline{OE} \times \overline{EE_0} \cos OEE_0.$$

Since

$$\overline{EE_0} = \overline{OE_3} = Iz,$$
$$OEE_0 = 180 - \theta_1 + \theta,$$

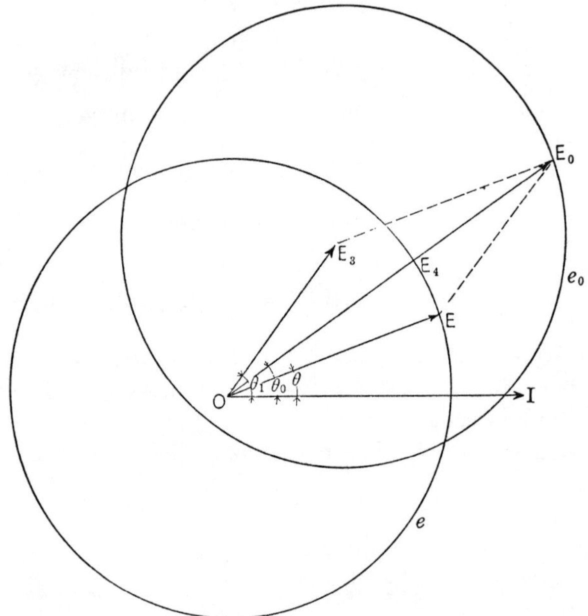

Fig. 28.—Locus of the generator and receiver e.m.fs. in a transmission line with varying load phase angle.

we have

$$E_0^2 = E^2 + I^2z^2 + 2\,EIz \cos(\theta_1 - \theta)$$
$$= (E + Iz)^2 - 4\,EIz \sin^2 \frac{\theta_1 - \theta}{2},$$

and

$$E_0 = \sqrt{(E + Iz)^2 - 4\,EIz \sin^2 \frac{\theta_1 - \theta}{2}},$$

and the drop of voltage in the line,

$$E_0 - E = \sqrt{(E + Iz)^2 - 4\,EIz \sin^2 \frac{\theta_1 - \theta}{2}} - E.$$

55. That is, the voltage E_0 required at the sending end of a line of resistance r and reactance x, delivering current I at voltage E, and the voltage drop in the line, do not depend upon current and line constants only, but depend also upon the angle of phase displacement of the current delivered over the line.

If $\theta = 0$, that is, non-inductive receiving circuit,

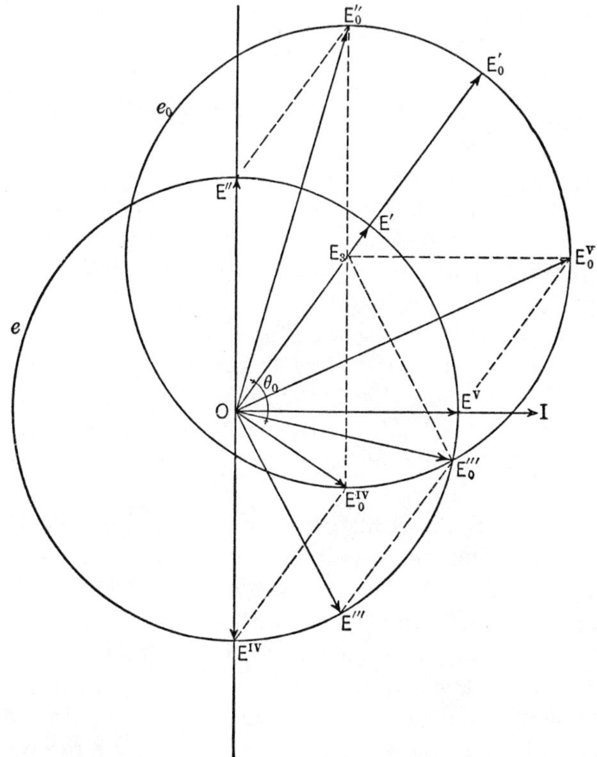

Fig. 29.—Locus of the generator and receiver e.m.fs. in a transmission line with varying load phase angle.

$$E_0 = \sqrt{(E + Iz)^2 - 4\ EIz \sin^2 \frac{\theta_1}{2}};$$

that is, less than $E + Iz$, and thus the line drop is less than Iz.

If $\theta = \theta_1$, E_0 is a maximum, $= E + Iz$, and the line drop is the impedance voltage.

With decreasing θ, E_0 decreases, and becomes $= E$; that is, no drop of voltage takes place in the line at a certain negative value of θ which depends not only on z and θ_1 but on E and I. Beyond this value of θ, E_0 becomes smaller than E; that is, a rise of voltage takes place in the line, due to its reactance. This can be seen best graphically.

Choosing the current vector $\overline{OI}$ as the horizontal axis, for the same e.m.f. E received, but different phase angles θ, all vectors $\overline{OE}$ lie on a circle e with O as center. Fig. 28. Vector $\overline{OE_3}$ is constant for a given line and given current I.

Since $E_3E_0 = \overline{OE}$ = constant, E_0 lies on a circle e_0 with E_3 as center and $\overline{OE} = E$ as radius.

To construct the diagram for angle θ, $\overline{OE}$ is drawn at the angle θ with $\overline{OI}$, and $\overline{EE_0}$ parallel to $\overline{OE_3}$.

The distance E_4E_0 between the two circles on vector $\overline{OE_0}$ is the drop of voltage (or rise of voltage) in the line.

As seen in Fig. 29, E_0 is maximum in the direction $\overline{OE_3}$ as $\overline{OE'_0}$, that is, for $\theta = \theta_0$, and is less for greater as well, $\overline{OE''_0}$, as smaller angles θ. It is $= E$ in the direction $\overline{OE'''_0}$, in which case $\theta < 0$, and minimum in the direction $\overline{OE^{IV}_0}$.

The values of E corresponding to the generator voltages E'_0, E''_0, E'''_0, E^{IV}_0 are shown by the points E' E'' E''' E^{IV} respectively. The voltages E''_0 and E^{IV}_0 correspond to a wattless receiver circuit E'' and E^{IV}. For non-inductive receiver circuit E^V the generator voltage is $\overline{OE^V_0}$.

56. That is, in an inductive transmission line the drop of voltage is maximum and equal to Iz if the phase angle θ of the receiving circuit equals the phase angle θ_0 of the line. The drop of voltage in the line decreases with increasing difference between the phase angles of line and receiving circuit. It becomes zero if the phase angle of the receiving circuit reaches a certain negative value (leading current). In this case no drop of voltage takes place in the line. If the current in the receiving circuit leads more than this value a rise of voltage takes place in the line. Thus by varying phase angle θ of the receiving circuit the drop of voltage in a transmission line with current I can be made anything between Iz and a certain negative value. Or inversely the same drop of voltage can be produced for different values of the current I by varying the phase angle.

Thus, if means are provided to vary the phase angle of the receiving circuit, by producing lagging and leading currents at will (as can be done by synchronous motors or converters), the voltage at the receiving circuit can be maintained constant within a certain range irrespective of the load and generator voltage.

In Fig. 30 let $\overline{OE} = E$, the receiving voltage; I, the power component of the line current; thus $\overline{OE_3} = E_3 = Iz$, the e.m.f. consumed by the power component of the current in the impedance. This e.m.f. consists of the e.m.f consumed by resistance $\overline{OE_1}$ and the e.m.f. consumed by reactance $\overline{OE_2}$.

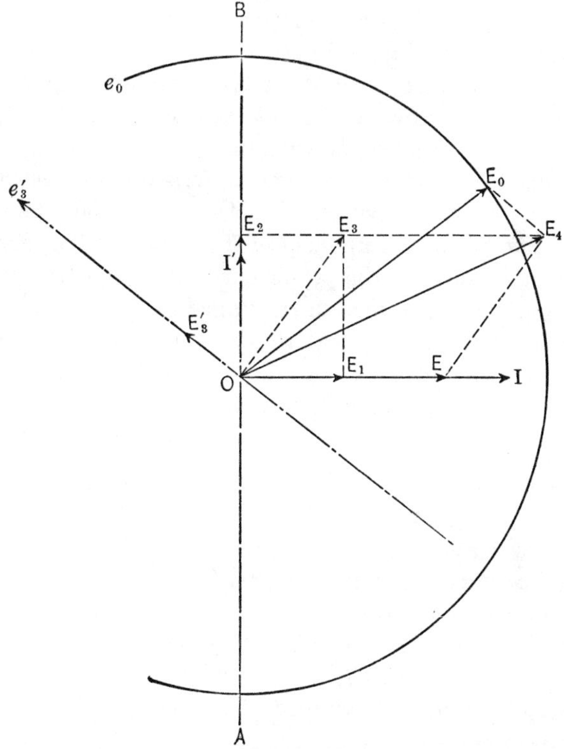

Fig. 30.—Regulation diagram for transmission line.

Reactive components of the current are represented in the diagram in the direction $\overline{OA}$ when lagging and $\overline{OB}$ when leading. The e.m.f. consumed by these reactive components of the current in the impedance is thus in the direction e'_3, perpendicular to $\overline{OE_3}$. Combining $\overline{OE_3}$ and $\overline{OE}$ gives the e.m.f. $\overline{OE_4}$ which would be required for non-inductive load. If E_0 is the generator voltage, E_0 lies on a circle e_0 with $\overline{OE_0}$ as radius. Thus drawing $\overline{E_4E_0}$ parallel to e'_3 gives $\overline{OE_0}$, the generator voltage; $OE'_3 = \overline{E_4E_0}$, the e.m.f. consumed in the impedance by the reactive component of the current; and as proportional thereto, $\overline{OI'} = I'$, the reactive current required to give at generator voltage E_0 and power current I the receiver voltage E. This reactive current I' lags behind E'_3 by less than 90 and more than zero degrees.

57. In calculating numerical values, we can proceed either trigonometrically as in the preceding, or algebraically by resolving all sine waves into two rectangular components; for instance, a horizontal and a vertical component, in the same way as in mechanics when combining forces.

Let the horizontal components be counted positive toward the right, negative toward the left, and the vertical components positive upward, negative downward.

Assuming the receiving voltage as zero line or positive horizontal line, the power current I is the horizontal, the wattless current I' the vertical component of the current. The e.m.f. consumed in resistance by the power current I is a horizontal component, and that consumed in resistance by the reactive current I' a vertical component, and the inverse is true of the e.m.f. consumed in reactance.

We have thus, as seen from Fig. 30:

	Horizontal component	Vertical component
Receiver voltage, E,	$+E$	0
Power current, I,	$+I$	0
Reactive current, I',	0	$\mp I'$
E.m.f. consumed in resistance r by the power current, Ir,	$+Ir$	0
E.m.f. consumed in resistance r by the reactive current, $I'r$,	0	$\mp I'r$
E.m.f. consumed in reactance x by the power current, Ix,	0	$+Ix$
E.m.f. consumed in reactance x by the reactive current, $I'x$,	$\pm I'x$	0
Thus, total e.m.f. required, or impressed e.m.f., E_0,	$E + Ir \pm I'x$	$\mp I'r + Ix$;

hence, combined,

$$E_0 = \sqrt{(E + Ir \pm I'x)^2 + (\mp I'r + Ix)^2};$$

or, expanded,

$$E_0 = \sqrt{E^2 + 2E(Ir \pm I'x) + (I^2 + I'^2)z^2}.$$

From this equation I' can be calculated; that is, the reactive current found which is required to give E_0 and E at energy current I.

The lag of the total current in the receiver circuit behind the receiver voltage is

$$\tan \theta = \frac{I'}{I}.$$

The lead of the generator voltage ahead of the receiver voltage is

$$\tan \theta_1 = \frac{\text{vertical component of } E_0}{\text{horizontal component of } E_0}$$

$$= \frac{\pm I'r - Ix}{E + Ir \pm I'x},$$

and the lag of the total current behind the generator voltage is

$$\theta_0 = \theta + \theta_1.$$

As seen, by resolving into rectangular components the phase angles are directly determined from these components.

The resistance voltage is the same component as the current to which it refers.

The reactance voltage is a component 90 time degrees ahead of the current.

The same investigation as made here on long-distance transmission applies also to distribution lines, reactive coils, transformers, or any other apparatus containing resistance and reactance inserted in series into an alternating-current circuit.

EXAMPLES

58. (1) An induction motor has 2000 volts impressed upon its terminals; the current and the power-factor, that is, the cosine of the angle of lag, are given as functions of the output in Fig. 31.

The induction motor is supplied over a line of resistance $r = 2.0$ and reactance $x = 4.0$.

(*a*) How must the generator voltage e_0 be varied to maintain constant voltage $e = 2000$ at the motor terminals, and

(*b*) At constant generator voltage $e_0 = 2300$, how will the voltage at the motor terminals vary?

We have

$$e_0 = \sqrt{(e + iz)^2 - 4\,eiz \sin^2 \frac{\theta_1 - \theta}{2}}. \qquad e = 2000.$$

$$z = \sqrt{r^2 + x^2} = 4.472.$$

$$\tan \theta_1 = \frac{x}{r} = 2. \qquad \theta_1 = 63.4°.$$

$$\cos \theta = \text{power-factor}.$$

Taking i from Fig. 31 and substituting, gives (*a*) the values of e_0 for $e = 2000$, which are recorded in the table, and plotted in Fig. 31.

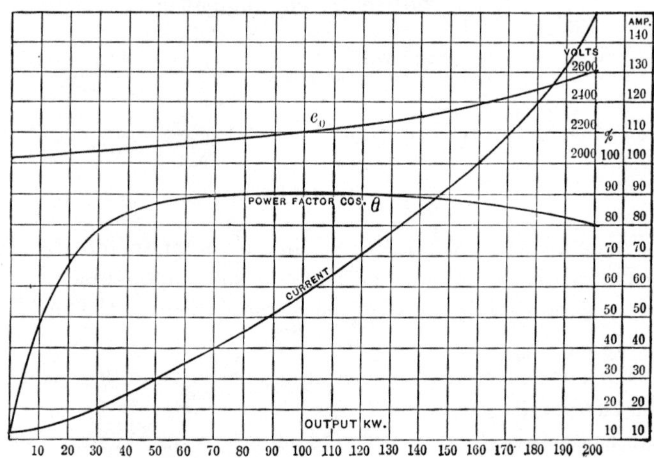

FIG. 31.—Characteristics of induction motor and variation of generator e.m.f. necessary to maintain constant the e.m.f. impressed upon the motor.

(*b*) At the terminal voltage of the motor $e = 2000$, the current is i, the output P, the generator voltage e_0. Thus at generator voltage $e'_0 = 2300$, the terminal voltage of the motor is

$$e' = \frac{2300}{e_0}\,e = \frac{2300}{e_0}\,2000;$$

the current is

$$i' = \frac{2300}{e_0}\,i$$

and the power is

$$P' = \left(\frac{2300}{e_0}\right)^2 P.$$

The values of e', i', P' are recorded in the second part of the table under (*b*) and plotted in Fig. 32.

(a) At e = 2000			Thus, e_0	(b) Hence, at e_0 = 2300		
Output, P = kw.	Current, i	Lag, θ		Output, P'	Current, i'	Voltage, e'
0	12.0	84.3°	2048	0.	13.45	2240
5	12.6	72.6°	2055	6.25	14.05	2234
10	13.5	62.6°	2060	12.4	15.00	2230
15	14.8	54.6°	2065	18.6	16.4	2220
20	16.3	47.9°	2071	24.4	18.0	2216
30	20.0	37.8°	2084	36.3	22.0	2200
40	25.0	32.8°	2093	48.0	27.5	2198
50	30.0	29.0°	2110	59.5	32.7	2180
69	40.0	26.3°	2146	78.5	42.8	2160
102	60.0	24.5°	2216	110.2	62.6	2080
132	80.0	25.8°	2294	131.0	79.5	1990
160	100.0	28.4°	2382	149.0	96.4	1928
180	120.0	31.8°	2476	156.5	111.5	1860
200	150.0	36.9°	2618	155.0	132.0	1760

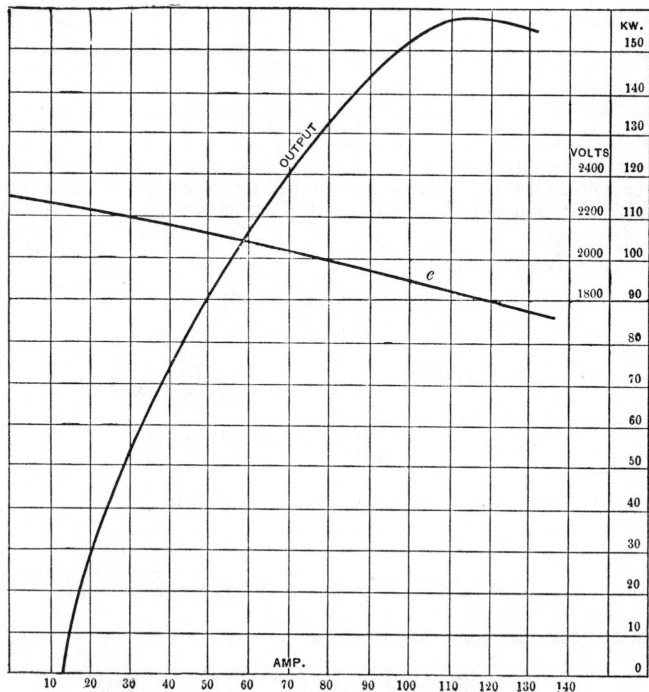

FIG. 32.—Characteristics of induction motor, constant generator e.m.f.

59. (2) Over a line of resistance $r = 2.0$ and reactance $x = 6.0$ power is supplied to a receiving circuit at a constant voltage of $e = 2000$. How must the voltage at the beginning of the line, or generator voltage, e_0, be varied if at no load the receiving circuit consumes a reactive current of $i_2 = 20$ amp., this reactive current decreases with the increase of load, that is, of power current i_1, becomes $i_2 = 0$ at $i_1 = 50$ amp., and then as leading current increases again at the same rate?

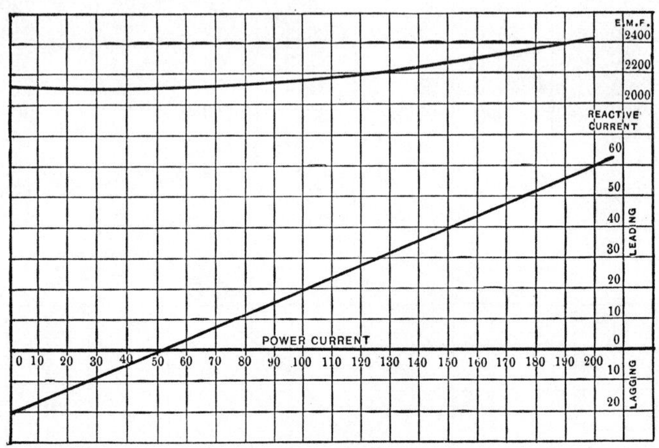

FIG. 33.—Variation of generator e.m.f. necessary to maintain constant receiver voltage if the reactive component of receiver current varies proportional to the change of power component of the current.

The reactive current,

$$i_2 = 20 \text{ at } i_1 = 0,$$
$$i_2 = 0 \text{ at } i_1 = 50,$$

and can be represented by

$$i_2 = \left(1 - \frac{i_1}{50}\right) 20 = 20 - 0.4\, i_1;$$

the general equation of the transmission line is

$$e_0 = \sqrt{(e + i_1 r + i_2 x)^2 + (i_2 r - i_1 x)^2}$$
$$= \sqrt{(2000 + 2\, i_1 + 6\, i_2)^2 + (2\, i_2 - 6\, i_1)^2};$$

hence, substituting the value of i_2,

$$e_0 = \sqrt{(2120 - 0.4\, i_1)^2 + (40 - 6.8\, i_1)^2}$$
$$= \sqrt{4{,}496{,}000 + 46.4\, i_1^2 - 2240\, i_1}.$$

Substituting successive numerical values for i_1 gives the values recorded in the following table and plotted in Fig. 33.

i_1	e_0
0	2120
20	2114
40	2116
60	2126
80	2148
100	2176
120	2213
140	2256
160	2308
180	2365
200	2430

13. ALTERNATING-CURRENT TRANSFORMER

60. The alternating-current transformer consists of one magnetic circuit interlinked with two electric circuits, the primary circuit which receives energy, and the secondary circuit which delivers energy.

Let r_1 = resistance, $x_1 = 2\pi f S_2$ = self-inductive or leakage reactance of secondary circuit,

r_0 = resistance, $x_0 = 2\pi f S_1$ = self-inductive or leakage reactance of primary circuit,

where S_2 and S_1 refer to that magnetic flux which is interlinked with the one but not with the other circuit.

Let a = ratio of $\frac{\text{secondary}}{\text{primary}}$ turns (ratio of transformation).

An alternating e.m.f. E_0 impressed upon the primary electric circuit causes a current, which produces a magnetic flux Φ interlinked with primary and secondary circuits. This flux Φ generates e.m.fs. E_1 and E_i in secondary and in primary circuit, which are to each other as the ratio of turns, thus $E_i = \frac{E_1}{a}$.

Let E = secondary terminal voltage, I_1 = secondary current, θ_1 = lag of current I_1 behind terminal voltage E (where $\theta_1 < 0$ denotes leading current).

Denoting then in Fig. 34 by a vector $\overline{OE} = E$ the secondary terminal voltage, $\overline{OI_1} = I_1$ is the secondary current lagging by the angle $EOI = \theta_1$.

The e.m.f. consumed by the secondary resistance r_1 is $\overline{OE'_1} = E'_1 = I_1 r_1$ in phase with I_1.

The e.m.f. consumed by the secondary reactance x_1 is $\overline{OE''_1} = E''_1 = I_1 x_1$, 90 degrees ahead of I_1. Thus the e.m.f. consumed by the secondary impedance $z_1 = \sqrt{r_1^2 + x_1^2}$ is the resultant of $\overline{OE'_1}$ and $\overline{OE''_1}$, or $\overline{OE'''_1} = E'''_1 = I_1 z_1$.

$\overline{OE'''_1}$ combined with the terminal voltage $\overline{OE} = E$ gives the secondary e.m.f. $\overline{OE_1} = E_1$.

Proportional thereto by the ratio of turns and in phase there-

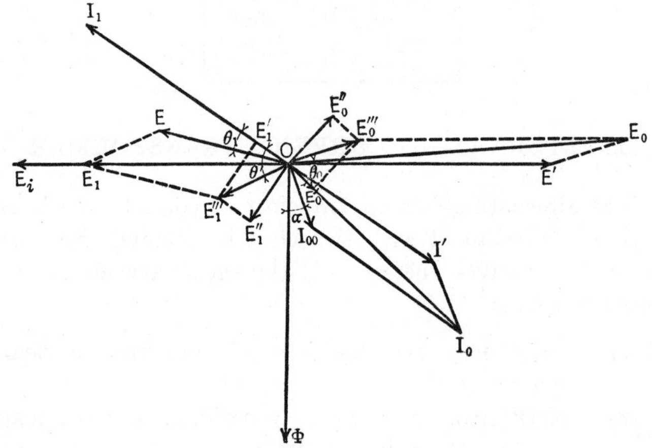

FIG. 34.—Vector diagram of e.m.fs. and currents in a transformer.

with is the e.m.f. generated in the primary $\overline{OE_i} = E_i$ where $E_i = \frac{E_1}{a}$.

To generate e.m.f. E_1 and E_i, the magnetic flux $\overline{O\Phi} = \Phi$ is required, 90 time degrees ahead of $\overline{OE_1}$ and $\overline{OE_i}$. To produce flux Φ the m.m.f. of F ampere-turns is required, as determined from the dimensions of the magnetic circuit, and thus the primary current I_{00}, represented by vector $\overline{OI_{00}}$, leading $\overline{O\Phi}$ by the angle α.

Since the total m.m.f. of the transformer is given by the primary exciting current I_{00}, there must be a component of primary current I', corresponding to the secondary current I_1, which may be called the primary load current, and which is opposite thereto and of the same m.m.f.; that is, of the intensity $I' = aI_1$, thus represented by vector $\overline{OI'} = I' = aI_1$.

$\overline{OI}_{00}$, the primary exciting current, and the primary load current $\overline{OI'}$, or component of primary current corresponding to the secondary current, combined, give the total primary current $\overline{OI}_0 = I_0$.

The e.m.f. consumed by resistance in the primary is $\overline{OE'}_0 = E'_0 = I_0r_0$ in phase with I_0.

The e.m.f. consumed by the primary reactance is $\overline{OE''}_0 = E''_0 = I_0x_0$, 90 degrees ahead of $\overline{OI}_0$.

$\overline{OE'}_0$ and $\overline{OE''}_0$ combined gives $\overline{OE'''}_0$, the e.m.f. consumed by the primary impedance.

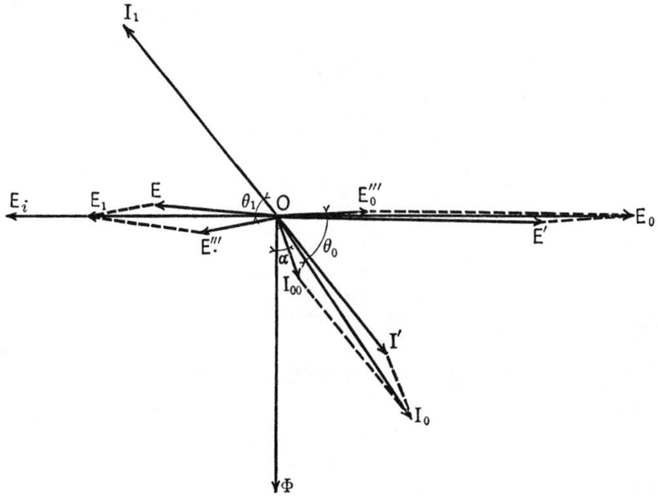

FIG. 35.—Vector diagram of transformer with lagging load current.

Equal and opposite to the primary counter-generated e.m.f. $\overline{OE}_i$ is the component of primary e.m.f., $\overline{OE'}$, consumed thereby.

$\overline{OE'}$ combined with $\overline{OE'''}_0$ gives $\overline{OE}_0 = E_0$, the primary impressed e.m.f., and angle $\theta_0 = E_0OI_0$, the phase angle of the primary circuit.

Figs. 35, 36, and 37 give the polar diagrams of $\theta_1 = 45°$ or lagging current, $\theta_1 =$ zero or non-inductive circuit, and $\theta = - 45°$ or leading current.

61. As seen, the primary impressed e.m.f. E_0 required to produce the same secondary terminal voltage E at the same current I_1 is larger with lagging or inductive and smaller with leading current than on a non-inductive secondary circuit; or, inversely, at the same secondary current I_1 the secondary terminal voltage E with lagging current is less and with leading current more than with non-inductive secondary circuit, at the same primary impressed e.m.f. E_0.

The calculation of numerical values is not practicable by measurement from the diagram, since the magnitudes of the different quantities are too different, $E'_1:E''_1:E_1:E_0$ being frequently in the proportion 1:10:100:2000.

Trigonometrically, the calculation is thus:

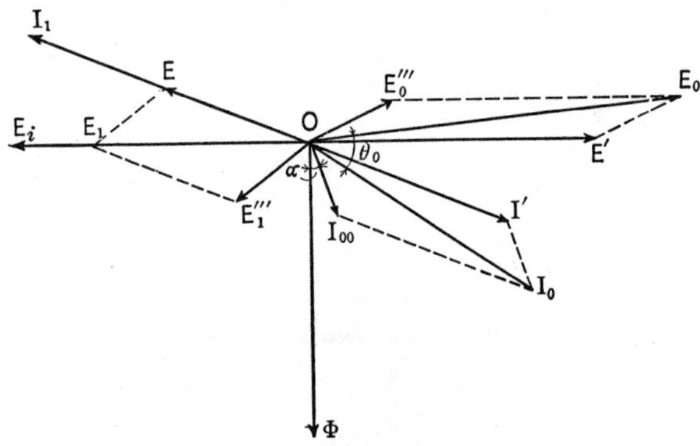

FIG. 36.—Vector diagram of transformer with non-inductive loading.

In triangle OEE_1, Fig. 34, writing

$$\tan \theta' = \frac{x_1}{r_1},$$

we have,

$$\overline{OE_1}^2 = \overline{OE}^2 + \overline{EE_1}^2 - 2\,\overline{OE}\,\overline{EE_1} \cos OEE_1;$$

also,

$$\overline{EE_1} = I_1 z_1$$

$$\sphericalangle OEE_1 = 180 - \theta' + \theta_1,$$

hence,

$$E_1^2 = E^2 + I_1^2 z_1^2 + 2\,E I_1 z_1 \cos(\theta' - \theta_1).$$

This gives the secondary e.m.f., E_1, and therefrom the primary counter-generated e.m.f.

$$E_i = \frac{E_1}{a}.$$

In triangle EOE_1 we have

$$\sin E_1OE \div \sin E_1EO = \overline{EE_1} \div \overline{E_1O}$$

thus, writing

$$\sphericalangle E_1OE = \theta'',$$

we have

$$\sin \theta'' \div \sin (\theta' - \theta_1) = I_1z \div E_1,$$

wherefrom we get

$$\sphericalangle \theta'', \text{ and } \sphericalangle E_1OI_1 = \theta = \theta_1 + \theta'',$$

the phase displacement between secondary current and secondary e.m.f.

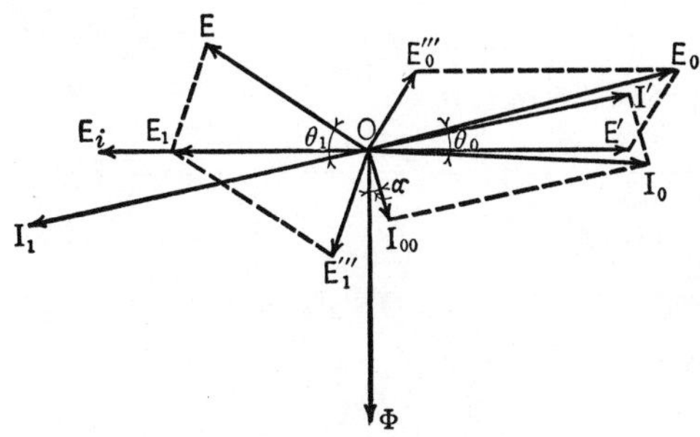

FIG. 37.—Vector diagram of transformer with leading load current.

In triangle $OI_{00}I_0$ we have

$$\overline{OI_0}^2 = \overline{OI_{00}}^2 + \overline{I_{00}I_0}^2 - 2\,\overline{OI_{00}}\,\overline{I_{00}I_0} \cos OI_{00}I_0,$$

since

$$\sphericalangle E_1O\phi = 90°,$$
$$\sphericalangle OI_{00}I_0 = 90 + \theta + \alpha,$$

and

$$\overline{I_{00}I_0} = I' = aI_1,$$
$$\overline{OI_{00}} = I_{00} = \text{exciting current},$$

calculated from the dimensions of the magnetic circuit. Thus the primary current is

$$I_0^2 = I_{00}^2 + a^2I_1^2 + 2\,aI_1I_{00} \sin (\theta + \alpha).$$

In triangle $OI_{00}I_0$ we have

$$\sin I_{00}OI_0 \div \sin OI_{00}I_0 = \overline{I_{00}I_0} \div \overline{OI_0};$$

writing

$$\sphericalangle I_{00}OI_0 = \theta''$$

this becomes

$$\sin \theta''_0 \div \sin (\theta + \alpha) = aI_1 \div I_0;$$

therefrom we get θ''_0, and thus

$$\sphericalangle E'OI_0 = \theta_2 = 90° - \alpha - \theta''_0.$$

In triangle $OE'E_0$ we have

$$\overline{OE_0}^2 = \overline{OE'}^2 + \overline{E'E_0}^2 - 2\,\overline{OE'}\;\overline{E'E_0}\cos \overline{OE'}\;E_0;$$

writing

$$\tan \theta'_0 = \frac{x_0}{r_0},$$

we have

$$\measuredangle\, OE'E_0 = 180° - \theta' + \theta_2,$$

$$\overline{OE'} = E_i = \frac{E_1}{a},$$

$$\overline{E'E_0} = I_0 z_0;$$

thus the impressed e.m.f. is

$$E_0^2 = \frac{E_1^2}{a^2} + I_0^2 z_0^2 + \frac{2E_1 I_0 z_0}{a}\cos(\theta_0' - \theta_2).$$

In triangle $OE'E_0$

$$\sin E'OE_0 \div \sin OE'E_0 = \overline{E'E_0} \div OE_0;$$

thus, writing

$$\measuredangle\, E'OE_0 = \theta_1'',$$

we have

$$\sin \theta''_1 \div \sin(\theta'_0 - \theta_2) = I_0 z_0 \div E_0;$$

herefrom we get $\measuredangle\, \theta''_1$, and

$$\measuredangle\, \theta_0 = \theta_2 + \theta''_1,$$

the phase displacement between primary current and impressed e.m.f.

As seen, the trigonometric method of transformer calculation is rather complicated.

62. Somewhat simpler is the algebraic method of resolving into rectangular components.

Considering first the secondary circuit, of current I_1 lagging behind the terminal voltage E by angle θ_1.

The terminal voltage E has the components $E \cos \theta_1$ in phase, $E \sin \theta_1$ in quadrature with and ahead of the current I_1.

The e.m.f. consumed by resistance r_1, $I_1 r_1$, is in phase.

The e.m.f. consumed by reactance x_1, $I_1 x_1$, is in quadrature ahead of I_1.

Thus the secondary e.m.f. has the components

$E \cos \theta_1 + I_1 r_1$ in phase,

$E \sin \theta_1 + I_1 x_1$ in quadrature ahead of the current I_1, and the total value,

$$E_1 = \sqrt{(E\cos\theta_1 + I_1 r_1)^2 + (E\sin\theta_1 + I_1 x_1)^2},$$

and the tangent of the phase angle of the secondary circuit is

$$\tan \theta = \frac{E \sin \theta_1 + I_1 x_1}{E \cos \theta_1 + I_1 r_1}.$$

Resolving all quantities into components in phase and in quadrature with the secondary e.m.f. E_1, or in horizontal and in vertical components, choosing the magnetism or mutual flux as vertical axis, and denoting the direction to the right and upward as positive, to the left and downward as negative, we have

	Horizontal component	Vertical component
Secondary current, I_1,	$-I_1 \cos \theta$	$+I_1 \sin \theta$
Secondary e.m.f., E_1,	$-E_1$	0
Primary counter-generated e.m.f., $E_1 = \frac{E_1}{a}$,	$-\frac{E_1}{a}$	0
Primary e.m.f. consumed thereby, $E' = -E_i$,	$+\frac{E_1}{a}$	0
Primary load current, $I' = -aI_1$,	$+aI_1 \cos \theta$	$-aI_1 \sin \theta$
Magnetic flux, Φ,	0	$-\Phi$
Primary exciting current, I_{00}, consisting of core loss current,	$I_{00} \sin \alpha$	
magnetizing current,		$-I_{00} \cos \alpha$

hence, total primary current, I_0,

Horizontal component	Vertical component
$aI_1 \cos \theta_1 + I_{00} \sin \alpha$	$-(aI_1 \sin \theta_1 + I_{00} \cos \alpha)$

E.m.f. consumed by primary resistance r_0, $E'_0 = I_0 r_0$ in phase with I_0,

Horizontal component	Vertical component
$r_0 aI_1 \cos \theta + r_0 I_{00} \sin \alpha$	$-(r_0 aI_1 \sin \theta + r_0 I_{00} \cos \alpha)$

E.m.f. consumed by primary reactance x_0, $E_0 = I_0 x_0$, 90° ahead of I_0,

Horizontal component	Vertical component
$x_0 aI_1 \sin \theta + x_0 I_{00} \cos \alpha$	$+x_0 aI_1 \cos \theta + x_0 I_{00} \sin \alpha$

E.m.f. consumed by primary generated e.m.f., $E' = \frac{E_1}{a}$ horizontal.

The total primary impressed e.m.f., E_0,

Horizontal component

$$\frac{E_1}{a} + aI_1 (r_0 \cos \theta + x_0 \sin \theta) + I_{00} (r_0 \sin \alpha + x_0 \cos \alpha).$$

Vertical component

$$aI_1 (r_0 \sin \theta - x_0 \cos \theta) + I_{00} (r_0 \cos \alpha - x_0 \sin \alpha),$$

or writing
$$\tan \theta'_0 = \frac{x_0}{r_0},$$

since

$$\sqrt{r_0^2 + x_0^2} = z_0, \sin \theta'_0 = \frac{x_0}{z_0}, \text{ and } \cos \theta'_0 = \frac{r_0}{z_0}.$$

Substituting this value, the horizontal component of E_0 is

$$\frac{E_1}{a} + az_0 I_1 \cos (\theta - \theta'_0) + z_0 I_{00} \sin (\alpha + \theta'_0);$$

the vertical component of E_0 is

$$az_0 I_1 \sin (\theta - \theta'_0) + z_0 I_{00} \cos (\alpha + \theta'_0),$$

and, the total primary impressed e.m.f. is

$$E_0 = \sqrt{\left[\frac{E_1}{a} + az_0 I_1 \cos(\theta - \theta'_0) + z_0 I_{00} \sin(\alpha + \theta'_0)\right]^2 + \left[az_0 I_1 \sin(\theta - \theta'_0) + z_0 I_{00} \cos(\alpha + \theta'_0)\right]^2}$$

$$= \frac{E_1}{a}\sqrt{1 + \frac{2a^2 z_0 I_1}{E_1}\cos(\theta - \theta'_0) + \frac{2az_0 I_{00}}{E_1}\sin(\alpha + \theta'_0) + \frac{a^4 z_0^2 I_1^2}{E_1^2} + \frac{a^2 z_0^2 I_{00}^2}{E_1^2} + \frac{2a^3 z_0^2 I_1 I_{00}}{E_1^2}\sin(\theta + \alpha)}.$$

Combining the two components, the total primary current is

$$I_0 = \sqrt{(aI_1 \cos \theta + I_{00} \sin \alpha)^2 + (aI_1 \sin \theta + I_{00} \cos \alpha)}$$

$$= aI_1 \sqrt{1 + \frac{2 I_{00}}{aI_1} \sin (\theta + \alpha) + \frac{I_{00}^2}{a^2 I_1^2}}.$$

Since the tangent of the phase angle is the ratio of vertical component to horizontal component, we have, primary e.m.f. phase,

$$\tan \theta' = \frac{az_0 I_1 \sin (\theta - \theta'_0) + z_0 I_{00} \cos (\alpha + \theta'_0)}{\frac{E_1}{a} + az_0 I_1 \cos (\theta - \theta'_0) + z_0 I_{00} \sin (\alpha - \theta'_0)}$$

primary current phase,

$$\tan \theta'' = \frac{aI_1 \sin \theta + I_{00} \cos \alpha}{aI_1 \cos \theta + I_{00} \sin \alpha}$$

and lag of primary current behind impressed e.m.f.,

$$\theta_0 = \theta'' - \theta'$$

EXAMPLES

63. (1) In a 20-kw. transformer the ratio of turns is 20 ÷ 1, and 100 volts is produced at the secondary terminals at full load. What is the primary current at full load, and the regulation, that is, the rise of secondary voltage from full load to no load, at constant primary voltage, and what is this primary voltage?

(*a*) at non-inductive secondary load,

(*b*) with 60 degrees time lag in the external secondary circuit,

(*c*) with 60 degrees time lead in the external secondary circuit.

The exciting current is 0.5 amp., the core loss 600 watts, the primary resistance 2 ohms, the primary reactance 5 ohms, the secondary resistance 0.004 ohm, the secondary reactance 0.01 ohm.

Exciting current and core loss may be assumed as constant.

600 watts at 2000 volts gives 0.3 amp. core loss current hence $\sqrt{0.5^2 - 0.3^2} = 0.4$ amp. magnetizing current.

We have thus

$$r_0 = 2 \qquad r_1 = 0.004 \qquad I_{00} \cos \alpha = 0.3 \qquad a = 0.05$$
$$x_0 = 5 \qquad x_1 = 0.01 \qquad I_{00} \sin \alpha = 0.4$$
$$I_{00} = 0.5$$

1. Secondary current as horizontal axis:

	Non-inductive, $\theta_1 = 0$		Lag, $\theta_1 = +60°$		Lead, $\theta_1 = -60°$	
	Hor.	Vert.	Hor.	Vert.	Hor.	Vert.
Secondary current, I_1..	200	0	200	0	200	0
Secondary terminal voltage, E..........	100	0	50	+86.6	50	−86.6
Resistance voltage, I_1r_1.	0.8	0	0.8	0	0.8	0
Reactance voltage, I_1x_1.	0	+2.0	0	+ 2.0	0	+ 2.0
Secondary e.m.f., E_1...	100.8	+2.0	50.8	+88.6	50.8	−84.6
Secondary e.m.f., total	100.80		102.13		98.68	
tan θ...............	+0.0198		+1.745		− 1.665	
θ..................	+1.1°		+60.2°		−59.0°	

2. Magnetic flux as vertical axis:

	Non-inductive, $\theta_1 = 0$		Lag, $\theta_1 = +60°$		Lead, $\theta_1 = -60°$	
	Hor.	Vert.	Hor.	Vert.	Hor.	Vert.
Secondary generated e.m.f., E	−100.80	0	−102.13	0	− 98.68	0
Secondary current, I_1	−200	+ 4	− 99.4	−172.8	−103	−171.4
Primary load current, $I' = -aI_1$	+ 10	− 0.2	+ 4.97	− 8.64	+ 5.15	+ 8.57
Primary exciting current, I_{00}	0.3	− 0.4	0.3	− 0.4	0.3	− 0.4
Total primary current, I_0	+ 10.3	− 0.6	+ 5.27	− 9.04	+ 5.45	+ 8.17
Primary resistance, voltage, I_0r_0	20.6	1.2	10.54	− 18.08	10.90	+16.34
Primary reactance, voltage, I_0x_0	3.0	+51.3	45.20	+ 26.35	− 40.85	+27.25
E.m.f. consumed by primary counter e.m.f., $\frac{-E_1}{a}$	2016	0	2042.6	0	1973.6	0
Total primary impressed e.m.f., E^0	2039.6	+50.1	2098.34	+ 8.27	1943.65	+43.59

Hence,

	Non-inductive, $\theta_1 = 0$	Lag, $\theta_1 = +60°$	Lead, $\theta_1 = -60°$
Resultant E_0	2040.1	2098.3	1944.2
Resultant I_0	10.32	10.47	9.82
Phase of E_0	−1.4°	− 0.2°	− 1.2°
Phase of I_0	+3.3°	+59.8°	−56.3°
Primary lag, θ_0	+4.7°	+60.0°	−55.1°
Regulation $\frac{E_0}{2000}$	1.02005	1.04915	0.9721
Drop of voltage, per cent	2.005	4.915	− 2.79
Change of phase, $\theta_0 - \theta_1$	4.7°	0	4.9°

14. RECTANGULAR COORDINATES

64. The vector diagram of sine waves gives the best insight into the mutual relations of alternating currents and e.m.fs.

For numerical calculation from the vector diagram either the trigonometric method or the method of rectangular components is used.

The method of rectangular components, as explained in the above paragraphs, is usually simpler and more convenient than the trigonometric method.

In the method of rectangular components it is desirable to distinguish the two components from each other and from the resultant or total value by their notation.

To distinguish the components from the resultant, small letters are used for the components, capitals for the resultant. Thus in the transformer diagram of Section 13 the secondary current I_1 has the horizontal component $i_1 = - I_1 \cos \theta_1$, and the vertical component $i'_1 = + I_1 \sin \theta_1$.

To distinguish horizontal and vertical components from each other, either different types of letters can be used, or indices, or a prefix or coefficient.

Different types of letters are inconvenient, indices distinguishing the components undesirable, since indices are reserved for distinguishing different e.m.fs., currents, etc., from each other.

Thus the most convenient way is the addition of a prefix or coefficient to one of the components, and as such the letter j is commonly used with the vertical component.

Thus the secondary current in the transformer diagram, Section 13, can be written

$$i_1 + ji_2 = I_1 \cos \theta_1 + jI_1 \sin \theta_1. \qquad (1)^*$$

This method offers the further advantage that the two components can be written side by side, with the plus sign between them, since the addition of the prefix j distinguishes the value ji_2 or $jI_1 \sin \theta_1$ as vertical component from the horizontal component i_1 or $I_1 \cos \theta_1$.

$$\dot{I}_1 = i_1 + ji_2 \qquad (2)$$

thus means that I_1 consists of a horizontal component i_1 and a vertical component i_2, and the plus sign signifies that i_1 and i_2 are combined by the parallelogram of sine waves.

* Numbers of the form $i_1 + ji_2$ are called *complex numbers.*

The secondary e.m.f. of the transformer in Section 13, Fig. 34, is written in this manner, $\underset{\cdot}{E}_1 = -e_1$, that is, it has the horizontal component $-e_1$ and no vertical component.

The primary generated e.m.f. is

$$\underset{\cdot}{E}_i = \frac{-e_1}{a}, \tag{3}$$

and the e.m.f. consumed thereby

$$\underset{\cdot}{E}' = +\frac{e_1}{a}. \tag{4}$$

The secondary current is

$$\underset{\cdot}{I}_1 = -i_1 + ji_2, \tag{5}$$

where

$$i_1 = I_1 \cos \theta_1, \quad i_2 = I_1 \sin \theta_1, \tag{6}$$

and the primary load current corresponding thereto is

$$\underset{\cdot}{I}' = -a\underset{\cdot}{I}_1 = ai_1 - jai_2. \tag{7}$$

The primary exciting current,

$$\underset{\cdot}{I}_{00} = h - jg, \tag{8}$$

where $h = I_{00} \sin \alpha$ is the hysteresis current, $g = I_{00} \cos \alpha$ the reactive magnetizing current.

Thus the total primary current is

$$\underset{\cdot}{I}_0 = \underset{\cdot}{I}' + \underset{\cdot\cdot}{I}_{00} = (ai_1 + h) - j(ai_2 + g). \tag{9}$$

The e.m.f. consumed by primary resistance r_0 is

$$r_0\underset{\cdot}{I}_0 = r_0(ai_1 + h) - jr_0(ai_2 + g). \tag{10}$$

The horizontal component of primary current $(ai_1 + h)$ gives as e.m.f. consumed by reactance x_0 a negative vertical component, denoted by $jx_0(ai_1 + h)$. The vertical component of primary current $j(ai_2 + g)$ gives as e.m.f. consumed by reactance x_0 a positive horizontal component, denoted by $x_0(ai_2 + g)$.

Thus the total e.m.f. consumed by primary reactance x_0 is

$$x_0(ai_2 + g) + jx_0(ai_1 + h), \tag{11}$$

and the total e.m.f. consumed by primary impedance is

$$r_0(ai_1 + h) + x_0(ai_2 + g) - j[r_0(ai_2 + g) - x_0(ai_1 + h)]. \tag{12}$$

Thus, to get from the current the e.m.f. consumed in reactance x_0 by the horizontal component of current, the coefficient j has to be added; in the vertical component the coefficient $-j$ omitted; or, we can say the reactance is denoted by jx_0 for the horizontal and by $-\frac{x_0}{j}$ for the vertical component of current. In other words, if $\dot{I} = i - ji'$ is a current, x the reactance of its circuit, the e.m.f. consumed by the reactance is

$$jxi + xi' = xi' + jxi.$$

65. If instead of omitting $-j$ in deriving the reactance e.m.f. for the vertical component of current we would add j also (as done when deriving the reactance e.m f. for the horizontal component of current), we get the reactance e.m.f.

$$jxi - j^2xi',$$

which gives the correct value $jxi + xi'$, if

$$j^2 = -1; \tag{13}$$

that is, we can say, in deriving the e.m.f. consumed by reactance, x, from the current, we multiply the current by jx, and substitute $j^2 = -1$.

By defining, and substituting, $j^2 = -1$, jx can thus be called the reactance in the representation in rectangular coordinates and $r + jx$ the impedance.

The primary impedance voltage of the transformer in the preceding could thus be derived directly by multiplying the current,

$$\dot{I}_0 = (ai_1 + h) - j(ai_2 + g), \tag{9}$$

by the impedance,

$$Z_0 = r_0 + jx_0,$$

which gives

$$\begin{aligned} \dot{E}'_0 = Z_0\dot{I}_0 &= (r_0 + jx_0)[(ai_1 + h) - j(ai_2 + g)] \\ &= r_0(ai_1 + h) - jr_0(ai_2 + g) + jx_0(ai_1 + h) - j^2x_0(ai_2 + g), \end{aligned}$$

and substituting $j^2 = -1$,

$$\dot{E}'_0 = [r_0(ai_1 + h) + x_0(ai_2 + g)] - j[r_0(ai_2 + g) - x_0(ai_1 + h)], \tag{14}$$

and the total primary impressed e.m.f. is thus

$$\underset{\cdot}{E}_0 = \underset{\cdot}{E}' + \underset{\cdot}{E}'_0$$

$$= \left[\frac{e_1}{a} + r_0 (ai_1 + h) + x_0 (ai_2 + g)\right] - j\left[r_0 (ai_2 + g) - x_0 (ai + h)\right]. \tag{15}$$

66. Such an expression in rectangular coordinates as

$$\underset{\cdot}{I} = i + ji' \tag{16}$$

represents not only the current strength but also its phase.

It means, in Fig. 38, that the total current $\overline{OI}$ has the two rectangular components, the horizontal component $I \cos \theta = i$ and the vertical component $I \sin \theta = i'$.

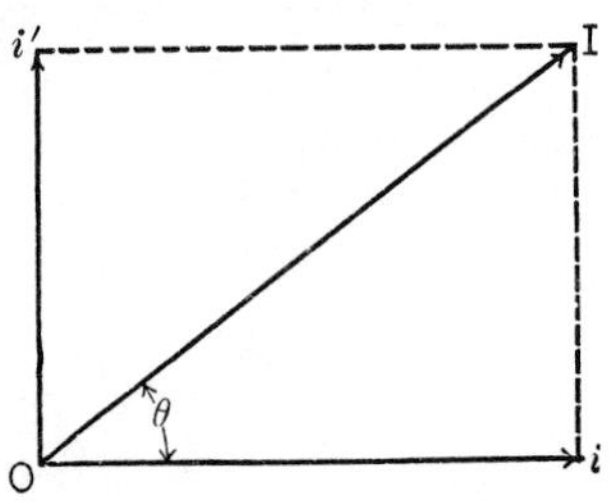

Fig. 38.—Magnitude and phase in rectangular coordinates.

Thus,

$$\tan \theta = \frac{i'}{i}; \tag{17}$$

that is, the tangent function of the phase angle is the vertical component divided by the horizontal component, or the term with prefix j divided by the term without j.

The total current intensity is obviously

$$I = \sqrt{i^2 + i'^2}. \tag{18}$$

The capital letter I in the symbolic expression $\underset{\cdot}{I} = i + ji'$ thus represents more than the I used in the preceding for total current, etc., and gives not only the intensity but also the phase. It is thus necessary to distinguish by the type of the latter the capital letters denoting the resultant current in symbolic expression (that is, giving intensity and phase) from the capital letters giving merely the intensity regardless of phase; that is,

$$\underset{\cdot}{I} = i + ji'$$

denotes a current of intensity

$$I = \sqrt{i^2 + i'^2}$$

and phase

$$\tan \theta = \frac{i'}{i}.$$

In the following, dotted italics will be used for the symbolic expressions and plain italics for the absolute values of alternating waves.

In the same way $z = \sqrt{r^2 + x^2}$ is denoted in symbolic representation of its rectangular components by

$$Z = r + jx. \tag{91}$$

When using the symbolic expression of rectangular coordinates it is necessary ultimately to reduce to common expressions.

Thus in the above discussed transformer the symbolic expression of primary impressed e.m.f.

$$\dot{E}_0 = \left[\frac{e_1}{a} + r_0\,(ai_1 + h) + x_0(ai_2 + g)\right] - j\left[r_0(ai_2 + g) - x_0(ai_1 + h)\right] \tag{15}$$

means that the primary impressed e.m.f. has the intensity

$$E_0 = \sqrt{\left[\frac{e_1}{a} + r_0\,(ai_1 + h) + x_0\,(ai_2 + g)\right]^2 + \left[r_0(ai_2 + g) - x_0(ai_1 + h)\right]}, \tag{20}$$

and the phase

$$\tan\theta_0 = \frac{r_0\,(ai_2 + g) - x_0\,(ai_1 + h)}{\frac{e_1}{a} + r_0\,(ai_1 + h) + x_0\,(ai_2 + g)}. \tag{21}$$

This symbolism of rectangular components is the quickest and simplest method of dealing with alternating-current phenomena, and is in many more complicated cases the only method which can solve the problem at all, and therefore the reader must become fully familiar with this method.

EXAMPLES

67. (1) In a 20-kw. transformer the ratio of turns is 20 : 1, and 100 volts are required at the secondary terminals at full load. What is the primary current, the primary impressed e.m.f., and the primary lag,

(*a*) at non-inductive load, $\theta_1 = 0$;

(*b*) with $\theta_1 = 60$ degrees time lag in the external secondary circuit;

(*c*) with $\theta_1 = -60$ degrees time lead in the external secondary circuit?

	Non-inductive	60° Lag	60° Lead
Secondary current, I_1 =	200	200	200
Secondary impedance voltage, $E' = I_1 Z_1$ =	$0.8 + 2j$	$0.8 + 2j$	$0.8 + 2j$
Secondary terminal voltage, $E = 100 (\cos \theta_1 + j \sin \theta_1)$ =	100	$50 + 86.6j$	$50 - 86.6j$
Thus, secondary counter-generated e.m.f., $E_1 = E + E'_1$ =	$100.8 + 2j$	$50.8 + 88.6j$	$50.8 - 84.6j$
Primary counter-generated e.m.f., $Ei = 20 E_1$ =	$2016 + 40j$	$1016 + 1772j$	$1016 - 1692j$
Primary load current' $= I \frac{I_1}{20}$ =	10	10	10
Primary exciting current, at $e = 2000$ volts impressed, I_{00} =	$0.3 - 0.4j$	$0.3 - 0.4j$	$0.3 - 0.4j$
Thus, at primary counter-generated e.m.f. Ei the exciting current is $\frac{E_i I_{00}}{2000}$ =	$\frac{(0.3 - 0.4j)\ (2016 + 40j)}{2000}$	$\frac{(0.3 - 0.4j)\ (1016 + 1772j)}{2000}$	$\frac{(0.3 - 0.4j)\ (1016 - 1692j)}{2000}$
Hence, expanded, I_{00} =	$0.310 - 397j$	$0.507 + 0.063j$	$-0.186 - 0.407j$
Total primary current, $I_0 = I_{00} + I'$	$10.31 - 0.397j$	$10.507 + 0.063j$	$9.814 - 407j$
Primary impedance voltage, $E'_0 = I_0 Z_0$ =	$(2 + 5j)\ 10.31 - 0.397j)$	$(2 + 5j)\ (10.507 + 0.063j)$	$(2 + 5j)\ (9.814 - 0.407j)$
Hence, expanded, E'_0 =	$22.6 + 50.76j$	$20.7 + 52.66j$	$21.66 + 48.26j$
Thus, primary impressed e.m.f., $E_0 = Ei + E'_0$ =	$2038.6 + 90.8j$	$1036.7 + 1824.7j$	$1037.7 - 1643.7j$
Hence, primary e.m.f. phase, $\tan \theta^1$ =	$-\frac{90.8}{2038.6}$	$-\frac{1824.7}{1036.7}$	$+\frac{1643.7}{1037.7}$
θ' =	$-2.6°$	$-60.4°$	$+57.7°$
Primary current phase, $\tan \theta''$ =	$+\frac{0.397}{10.31}$	$-\frac{0.063}{10.507}$	$+\frac{0.407}{9.814}$
θ'' =	$+2.2°$	$-0.4°$	$+2.4°$
Primary lag, $\theta_0 = \theta'' - \theta'$ =	$+4.8°$	$+60.0°$	$-55.3°$
And, reduced, primary impressed e.m.f., E_0 =	$\sqrt{2038.6^2 + 90.8^2} = 2041$	$\sqrt{1036.7^2 + 1824.7^2} = 2099$	$\sqrt{1037.7^2 + 1643.7^2} = 1943$
Primary current, I_0	$\sqrt{10.31^2 + 0.397^2} = 10.32$	$\sqrt{10.507^2 + 0.063^2} = 10.51$	$\sqrt{9.814^2 + 0.407^2} = 9.82$

	Non-inductive	60° Lag	60° Lead
We then have	$Z = r = 0.5$	$Z = 0.3 + 0.4j$	$Z = 0.3 - 0.4j$
Secondary current $I_1 = \frac{e}{Z}$	$= \frac{e}{0.5}$	$\frac{e}{0.3 + 0.4j}$	$\frac{e}{0.3 - 0.4j}$
Expanded by the associate term of the denominator, and substitute, $j^2 = -1$, $I_1 =$	$2e$ $= 2e$	$\frac{e(0.3 - 0.4j)}{(0.3 + 0.4j)(0.3 - 0.4j)}$ $= 4e(0.3 - 0.4j)$	$\frac{e(0.3 + 0.4j)}{(0.3 - 0.4j)(0.3 + 0.4j)}$ $= 4e(0.3 + 0.4j)$
Secondary impedance voltage, $E'_1 = I_1 Z_1 =$	$2e(0.004 + 0.01j)$ $= e(0.008 + 0.02j)$	$4e(0.3 - 0.4j)(0.004 + 0.01j)$ $= e(0.0208 + 0.0056j)$	$4e(3 - 4j)(0.004 + 0.01j)$ $= e(-0.0112 + 0.0184j)$
Secondary terminal voltage, $E = e - E'_1 =$	$e(0.992 - 0.02j)$	$e(0.9792 - 0.0056j)$	$e(1.0112 - 0.0184j)$
Or, reduced, $E =$	$e\sqrt{0.992^2 + 0.02^2}$ $= 0.992e$	$e\sqrt{0.9792^2 + 0.0056^2}$ $= 0.9792e$	$e\sqrt{1.011^2 + 0.0184^2}$ $= 1.0114e$
Primary counter-generated e.m.f., $E_i =$	$20e$	$20e$	$20e$
Primary load current, $I' = \frac{1}{20} I_1 =$	$0.1e$	$0.2e(0.3 - 0.4j)$	$0.2e(0.3 + 0.4j)$
Primary exciting current, $I_{00} = EY =$	$e(3 - 4j)\ 10^{-3}$	$e(3 - 4j)\ 10^{-3}$	$e(0.3 - 0.4j)\ 10^2$
Thus, total primary current $I_0 = I' + I_{00} =$	$e(0.103 - 0.004j)$	$e(0.063 - 0.084j)$	$e(0.063 + 0.076j)$
Primary impedance voltage, $E'_0 = Z_0 I_0 =$	$e(0.103 - 0.004j)(2 + 5j)$	$e(0.063 - 0.084j)\ 2 - 5j)$	$e(0.063 + 0.076j)(2 + 5j)$
Expanded =	$e(0.226 + 505j)$	$e(0.546 + 0.147j)$	$e(-0.254 + 0.467j)$
Thus, primary impressed e.m.f., $E_0 = E_i + E'_0 =$	$e(20.226 + 0.505j)$	$e(20.546 + 0.147j)$	$e(19.746 + 0.467j)$
Or, reduced, $e_0 =$	$e\sqrt{20.226^2 + 0.505^2}$ $= 20.23e$	$e\sqrt{20.564^2 + 0.147^2}$ $= 20.55e$	$e\sqrt{19.746^2 + 0.467^2}$ $= 19.75e$
Or, $e =$	$\frac{e_0}{20.23}$	$\frac{e_0}{20.55}$	$\frac{e_0}{19.75}$
Since $e_0 = 2000$, $e =$	98.85	97.32	101.25
Substituting e gives			
Secondary current, $I_1 =$	197.7	$116.8 - 155.6j$	$121.8 + 162j$
Reduced, $I_1 =$	197.7	194.6	202.5
Secondary terminal voltage, $E_1 =$	$98.1 - 2j$	$95.3 - 0.54j$	$102.4 - 1.86j$
Reduced, $E_1 =$	98.1	95.3	102.4
Primary current, $I_0 =$	$10.18 - 0.004j$	$6.13 - 8.17j$	$6.38 + 7.70j$
Reduced, $I_0 =$	10.18	10.22	10.00

The exciting current is $I'_{00} = 0.3 - 0.4\,j$ amp. at $e = 2000$ volts impressed, or rather, primary counter-generated e.m.f.

The primary impedance, $Z_0 = 2 + 5\,j$ ohms.

The secondary impedance, $Z_1 = 0.004 + 0.01\,j$ ohm.

We have, in symbolic expression, choosing the secondary current I_1 as real axis, the results calculated in tabulated form on page 82.

68. (2) $e_0 = 2000$ volts are impressed upon the primary circuit of a transformer of ratio of turns 20:1. The primary impedance is $Z_0 = 2 + 5\,j$, the secondary impedance, $Z_1 = 0.004 + 0.01\,j$, and the exciting current at $e' = 2000$ volts counter-generated e.m.f. is $\dot{I}_{00} = 0.3 - 0.4\,j$; thus the exciting admittance, $Y = \frac{I'_{00}}{e'} = (0.15 - 0.2\,j)10^{-3}$.

What is the secondary current and secondary terminal voltage and the primary current if the total impedance of the secondary circuit (internal impedance plus external load) consists of

(*a*) resistance,
$Z = r = 0.5$ – non-inductive circuit.

(*b*) impedance,
$Z = r + jx = 0.3 + 0.4\,j$ – inductive circuit.

(*c*) impedance,
$Z = r + jx = 0.3 - 0.4\,j$ – anti-inductive circuit.

Let e = secondary e.m.f.,

assumed as real axis in symbolic expression, and carrying out the calculation in tabulated form, on page 83.

69. (3) A transmission line of impedance $Z = r + jx = 20 + 50\,j$ ohms feeds a receiving circuit. At the receiving end an apparatus is connected which produces reactive lagging or leading currents at will (synchronous machine); 12,000 volts are impressed upon the line. How much lagging and leading currents respectively must be produced at the receiving end of the line to get 10,000 volts (*a*) at no load, (*b*) at 50 amp. power current as load, (*c*) at 100 amp. power current as load?

Let $e = 10{,}000$ = e.m.f. received at end of line, i_1 = power current, and i_2 = reactive lagging current; then

$$\dot{I} = i_1 - ji_2 = \text{total line current.}$$

The voltage at the generator end of the line is then

$$\begin{aligned} \dot{E}_0 &= e + Z\dot{I} \\ &= e + (r + jx)(i_1 - ji_2) \\ &= (e + ri_1 + xi_2) - j\,(ri_2 - xi_1) \\ &= (10{,}000 + 20\,i_1 + 50\,i_2) - j\,(20\,i_2 - 50\,i_1); \end{aligned}$$

or, reduced,

$$E_0 = \sqrt{(e + ri_1 + xi_2)^2 + (ri_2 - xi_1)^2};$$

thus, since $E_0 = 12{,}000$,

$$12{,}000 = \sqrt{(10{,}000 + 20\,i_1 + 50\,i_2)^2 + (20\,i_2 - 50\,i_1)^2}.$$

(*a*) At no load $i_1 = 0$, and

$$12{,}000 = \sqrt{(10{,}000 + 50\,i_2)^2 + 400\,i_2{}^2};$$

hence,

$$i_2 = +\,39.5 \text{ amp., reactive lagging current, } \dot{I} = -\,39.5\,j.$$

(*b*) At half load $i_1 = 50$, and

$$12{,}000 = \sqrt{(11{,}000 + 50\,i_2)^2 + (20\,i_2 - 2500)^2};$$

hence,

$$i_2 = +\,16 \text{ amp., lagging current, } \dot{I} = 50 - 16\,j.$$

(*c*) At full load $i_1 = 100$, and

$$12{,}000 = \sqrt{(12{,}000 + 50\,i_2)^2 + (20\,i - 5000)^2};$$

hence,

$$i_2 = -\,27.13 \text{ amp., leading current, } \dot{I} = 100 + 27.13\,j.$$

15. LOAD CHARACTERISTIC OF TRANSMISSION LINE

70. The load characteristic of a transmission line is the curve of volts and watts at the receiving end of the line as function of the amperes, and at constant e.m.f. impressed upon the generator end of the line.

Let r = resistance, x = reactance of the line. Its impedance $z = \sqrt{r^2 + x^2}$ can be denoted symbolically by

$$Z = r + jx.$$

Let E_0 = e.m.f. impressed upon the line.

Choosing the e.m.f. at the end of the line as horizontal component in the vector diagram, it can be denoted by $\dot{E} = e$.

At non-inductive load the line current is in phase with the e.m.f. e, thus denoted by $\dot{I} = i$.

The e.m.f. consumed by the line impedance $Z = r + jx$ is

$$\dot{E}_1 = Z\dot{I} = (r + jx)\, i$$
$$= ri + jx\, i. \qquad (1)$$

Thus the impressed voltage,

$$\dot{E}_0 = \dot{E} + \dot{E}_1 = e + ri + jxi. \qquad (2)$$

or, reduced,

$$E_0 = \sqrt{(e + ri)^2 + x^2 i^2}, \qquad (3)$$

and

$$e = \sqrt{E_0{}^2 - x^2 i^2} - ri, \text{ the e.m.f.} \qquad (4)$$

$$P = ei = i\sqrt{E_0{}^2 - x^2 i^2} - ri^2, \qquad (5)$$

the power received at end of the line.

The curve of e.m.f. e is an arc of an ellipse.

With open circuit $i = 0$, $e = E_0$ and $P = 0$, as is to be expected.

At short circuit, $e = 0$, $0 = \sqrt{E_0{}^2 - x^2 i^2} - ri$, and

$$i = \frac{E_0}{\sqrt{r^2 + x^2}} = \frac{E_0}{z}; \qquad (6)$$

that is, the maximum line current which can be established with a non-inductive receiver circuit and negligible line capacity.

71. The condition of maximum power delivered over the line is

$$\frac{dP}{di} = 0; \qquad (7)$$

that is,

$$\sqrt{E_0{}^2 - x^2 i^2} + \frac{\tfrac{1}{2}\, i\,(-2x^2 i)}{\sqrt{E_0{}^2 - x^2 i^2}} - 2ri = 0;$$

substituting (3):

$$\sqrt{E_0{}^2 - x^2 i^2} = e + ri,$$

and expanding, gives

$$e^2 = (r^2 + x^2)\, i^2$$
$$= z^2 i^2; \qquad (8)$$

hence,

$$e = zi, \text{ and } \frac{e}{i} = z. \qquad (9)$$

$\frac{e}{i} = r_1$ is the resistance or effective resistance of the receiving circuit; that is, the maximum power is delivered into a non-inductive receiving circuit over an inductive line upon which is impressed a constant e.m.f., if the resistance of the receiving circuit equals the impedance of the line, $r_1 = z$.

In this case the total impedance of the system is

$$Z_0 = Z + r_1 = r + z + jx, \tag{10}$$

or,

$$z_0 = \sqrt{(r + z)^2 + x^2}. \tag{11}$$

Thus the current is

$$i_1 = \frac{E_0}{z_0} = \frac{E_0}{\sqrt{(r + z)^2 + x^2}}, \tag{12}$$

and the power transmitted is

$$\begin{aligned} P_1 = i_1^2 r_1 &= \frac{E_0^2 z}{(r + z)^2 + x^2} \\ &= \frac{E_0^2}{2\,(r + z)}; \end{aligned} \tag{13}$$

that is, the maximum power which can be transmitted over a line of resistance r and reactance x is the square of the impressed e.m.f. divided by twice the sum of resistance and impedance of the line.

At $x = 0$, this gives the common formula,

$$P_1 = \frac{E_0^2}{4\,r}. \tag{14}$$

Inductive Load

72. With an inductive receiving circuit of lag angle θ, or power-factor $p = \cos\,\theta$, and inductance factor $q = \sin\,\theta$, at e.m.f. $\underset{\cdot}{E} = e$ at receiving circuit, the current is denoted by

$$\underset{\cdot}{I} = I\,(p - jq); \tag{15}$$

thus the e.m.f. consumed by the line impedance $Z = r + jx$ is

$$\begin{aligned} \underset{\cdot}{E_1} = Z\underset{\cdot}{I} &= I\,(p - jq)(r + jx) \\ &= I\,[(rp + xq) - j\,(rq - xp)], \end{aligned}$$

and the generator voltage is

$$\begin{aligned} \underset{\cdot}{E_0} &= \underset{\cdot}{E} + \underset{\cdot}{E_1} \\ &= [e + I\,(rp + xq)] - jI\,(rq - xp); \end{aligned} \tag{16}$$

or, reduced,

$$E_0 = \sqrt{[e + I(rp + xq)]^2 + I^2(rq - xp)^2}, \tag{17}$$

and

$$e = \sqrt{E_0^2 - I^2(rq - xp)^2} - I(rp + xq). \tag{18}$$

The power received is the e.m.f. times the power component of the current; thus

$$\begin{aligned} P &= eIp \\ &= Ip\sqrt{E_0^2 - I^2(rq - xp)^2} - I^2p(rp + xq). \end{aligned} \tag{19}$$

The curve of e.m.f., e, as function of the current I is again an arc of an ellipse.

At short circuit $e = 0$; thus, substituted,

$$I = \frac{E_0}{z}, \tag{20}$$

the same value as with non-inductive load, as is obvious.

73. The condition of maximum output delivered over the line is

$$\frac{dP}{dI} = 0; \tag{21}$$

that is, differentiated,

$$\sqrt{E_0^2 - I^2(rq - xp)^2} = e + I(rp + xq); \tag{22}$$

substituting and expanding,

$$\begin{aligned} e^2 &= I^2(r^2 + x^2) \\ &= I^2z^2; \\ e &= Iz; \end{aligned}$$

or

$$\frac{e}{I} = z. \tag{23}$$

$z_1 = \dfrac{e}{I}$ is the impedance of the receiving circuit; that is, the power received in an inductive circuit over an inductive line is a maximum if the impedance of the receiving circuit, z_1, equals the impedance of the line, z.

In this case the impedance of the receiving circuit is

$$Z_1 = z(p + jq), \tag{24}$$

and the total impendance of the system is

$$\begin{aligned} Z_0 &= Z + Z_1 \\ &= r + jx + z(p + jq) \\ &= (r + pz) + j(x + qz). \end{aligned}$$

Thus, the current is

$$I_1 = \frac{E_0}{\sqrt{(r + pz)^2 + (x + qz)^2}}, \qquad (25)$$

and the power is

$$P_1 = I_1^2 zp = \frac{E_0^2 zp}{(r + pz)^2 + (x + qz)^2}$$
$$= \frac{E_0^2 p}{2(z + rp + xq)}. \qquad (26)$$

EXAMPLES

74. (1) 12,000 volts are impressed upon a transmission line of impedance $Z = r + jx = 20 + 50j$. How do the voltage

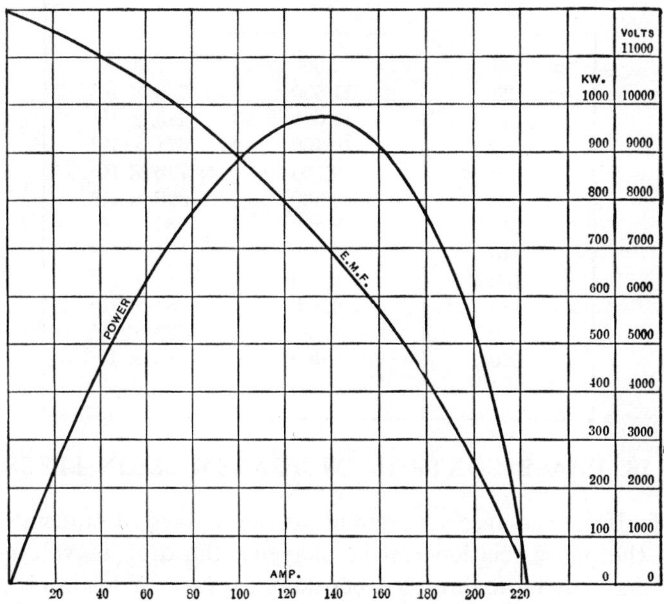

FIG. 39.—Non-reactive load characteristids of a transmission line. Constant impressed e.m.f.

and the output in the receiving circuit vary with the current with non-inductive load?

Let e = voltage at the receiving end of the line, i = current: thus = ei = power received. The voltage impressed upon the line is then

$$E_0 = e + Zi$$
$$= e + ri + jxi;$$

or, reduced,

$$E_0 = \sqrt{(e + ri)^2 + x^2 i^2}.$$

Since $E_0 = 12{,}000$,

$$12{,}000 = \sqrt{(e + ri)^2 + x^2 i^2} = \sqrt{(e + 20\,i)^2 + 2500\,i^2},$$

$$e = \sqrt{12{,}000^2 - x^2 i^2} - ri = \sqrt{12{,}000^2 - 2500\,i^2} - 20\,i.$$

The maximum current for $e = 0$ is

$$0 = \sqrt{12{,}000^2 - 2500\,i^2} - 20\,i;$$

thus,

$$i = 223.$$

Substituting for i gives the values plotted in Fig. 39.

i	e	$p = ei$
0	12,000	0
20	11,500	230×10^3
40	11,000	440×10^3
60	10,400	624×10^3
80	9,700	776×10^3
100	8,900	890×10^3
120	8,000	960×10^3
140	6,940	971×10^3
160	5,750	920×10^3
180	4,340	784×10^3
200	2,630	526×10^3
220	400	88×10^3
223	0	0

16. PHASE CONTROL OF TRANSMISSION LINES

75. If in the receiving circuit of an inductive transmission line the phase relation can be changed, the drop of voltage in the line can be maintained constant at varying loads or even decreased with increasing load; that is, at constant generator voltage the transmission can be compounded for constant voltage at the receiving end, or even over-compounded for a voltage increasing with the load.

1. Compounding of Transmission Lines for Constant Voltage

Let r = resistance, x = reactance of the transmission line, e_0 = voltage impressed upon the beginning of the line, e = voltage received at the end of end line.

Let i = power current in the receiving circuit; that is, $P = ei$ = transmitted power, and i_1 = reactive current produced in the system for controlling the voltage. i_1 shall be considered positive as lagging, negative as leading current.

Then the total current, in symbolic representation, is

$$I = i - ji_1;$$

the line impedance is

$$Z = r + jx,$$

and thus the e.m.f. consumed by the line impedance is

$$\begin{aligned} \underset{\cdot}{E}_1 &= \underset{\cdot}{Z}I = (r + jx)\,(i - ji_1) \\ &= ri + jri_1 + jxi - j^2xi_1; \end{aligned}$$

and substituting $j^2 = -1$,

$$\underset{\cdot}{E}_1 = (ri + xi_1) - j\,(ri_1 - xi).$$

Hence the voltage impressed upon the line

$$\begin{aligned} \underset{\cdot}{E}_0 &= e + \underset{\cdot}{E}_1 \\ &= (e + ri + xi_1) - j\,(ri_1 - xi); \qquad (1) \end{aligned}$$

or, reduced,

$$e_0 = \sqrt{(e + ri + xi_1)^2 + (ri_1 - xi)^2}. \qquad (2)$$

If in this equation e and e_0 are constant, i_1, the reactive component of the current, is given as a function of the power component current i and thus of the load ei.

Hence either e_0 and e can be chosen, or one of the e.m.fs. e_0 or e and the reactive current i_1 corresponding to a given power current i.

76. If $i_1 = 0$ with $i = 0$, and e is assumed as given, $e_0 = e$. Thus,

$$\begin{aligned} e &= \sqrt{(e + ri + xi_1)^2 + (ri_1 - xi)^2}; \\ 2\,e\,(ri + xi_1) &+ (r^2 + x^2)\,(i^2 + i_1^2) = 0. \end{aligned}$$

From this equation it follows that

$$i_1 = -\frac{ex \pm \sqrt{e^2x^2 - 2\,eriz^2 - i^2z^4}}{z^2}. \qquad (3)$$

Thus, the reactive current i_1 must be varied by this equation to maintain constant voltage $e = e_0$ irrespective of the load ei.

As seen, in this equation, i_1 must always be negative, that is, the current leading.

i_1 becomes impossible if the term under the square root becomes negative, that is, at the value

$$e^2x^2 - 2\,eriz^2 - i^2z^4 = 0;$$

or,

$$i = \frac{e\,(z - r)}{z^2}. \tag{4}$$

At this point the power transmitted is

$$P = ei = \frac{e^2\,(z - r)}{z^2}. \tag{5}$$

This is the maximum power which can be transmitted without drop of voltage in the line, with a power current $i = \frac{e\,(z-r)}{z^2}$.

The reactive current corresponding hereto, since the square root becomes zero, is

$$i_1 = \frac{ex}{z^2}; \tag{6}$$

thus the ratio of reactive to power current, or the tangent of the phase angle of the receiving circuit, is

$$\tan\theta_1 = \frac{i_1}{i} = -\frac{x}{z - r}. \tag{7}$$

A larger amount of power is transmitted if e_0 is chosen $> e$, a smaller amount of power if $e_0 < e$.

In the latter case i_1 is always leading; in the former case i_1 is lagging at no load, becomes zero at some intermediate load, and leading at higher load.

77. If the line impedance $Z = r + jx$ and the received voltage e is given, and the power current i_0 at which the reactive current shall be zero, the voltage at the generator end of the line is determined hereby from the equation (2):

$$e_0 = \sqrt{(e + ri + xi_1)^2 + (ri_1 - xi)^2},$$

by substituting $i_1 = 0$, $i = i_0$,

$$e_0 = \sqrt{(e + ri_0)^2 + x^2i_0^2}. \tag{8}$$

Substituting this value in the general equation (2):

$$e_0 = \sqrt{(e + ri + xi_1)^2 + (ri_1 - xi)^2}$$

gives

$$(e + ri_0)^2 + x^2i_0^2 = (e + ri + xi_1)^2 + (ri_1 - xi)^2 \tag{9}$$

as equation between i and i_1.

If at constant generator voltage e_0:
at no load,

$$\left.\begin{aligned} i = 0,\ e = e_0,\ i_1 = i'_0, \\ i = i_0,\ e = e_0,\ i_1 = 0 \end{aligned}\right\} \tag{10}$$

and at the load,

it is, substituted:
no load,

$$e_0 = \sqrt{(e_0 + x i_0)^2 + r^2 i'_0{}^2}, \tag{11}$$

load i_0,

$$e_0 = \sqrt{(e_0 + r i_0)^2 + x^2 i_0{}^2}. \tag{12}$$

Thus,

$$(e_0 + x i'_0) + x^2 i'^2_0 = (e_0 + r i_0)^2 + x^2 i_0{}^2;$$

or, expanded,

$$i'_0{}^2(r^2 + x^2) + 2\, i'_0\, x e_0 = i_0{}^2\,(r^2 + x^2) + 2\, i_0 r e_0. \tag{13}$$

This equation gives i'_0 as function of i_0, e_0, r, x.

If now the reactive current i_1 varies as linear function of the power current i, as in case of compounding by rotary converter with shunt and series field, it is

$$i_1 = \frac{(i_0 - i)}{i_0}\, i'_0. \tag{14}$$

Substituting this value in the general equation

$$(e_0 + r i_0)^2 + x^2 i_0{}^2 = (e + ri + x i_1)^2 + (r i_1 - xi)^2$$

gives e as function of i; that is, gives the voltage at the receiving end as function of the load, at constant voltage e_0 at the generating end, and $e = e_0$ for no load,

$$i = 0,\ i_1 = i'_0,$$

and $e = e_0$ for the load,

$$i = i_0,\ i_1 = 0.$$

Between $i = 0$ and $i = i_0$, $e > e_0$, and the current is lagging. Above $i = i_0$, $e < e_0$, and the current is leading.

By the reaction of the variation of e from e_0 upon the receiving apparatus producing reactive current i_1, and by magnetic saturation in the receiving apparatus, the deviation of e from e_0 is reduced, that is, the regulation improved.

2. Over-compounding of Transmission Lines

78. The impressed voltage at the generator end of the line was found in the preceding,

$$e_0 = \sqrt{(e + ri + x i_1)^2 + (r i_1 - xi)^2}. \tag{2}$$

If the voltage at the end of the line e shall rise proportionally to the power current i, then

$$e = e_1 + ai; \tag{15}$$

thus,

$$e_0 = \sqrt{[e_1 + (a + r)\, i + xi_1]^2 + (ri_1 - xi)^2}, \tag{16}$$

and herefrom in the same way as in the preceding we get the characteristic curve of the transmission.

If $e_0 = e_1$, $i_1 = 0$ at no load, and is leading at load. If $e_0 < e_1$, i_1 is always leading, the maximum output is less than before.

If $e_0 > e_1$, i_1 is lagging at no load, becomes zero at some intermediate load, and leading at higher load. The maximum output is greater than at $e_0 = e_1$.

The greater a, the less is the maximum output at the same e_0 and e_1.

The greater e_0, the greater is the maximum output at the same e_1 and a, but the greater at the same time the lagging current (or less the leading current) at no load.

EXAMPLES

79. (1) A constant voltage of e_0 is impressed upon a transmission line of impedance $Z = r + jx = 10 + 20\, j$. The voltage at the receiving end shall be 10,000 at no load as well as at full load of 75 amp. power current. The reactive current in the receiving circuit is raised proportionally to the load, so as to be lagging at no load, zero at full load or 75 amp., and leading beyond this. What voltage e_0 has to be impressed upon the line, and what is the voltage e at the receiving end at ⅓, ⅔, and 1⅓ load?

Let $\dot{I} = i_1 - ji_2$ = current, $E = e$ voltage in receiving circuit. The generator voltage is then

$$\begin{aligned} \dot{E}_0 &= e + Z\dot{I} \\ &= e + (r + jx)\,(i_1 - ji_2) \\ &= (e + ri_1 + xi_2) - j\,(ri_2 - xi_1) \\ &= (e + 10\, i_1 + 20\, i_2) - j\,(10\, i_2 - 20\, i_1); \end{aligned}$$

or, reduced,

$$\begin{aligned} e_0^2 &= (e + ri_1 + xi_2)^2 + (ri_2 - xi_1)^2; \\ &= (e + 10\, i_1 + 20\, i_2)^2 + (10\, i_2 - 20\, i_1)^2. \end{aligned}$$

When

$$i_1 = 75,\ i_2 = 0,\ e = 10{,}000;$$

substituting these values,

$$e_0^2 = 10{,}750^2 + 1500^2 = 117.81 \times 10^6;$$

hence,

$$e_0 = 10{,}860 \text{ volts is the generator voltage.}$$

When

$$i_1 = 0,\ e = 10{,}000,\ e_0 = 10{,}860, \text{ let } i_2 = i;$$

these values substituted give

$$\begin{aligned} 117.81 \times 10^6 &= (10{,}000 + 20\,i)^2 + 100\,i^2 \\ &= 100 \times 10^6 + 400\,i \times 10^3 + 500\,i^2, \end{aligned}$$

or,

$$i = 44.525 - 1.25\,i^2\,10^{-3};$$

this equation is best solved by approximation, and then gives

$$p = 42.3 \text{ amp. reactive lagging current at no load.}$$

Since

$$e_0^2 = (e + ri_1 + xi_2)^2 + (ri_2 - xi_1)^2,$$

it follows that

$$e = \sqrt{e_0^2 - (ri_2 - xi_1)^2} - (ri_1 + xi_2);$$

or,

$$e = \sqrt{117.81 + 10^6 - (10\,i_2 - 20\,i_1)^2} - (10\,i_1 + 20\,i_2).$$

Substituting herein the values of i_1 and i_2 gives e.

i_1	i_2	e
0	42.3	10,000
25	28.2	10,038
50	14.1	10,038
75	0	10,000
100	−14.1	9,922
125	−28.2	9,803

80. (2) A constant voltage e_0 is impressed upon a transmission line of impedance $Z = r + jx = 10 + 10\,j$. The voltage at the receiving end shall be 10,000 at no load as well as at full load of 100 amp. power current. At full load the total current shall be in phase with the e.m.f. at the receiving end, and at no load a lagging current of 50 amp. is permitted. How much additional reactance x_0 is to be inserted, what must be the generator voltage e_0, and what will be the voltage e at the receiving end at ½ load and at 1½ load, if the reactive current varies proportionally with the load?

Let x_0 = additional reactance inserted in circuit.

Let $\dot{I} = i_1 - ji_2$ = current.

Then

$$e_0^2 = (e + ri_1 + x_1i_2)^2 + (ri_2 - x_1i_1)^2 = (e + 10\,i_1 + x_1i_2)^2 + (10\,i_2 - x_1i_1)^2,$$

where

$x_1 = x + x_0$ = total reactance of circuit between e and e_0.

At no load,

$$i_1 = 0,\ i_2 = 50,\ e = 10{,}000;$$

thus, substituting,

$$e_0^2 = (10{,}000 + 50\,x_1)^2 + 250{,}000.$$

At full load,

$$i_1 = 100,\ i_2 = 0,\ e = 10{,}000;$$

thus, substituting,

$$e_0^2 = 121 \times 10^6 + 10{,}000\,x_1^2.$$

Combining these gives

$$(10{,}000 + 50\,x_1)^2 + 250{,}000 = 121 \times 10^6 + 10{,}000\,x_1^2;$$

hence,

$$x_1 = 66.5 \pm 40.8$$
$$= 107.3 \text{ or } 25.7;$$

thus

$x_0 = x_1 - x$ = 97.3 or 15.7 ohms additional reactance.

Substituting

$$x_1 = 25.7$$

gives

$$e_0^2 = (e + 10\,i_1 + 25.7\,i_2)^2 + (10\,i_2 - 25.7\,i_1)^2,$$

but at full load

$$i_1 = 100,\ i_2 = 0,\ e = 10{,}000,$$

which values substituted give

$$e_0^2 = 121 \times 10^6 + 6.605 \times 10^6 = 127.605 \times 10^6,$$

e_0 = 11,300, generator voltage.

Since

$$e = \sqrt{e_0^2 - (10\,i_2 - 25.7\,i_1)^2} - (10\,i_1 + 25.7\,i_2),$$

it follows that

$$e = \sqrt{127.605 \times 10^6 - (10\,i_2 - 25.7\,i_1)^2} - (10\,i_1 + 25.7\,i_2).$$

Substituting for i_1 and i_2 gives e.

i_1	i_2	e
0	50	10,000
50	25	10,105
100	0	10,000
150	−25	9,658

81. (3) In a circuit whose voltage e_0 fluctuates by 20 per cent. between 1800 and 2200 volts, a synchronous motor of internal impedance $Z_0 = r_0 + jx_0 = 0.5 + 5\,j$ is connected through a reactive coil of impedance $Z_1 = r_1 + jx_1 = 0.5 + 10j$ and run light, as compensator (that is, generator of reactive currents). How will the voltage at the synchronous motor terminals e_1, at constant excitation, that is, constant counter e.m.f. $e = 2000$, vary as function of e_0 at no load and at a load of $i = 100$ amp. power current, and what will be the reactive current in the synchronous motor?

Let $\underset{\cdot}{I} = i_1 - ji_2$ = current in receiving circuit of voltage e_1. Of this current $\underset{\cdot}{I}, -ji_2$ is taken by the synchronous motor of counter e.m.f. e, and thus

$$\begin{aligned} \underset{\cdot}{E}_1 &= e - Z_0 ji_2 \\ &= e + x_0 i_2 - jr_0 i_2; \end{aligned}$$

or, reduced,

$$e_1^2 = (e + x_0 i_2)^2 + r_0^2 i_2^2.$$

In the supply circuit the voltage is

$$\begin{aligned} \underset{\cdot}{E}_0 &= \underset{\cdot}{E}_1 + \underset{\cdot}{I} Z_1 \\ &= e + x_0 i_2 - jr_0 i_2 + (i_1 - ji_2)\,(r_1 + jx_1) \\ &= [e + r_1 i_1 + (x_0 + x_1)\,i_2] - j\,[(r_0 + r_1)\,i_2 - x_1 i_1]; \end{aligned}$$

or, reduced,

$$e_0^2 = [e + r_1 i_1 + (x_0 + x_1)\,i_2]^2 + [(r_0 + r_1)\,i_2 - x_1 i_1]^2.$$

Substituting in the equations for e_1^2 and e_0^2 the above values of r_0 and x_0: at no load, $i_1 = 0$, we have

$$e_1^2 = (e + 5\,i_2)^2 + 0.25\,i_2^2 \text{ and } e_0^2 = (e + 15\,i_2)^2 + i_2^2;$$

at full load, $i_1 = 100$, we have

$$\begin{aligned} e_1^2 &= (e + 5\,i_2)^2 + 0.25\,i_2^2, \\ e_0^2 &= (e + 50 + 15\,i_2)^2 + (i_2 - 1000)^2, \end{aligned}$$

and at no load, $i_1 = 0$, substituting $e = 2000$, we have

$$e_1^2 = (2000 + 5\,i_2)^2 + 0.25\,i_2^2,$$
$$e_0^2 = (2000 + 15\,i_2)^2 + i_2^2;$$

at full load, $i_1 = 100$, we have

$$e_1^2 = (2000 + 5\,i_2)^2 + 0.25\,i_2^2,$$
$$e_0^2 = (2050 + 15\,i_2)^2 + (i_2 - 1000)^2.$$

Substituting herein e_0 = successively 1800, 1900, 2000, 2100, 2200, gives values of i_2, which, substituted in the equation for e_1^2, give the corresponding values of e_1 as recorded in the following table.

As seen, in the local circuit controlled by the synchronous compensator, and separated by reactance from the main circuit of fluctuating voltage, the fluctuations of voltage appear in a greatly reduced magnitude only, and could be entirely eliminated by varying the excitation of the synchronous compensator.

$e = 2000$				
	No load, $i_1 = 0$		Full load, $i_1 = 100$	
e_0	i_2	e_1	i_2	e_1
1,800	−13.3	1,937	−39	1,810
1,900	− 6.7	1,965	−30.1	1,850
2,000	0	2,000	−22	1,885
2,100	+ 6.7	2,035	−13.5	1,935
2,200	+13.3	2,074	− 6.5	1,970

17. IMPEDANCE AND ADMITTANCE

82. In direct-current circuits the most important law is Ohm's law,

$$i = \frac{e}{r}, \quad \text{or } e = ir, \quad \text{or } r = \frac{e}{i},$$

where e is the e.m.f. impressed upon resistance r to produce current i therein.

Since in alternating-current circuits a current i through a resistance r may produce additional e.m.fs. therein, when applying Ohm's law, $i = \frac{e}{r}$ to alternating-current circuits, e is the total e.m.f. resulting from the impressed e.m.f. and all e.m.fs. produced by the current i in the circuit.

Such counter e.m.fs. may be due to inductance, as self-inductance, or mutual inductance, to capacity, chemical polarization, etc.

The counter e.m.f. of self-induction, or e.m.f. generated by the magnetic field produced by the alternating current i, is represented by a quantity of the same dimensions as resistance, and measured in ohms: reactance x. The e.m.f. consumed by reactance x is in quadrature with the current, that consumed by resistance r in phase with the current.

Reactance and resistance combined give the impedance,

$$z = \sqrt{r^2 + x^2};$$

or, in symbolic or vector representation,

$$Z = r + jx.$$

In general in an alternating-current circuit of current i, the e.m.f. e can be resolved in two components, a power component e_1 in phase with the current, and a wattless or reactive component e_2 in quadrature with the current.

The quantity

$$\frac{e_1}{i} = \frac{\text{power e.m.f., or e.m.f. in phase with the current}}{\text{current}} = r_1$$

is called the *effective resistance.*

The quantity

$$\frac{e_2}{i} = \frac{\text{reactive e.m.f., or e.m.f. in quadrature with the current}}{\text{current}} = x_1$$

is called the *effective reactance* of the circuit.

And the quantity

$$z_1 = \sqrt{r_1^2 + x^2}$$

or, in symbolic representation,

$$Z_1 = r_1 + jx_1$$

is the impedance of the circuit.

If power is consumed in the circuit only by the ohmic resistance r, and counter e.m.f. produced only by self-inductance, the effective resistance r_1 is the true or ohmic resistance r, and the effective reactance x_1 is the true or inductive reactance x.

By means of the terms effective resistance, effective reactance, and impedance, Ohm's law can be expressed in alternating-current circuits in the form

$$i = \frac{e}{z_1} = \frac{e}{\sqrt{r_1^2 + x_1^2}}; \tag{1}$$

or,
$$e = iz_1 = i\sqrt{r_1^2 + x_1^2}; \tag{2}$$

or,
$$z_1 = \sqrt{r_1^2 + x_1^2} = \frac{e}{i}; \tag{3}$$

or, in symbolic or vector representation,

$$\dot{I} = \frac{\dot{E}}{Z_1} = \frac{\dot{E}}{r_1 + jx}; \tag{4}$$

or,
$$\dot{E} = \dot{I}Z_1 = \dot{I}(r_1 + jx_1); \tag{5}$$

or,
$$Z_1 = r_1 + jx_1 = \frac{\dot{E}}{\dot{I}}. \tag{6}$$

In this latter form Ohm's law expresses not only the intensity but also the phase relation of the quantities; thus

$$e_1 = ir_1 = \text{power component of e.m.f.},$$
$$e_2 = ix_1 = \text{reactive component of e.m.f.}$$

83. Instead of the term impedance $z = \frac{e}{i}$ with its components, the resistance and reactance, its reciprocal can be introduced.

$$\frac{i}{e} = \frac{1}{z},$$

which is called the *admittance.*

The components of the admittance are called the *conductance* and the *susceptance.*

Resolving the current i into a power component i_1 in phase with the e.m.f. and a wattless component i_2 in quadrature with the e.m.f., the quantity

$$\frac{i_1}{e} = \frac{\text{power current, or current in phase with e.m.f.}}{\text{e.m.f.}} = g$$

is called the *conductance.*

The quantity

$$\frac{i_2}{e} = \frac{\text{reactive current, or current in quadrature with e.m.f.}}{\text{e.m.f.}} = b$$

is called the *susceptance* of the circuit.

The conductance represents the current in phase with the

e.m.f., or power current, the susceptance the current in quadrature with the e.m.f., or reactive current.

Conductance g and susceptance b combined give the admittance

$$y = \sqrt{g^2 + b^2}; \tag{7}$$

or, in symbolic or vector representation,

$$Y = g - jb. \tag{8}$$

Thus Ohm's law can also be written in the form

$$i = ey = e\sqrt{g^2 + b^2}; \tag{9}$$

or,

$$e = \frac{i}{y} = \frac{i}{\sqrt{g^2 + b^2}}; \tag{10}$$

or,

$$y = \sqrt{g^2 + b^2} = \frac{i}{e}; \tag{11}$$

or, in symbolic or vector representation,

$$\underset{\cdot}{I} = \underset{\cdot}{E}Y = \underset{\cdot}{E}\,(g - jb); \tag{12}$$

or,

$$\underset{\cdot}{E} = \frac{\underset{\cdot}{I}}{Y} = \frac{\underset{\cdot}{I}}{g - jb}; \tag{13}$$

or,

$$Y = g - jb = \frac{\underset{\cdot}{I}}{\underset{\cdot}{E}}. \tag{14}$$

and $\quad i_1 = eg =$ power component of current,
$\quad\quad\; i_2 = eb =$ reactive component of current.

84. According to circumstances, sometimes the use of the terms impedance, resistance, reactance, sometimes the use of the terms admittance, conductance, susceptance, is more convenient.

Since, in a number of series-connected circuits, the total e.m.f., in symbolic representation, is the sum of the individual e.m.fs., it follows that in a number of series-connected circuits the total impedance, in symbolic expression, is the sum of the impedances of the individual circuits connected in series.

Since, in a number of parallel-connected circuits, the total current, in symbolic representation, is the sum of the individual currents, it follows that in a number of parallel-connected circuits the total admittance, in symbolic expression, is the sum of the admittances of the individual circuits connected in parallel.

Thus in series connection the use of the term impedance, in parallel connection the use of the term admittance, is generally more convenient.

Since in symbolic representation

$$Y = \frac{1}{Z}; \tag{15}$$

or,

$$ZY = 1; \tag{16}$$

that is,

$$(r + jx)(g - jb) = 1; \tag{17}$$

it follows that

$$(rg + xb) - j\,(rb - xg) = 1;$$

that is

$$rg + zb = 1,$$
$$rb - xg = 0.$$

Thus,

$$r = \frac{g}{g^2 + b^2} = \frac{g}{y^2}, \tag{18}$$

$$x = \frac{b}{g^2 + b^2} = \frac{b}{y^2}, \tag{19}$$

$$g = \frac{r}{r^2 + x^2} = \frac{r}{z^2} \tag{20}$$

$$b = \frac{r}{r^2 + x^2} = \frac{x}{z^2}, \tag{21}$$

or, in absolute values,

$$y = \frac{1}{z}, \tag{22}$$

$$zy = 1, \tag{23}$$

$$(r^2 + x^2)(g^2 + b^2) = 1. \tag{24}$$

Thereby the admittance with its components, the conductance and susceptance, can be calculated from the impedance and its components, the resistance and reactance, and inversely.

If $x = 0$, $z = r$ and $g = \frac{1}{r}$, that is, g is the reciprocal of the resistance in a non-inductive circuit; not so, however, in an inductive circuit.

EXAMPLES

85. (1) In a quarter-phase induction motor having an impressed e.m.f. $e = 110$ volts per phase, the current is $\dot{I}_0 = i_1 - ji_2 = 100 - 100\,j$ at standstill, the torque $= D_0$.

The two phases are connected in series in a single-phase circuit of e.m.f. $e = 220$, and one phase shunted by a condenser of 1 ohm capacity reactance.

What is the starting torque D of the motor under these conditions, compared with D_0, the torque on a quarter-phase circuit, and what the relative torque per volt-ampere input, if the torque is proportional to the product of the e.m.fs. impressed upon the two circuits and the sine of the angle of phase displacement between them?

In the quarter-phase motor the torque is

$$D_0 = ae^2 = 12{,}100\ a,$$

where a is a constant. The volt-ampere input is

$$Q_0 = 2\,e\,\sqrt{i_1^2 + i_2^2} = 31{,}200;$$

hence, the "apparent torque efficiency," or torque per volt-ampere input,

$$\eta_0 = \frac{D_0}{Q_0} = 0.388\ a.$$

The admittance per motor circuit is

$$Y = \frac{I}{e} = 0.91 - 0.91\,j,$$

the impedance is

$$Z = \frac{e}{I} = \frac{110}{100 - 100j} = \frac{110\,(100 + 100j)}{(100 - 100j)\,(100 + 100j)} = 0.55 + 0.55\,j.$$

the admittance of the condenser is

$$Y_0 = j;$$

thus, the joint admittance of the circuit shunted by the condenser is

$$\begin{aligned} Y_1 &= Y + Y_0 = 0.91 - 0.91\,j + j \\ &= 0.91 + 0.09\,j; \end{aligned}$$

its impedance is

$$Z_1 = \frac{1}{Y_1} = \frac{1}{0.91 + 0.09\,j} = \frac{0.91 - 0.09\,j}{0.91^2 + 0.09^2} = 1.09 - 0.11\,j,$$

and the total impedance of the two circuits in series is

$$\begin{aligned} Z_2 &= Z + Z_1 \\ &= 0.55 + 0.55\,j + 1.09 - 0.11\,j \\ &= 1.64 + 0.44\,j. \end{aligned}$$

Hence, the current, at impressed e.m.f. $e = 220$,

$$\begin{aligned} I = i_1 - ji_2 &= \frac{e}{Z_2} = \frac{220}{1.64 + 0.44\,j} = \frac{220\,(1.64 - 0.44\,j)}{1.64^2 + 0.44^2} \\ &= 125 - 33.5\,j; \end{aligned}$$

or, reduced,

$$\begin{aligned} I &= \sqrt{125^2 + 33.5^2} \\ &= 129.4 \text{ amp.} \end{aligned}$$

Thus, the volt-ampere input,

$$\begin{aligned} Q &= eI = 220 \times 129.4 \\ &= 28{,}470. \end{aligned}$$

The e.m.fs. acting upon the two motor circuits respectively are

$$\underset{\cdot}{E}_1 = \underset{\cdot}{I}Z_1 = (125 - 33.5\,j)\,(1.09 - 0.11\,j) = 132.8 - 50.4\,j$$

and

$$\underset{\cdot}{E}' = \underset{\cdot}{I}Z = (125 - 33.5\,j)(0.55 + 0.55\,j) = 87.2 + 50.4\,j.$$

Thus, the tangents of their phase angles are

$$\tan\theta_1 = +\frac{50.4}{132.8} = +\,0.30;\ \text{hence, } \theta_1 = +\,21°;$$

$$\tan\theta' = -\frac{50.4}{87.2} = -\,0.579;\ \text{hence, } \theta' = -\,30°;$$

and the phase difference,

$$\theta = \theta_1 - \theta' = 51°.$$

The absolute values of these e.m.fs. are

$$e_1 = \sqrt{132.8 + 50.4^2} = 141.5$$

and

$$e' = \sqrt{87.2^2 - 50.4^2} = 100.7;$$

thus, the torque is

$$\begin{aligned} D &= ae_1e' \sin\theta \\ &= 11{,}100\,a; \end{aligned}$$

and the apparent torque efficiency is

$$\eta_t = \frac{D}{Q}\ \frac{11{,}100\,a}{28{,}470} = 0.39\,a.$$

Hence, comparing this with the quarter-phase motor, the relative torque is

$$\frac{D}{D_0} = \frac{11{,}100\,a}{12{,}100\,a} = 0.92,$$

and the relative torque per volt-ampere, or relative apparent torque efficiency, is

$$\frac{\eta_t}{\eta_0} = \frac{0.39\,a}{0.388\,a} = 1.005.$$

86. (2) At constant field excitation, corresponding to a nominal generated e.m.f. $e_0 = 12{,}000$, a generator of synchronous impedance $Z_0 = r_0 + jx_0 = 0.6 + 60\,j$ feeds over a transmission line of impedance $Z_1 = r_1 + jx_1 = 12 + 18j$, and of capacity susceptance 0.003, a non-inductive receiving circuit. How will the voltage at the receiving end, e, and the voltage at the generator terminals, e_1, vary with the load if the line capacity is represented by a condenser shunted across the middle of the line?

Let $\dot{I} = i$ = current in receiving circuit, in phase with the e.m.f., $\dot{E} = e$.

The voltage in the middle of the line is

$$\dot{E}_2 = \dot{E} + \frac{Z_1}{2}\dot{I}$$
$$= e + 6\,i + 9\,ij.$$

The capacity susceptance of the line is, in symbolic expression, $Y = 0.003\,j$; thus the charging current is

$$\dot{I}_2 = \dot{E}_2 Y = 0.003\,j\,(e + 6\,i + 9\,ij)$$
$$= 0.027\,i + j\,(0.003\,e + 0.018\,i),$$

and the total current is

$$\dot{I}_1 = \dot{I} + \dot{I}_2 = 0.973\,i + j\,(0.003\,e + 0.018\,i).$$

Thus, the voltage at the generator end of the line is

$$\dot{E}_1 = \dot{E}_2 + \frac{Z_1}{2}\dot{I}_1$$
$$= e + 6\,i + 9\,ij + (6 + 9\,j)[0.973\,i + j\,(0.003\,e + 0.018\,i)]$$
$$= (0.973\,e + 11.68\,i) + j\,(17.87\,i + 0.018\,e),$$

and the nominal generated e.m.f. of the generator is

$$\dot{E}_0 = \dot{E}_1 + Z_0\dot{I}_1$$
$$= (0.973\,e + 11.68\,i) + j\,(17.87\,i + 0.018\,e) + (0.6 + 60\,j)[0.973\,i + j\,(0.003\,e + 0.018\,i)]$$
$$= (0.793\,e + 11.18\,i) + j\,(76.26\,i + 0.02\,e);$$

or, reduced, and $e_0 = 12{,}000$ substituted,

$$e_0^2 = 144 \times 10^6 = (0.793\,e + 11.18\,i)^2 + (76.26\,i + 0.02\,e)^2;$$

thus,

$$e^2 + 33\,ei + 9450\,i^2 = 229 \times 10^6,$$
$$e = -\,16.5\,i + \sqrt{229 \times 10^6 - 9178\,i^2},$$

and

$$e_1 = \sqrt{(0.973\,e + 11.68\,i)^2 + (17.87\,i + 0.018\,e)^2};$$

at

$$i = 0,\ e = 15{,}133,\ e_1 = 14{,}700;$$

at

$$e = 0,\ i = 155.6,\ e_1 = 3327.$$

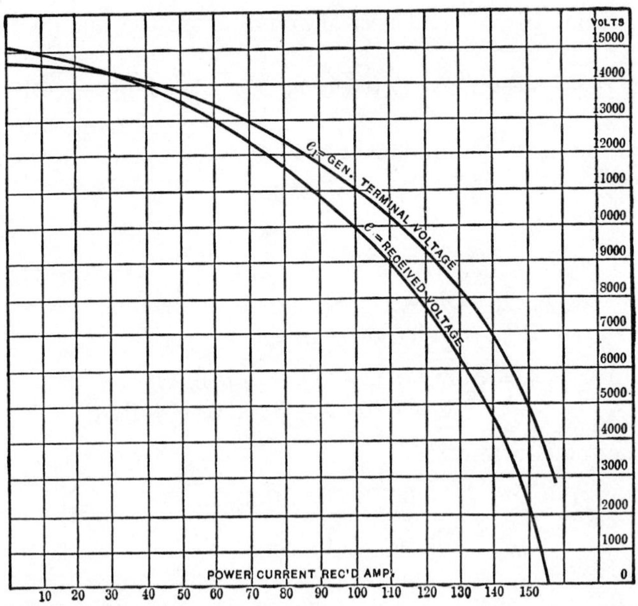

Fig. 40.—Reactive load characteristics of a transmission line fed by synchronous generator with constant field excitation.

Substituting different values for i gives

i	e	e_1	i	e	e_1
0	15,133	14,700	100	10,050	11,100
25	14,488	14,400	125	7,188	8,800
50	13,525	13,800	150	2,325	4,840
75	12,063	12,730	155.6	0	3,327

which values are plotted in Fig. 40.

18. EQUIVALENT SINE WAVES

87. In the preceding chapters, alternating waves have been assumed and considered as sine waves.

The general alternating wave is, however, never completely, frequently not even approximately, a sine wave.

A sine wave having the same effective value, that is, the same square root of mean squares of instantaneous values, as a general alternating wave, is called its corresponding "equivalent sine wave." It represents the same effect as the general wave.

With two alternating waves of different shapes, the phase relation or angle of lag is indefinite. Their equivalent sine waves, however, have a definite phase relation, that which gives the same effect as the general wave, that is, the same mean (ei).

Hence if e = e.m.f. and i = current of a general alternating wave, their equivalent sine waves are defined by

$$e_0 = \sqrt{\text{mean } (e^2)},$$
$$i_0 = \sqrt{\text{mean } (i^2)};$$

and the power is

$$p_0 = e_0 i_0 \cos \widehat{e_0 i_0} = \text{mean } (ei);$$

thus,

$$\cos \widehat{e_0 i_0} = \frac{\text{mean } (ei)}{\sqrt{\text{mean } (e^2)} \sqrt{\text{mean } (i^2)}}.$$

Since by definition the equivalent sine waves of the general alternating waves have the same effective value or intensity and the same power or effect, it follows that in regard to intensity and effect the general alternating waves can be represented by their equivalent sine waves.

Considering in the preceding the alternating currents as equivalent sine waves representing general alternating waves, the investigation becomes applicable to any alternating circuit irrespective of the wave shape.

The use of the terms reactance, impedance, etc., implies that a wave is a sine wave or represented by an equivalent sine wave.

Practically all measuring instruments of alternating waves (with exception of instantaneous methods) as ammeters, voltmeters, wattmeters, etc., give not general alternating waves but their corresponding equivalent sine waves.

EXAMPLES

88. In a 25-cycle alternating-current transformer, at 1000 volts primary impressed e.m.f., of a wave shape as shown in

TABLE I

(1) Degrees	(2) e	(3) e^2	(4) $e\frac{1000}{37.22} = e_0$	(5) Σe_0	(6) $\Sigma e_0 - 7324 = B'$	(7) $B'\frac{15,000}{7324} = B$	(8) From hysteresis cycle, f	(9) $50f = F$	(10) $\frac{F}{500} = i$	(11) i^2	(12) $p = e_0$
0	0	0	0	0	−7,324	−15,000	−20	−1,000	−2.00	4.00	0
10	1	1	27	27	−7,296	−14,950	−19.5	− 975	−1.95	3.80	− 53
20	3.5	12	94	121	−7,203	−14,800	−18	− 900	−1.80	3.24	− 169
30	8	64	216	337	−6,987	−14,300	−14.3	− 715	−1.43	2.04	− 308
40	14	196	377	714	−6,610	−13,550	−10	− 500	−1.00	1.00	− 377
50	22	484	591	1,305	−6,019	−12,350	− 5.5	− 275	−0.55	0.30	− 325
60	31	961	835	2,140	−5,184	−10,600	− 2.3	− 115	−0.23	0.05	+ 191
70	41	1,681	1,100	3,240	−4,084	− 8,370	− 0.2	− 10	−0.02	0.00	− 22
80	50	2,500	1,345	4,585	−2,739	− 5,600	+ 1.0	+ 50	+0.10	0.01	+ 134
90	55	3,025	1,480	6,065	−1,259	− 2,580	+ 1.9	+ 95	+0.19	0.04	+ 281
100	57	3,249	1,535	7,600	+ 276	+ 570	+ 2.6	+ 130	+0.26	0.07	+ 398
110	58	3,364	1,560	9,160	+1,836	+ 3,550	+ 3.3	+ 165	+0.33	0.11	+ 514
120	58	3,364	1,560	10,720	+3,396	+ 6,970	+ 4.5	+ 225	+0.45	0.20	+ 700
130	56	3,136	1,508	12,228	+4,904	+10,050	+ 6.6	+ 330	+0.66	0.44	+ 995
140	43	1,849	1,155	13,383	+6,059	+12,400	+10.0	+ 500	+1.00	1.00	+1,155
150	29	841	780	14,163	+6,839	+14,000	+14.2	+ 710	+1.42	2.02	+1,108
160	14	196	377	14,540	+7,216	+14,800	+18.8	+ 940	+1.88	3.53	+ 710
170	4	16	108	14,648	+7,324	+15,000	+20.0	+1,000	+2.00	4.09	+ 216
180	0	0	0	14,648	+7,324	+15,000					
		$\Sigma = 24,939$								$\Sigma = 25.85$	$\Sigma = 4,766$

$$\text{mean } e^2 = \frac{24,939}{18} = 1385.5$$

$$e' = \text{eff. } e = \sqrt{1385.5} = 37.22.$$

$$\text{mean } i^2 = \frac{25.85}{18} = 1.436$$

$$i' = \text{eff. } i = \sqrt{1.436} = 1.198$$

$$p' = \text{mean } p = \frac{4766}{18} = 264.8$$

Fig. 41 and Table I, the number of primary turns is 500, the length of the magnetic circuit 50 cm., and its section shall be chosen so as to give a maximum density $B = 15{,}000$.

At this density the hysteretic cycle is as shown in Fig. 42 and Table II.

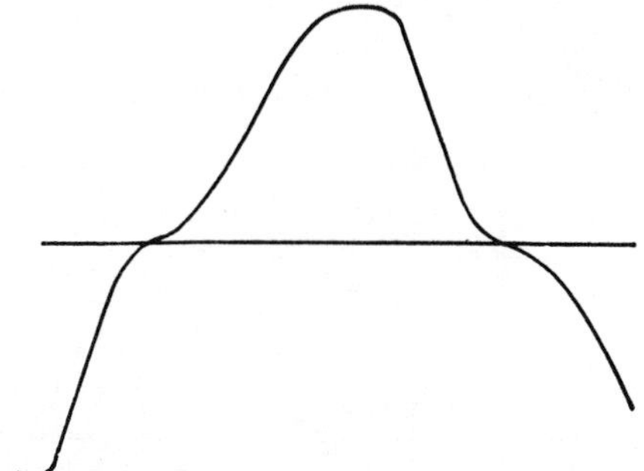

FIG. 41.—Wave-shape of e.m.f. in example 88.

What is the shape of current wave, and what the equivalent sine waves of e.m.f., magnetism, and current?

The calculation is carried out in attached table.

TABLE II

f	B
0	± 8,000
2	+ 10,400 − 2,500
4	+ 11,700 + 5,800
6	+ 12,400 + 9,300
8	+ 13,000 + 11,200
10	+ 13,500 + 12,400
12	+ 13,900 + 13,200
14	+ 14,200 + 13,800
16	+ 14,500 + 14,300
18	+ 14,800 + 14,700
20	+ 15,000

In column (1) are given the degrees, in column (2) the relative values of instantaneous e.m.fs., e corresponding thereto, as taken from Fig. 41.

Column (3) gives the squares of e. Their sum is 24,939; thus the mean square, $\frac{24{,}939}{18} = 1385.5$, and the effective value,

$$e' = \sqrt{1385.5} = 37.22.$$

Since the effective value of impressed e.m.f. is = 1000, the instantaneous values are $e_0 = e\dfrac{1000}{37.22}$ as given in column (4).

Since the e.m.f. e_0 is proportional to the rate of change of magnetic flux, that is, to the differential coefficient of B, B is proportional to the integral of the e.m.f., that is, to Σe_0 plus an integration constant. Σe_0 is given in column (5), and the integration constant follows from the condition that B at 180°

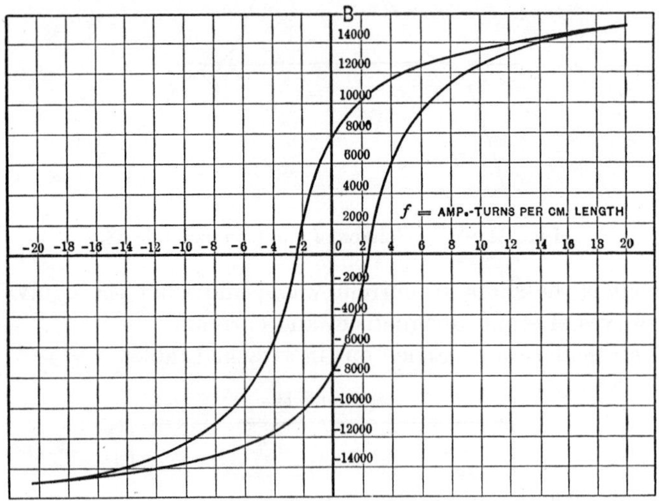

FIG. 42.—Hysteretic cycle in example 88.

must be equal, but opposite in sign, to B at 0°. The integration constant is, therefore,

$$-\tfrac{1}{2}\sum_0^{180} e_0 = -\frac{14,648}{2} = -7324,$$

and by subtracting 7324 from the values in column (5) the values of B' of column (6) are found as the relative instantaneous values of magnetic flux density.

Since the maximum magnetic flux density is 15,000 the instantaneous values are $B = B'\dfrac{15,000}{7324}$, plotted in column (7).

From the hysteresis cycle in Fig. 42 are taken the values of magnetizing force f, corresponding to magnetic flux density B. They are recorded in column (8), and in column (9) the instantaneous values of m.m.f. $F = lf$, where $l = 50$ = length of magnetic circuit.

$i = \frac{F}{n}$, where $n = 500$ = number of turns of the electric circuit, gives thus the exciting current in column (10).

Column (11) gives the squares of the exciting current, i^2. Their sum is 25.85; thus the mean square, $\frac{25.85}{18} = 1.436$, and the effective value of exciting current, $i' = \sqrt{1.436} = 1.198$ amp.

Column (12) gives the instantaneous values of power, $p = ie_0$. Their sum is 4766; thus the mean power, $p' = \frac{4766}{18} = 264.8$.

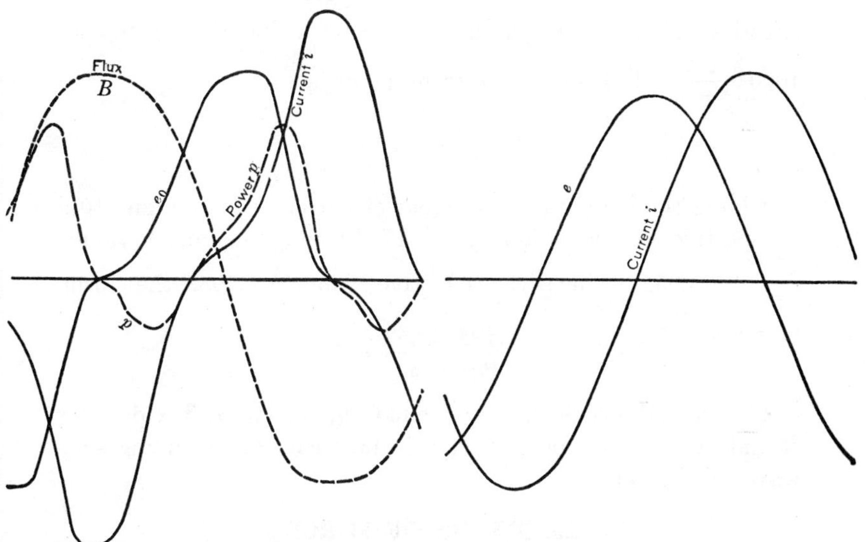

FIG. 43.—Waves of exciting current. Power and flux density corresponding to e.m.f. in Fig. 41 and hysteretic cycle in Fig. 42.

FIG. 44.—Corresponding sine waves for e.m.f. and exciting current in Fig. 43.

Since
$$p' = i'e'_0 \cos \theta,$$
where e'_0 and i' are the equivalent sine waves of e.m.f. and of current respectively, and θ their phase displacement, substituting these numerical values of p', e', and i', we have

$$264.8 = 1000 \times 1.198 \cos \theta.$$

hence,

$$\cos \theta = 0.221,$$
$$\theta = 77.2°,$$

and the angle of hysteretic advance of phase,

$$\alpha = 90° - \theta = 12.8°.$$

The hysteresis current is then

$$i' \cos \theta = 0.265,$$

and the magnetizing current,

$$i' \sin \theta = 1.165.$$

Adding the instantaneous values of e.m.f. e_0 in column (4) gives 14,648; thus the mean value, $\frac{14{,}648}{18} = 813.8$. Since the effective value is 1000, the mean value of a sine wave would be $1000\frac{2\sqrt{2}}{\pi} = 904$; hence the form factor is

$$\gamma = \frac{904}{813.8} = 1.11.$$

Adding the instantaneous values of current i in column (10), irrespective of their sign, gives 17.17; thus the mean value, $\frac{17.17}{18} = 0.954$. Since the effective value = 1.198, the form factor is

$$\gamma = \frac{1.198}{0.954}\frac{2\sqrt{2}}{\pi} = 1.12.$$

The instantaneous values of e.m.f. e_0, current i, flux density B and power p are plotted in Fig. 43, their corresponding sine waves in Fig. 44.

19. FIELDS OF FORCE

89. When an electric current flows through a conductor, power is consumed and heat produced inside of the conductor. In the space outside and surrounding the conductor, a change has taken place also, and this space is not neutral and inert any more, but if we try to move a solid mass of metal rapidly through it, the motion is resisted, and heat produced in the metal by induced currents. Materials of high permeability, as iron filings, brought into this space arrange themselves in chains; a magnetic needle is moved and places itself in a definite direction. Due to the passage of the current in the conductor, there are therefore in the spaces outside of the conductor—where the current does not flow—forces exerted, and this space then is not neutral space, but has become a *field of force*, and the cause of the field, in this case the electric current in the conductor, is its "*motive force*." As in this case the actions exerted in the field of force are magnetic, the space surrounding a conductor traversed by a current is a *field of magnetic force*, and the current in the conductor is the *magnetomotive force*.

In the space surrounding a ponderable mass, as our earth, forces are exerted on other masses—which cause the stone to fall toward the earth, and water to run down hill—and this space thus is a *field of gravitational force*, the earth the *gravimotive force*.

In the space surrounding conductors having a high potential difference, we observe a *field of dielectric force*, that is, electrostatic or dielectric forces are exerted, and the potential difference between the conductors is the *electromotive force* of the dielectric field.

The force exerted by the earth as *gravimotive force*, on any mass in the gravitational field of the earth, causes the mass to move with increasing rapidity. The direction of motion then shows the direction in which the force acts, that is, the direction of the gravitational field. The force g, which the field exerts on unit mass, that is, the acceleration of the mass, measures the intensity of the field: in the gravitational field of the earth 981 cm g sec. The force acting upon a mass m, then, is: $F = gm$, and is called the *weight* of the mass.

In the same manner, in the magnetic field of a current as magnetomotive force, the intensity H of the magnetic field is measured by the force F which the field exerts on a magnetic mass or pole strength m: $F = Hm$; the intensity K of the dielectric field of a potential difference as electromotive force is measured by the force F exerted upon an electric pole strength e: $F = Ke$; the direction of the force represents the direction of the field of force.

90. This conception of the field of force is one of the most important and fundamental ones of all sciences and applied sciences: a condition of space, brought about by some exciting cause or motive force, whereby the space is not neutral any more, but capable of exerting forces on anything susceptible to these forces: mechanical forces on masses in a gravitational field, magnetic forces on magnetic materials in a magnetic field,

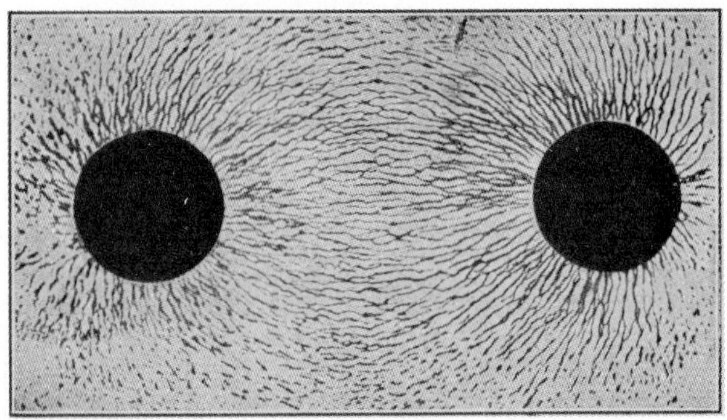

A.—A photograph of a mica-filing map of the dielectric lines of force between two cylinders.

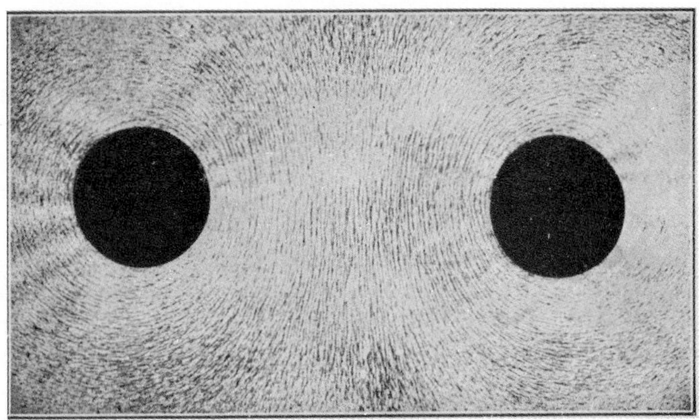

B.—A photograph of an iron-filing map of the magnetic lines of force about two cylinders.

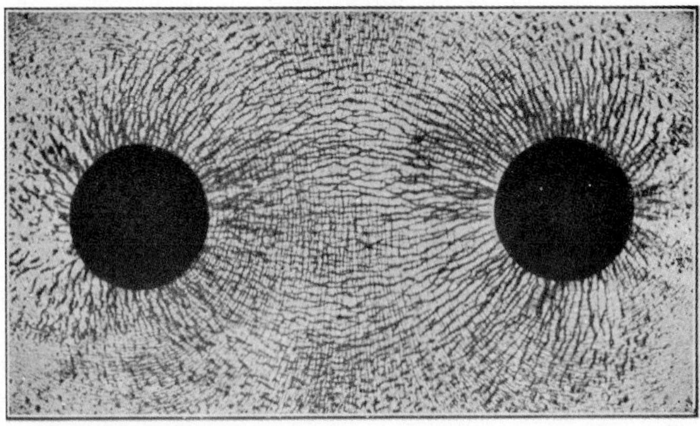

C.—A photographic superposition of A and B representing the magnetic and dielectric fields of the space surrounding two conductors which are carrying energy.

FIG. 45.

dielectric forces on dielectrics in a dielectric field, etc. The field of force then is characterized by having, at any point, a definite direction—the direction in which the force acts—and a definite intensity, to which the forces are proportional.

Such fields of force can be graphically represented by lines showing the direction in which the force acts: the lines of force and, at right angles thereto, the equipotential lines or surfaces, as the direction in which no force acts. Thus the lines of gravitational force of the earth are the verticals, the equipotential surfaces, or level surfaces, are the horizontals. Such pictures of a field of force also illustrate the intensity: where the lines of force and therefore the equipotential lines come closer together, the field is more intense, that is, the forces greater.

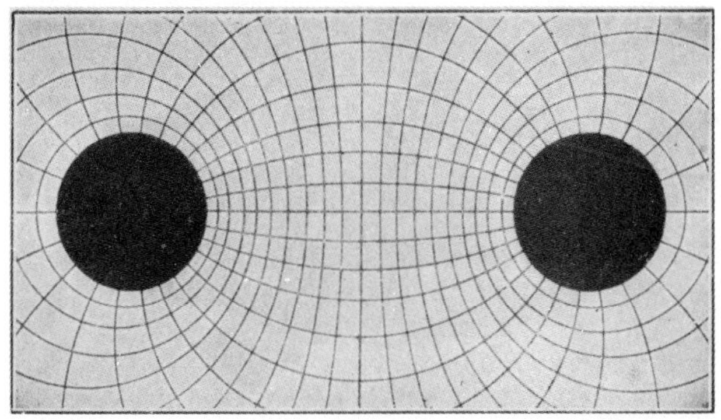

Fig. 46.—A mathematical plot of fields shown in C.

Magnetic fields may be demonstrated by iron filings brought into the field; dielectric fields by particles of a material of high specific capacity, such as mica. Fig. 45 shows the dielectric field of a pair of parallel conductors, the magnetic field between these conductors, and their combination. Fig. 46 shows the same as calculated.

As further illustration, Fig. 47 shows, from observation, half of the dielectric field between a rod with circular disc, as one terminal, passing symmetrically through the center of a cylinder placed in a circular hole in a plate as other terminal: the lines of force pass from terminal to terminal; the equipotential surfaces intersect at right angles (A 10,292).

91. In electrical engineering we have to deal with the electrical quantities: voltage, current, resistance, etc.; the magnetic quantities: magnetic flux, field intensity, permeability, etc.; and the dielectric quantities: dielectric flux, field intensity, permittivity, etc.

The electric current is the *magnetomotive force* F which produces the magnetic field, acting upon space. It is expressed in amperes, or rather in ampere-turns, and thus is an electrical quantity, its

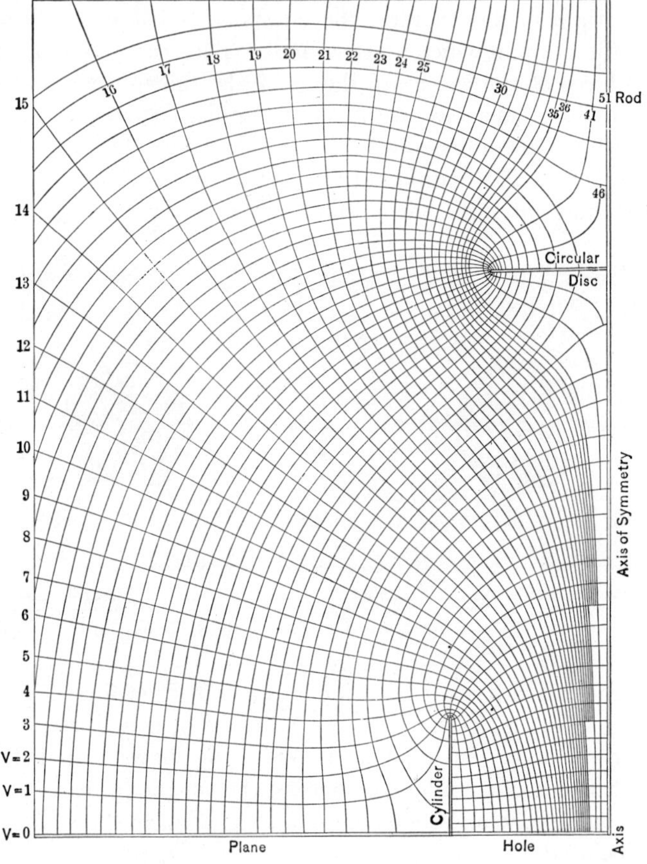

FIG. 47.—Observed dielectric field.

unit being determined by the unit of current, as the ampere-turn equal to 10^{-1} absolute units.

The magnetomotive force per unit length of the magnetic circuit then is the *magnetizing force* or *magnetic gradient* f, in ampere-turns per centimeter, hence still an electrical quantity.

Proportional thereto, and of the same dimension, is the

magnetic field intensity H. It differs from the magnetic gradient merely by a numerical factor 4π; $H = 4\pi f\, 10^{-1}$. Magnetic field intensity is a magnetic quantity, and its unit defined by the magnetic forces exerted in the field, thus different from the unit of magnetic gradient, which is determined by the unit of electric current; hence the factor 4π. The factor 10^{-1} merely reduces from amperes to absolute unit.

If then μ is the magnetic conductivity of the material in the magnetic field, called its *permeability,* $B = \mu H$ is the magnetic *flux density,* and the total magnetic flux Φ is given by the density B times the area or section of the flux.

Or, passing directly from the magnetomotive force F to the magnetic flux, by the conception of the magnetic circuit: $\Phi = \frac{F}{R}$, where R is the magnetic resistance, or *reluctance* of the magnetic circuit.

R is an electric quantity, and does not contain the 4π.

In the dielectric field, the potential difference e is the electromotive force expressed in volts. The electromotive force per unit length of the dielectric circuit is the *electrifying force* or *voltage gradient* or *dielectric gradient* g, expressed in volts per centimeter. This is still an electric quantity.

Proportional thereto by a numerical factor is the dielectric quantity: *dielectric field intensity* $K = \frac{g}{4\pi v^2}$, and if k is the dielectric conductivity of the medium in the dielectric field, called *specific capacity* or *permittivity,* the *dielectric flux density* is $D = kK$, and the total *dielectric flux* Ψ is flux density times area.

Here again, at the transition from the electric quantity "gradient" to the dielectric quantity "field intensity," a numerical factor $4\pi v^2$ enters, the one quantity being based on the volt as unit, the other on unit force action. v is the velocity of light, 3×10^{10}, and the factor v^2 the result of the convention of assuming the permittivity of empty space as unity.

It is now easy to remember, where in the electromagnetic system of units the factor 4π enters: it is at the transition from the electrical quantities to the magnetic or dielectric quantities, from *gradient* to *field intensity.*

92. The dielectric field and the magnetic field are analogous, and to magnetic flux, magnetic field intensity, permeability, as used in dealing with magnetic circuits, correspond the terms dielectric flux, dielectric field intensity, permittivity, as used in dealing with the electrostatic fields of high potential apparatus, as transmission insulators, transformer bushings, etc. The foremost difference is that in the magnetic field, a line of force must always return into itself in a closed circuit, while in the electrostatic or dielectric field, a line of force may terminate in a conductor. The terminals of the lines of electrostatic flux, Ψ at the conductor, then are represented by the conception of a *quantity of electricity* or *electric charge*, Q, being located on the conductor. Thus, at the terminal of the line of unit dielectric flux, unit electric quantity is located on the conductor.

Dielectric flux Ψ and electric quantity or charge Q thus are identical, and merely different conceptions of the dielectric circuit:

$$Q = \Psi.$$

In using the conception of electric quantity Q, we consider only the terminals of the lines of dielectric flux, that is, deal merely with the effect of the dielectric flux on the electric circuit which produced it. This conception is in many cases more convenient, but it necessarily fails, when the distribution of the dielectric flux in the dielectric field is of importance, such as is the case when dealing with high dielectric field intensities, approaching the possibility of disruptive effects in the field of force, or when dealing with the effect produced by the introduction of materials of different permittivity into the dielectric field. Therefore, with the increasing importance of the dielectric field in engineering, the conception of electric quantity, or charge, is gradually being replaced by the conception of the dielectric flux and the dielectric field, analogous to the magnetic field, which has replaced the previous conception of "magnetic poles."

20. NOMENCLATURE

93. The following nomenclature and symbols of the quantities most frequently used in electrical engineering appears most satisfactory, and is therefore recommended. It is in agreement with the Standardization Rules of the A. I. E. E., but as far as possible standard letters have been used, and script letters avoided as impracticable or at least inconvenient in writing and still more in typewriting. Therefore F has been chosen for m.m.f., and dielectric field intensity changed to K. Also, a few symbols not contained in the Standardization Rules had to be added.

TABLE OF SYMBOLS

Symbol	Name	Unit	Character
E, e	Voltage Potential difference Electromotive force	Volt	Electrical
I, i	Current	Ampere	Electrical
R, r	Resistance	Ohm	Electrical
x	Reactance	Ohm	Electrical
Z, z	Impedance	Ohm	Electrical
g	Conductance	Mho	Electrical
b	Susceptance	Mho	Electrical
Y, y	Admittance	Mho	Electrical
ρ	Resistivity	Ohm-centimeter	Electrical
γ	Conductivity	Mho-centimeter	Electrical
Φ	Magnetic flux[1]	Line; kiloline; megaline	Magnetic
B	Magnetic density[2]	Lines per cm.^{2}; kilolines per cm.^{2}	Magnetic
H	Magnetic field intensity	Lines per cm.^{2}	Magnetic
μ	Permeability (magnetic conductivity)	....	Magnetic
f	Magnetic gradient Magnetizing force[3]	Ampere-turns per centimeter.	Electrical
F	Magnetomotive force	Ampere-turns	Electrical
R	Reluctance (magnetic resistance)	....	Electrical
L	Inductance	Henry; milhenry	Magnetic
M	Mutual inductance	Henry; milhenry	Magnetic
S	Self-inductance Leakage inductance	Henry; milhenry	Magnetic
Ψ, Q	Dielectric flux Electric quantity or charge	Lines of dielectric force Coulombs	Dielectric
D	Dielectric density	Dielectric lines per cm.^{2}	Dielectric
K	Dielectric field intensity	Coulombs per cm.^{2}	Dielectric
k	Permittivity Specific capacity	....	Dielectric

[1] 10^8 lines = 1 *weber*.
[2] 1 line per sq. cm. = 1 *gauss*.
[3] 0.4 amp-turn per cm. = 1 *oersted*.

TABLE OF SYMBOLS. *Continued*

Symbol	Name	Unit	Character
g........	Dielectric gradient Voltage gradient Electrifying force	Volts per centimeter	Electrical
C.......	Capacity	Farad; microfarad	Dielectric
P, p.....	Power, effect	Watt; kilowatt	General
W, w....	Energy, work	Joule; kilojoule	General
T, ϑ.....	Temperature	Degrees Centigrade	General
t........	Time	Seconds	General
θ, ϕ, β...	Time angle	Degrees or radians	General
α, τ.....	Space angle	Degrees or radians	General
f........	Frequency	Cycles per second	General

PART II

ALTERNATING-CURRENT PHENOMENA

CHAPTER I

INTRODUCTION

1. In the practical applications of electrical energy, we meet with two different classes of phenomena, due respectively to the continuous current and to the alternating current.

The continuous-current phenomena have been brought within the realm of exact analytical calculation by a few fundamental laws:

1. Ohm's law: $i = \frac{e}{r}$, where r, the resistance, is a constant of the circuit.

2. Joule's law: $P = i^2r$, where P is the power, or the rate at which energy is expended by the current, i, in the resistance, r.

3. The power equation: $P_0 = ei$, where P_0 is the power expended in the circuit of e.m.f., e, and current, i.

4. Kirchhoff's laws:

(*a*) The sum of all the e.m.fs. in a closed circuit = 0, if the e.m.f. consumed by the resistance, ir, is also considered as a counter e.m.f., and all the e.m.fs. are taken in their proper direction.

(*b*) The sum of all the currents directed toward a distributing point = 0.

In alternating-current circuits, that is, in circuits in which the currents rapidly and periodically change their direction, these laws cease to hold. Energy is expended, not only in the conductor through its ohmic resistance, but also outside of it; energy is stored up and returned, so that large currents may exist simultaneously with high e.m.fs., without representing any considerable amount of expended energy, but merely a surging to and fro of energy; the ohmic resistance ceases to be the determining factor of current value; currents may divide into components, each of which is larger than the undivided current, etc.

2. In place of the above-mentioned fundamental laws of continuous currents, we find in alternating-current circuits the following:

Ohm's law assumes the form $i = \frac{e}{z}$, where z, the apparent resistance, or *impedance*, is no longer a constant of the circuit, but depends upon the frequency of the currents; and in circuits containing iron, etc., also upon the e.m.f.

Impedance, z, is, in the system of absolute units, of the same dimension as resistance (that is, of the dimension lt^{-1} = velocity), and is expressed in ohms.

It consists of two components, the resistance, r, and the reactance, x, or

$$z = \sqrt{r^2 + x^2}.$$

The resistance, r, in circuits where energy is expended only in heating the conductor, is the same as the ohmic resistance of continuous-current circuits. In circuits, however, where energy is also expended outside of the conductor by magnetic hysteresis, mutual inductance, dielectric hysteresis, etc., r is larger than the true ohmic resistance of the conductor, since it refers to the total expenditure of energy. It may be called then the *effective resistance.* It may no longer be a constant of the circuit.

The reactance, x, does not represent the expenditure of energy as does the effective resistance, r, but merely the surging to and fro of energy. It is not a constant of the circuit, but depends upon the frequency, and frequently, as in circuits containing iron, or in electrolytic conductors, upon the e.m.f. also. Hence while the effective resistance, r, refers to the power or active component of e.m.f., or the e.m.f. in phase with the current, the reactance, x, refers to the wattless or reactive component of e.m.f., or the e.m.f. in quadrature with the current.

3. The principal sources of reactance are electromagnetism and capacity.

Electromagnetism

An electric current, i, in a circuit produces a magnetic flux surrounding the conductor in lines of magnetic force (or more correctly, lines of magnetic induction), of closed, circular, or other form, which alternate with the alternations of the current, and thereby generate an e.m.f. in the conductor. Since the magnetic flux is in phase with the current, and the generated e.m.f. 90°, or a quarter period, behind the flux, this *e.m.f. of self-induction* lags 90°, or a quarter period, behind the current; that is, is in quadrature therewith, and therefore wattless.

If now Φ = the magnetic flux produced by, and interlinked with, the current, i (where those lines of magnetic force which are interlinked n-fold, or pass around n turns of the conductor, are counted n times), the ratio, $\frac{\Phi}{i}$, is denoted by L, and called the *inductance* of the circuit. It is numerically equal, in absolute units, to the interlinkages of the circuit with the magnetic flux produced by unit current, and is, in the system of absolute units, of the dimension of length. Instead of the inductance, L, sometimes its ratio with the ohmic resistance, r, is used, and is called the *time-constant* of the circuit,

$$T = \frac{L}{r}.$$

If a conductor surrounds with n turns a magnetic circuit of reluctance, $\mathcal{R}$, the current, i, in the conductor represents the m.m.f. of ni ampere-turns, and hence produces a magnetic flux of $\frac{ni}{\mathcal{R}}$ lines of magnetic force, surrounding each n turns of the conductor, and thereby giving $\Phi = \frac{n^2 i}{\mathcal{R}}$ interlinkages between the magnetic and electric circuits. Hence the inductance is

$$L = \frac{\Phi}{i} = \frac{n^2}{\mathcal{R}}.$$

The fundamental law of electromagnetic induction is, that the e.m.f. generated in a conductor by a magnetic field is proportional to the rate of cutting of the conductor through the magnetic field.

Hence, if i is the current and L is the inductance of a circuit, the magnetic flux interlinked with a circuit of current, i, is Li, and $4fLi$ is consequently the average rate of cutting; that is, the number of lines of force cut by the conductor per second, where f = frequency, or number of complete periods (double reversals) of the current per second, i = maximum value of current.

Since the maximum rate of cutting bears to the average rate the same ratio as the quadrant to the radius of a circle (a sinusoidal variation supposed), that is, the ratio $\frac{\pi}{2} \div 1$, the maximum rate of cutting is $2\pi f$, and, consequently, the maximum value of e.m.f. generated in a circuit of maximum current value, i, and inductance, L, is

$$e = 2\pi f L i.$$

Since the maximum values of sine waves are proportional (by factor $\sqrt{2}$) to the effective values (square root of mean squares), if i = effective value of alternating current, $e = 2\pi f L i$ is the effective value of e.m.f. of self-induction, and the ratio, $\frac{e}{i} = 2\pi f L$, is the *inductive reactance*,

$$x_m = 2\pi f L.$$

Thus, if r = resistance, x_m = reactance, z = impedance,
the e.m.f. consumed by resistance is $e_1 = ir$;
the e.m.f. consumed by reactance is $e_2 = ix_m$;
and, since both e.m.fs. are in quadrature to each other, the total e.m.f. is

$$e = \sqrt{e_1^2 + e_2^2} = i\sqrt{r^2 + x_m^2} = iz;$$

that is, the impedance, z, takes in alternating-current circuits the place of the resistance, r, in continuous-current circuits.

Capacity

4. If upon a condenser of capacity C an e.m.f., e, is impressed, the condenser receives the electrostatic charge, Ce.

If the e.m.f., e, alternates with the frequency, f, the average rate of charge and discharge is $4f$, and $2\pi f$ the maximum rate of charge and discharge, sinusoidal waves supposed; hence, $i = 2\pi f C e$, the current to the condenser, which is in quadrature to the e.m.f. and leading.

It is then

$$x_c = \frac{e}{i} = \frac{1}{2\pi f C},$$

the "*condensive reactance.*"

Polarization in electrolytic conductors acts to a certain extent like capacity.

The condensive reactance is inversely proportional to the frequency and represents the leading out-of-phase wave; the inductive reactance is directly proportional to the frequency, and represents the lagging out-of-phase wave. Hence both are of opposite sign with regard to each other, and the total reactance of the circuit is their difference, $x = x_m - x_c$.

The total resistance of a circuit is equal to the sum of all the resistances connected in series; the total reactance of a circuit is equal to the algebraic sum of all the reactances connected in series; the total impedance of a circuit, however, is not equal to the sum of all the individual impedances, but in general less, and is the resultant of the total resistance and the total reactance. Hence it is not permissible directly to add impedances, as it is with resistances or reactances.

A further discussion of these quantities will be found in the later chapters.

5. In Joule's law, $P = i^2r$, r is not the true ohmic resistance, but the "effective resistance;" that is, the ratio of the power component of e.m.f. to the current. Since in alternating-current circuits, in addition to the energy expended in the ohmic resistance of the conductor, energy is expended, partly outside, partly inside of the conductor, by magnetic hysteresis, mutual induction, dielectric hysteresis, etc., the effective resistance, r, is in general larger than the true resistance of the conductor, sometimes many time larger, as in transformers at open secondary circuit, and is no longer a constant of the circuit. It is more fully discussed in Chapter VIII (page 174).

In alternating-current circuits the power equation contains a third term, which, in sine waves, is the cosine of the angle of the difference of phase between e.m.f. and current:

$$P_0 = ei \cos \theta.$$

Consequently, even if e and i are both large, P_0 may be very small, if $\cos \theta$ is small, that is, θ near 90°.

Kirchhoff's laws become meaningless in their original form, since these laws consider the e.m.fs. and currents as directional quantities, counted positive in the one, negative in the opposite direction, while the alternating current has no definite direction of its own.

6. The alternating waves may have widely different shapes; some of the more frequent ones are shown in a later chapter.

The simplest form, however, is the sine wave, shown in Fig. 1, or, at least, a wave very near sine shape, which may be represented analytically by

$$i = I \sin \frac{2\pi}{t_0} (t - t_1) = I \sin 2\pi f(t - t_1),$$

where I is the maximum value of the wave, or its *amplitude;* t_0 is the time of one complete cyclic repetition, or the *period* of the wave, or $f = \frac{1}{t_0}$ is the *frequency* or number of complete periods per second; and t_1 is the time, where the wave is zero, or the *epoch* of the wave, generally called the *phase.*[1]

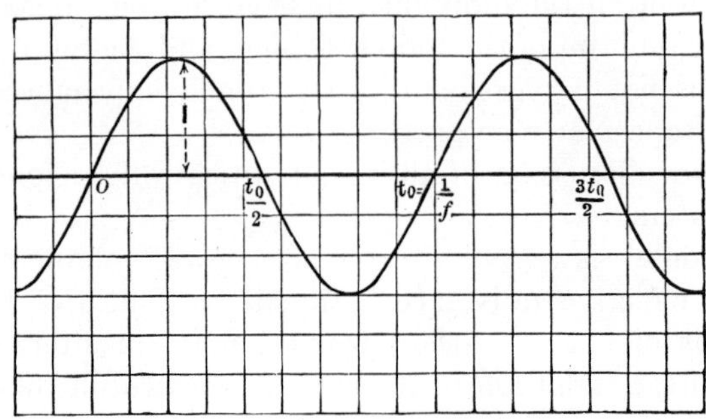

FIG. 1.—Sine wave.

Obviously, "phase" or "epoch" attains a practical meaning only when several waves of different phases are considered, as "difference of phase." When dealing with one wave only, we may count the time from the moment when the wave is zero, or from the moment of its maximum, representing it respectively by

$$i = I \sin 2\pi ft,$$

and

$$i = I \cos 2\pi ft.$$

Since it is a univalent function of time, that is, can at a given instant have one value only, by Fourier's theorem, any alternating wave, no matter what its shape may be, can be represented by a series of sine functions of different frequencies and different phases, in the form

$$i = I_1 \sin 2\pi f(t - t_1) + I_2 \sin 4\pi f(t - t_2) + I_3 \sin 6\pi f(t - t_3) + \ldots$$

where $I_1, I_2, I_3, \ldots$ are the maximum values of the different components of the wave, $t_1, t_2, t_3 \ldots$ the times, where the respective components pass the zero value.

[1] "Epoch" is the time where a periodic function reaches a certain value, for instance, zero; and "phase" is the angular position, with respect to a datum position, of a periodic function at a given time. Both are in alternate-current phenomena only different ways of expressing the same thing.

The first term, $I_1 \sin 2\pi f(t - t_1)$, is called the *fundamental wave*, or the *first harmonic;* the further terms are called the *higher harmonics*, or "overtones," in analogy to the overtones of sound waves. $I_n \sin 2n\pi f(t - t_n)$ is the nth harmonic.

By resolving the sine functions of the time differences, $t - t_1$, $t - t_2 \ldots$, we reduce the general expression of the wave to the form:

$$i = A_1 \sin 2\pi ft + A_2 \sin 4\pi ft + A_3 \sin 6\pi ft + \ldots$$
$$+ B_1 \cos 2\pi ft + B_2 \cos 4\pi ft + B_3 \cos 6\pi ft + \ldots$$

The two half-waves of each period, the *positive wave* and the *negative wave* (counting in a definite direction in the circuit), are usually identical, because, for reasons inherent in their construction, practically all alternating-current machines generate e.m.fs. in which the negative half-wave is identical with the positive. Hence the even higher harmonics, which cause a difference in the shape of the two half-waves, disappear, and only the odd harmonics exist, except in very special cases.

Hence the general alternating-current wave is expressed by:

$$i = I_1 \sin 2\pi f(t - t_1) + I_3 \sin 6\pi f(t - t_3)$$
$$+ I_5 \sin 10\pi f(t - t_5) + \ldots$$

or,

$$i = A_1 \sin 2\pi ft + A_3 \sin 6\pi ft + A_5 \sin 10\pi ft + \ldots$$
$$+ B_1 \cos 2\pi ft + B_3 \cos 6\pi ft + B_5 \cos 10\pi ft + \ldots$$

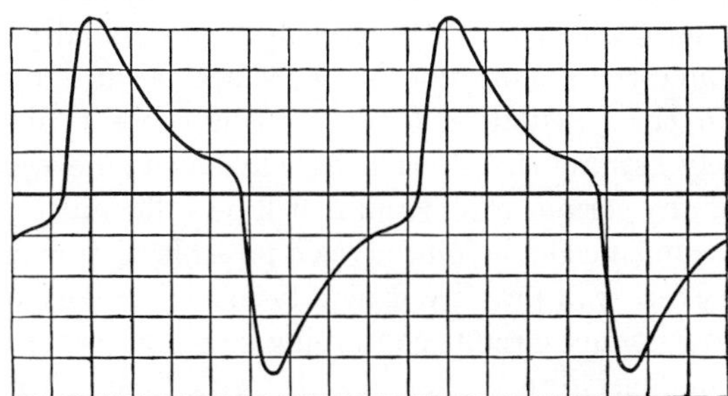

FIG. 2.—Wave without even harmonics.

Such a wave is shown in Fig. 2, while Fig. 3 shows a wave whose half-waves are different. Figs. 2 and 3 represent the secondary currents of a Ruhmkorff coil, whose secondary coil is closed by a high external resistance; Fig. 3 is the coil operated in the usual way, by make and break of the primary battery current; Fig. 2 is the coil fed with reversed currents by a commutator from a battery.

7. Inductive reactance, or electromagnetic momentum, which is always present in alternating-current circuits—to a large extent in generators, transformers, etc.—tends to suppress the higher harmonics of a complex harmonic wave more than the

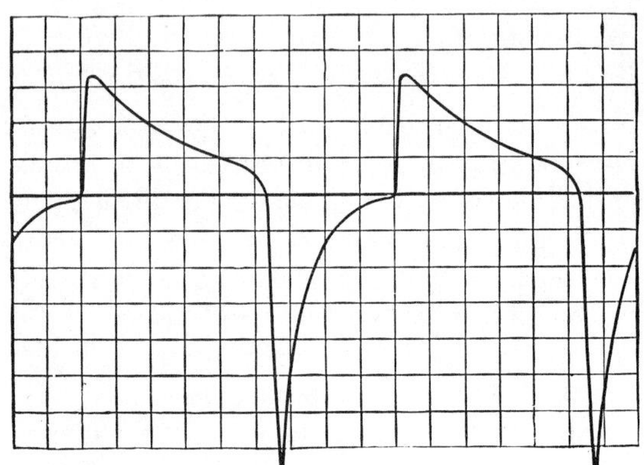

FIG. 3.—Wave with even harmonics.

fundamental harmonic, since the inductive reactance is proportional to the frequency, and is thus greater with the higher harmonics, and thereby causes a general tendency toward simple sine shape, which has the effect that, in general, the alternating currents in our light and power circuits are sufficiently near sine waves to make the assumption of sine shape permissible.

Hence, in the calculation of alternating-current phenomena, we can safely assume the alternating wave as a sine wave, without making any serious error; and it will be sufficient to keep the distortion from sine shape in mind as a possible disturbing factor, which, however, is in practice generally negligible—except in the case of low-resistance circuits containing large inductive reactance and large condensive reactance in series with each other, so as to produce resonance effects of these higher harmonics, and also under certain conditions of long-distance power transmission and high-potential distribution.

8. Experimentally, the impedance, effective resistance, inductance, capacity, etc., of a circuit or a part of a circuit are conveniently determined by impressing a sine wave of alternating e.m.f. upon the circuit and measuring with alternating-current ammeter, voltmeter and wattmeter the current, i, in the circuit, the potential difference, e, across the circuit, and the power, p, consumed in the circuit.

Then,

$$\text{The impedance, } z = \frac{e}{i};$$

$$\text{The phase angle, } \cos\theta = \frac{p}{ei};$$

$$\text{The effective resistance, } r = \frac{P}{i^2}.$$

From these equations,

$$\text{The reactance, } x = \sqrt{z^2 - r^2}.$$

If the reactance is inductive, the inductance is

$$L = \frac{x}{2\,\pi f}.$$

If the reactance is condensive, the capacity or its equivalent is

$$C = \frac{1}{2\,\pi f x},$$

wherein f = the frequency of the impressed e.m.f. If the reactance is the resultant of inductive and condensive reactances connected in series, it is

$$x = 2\,\pi f L - \frac{1}{2\,\pi f C};$$

L and C can be found by measuring the reactance at two different frequencies, f_1 and f_2, as follows;

$$x_1 = 2\,\pi f_1 L - \frac{1}{2\pi f_1 C},$$

$$x_2 = 2\,\pi f_2 L - \frac{1}{2\pi f_2 C},$$

then,

$$L = \frac{x_1 f_1 - x_2 f_2}{2\,\pi(f_1^2 - f_2^2)},$$

and

$$C = \frac{f_1^2 - f_2^2}{2\,\pi f_1 f_2 (x_1 f_2 - x_2 f_1)}.$$

A moderate deviation of the wave of alternating impressed e.m.f. from sine shape does not cause any serious error as long as the circuit contains no capacity.

In the presence of capacity, however, even a very slight distortion of wave shape may cause an error of some hundred per cent.

To measure capacity and condensive reactance by ordinary alternating currents it is, therefore, advisable to insert in series with the condensive reactance a non-inductive resistance or inductive reactance which is larger than the condensive reactance, or to use a source of alternating current, in which the higher harmonics are suppressed, as the *T*-connection of Constant Potential —Constant-current Transformation, paragraph 64, page 196.

In iron-clad inductive reactances, or reactances containing iron in the magnetic circuit, the reactance varies with the magnetic induction in the iron, and thereby with the current and the impressed e.m.f. Therefore the impressed e.m.f. or the magnetic induction must be given, to which the ohmic reactance refers, or preferably a curve is plotted from test (or calculation), giving the ohmic reactance, or, as usually done, the impressed e.m.f. as function of the current. Such a curve is called an *excitation curve* or *impedance curve,* and has the general character of the magnetic characteristic. The same also applies to electrolytic reactances, etc.

The calculation of an inductive reactance is accomplished by calculating the magnetic circuit, that is, determining the ampere-turns m.m.f. required to send the magnetic flux through the magnetic reluctance. In the air part of the magnetic circuit, unit permeability (or, referred to ampere-turns as m.m.f., reluctivity $\frac{10}{4\pi}$) is used; for the iron part, the ampere-turns are taken from the curve of the magnetic characteristic, as discussed in the following.

CHAPTER II

INSTANTANEOUS VALUES AND INTEGRAL VALUES

9. In a periodically varying function, as an alternating current, we have to distinguish between the *instantaneous value,* which varies constantly as function of the time, and the *integral value,* which characterizes the wave as a whole.

As such integral value, almost exclusively the *effective value* is used, that is, the square root of the mean square; and wherever the intensity of an electric wave is mentioned without further reference, the effective value is understood.

The *maximum value* of the wave is of practical interest only in few cases, and may, besides, be different for the two half-waves, as in Fig. 3.

As *arithmetic mean,* or *average value,* of a wave as in Figs. 4 and 5, the arithmetical average of all the instantaneous values during one complete period is understood.

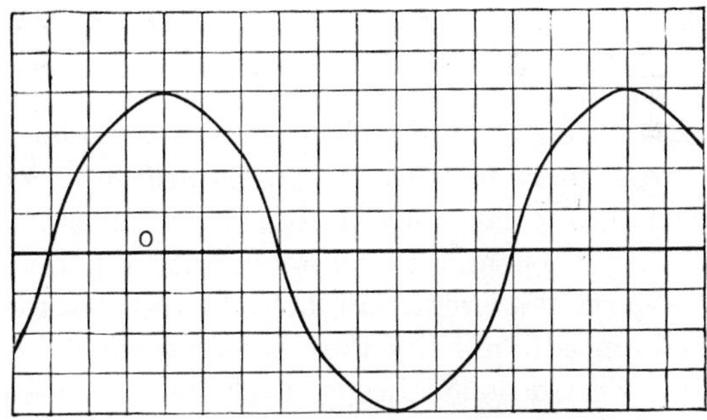

FIG. 4.—Alternating wave.

This arithmetic mean is either = 0, as in Fig. 4, or it differs from 0, as in Fig. 5. In the first case, the wave is called an *alternating wave,* in the latter a *pulsating wave.*

Thus, an alternating wave is a wave whose positive values give the same sum total as the negative values; that is, whose two half-waves have in rectangular coördinates the same area, as shown in Fig. 4.

A pulsating wave is a wave in which one of the half-waves preponderates, as in Fig. 5.

By electromagnetic induction, pulsating waves are produced only by commutating and unipolar machines (or by the superposition of alternating upon direct currents, etc.).

All inductive apparatus without commutation give exclusively alternating waves, because, no matter what conditions may exist in the circuit, any line of magnetic force which during a complete period is cut by the circuit, and thereby generates an e.m.f., must during the same period be cut again in the opposite direction, and thereby generate the same total amount of e.m.f. (Obviously, this does not apply to circuits consisting of different

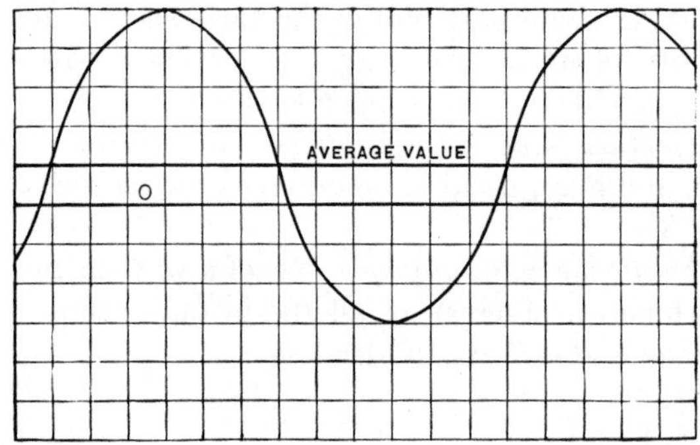

FIG. 5.—Pulsating wave.

parts movable with regard to each other, as in unipolar machines.) A direct-current machine without commutator or collector rings, or a coil-wound unipolar machine, thus is an impossibility.

Pulsating currents, and therefore pulsating potential differences across parts of a circuit can, however, be produced from an alternating induced e.m.f. by the use of asymmetrical circuits, as arcs, some electrochemical cells, as the aluminum-carbon cell, etc. Most of the alternating-current rectifiers are based on the use of such asymmetrical circuits.

In the following we shall almost exclusively consider the alternating wave, that is, the wave whose true arithmetic mean value $= 0$.

Frequently, by mean value of an alternating wave, the average of one half-wave only is denoted, or rather the average of all instantaneous values without regard to their sign. This *mean value* of one half-wave is of importance mainly in the rectification of alternating e.m.fs., since it determines the unidirectional value derived therefrom.

10. In a sine wave, the relation of the mean to the maximum value is found in the following way:

Let, in Fig. 6, AOB represent a quadrant of a circle with radius 1.

Then, while the angle θ traverses the arc $\frac{\pi}{2}$ from A to B, the sine varies from 0 to $OB = 1$. Hence the average variation of the sine bears to that of the corresponding arc the ratio $1 \div \frac{\pi}{2}$, or $\frac{2}{\pi} \div 1$. The maximum variation of the sine takes place about its zero value, where the sine is equal to the arc. Hence the maximum variation of the sine is equal to the variation of the

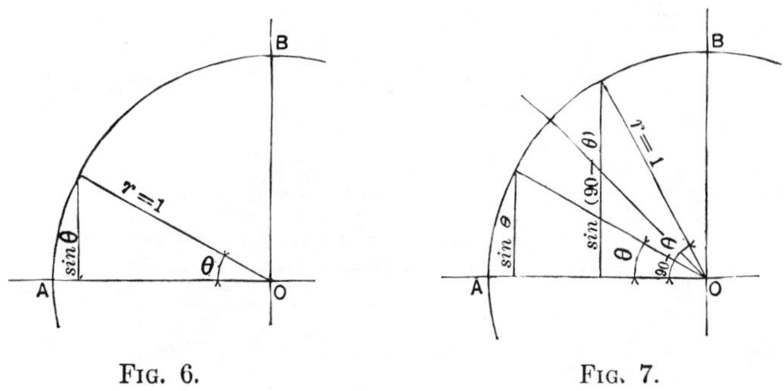

FIG. 6. FIG. 7.

corresponding arc, and consequently the maximum variation of the sine bears to its average variation the same ratio as the average variation of the arc to that of the sine, that is, $1 \div \frac{2}{\pi}$, and since the variations of a sine function are sinusoidal also, we have

Mean value of sine wave $\div$ maximum value $= \frac{2}{\pi} \div 1 = 0.63663.$

The quantities, "current," "e.m.f.," "magnetism," etc., are in reality mathematical fictions only, as the components of the entities, "energy," "power," etc.; that is, they have no independent existence, but appear only as squares or products.

Consequently, the only integral value of an alternating wave which is of practical importance, as directly connected with the mechanical system of units, is that value which represents the same *power* or *effect* as the periodical wave. This is called the *effective value.* Its square is equal to the mean square of the periodic function, that is:

The effective value of an alternating wave, or the value representing the same effect as the periodically varying wave, is the square root of the mean square.

In a sine wave, its relation to the maximum value is found in the following way:

Let, in Fig. 7, AOB represent a quadrant of a circle with radius 1.

Then, since the sines of any angle, θ, and its complementary angle, $90° - \theta$, fulfill the condition,

$$\sin^2 \theta + \sin^2 (90 - \theta) = 1,$$

the sines in the quadrant, AOB, can be grouped into pairs, so that the sum of the squares of any pair $= 1$; or, in other words, the mean square of the sine $= \frac{1}{2}$, and the square root of the mean square, or the effective value of the sine, $= \frac{1}{\sqrt{2}}$. That is:

The effective value of a sine function bears to its maximum value the ratio,

$$\frac{1}{\sqrt{2}} \div 1 = 0.70711.$$

Hence, we have for the sine wave the following relations:

Max.	Eff.	Arith. mean	
		Half period	Whole period
1	$\frac{1}{\sqrt{2}}$	$\frac{2}{\pi}$	0
1.0	0.7071	0.63663	0
1.4142	1.0	0.90034	0
1.5708	1.1107	1.0	0

11. Coming now to the general alternating wave,

$$i = A_1 \sin 2\pi ft + A_2 \sin 4\pi ft + A_3 \sin 6\pi ft + \ldots$$
$$+ B_1 \cos 2\pi ft + B_2 \cos 4\pi ft + B_3 \cos 6\pi ft + \ldots,$$

we find, by squaring this expression and cancelling all the products which give 0 as mean square, the *effective value*

$$I = \sqrt{\tfrac{1}{2}(A_1^2 + A_2^2 + A_3^2 + \ldots + B_1^2 + B_2^2 + B_3^2 \ldots)}.$$

The *mean value* does not give a simple expression, and is of no general interest.

12. All alternating-current instruments, as ammeter, voltmeter, etc., measure and indicate the *effective value*. The maximum value and the mean value can be derived from the curve of instantaneous values, as determined by wave-meter or oscillograph.

Measurement of the alternating wave after rectification by a unidirectional conductor, as an arc, gives the *mean* value with direct-current instruments, that is, instruments employing a permanent magnetic field, and the *effective value* with alternating-current instruments.

Voltage determination by spark-gap, that is, by the striking distance, gives a value approaching the *maximum*, especially with spheres as electrodes of a diameter larger than the spark-gap.

CHAPTER III

LAW OF ELECTROMAGNETIC INDUCTION

13. If an electric conductor moves relatively to a magnetic field, an e.m.f. is generated in the conductor which is proportional to the intensity of the magnetic field, to the length of the conductor, and to the speed of its motion perpendicular to the magnetic field and the direction of the conductor; or, in other words, proportional to the number of lines of magnetic force cut per second by the conductor.

As a practical unit of e.m.f., the *volt* is defined by the e.m.f. generated in a conductor, which cuts $10^8 = 100{,}000{,}000$ lines of magnetic flux per second.

If the conductor is closed upon itself, the e.m.f. produces a current.

A closed conductor may be called a turn or a convolution. In such a turn, the number of lines of magnetic force cut per second is the increase or decrease of the number of lines inclosed by the turn, or n times as large with n turns.

Hence the e.m.f. in volts generated in n turns, or convolutions, is n times the increase or decrease, per second, of the flux inclosed by the turns, times 10^{-8}.

If the change of the flux inclosed by the turn, or by n turns, does not take place uniformly, the product of the number of turns times change of flux per second gives the average e.m.f.

If the magnetic flux, Φ, alternates relatively to a number of turns, n—that is, when the turns either revolve through the flux or the flux passes in and out of the turns—the total flux is cut four times during each complete period or cycle, twice passing into, and twice out of, the turns.

Hence, if f = number of complete cycles per second, or the frequency of the flux, Φ, the average e.m.f. generated in n turns is

$$E_{avg.} = 4\, n\Phi f\, 10^{-8} \text{ volts.}$$

This is the fundamental equation of electrical engineering, and applies to continuous-current, as well as to alternating-current, apparatus.

14. In continuous-current machines and in many alternators, the turns revolve through a constant magnetic field; in other alternators and in induction motors, the magnetic field revolves; in transformers, the field alternates with respect to the stationary turns; in other apparatus, alternation and rotation occur simultaneously, as in alternating-current commutator motors.

Thus, in the continuous-current machine, if n = number of turns in series from brush to brush, Φ = flux inclosed per turn, and f = frequency, the e.m.f. generated in the machine is $E = 4\,n\Phi f\,10^{-8}$ volts, independent of the number of poles, of series or multiple connection of the armature, whether of the ring, drum, or other type.

In an alternator or transformer, if n is the number of turns in series, Φ the maximum flux inclosed per turn, and f the frequency, this formula gives

$$E_{avg.} = 4\,n\Phi f\,10^{-8} \text{ volts.}$$

Since the maximum e.m.f. is given by

$$E_{max.} = \frac{\pi}{2} E_{avg.},$$

we have

$$E_{max.} = 2\,\pi n\Phi f\,10^{-8} \text{ volts.}$$

And since the effective e.m.f. is given by

$$E_{eff.} = \frac{E_{max.}}{\sqrt{2}}$$

we have

$$\begin{aligned} E_{eff.} &= \sqrt{2}\,\pi n\Phi f\,10^{-8} \\ &= 4.44\,nf\Phi\,10^{-8} \text{ volts,} \end{aligned}$$

which is the fundamental formula of alternating-current induction by sine waves.

15. If, in a circuit of n turns, the magnetic flux, Φ, inclosed by the circuit is produced by the current in the circuit, the ratio,

$$\frac{\text{flux} \times \text{number of turns} \times 10^{-8}}{\text{current}},$$

is called the inductance, L, of the circuit, in henrys.

The product of the number of turns, n, into the maximum flux, Φ, produced by a current of I amperes effective, or $I\sqrt{2}$ amperes maximum, is therefore

$$n\Phi = LI\sqrt{2}\,10^{8};$$

and consequently the effective e.m.f. of self-induction is

$$E = \sqrt{2}\,\pi n \Phi f\, 10^{-8}$$
$$= 2\,\pi f L I \text{ volts.}$$

The product, $x = 2\,\pi f L$, is of the dimension of resistance, and is called the *inductive reactance* of the circuit; and the e.m.f. of self-induction of the circuit, or the reactance voltage, is

$$E = Ix,$$

and lags 90° behind the current, since the current is in phase with the magnetic flux produced by the current, and the e.m.f. lags 90° behind the magnetic flux. The e.m.f. lags 90° behind the magnetic flux, as it is proportional to the rate of change in flux; thus it is zero when the magnetism does not change, at its maximum value, and a maximum when the flux changes quickest, which is where it passes through zero.

CHAPTER IV

VECTOR REPRESENTATION

16. While alternating waves can be, and frequently are, represented graphically in rectangular coördinates, with the time as abscissæ, and the instantaneous values of the wave as ordinates, the best insight with regard to the mutual relation of different alternating waves is given by their representation as vectors, in the so-called *crank diagram.* A vector, equal in length to the maximum value of the alternating wave, revolves at uniform speed so as to make a complete revolution per period, and the projections of this revolving vector on the horizontal then denote the instantaneous values of the wave.

Obviously, by this diagram only sine waves can be represented or, in general, waves which are so near sine shape that they can be represented by a sine.

Let, for instance, $\overline{OI}$ represent in length the maximum value I of a sine wave of current. Assuming then a vector, $\overline{OI}$, to revolve, left handed or in counter-clockwise direction, so that it makes a complete revolution during each cycle or period t_0. If then at a certain moment of time, this vector stands in position $\overline{OI_1}$ (Fig. 8), the projection, $\overline{OA_1}$, of $\overline{OI_1}$ on the horizontal line $\overline{OA}$ represents the instantaneous value of the current at this moment. At a later moment, $\overline{OI}$ has moved farther, to $\overline{OI_2}$, and the projection, $\overline{OA_2}$, of $\overline{OI_2}$ on $\overline{OA}$ is the instantaneous value at this later moment. The diagram so shows the instantaneous condition of the sine wave: each sine wave reaches its maximum at the moment of time where its revolving vector passes the horizontal, and reaches zero at the moment where its revolving vector passes the vertical.

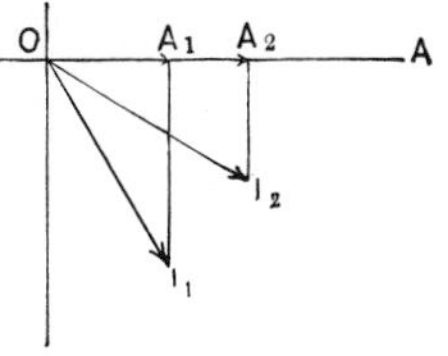

FIG. 8.

If now the time, t, and thus the angle, $\vartheta = IOA = 2\pi \frac{t}{t_0}$ (where t_0 = time of one complete cycle or period), is counted from the moment of time where the revolving vector $\overline{OI}$ in Fig. 8 stands in position $\overline{OI_1}$, then this sine wave would be represented by

$$i = I \cos (\vartheta - \vartheta_1),$$

where $\vartheta_1 = I_1OA$ may be called the *phase* of the wave, and $I = \overline{OI_1}$ the *amplitude* or *intensity*.

At the time, $\vartheta = \vartheta_1$, that is, the angle, ϑ_1, after the moment of time represented by position $\overline{OI_1}$, $i = I$, and $\overline{OI}$ passes through the horizontal $\overline{OA}$, that is, has its maximum value. The phase ϑ_1 thus is the angle representing the time, t_1, at which the wave reaches its maximum value.

If the time, t, and thus the angle, ϑ, are counted from the moment at which the revolving vector reaches position $\overline{OI_2}$, the equation of the wave would be

$$i = I \cos (\vartheta - \vartheta_2),$$

and $\vartheta_2 = I_2OA$ is the phase.

17. When dealing with one wave only, it obviously is immaterial from which moment of time as zero value the time and thus the angle, ϑ, is counted. That is, the phase ϑ_1 or ϑ_2 may be chosen anything desired. As soon, however, as several alternating waves enter the diagram, it is obvious that for all the waves of the same diagram the time must be counted from the same moment, and by choosing the phase angle of one of the waves, that of the others is determined.

Thus, let I = the maximum value of a current, lagging behind the maximum value of voltage E by time t_1, that is, angle of phase difference $\vartheta_1 = 2\pi \frac{t_1}{t_0}$.

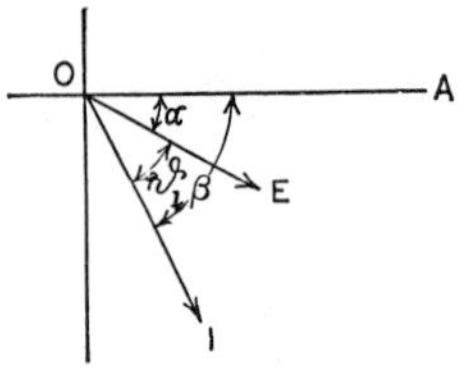

FIG. 9.

The phase of the voltage, E, then may be chosen as α, and the voltage represented, in Fig. 9, by vector $\overline{OE} = E$ at phase angle $EOA = \alpha$. As the current lags by phase difference ϑ_1, the phase of the current then must be $\beta = \alpha + \vartheta_1$, and the current is represented, in Fig. 9, by vector $\overline{OI} = I$, under phase angle $\beta = IOA$.

The equations of voltage and current then are:

$$\begin{aligned} e &= E \cos (\vartheta - \alpha) \\ i &= I \cos (\vartheta - \beta) \\ &= I \cos (\vartheta - \alpha - \vartheta_1). \end{aligned}$$

The voltage $\overline{OE} = E$, as the first vector, may be plotted in any desired direction, for instance, under angle $-\alpha' = EOA$ in Fig. 10. The current then would be represented by $\overline{OI} = I$, under phase angle $-\beta' = -(\alpha' - \vartheta_1) = IOA$, and the equations of voltage and current would be:

$$e = E \cos(\vartheta + \alpha')$$
$$i = I \cos(\vartheta + \beta')$$
$$= I \cos(\vartheta + \alpha' - \vartheta_1).$$

Or, the current $\overline{OI} = I$ may be chosen as the first vector, in Fig. 9, under phase angle $\beta = IOA$, and the voltage then would have the phase angle $\alpha = \beta - \vartheta_1$, and be represented by vector $\overline{OE} = E$, and the equations would be:

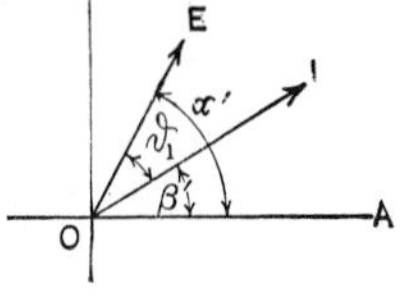

FIG. 10.

$$i = I \cos(\vartheta - \beta)$$
$$e = E \cos(\vartheta - \alpha)$$
$$= E \cos(\vartheta - \beta + \vartheta_1).$$

In this vector representation, a current *lagging* behind its voltage makes a *greater* angle with the horizontal, $\overline{OA}$, that is, the current vector, $\overline{OI}$, lags behind the voltage vector, $\overline{OE}$, in the direction of rotation, thus passes the zero line, $\overline{OA}$, of maximum value, at a later time.

Inversely, a leading current passes the zero line $\overline{OA}$ earlier, that is, is ahead in the direction of rotation.

Instead of the maximum value of the rotating vector, the effective value is commonly used, especially where the instantaneous values are not required, but the diagram intended to represent the relations of the different alternating waves to each other. With the length of the rotating vector equal to the effective value of the alternating wave, the maximum value obviously is $\sqrt{2}$ times the length of the vector, and the instantaneous values are $\sqrt{2}$ times the projections of the vectors on the horizontal.

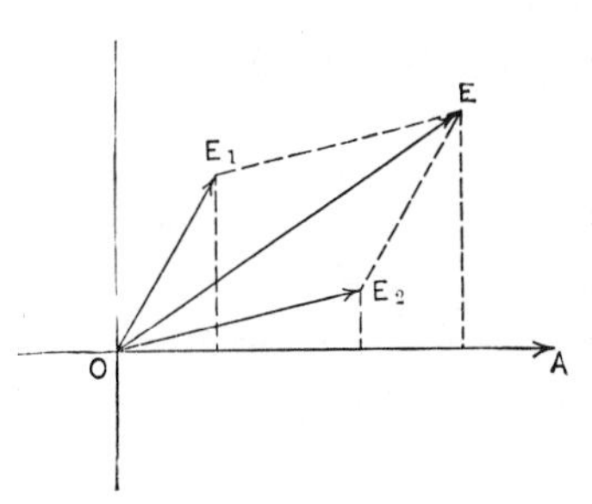

FIG. 11.

18. To combine different sine waves, their graphical representations as vectors, are combined by the parallelogram law.

If, for instance, two sine waves, $\overline{OE_1}$, and $\overline{OE_2}$ (Fig. 11), are superposed—as, for instance, two e.m.fs. acting in the same circuit—their resultant wave is represented by $\overline{OE}$, the diagonal of a parallelogram with $\overline{OE_1}$ and $\overline{OE_2}$ as sides. As the projection of the diagonal of a parallelogram equals the sum of the projections of the sides, during the rotation of the parallelogram OE_1EE_2, the projection of $\overline{OE}$ on the horizontal $\overline{OA}$, that is, the instantaneous value of the wave represented by vector $\overline{OE}$, is equal to the sum of the projection of the two sides $\overline{OE_1}$ and $\overline{OE_2}$, that is, the sum of the instantaneous values of the component vectors $\overline{OE_1}$ and $\overline{OE_2}$.

From the foregoing considerations we have the conclusions:

The sine wave is represented graphically in the crank diagram, by a vector, which by its length, $\overline{OE}$, denotes the intensity, and by its amplitude, AOE, the phase, of the sine wave.

Sine waves are combined or resolved graphically, in vector representation, by the law of the parallelogram or the polygon of sine waves.

Kirchhoff's laws now assume, for alternating sine waves, the form:

(*a*) The resultant of all the e.m.fs. in a closed circuit, as found by the parallelogram of sine waves, is zero if the counter e.m.fs. of resistance and of reactance are included.

(*b*) The resultant of all the currents toward a distributing point, as found by the parallelogram of sine waves, is zero.

The power equation expressed graphically is as follows:

The power of an alternating-current circuit is represented in vector representation by the product of the current, I, into the projection of the e.m.f., E, upon the current, or by the e.m.f., E, into the projection of the current, I, upon the e.m.f., or by $IE \cos \theta$, where θ = angle of phase displacement.

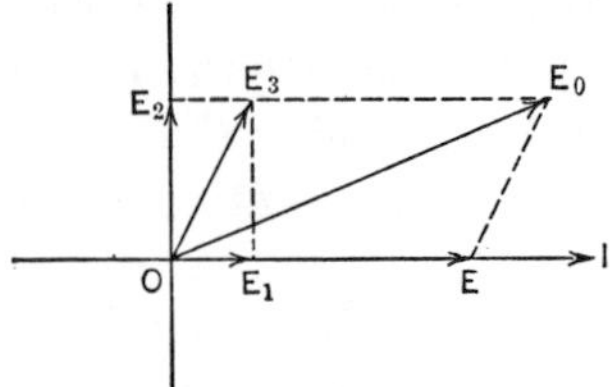

FIG. 12.

19. Suppose, as an example, that in a line having the resistance, r, and the reactance, $x = 2\,\pi f L$—where f = frequency and L = inductance—there exists a current of I amp., the line being connected to a non-inductive circuit operating at a voltage of E volts. What will be the voltage required at the generator end of the line?

In the vector diagram, Fig. 12, let the phase of the current be assumed as the initial or zero line, $\overline{OI}$. Since the receiving circuit is non-inductive, the current is in phase with its voltage. Hence the voltage, E, at the end of the line, impressed upon the receiving circuit, is represented by a vector, $\overline{OE}$. To overcome the resistance, r, of the line, a voltage, Ir, is required in phase with the current, represented by $\overline{OE_1}$ in the diagram. The inductive reactance of the line generates an e.m.f. which is proportional to the current, I, and the reactance, x, and lags a quarter of a period, or 90°, behind the current. To overcome this counter e.m.f. of inductive reactance, a voltage of the value Ix is required, in phase 90° ahead of the current, hence represented by vector $\overline{OE_2}$. Thus resistance consumes voltage in phase, and reactance voltage 90° ahead of the current. The voltage of the generator, E_0, has to give the three voltages E, E_1, E_2, hence it is determined as their resultant. Combining by the parallelogram law, $\overline{OE_1}$ and $\overline{OE_2}$, give OE_3, the voltage required to overcome the impedance of the line, and similarly $\overline{OE_3}$ and $\overline{OE}$ give $\overline{OE_0}$, the voltage required at the generator side of the line, to yield the voltage, E, at the receiving end of the line. Algebraically, we get from Fig. 12

$$E_0 = \sqrt{(E + Ir)^2 + (Ix)^2}$$

or

$$E = \sqrt{E_0^2 - (Ix)^2} - Ir.$$

In this example we have considered the voltage consumed by the resistance (in phase with the current) and the voltage consumed by the reactance (90° ahead of the current) as parts, or components, of the impressed voltage, E_0, and have derived E_0 by combining Er, Ex, and E.

20. We may, however, introduce the effect of the inductive reactance directly as an e.m.f., E'_2, the counter e.m.f. of inductive reactance $= Ix$, and lagging 90° behind the current; and the e.m.f. consumed by the resistance as a counter e.m.f., $E'_1 = Ir$, in opposition to the current, as is done in Fig. 13; and combine the three voltages E_0, E'_1, E'_2, to form a resultant voltage E, which is left at the end of the line. E'_1 and E'_2 combine to form E'_3, the counter e.m.f. of impedance; and since E'_3 and E_0 must combine to form E, E_0 is found as the side of a parallelogram, $OE_0EE'_3$, whose other side, $\overline{OE'_3}$, and diagonal $\overline{OE}$, are given.

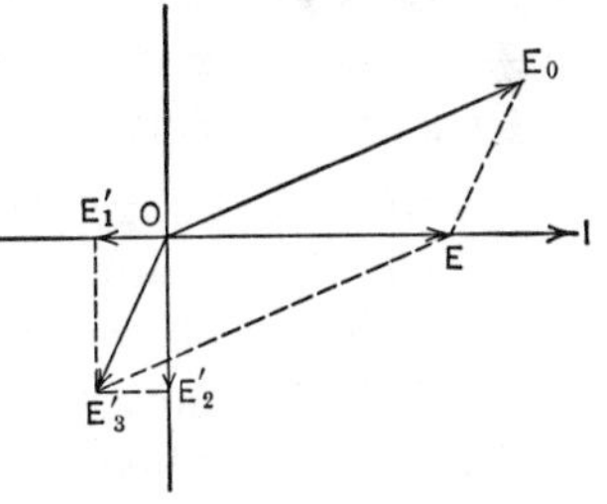

FIG. 13.

Or we may say (Fig. 14), that to overcome the counter e.m.f.

of impedance, $\overline{OE'_3}$, of the line, the component, $\overline{OE_3}$, of the impressed voltage is required which, with the other component, $\overline{OE}$, must give the impressed voltage, $\overline{OE_0}$.

As shown, we can represent the voltages produced in a circuit in two ways—either as counter e.m.fs., which combine with the impressed voltage, or as parts, or components, of the impressed voltage, in the latter case being of opposite phase. According to the nature of the problem, either the one or the other way may be preferable.

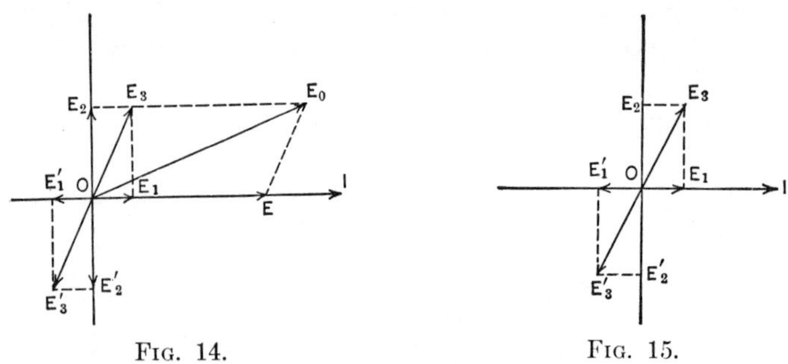

FIG. 14. FIG. 15.

As an example, the voltage consumed by the resistance is Ir, and in phase with the current; the counter e.m.f. of resistance is in opposition to the current. The voltage consumed by the reactance is Ix, and 90° ahead of the current, while the counter e.m.f. of reactance is 90° behind the current; so that, if, in Fig. 15, $\overline{OI}$ is the current.

$\overline{OE_1}$ = voltage consumed by resistance,
$\overline{OE'_1}$ = counter e.m.f. of resistance,
$\overline{OE_2}$ = voltage consumed by inductive reactance,
$\overline{OE'_2}$ = counter e.m.f. of inductive reactance,
$\overline{OE_3}$ = voltage consumed by impedance,
$\overline{OE'_3}$ = counter e.m.f. of impedance.

Obviously, these counter e.m.fs. are different from, for instance, the counter e.m.f. of a synchronous motor, in so far as they have no independent existence, but exist only through, and as long as the current exists. In this respect they are analogous to the opposing force of friction in mechanics.

21. Coming back to the equation found for the voltage at the generator end of the line,

$$E_0 = \sqrt{(E + Ir)^2 + (Ix)^2}$$

we find, as the drop of potential in the line,

$$e = E_0 - E = \sqrt{(E + Ir)^2 + (Ix)^2} - E.$$

This is different from, and less than, the e.m.f. of impedance,

$$E_3 = Iz = I\sqrt{r^2 + x^2}.$$

Hence it is wrong to calculate the drop of potential in a circuit by multiplying the current by the impedance; and the drop of potential in the line depends, with a given current fed over the line into a non-inductive circuit, not only upon the constants of the line, r and x, but also upon the voltage, E, at the end of line, as can readily be seen from the diagrams.

22. If the receiver circuit is inductive, that is, if the current, I, lags behind the voltage, E, by an angle, θ, and we choose again as the zero line, the current $\overline{OI}$ (Fig. 16), the voltage, $\overline{OE}$, is ahead of

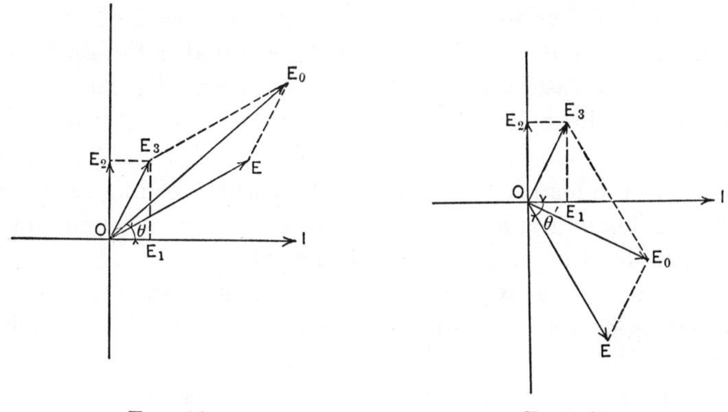

FIG. 16. FIG. 17.

the current by the angle, θ. The voltage consumed by the resistance, Ir, is in phase with the current, and represented by $\overline{OE_1}$ the voltage consumed by the reactance, Ix, is 90° ahead of the current, and represented by $\overline{OE_2}$. Combining $\overline{OE}$, $\overline{OE_1}$, and $\overline{OE_2}$, we get $\overline{OE_0}$, the voltage required at the generator end of the line. Comparing Fig. 16 with Fig. 12, we see that in the former $\overline{OE_0}$ is larger; or conversely, if E_0 is the same, E will be less with an inductive load. In other words, the drop of potential in an inductive line is greater if the receiving circuit is inductive than if it is non-inductive. From Fig. 16,

$$E_0 = \sqrt{(E\cos\theta + Ir)^2 + (E\sin\theta + Ix)^2}.$$

If, however, the current in the receiving circuit is leading, as is the case when feeding condensers or synchronous motors whose counter e.m.f. is larger than the impressed voltage, then the voltage will be represented, in Fig. 17, by a vector, $\overline{OE}$, lagging behind the current, $\overline{OI}$, by the angle of lead, θ'; and in this case we get, by combining $\overline{OE}$ with $\overline{OE_1}$, in phase with the current, and $\overline{OE_2}$, 90° ahead of the current, the generator voltage, $\overline{OE_0}$, which in this case is not only less than in Fig. 16 and in Fig. 12, but may be even less than E; that is, the voltage rises in the line. In other words, in a circuit with leading current, the inductive reactance of the line raises the voltage, so that the drop of voltage is less than with a non-inductive load, or may even be negative, and the voltage at the generator lower than at the other end of the line.

These diagrams, Figs. 12 to 17, can be considered vector diagrams of an alternating-current generator of a generated e.m.f., E_0, a resistance voltage, $E_1 = Ir$, a reactance voltage, $E_2 = Ix$, and a difference of potential, E, at the alternator terminals; and we see, in this case, that with the same generated e.m.f., with an inductive load the potential difference at the alternator terminals will be lower than with a non-inductive load, and that with a non-inductive load it will be lower than when feeding into a circuit with leading current, as for instance, a synchronous motor circuit under the circumstances stated above.

23. As a further example, we may consider the diagram of an alternating-current transformer, feeding through its secondary circuit an inductive load.

For simplicity, we may neglect here the magnetic hysteresis, the effect of which will be fully treated in a separate chapter on this subject.

Let the time be counted from the moment when the magnetic flux is zero and rising. The magnetic flux then passes its maximum at the time $\vartheta = 90°$, and the phase of the magnetic flux thus is $\vartheta = 90°$, the flux thus represented by the vector $\overline{O\Phi}$ in Fig. 18, vertically downward. The e.m.f. generated by this magnetic flux in the secondary circuit, E_1, lags 90° behind the flux; thus its vector, $\overline{OE_1}$, passes the zero line, $\overline{OA}$ 90°, later than the magnetic flux vector, or at the time $\vartheta = 180°$; that is, the e.m.f. generated in the secondary by the magnetic flux, $\overline{OE_1}$, has the phase $\vartheta = 180°$. The secondary current, I_1, lags behind the e.m.f., E_1, by an angle, θ_1, which is determined by the resistance and inductive reactance of the secondary circuit; that is, by the load in the secondary circuit, and is represented in the diagram by the vector, $\overline{OF}_1$, of phase $180 + \theta_1$.

Instead of the secondary current, I_1, we plot, however, the secondary m.m.f., $F_1 = n_1 I_1$, where n_1 is the number of secondary turns, and F_1 is given in ampere-turns. This makes us independent of the ratio of transformation.

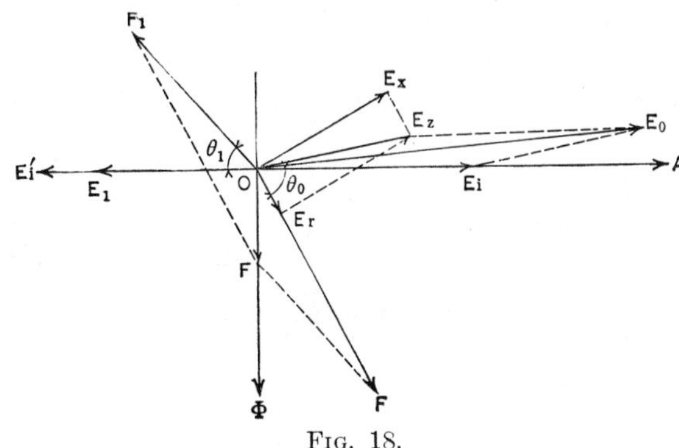

FIG. 18.

From the secondary e.m.f., E_1, we get the flux, Φ, required to induce this e.m.f., from the equation

$$E_1 = \sqrt{2}\,\pi n_1 f \Phi \, 10^{-8};$$

where

E_1 = secondary e.m.f., in effective volts,
f = frequency, in cycles per second,
n_1 = number of secondary turns,
Φ = maximum value of magnetic flux, in lines of magnetic force.

The derivation of this equation has been given in a preceding chapter.

This magnetic flux, Φ, is represented by a vector, $\overline{O\Phi}$, 90° in phase, and to produce it a m.m.f., F, is required, which is determined by the magnetic characteristic of the iron and the section and length of the magnetic circuit of the transformer; this m.m.f. is in phase with the flux, Φ, and is represented by the vector, $\overline{OF}$, in effective ampere-turns.

The effect of hysteresis, neglected at present, is to shift $\overline{OF}$ ahead of $\overline{O\Phi}$, by an angle, α, the angle of hysteretic lead. (See Chapter on Hysteresis.)

This m.m.f., F, is the resultant of the secondary m.m.f., F_1, and the primary m.m.f., F_0; or graphically, $\overline{OF}$ is the diagonal of a parallelogram with $\overline{OF_1}$ and $\overline{OF_0}$ as sides. $\overline{OF_1}$ and $\overline{OF}$ being known, we find $\overline{OF_0}$, the primary ampere-turns, and therefrom and the number of primary turns, n_0, the primary current, $I_0 = \frac{F_0}{n_0}$, which corresponds to the secondary, I_1.

To overcome the resistance, r_0, of the primary coil, a voltage, $E_r = I_0 r_0$, is required, in phase with the current, I_0, and represented by the vector, $\overline{OE_r}$.

To overcome the reactance, $x_0 = 2\pi f L_0$, of the primary coil, a voltage, $E_x = I_0 x_0$, is required, 90° ahead of the current, I_0, and represented by vector, $\overline{OE_x}$.

The resultant magnetic flux, Φ, which generates in the secondary coil the e.m.f., E_1, generates in the primary coil an e.m.f. proportional to E_1 by the ratio of turns $\frac{n_0}{n_1}$ and in phase with E_1, or,

$$E'_i = \frac{n_0}{n_1} E_1,$$

which is represented by the vector, $\overline{OE'_i}$. To overcome this counter e.m.f., E'_i, a primary voltage, E_i, is required, equal but in phase opposition to E'_i, and represented by the vector, $\overline{OE_i}$.

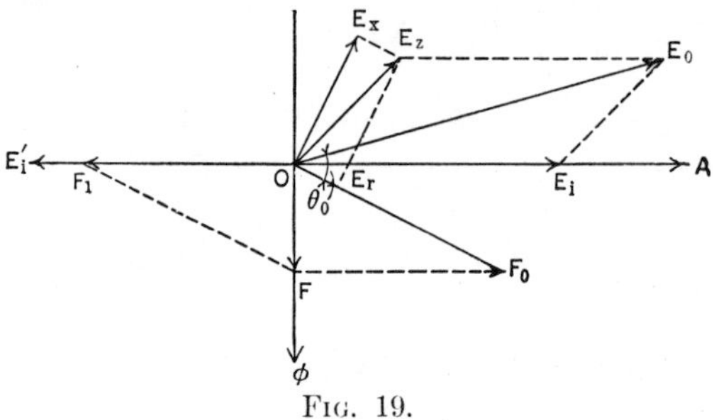

FIG. 19.

The primary impressed e.m.f., E_0, must thus consist of the three components $\overline{OE_i}$, $\overline{OE_r}$, and $\overline{OE_x}$, and is, therefore, their resultant $\overline{OE_0}$, while the difference of phase in the primary circuit is found to be

$$\theta_0 = E_0 O F_0.$$

24. Thus, in Figs. 18 to 20, the diagram of a transformer is drawn for the same secondary e.m.f., E_1, secondary current, I_1, and therefore secondary m.m.f., F_1, but with different conditions of secondary phase displacement:

In Fig. 18 the secondary current, I_1, lags 60° behind the secondary e.m.f., E_1.

In Fig. 19, the secondary current, I_1, is in phase with the secondary e.m.f., E_1.

In Fig. 20 the secondary current, I_1, leads by 60° the secondary e.m.f., E_1.

These diagrams show that lag of the current in the secondary circuit increases and lead decreases the primary current and primary impressed e.m.f. required to produce in the secondary circuit the same e.m.f. and current; or conversely, at a given primary impressed e.m.f., E_0, the secondary e.m.f., E_1, will be smaller with an inductive, and larger with a condensive (leading current), load than with a non-inductive load.

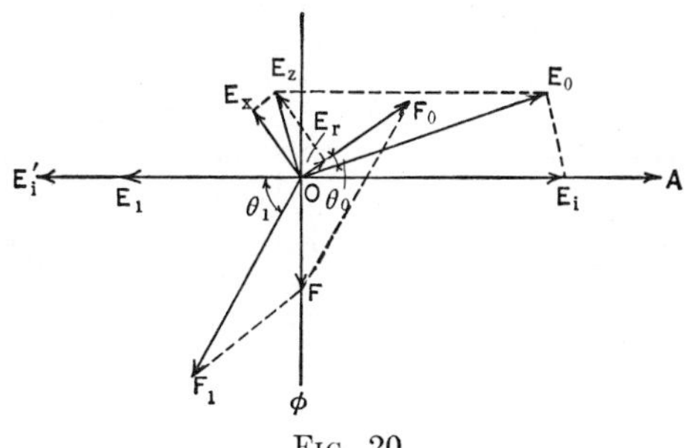

Fig. 20.

At the same time we see that a difference of phase existing in the secondary circuit of a transformer reappears in the primary circuit, somewhat decreased, if the current is leading, and slightly increased if lagging in phase. Later we shall see that hysteresis reduces the displacement in the primary circuit, so that, with an excessive lag in the secondary circuit, the lag in the primary circuit may be less than in the secondary.

A conclusion from the foregoing is that the transformer is not suitable for producing currents of displaced phase, since primary and secondary current are, except at very light loads, very nearly in phase, or rather in opposition, to each other.

CHAPTER V

SYMBOLIC METHOD

25. The graphical method of representing alternating-current phenomena affords the best means for deriving a clear insight into the mutual relation of the different alternating sine waves entering into the problem. For numerical calculation, however, the graphical method is generally not well suited, owing to the widely different magnitudes of the alternating sine waves represented in the same diagram, which make an exact diagrammatic determination impossible. For instance, in the transformer diagrams (*cf.* Figs. 18–20), the different magnitudes have numerical values in practice somewhat like the following: $E_1 = 100$ volts, and $I_1 = 75$ amp. For a non-inductive secondary load, as of incandescent lamps, the only reactance of the secondary circuit thus is that of the secondary coil, or $x_1 = 0.08$ ohms, giving a lag of $\theta_1 = 3.6°$. We have also,

$$
\begin{aligned}
n_1 &= 30 \text{ turns.} \\
n_0 &= 300 \text{ turns.} \\
F_1 &= 2250 \text{ ampere-turns.} \\
F &= 100 \text{ ampere-turns.} \\
E_r &= 10 \text{ volts.} \\
E_x &= 60 \text{ volts.} \\
E_i &= 1000 \text{ volts.}
\end{aligned}
$$

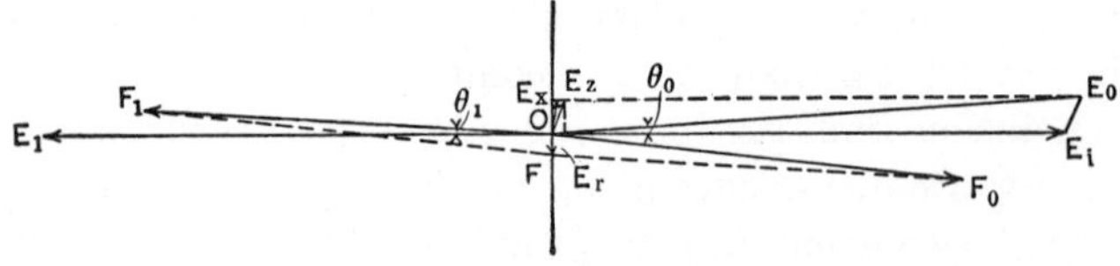

FIG. 21.—Vector diagram of transformer.

The corresponding diagram is shown in Fig. 21. Obviously, no exact numerical values can be taken from a parallelogram as flat as OF_1FF_0, and from the combination of vectors of the relative magnitudes 1 : 6 : 100.

Hence the importance of the graphical method consists not

so much in its usefulness for practical calculation as to aid in the simple understanding of the phenomena involved.

26. Sometimes we can calculate the numerical values trigonometrically by means of the diagram. Usually, however, this becomes too complicated, as will be seen by trying to calculate, from the above transformer diagram, the ratio of transformation. The primary m.m.f. is given by the equation

$$F_0 = \sqrt{F^2 + F_1{}^2 + 2\,FF_1 \sin \theta_1},$$

an expression not well suited as a starting-point for further calculation.

A method is therefore desirable which combines the exactness of analytical calculation with the clearness of the graphical representation.

27. We have seen that the alternating sine wave is represented in intensity, as well as phase, by a vector, $\overline{OI}$, which is determined analytically by two numerical quantities—the length, $\overline{OI}$, or intensity; and the amplitude, AOI, or phase, θ, of the wave, I.

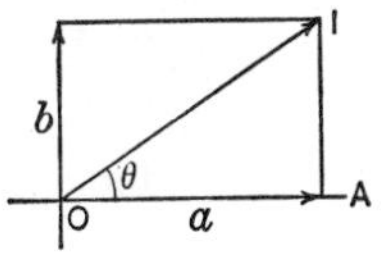

FIG. 22.

Instead of denoting the vector which represents the sine wave in the polar diagram by the polar coordinates, I and θ, we can represent it by its rectangular coordinates, a and b (Fig. 22), where

$a = I \cos \theta$ is the horizontal component,
$b = I \sin \theta$ is the vertical component of the sine wave.

This representation of the sine wave by its rectangular components is very convenient, in so far as it avoids the use of trigonometric functions in the combination or solution of sine waves.

Since the rectangular components, a and b, are the horizontal and the vertical projections of the vector representing the sine wave, and the projection of the diagonal of a parallelogram is equal to the sum of the projections of its sides, the combination of sine waves by the parallelogram law is reduced to the addition, or subtraction, of their rectangular components. That is:

Sine waves are combined, or resolved, by adding, or subtracting, their rectangular components.

For instance, if a and b are the rectangular components of a sine wave, I, and a' and b' the components of another sine wave,

I' (Fig. 23), their resultant sine wave, I_0, has the rectangular components $a_0 = (a + a')$, and $b_0 = (b + b')$.

To get from the rectangular components, a and b, of a sine wave its intensity, i, and phase, θ, we may combine a and b by the parallelogram, and derive

$$i = \sqrt{a^2 + b^2};$$

$$\tan \theta = \frac{b}{a}.$$

Hence we can analytically operate with sine waves, as with forces in mechanics, by resolving them into their rectangular components.

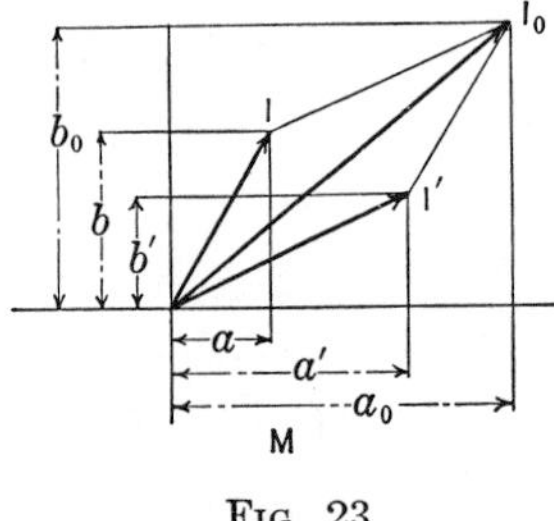

FIG. 23.

28. To distinguish, however, the horizontal and the vertical components of sine waves, so as not to be confused in lengthier calculation, we may mark, for instance, the vertical components by a distinguishing index, or the addition of an otherwise meaningless symbol, as the letter j, and thus represent the sine wave by the expression

$$\dot{I} = a + jb,$$

which now has the meaning that a is the horizontal and b the vertical component of the sine wave I, and that both components are to be combined in the resultant wave of intensity,

$$i = \sqrt{a^2 + b^2},$$

and of phase, $$\tan \theta = \frac{b}{a}.$$

Similarly, $a - jb$ means a sine wave with a as horizontal, and $-b$ as vertical, components, etc.

Obviously, the plus sign in the symbol, $a + jb$, does not imply simple addition, since it connects heterogeneous quantities—horizontal and vertical components—but implies combination by the parallelogram law.

For the present, j is nothing but a distinguishing index, and otherwise free for definition except that it is not an ordinary number.

29. A wave of equal intensity, and differing in phase from the wave, $a + jb$, by 180°, or one-half period, is represented in polar coördinates by a vector of opposite direction, and denoted by the symbolic expression, $-a-jb$. Or,

Multiplying the symbolic expression, $a+jb$, *of a sine wave by* -1 *means reversing the wave, or rotating it through* 180°, *or one-half period.*

A wave of equal intensity, but leading $a+jb$ by 90°, or one-quarter period, has (Fig. 24) the horizontal component, $-b$, and the vertical component, a, and is represented symbolically by the expression, $ja-b$.

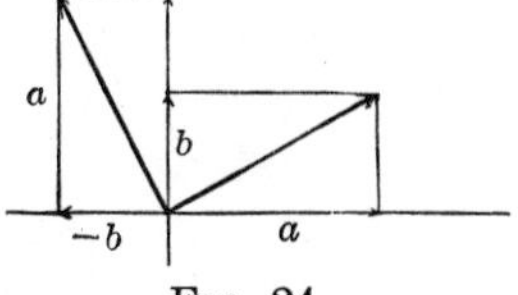

FIG. 24.

Multiplying, however, $a+jb$ by j, we get

$$ja + j^2b;$$

therefore, if we define the heretofore meaningless symbol, j, by the condition,

$$j^2 = -1,$$

we have

$$j(a + jb) = ja - b;$$

hence,

Multiplying the symbolic expression, $a+jb$, *of a sine wave by* j *means rotating the wave through* 90°, *or one-quarter period; that is, leading the wave by one-quarter period.*

Similarly—

Multiplying by $-j$ *means lagging the wave by one-quarter period.*

Since

$$j^2 = -1,$$

it is

$$j = \sqrt{-1};$$

and

j *is the imaginary unit, and the sine wave is represented by a complex imaginary quantity or general number,* $a+jb$.

As the imaginary unit, j, has no numerical meaning in the system of ordinary numbers, this definition of $j=\sqrt{-1}$ does not contradict its original introduction as a distinguishing index.

30. In the vector diagram, the sine wave is represented in intensity as well as phase by one complex quantity,

$$a + jb,$$

where a is the horizontal and b the vertical component of the wave; the intensity is given by

$$i = \sqrt{a^2 + b^2},$$

the phase by

$$\tan \theta = \frac{b}{a}.$$

and

$$a = i \cos \theta,$$
$$b = i \sin \theta;$$

hence the wave, $a + jb$, can also be expressed by

$$i(\cos \theta + j \sin \theta),$$

or, by substituting for $\cos \theta$ and $\sin \theta$ their exponential expressions, we obtain

$$i\epsilon^{j\theta}.$$ [1]

Since we have seen that sine waves may be combined or resolved by adding or subtracting their rectangular components, consequently,

Sine waves may be combined or resolved by adding or subtracting their complex algebraic expressions.

For instance, the sine waves,

$$a + jb$$

and

$$a' + jb',$$

combined give the sine wave,

$$\underset{\cdot}{I} = (a + a') + j(b + b').$$

It will thus be seen that the combination of sine waves is reduced to the elementary algebra of complex quantities.

31. If $I = i + ji'$ is a sine wave of alternating current, and r is the resistance, the voltage consumed by the resistance is in phase with the current, and equal to the product of the current and resistance. Or

$$r\underset{\cdot}{I} = ri + jri'.$$

If L is the inductance, and $x = 2\pi f L$ the inductive reactance, the e.m.f. produced by the reactance, or the counter e.m.f.

[1] In this representation of the sine wave by the exponential expression of the complex quantity, the angle θ necessarily must be expressed in *radians*, and not in degrees, that is, with one complete revolution or cycle as 2π, or with $\frac{180}{\pi} = 57.3°$ as unit.

of self-induction, is the product of the current and reactance, and lags in phase 90° behind the current; it is, therefore, represented by the expression

$$-jx\dot{I} = -jxi + xi'.$$

The voltage required to overcome the reactance is consequently 90° ahead of the current (or, as usually expressed, the current lags 90° behind the e.m.f.), and represented by the expression

$$jx\dot{I} = jxi - xi'.$$

Hence, the voltage required to overcome the resistance, r, and the reactance, x, is

$$(r + jx)\dot{I};$$

that is,

$Z = r + jx$ is the expression of the impedance of the circuit in complex quantities.

Hence, if $\dot{I} = i + ji'$ is the current, the voltage required to overcome the impedance, $Z = r + jx$, is

$$\dot{E} = Z\dot{I} = (r + jx)\,(i + ji')$$
$$= (ri + j^2xi') + j(ri' + xi);$$

hence, since $\quad j^2 = -1$

$$\dot{E} = (ri - xi') + j(ri' + xi);$$

or, if $\dot{E} = e + je'$ is the impressed voltage and $Z = r + jx$ the impedance, the current through the circuit is

$$\dot{I} = \frac{\dot{E}}{Z} = \frac{e + je'}{r + jx};$$

or, multiplying numerator and denominator by $(r - jx)$ to eliminate the imaginary from the denominator, we have

$$\dot{I} = \frac{(e + je')\,(r - jx)}{r^2 + x^2} = \frac{er + e'x}{r^2 + x^2} + j\,\frac{e'r - ex}{r^2 + x^2};$$

or, if $\dot{E} = e + je'$ is the impressed voltage and $\dot{I} = i + ji'$ the current in the circuit, its impedance is

$$Z = \frac{\dot{E}}{\dot{I}} = \frac{e + je'}{i + ji'} = \frac{(e + je')\,(i - ji')}{i^2 + i'^2} = \frac{ei + e'i'}{i^2 + i'^2} + j\,\frac{e'i - ei'}{i^2 + i'^2}.$$

32. If C is the capacity of a condenser in series in a circuit in which exists a current $\dot{I} = i + ji'$, the voltage impressed upon the terminals of the condenser is $\dot{E} = \dfrac{\dot{I}}{2\pi fC}$, 90° behind the current; and may be represented by $-\frac{j\dot{I}}{2\pi fC}$ or $-jx_1\underset{\cdot}{I}$, where $x_1 = \frac{1}{2\pi fC}$ is the *condensive reactance* or *condensance* of the condenser.

Condensive reactance is of opposite sign to inductive reactance; both may be combined in the name reactance.

We therefore have the conclusion that

If r = resistance and L = inductance,

thus $\qquad x = 2\pi fL$ = inductive reactance.

If C = capacity, $x_1 = \frac{1}{2\pi fC}$ = condensive reactance,

$Z = r + j(x - x_1)$ is the impedance of the circuit.

Ohm's law is then re-established as follows:

$$\underset{\cdot}{E} = Z\underset{\cdot}{I}, \quad \underset{\cdot}{I} = \frac{\dot{E}}{Z}, \quad Z = \frac{\dot{E}}{\underset{\cdot}{I}}.$$

The more general form gives not only the intensity of the wave but also its phase, as expressed in complex quantities.

33. Since the combination of sine waves takes place by the addition of their symbolic expressions, Kirchhoff's laws are now re-established in their original form:

(*a*) The sum of all the e.m.fs. acting in a closed circuit equals zero, if they are expressed by complex quantities, and if the resistance and reactance e.m.fs. are also considered as counter e.m.fs.

(*b*) The sum of all the currents directed toward a distributing point is zero, if the currents are expressed as complex quantities.

If a complex quantity equals zero, the real part as well as the imaginary part must be zero individually; thus, if

$$a + jb = 0, \qquad a = 0,\ b = 0.$$

Resolving the e.m.fs. and currents in the expression of Kirchhoff's law, we find:

(*a*) The sum of the components, in any direction, of all the e.m.fs. in a closed circuit equals zero, if the resistance and reactance are represented as counter e.m.fs.

(*b*) The sum of the components, in any direction, of all the currents at a distributing point equals zero.

Joule's law and the power equation do not give a simple expression in complex quantities, since the effect or power is a quantity of double the frequency of the current or e.m.f. wave, and therefore requires for its representation as a vector a transition from single to double frequency, as will be shown in Chapter XVI (page 299).

In what follows, complex vector quantities will always be denoted by dotted capitals when not written out in full; absolute quantities and real quantities by undotted letters.

34. Referring to the example given in the fourth chapter, of a circuit supplied with a voltage, E, and a current, I, over an inductive line, we can now represent the impedance of the line by $Z = r + jx$, where r = resistance, x = reactance of the line, and have thus as the voltage at the beginning of the line, or at the generator, the expression

$$\dot{E}_0 = \dot{E} + Z\dot{I}.$$

Assuming now again the current as the zero line, that is, $\dot{I} = i$, we have in general

$$\dot{E}_0 = \dot{E} + ir + jix;$$

hence, with non-inductive load, or $\dot{E} = e$,

$$\dot{E}_0 = (e + ir) + jix,$$

or $$e_0 = \sqrt{(e + ir)^2 + (ix)^2}, \qquad \tan \theta_0 = \frac{ix}{e + ir}.$$

In a circuit with lagging current, that is, with leading e.m.f., $\dot{E} = e + je'$, and

$$\dot{E}_0 = e + je' + (r + jx)i$$
$$= (e + ir) + j(e' + ix),$$

or $$e_0 = \sqrt{(e + ir)^2 + (e' + ix)^2}, \qquad \tan \theta_0 = \frac{e' + ix}{e + ir}.$$

In a circuit with leading current, that is, with lagging e.m.f., $\dot{E} = e - je'$, and

$$\dot{E}_0 = (e - je') + (r + jx)i$$
$$= (e + ir) - j(e' - ix),$$

or $$e_0 = \sqrt{(e + ir)^2 + (e' - ix)^2}, \qquad \tan \theta_0 = -\frac{e' - ix}{e + ir},$$

values which easily permit calculation.

35. When transferring from complex quantities to absolute values, it must be kept in mind that:

The absolute value of a product or a ratio of complex quantities is the product or ratio of their absolute values.

The phase angle of a product or a ratio of complex quantities is the sum or difference of their phase angles.

That is, if

$$\begin{aligned} \underset{\cdot}{A} &= a' + ja'' = a(\cos\alpha + j\sin\alpha) \\ \underset{\cdot}{B} &= b' + jb'' = b(\cos\beta + j\sin\beta) \\ \underset{\cdot}{C} &= c' + jc'' = c(\cos\gamma + j\sin\gamma) \end{aligned}$$

the absolute value of $\frac{\underset{\cdot}{A}\underset{\cdot}{B}}{\underset{\cdot}{C}}$ is given by $\frac{ab}{c}$, and its phase angle by $\alpha + \beta - \gamma$, that is, it is

$$\frac{\underset{\cdot}{A}\underset{\cdot}{B}}{\underset{\cdot}{C}} = \frac{ab}{c}[\cos(\alpha + \beta - \gamma) + j\sin(\alpha + \beta - \gamma)],$$

where

$$\begin{aligned} a &= \sqrt{a'^2 + a''^2} \\ b &= \sqrt{b'^2 + b''^2} \\ c &= \sqrt{c'^2 + c''^2} \end{aligned}$$

are the absolute values of $\underset{\cdot}{A}$, $\underset{\cdot}{B}$ and $\underset{\cdot}{C}$.

This rule frequently simplifies greatly the derivation of the absolute value and phase angle, from a complicated complex expression.

CHAPTER VI

TOPOGRAPHIC METHOD

36. In the representation of alternating sine waves by vectors, a certain ambiguity exists, in so far as one and the same quantity —voltage, for instance—can be represented by two vectors of opposite direction, according as to whether the e.m.f. is considered as a part of the impressed voltage or as a counter e.m.f. This is analogous to the distinction between action and reaction in mechanics.

Further, it is obvious that if in the circuit of a generator, G (Fig. 25), the current in the direction from terminal A over resistance R to terminal B is represented by a vector, $\overline{OI}$ (Fig. 26), or by $\dot{I} = i + ji'$, the same current can be considered as being

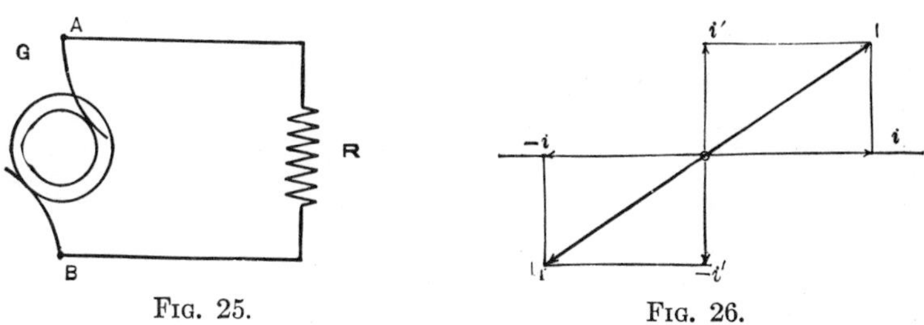

FIG. 25. FIG. 26.

in the opposite direction, from terminal B to terminal A in opposite phase, and therefore represented by a vector, $\overline{OI_1}$ (Fig. 26), or by $\dot{I}_1 = - i - ji'$.

Or, if the difference of potential from terminal B to terminal A is denoted by the $\dot{E} = e + je'$, the difference of potential from A to B is $E_1 = - e - je'$.

Hence, in dealing with alternating-current sine waves it is necessary to consider them in their proper direction with regard to the circuit. Especially in more complicated circuits, as interlinked polyphase systems, careful attention has to be paid to this point.

37. Let, for instance, in Fig. 27, an interlinked three-phase system be represented diagrammatically as consisting of three voltages, of equal intensity, differing in phase by one-third of a period. Let the voltages in the direction from the common connection, O, of the three branch circuits to the terminals, A_1, A_2, A_3, be represented by $\dot{E}_1$, $\dot{E}_2$, $\dot{E}_3$. Then the difference of potential from A_2 to A_1 is $\dot{E}_2 - \dot{E}_1$, since the two voltages, $\dot{E}_1$ and $\dot{E}_2$, are connected in circuit between the terminals, A_1 and A_2, in the direction $A_1—O—A_2$; that is, the one, E_2, in the direction, OA_2, from the common connection to terminal, the other, E_1, in the opposite direction, A_1O, from the terminal to common connection, and represented by $- E_1$. Conversely, the difference of potential from A_1 to A_2 is $\dot{E}_1 - \dot{E}_2$.

It is then convenient to go still a step farther, and drop the vector line altogether in the diagrammatic representation; that is, denote the sine wave by a point only, the end of the corresponding vector.

Looking at this from a different point of view, it means that we choose one point of the system—for instance, the common

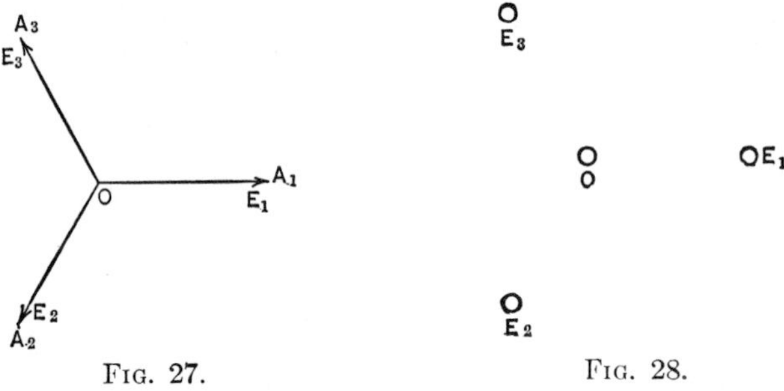

FIG. 27. FIG. 28.

connection, or neutral O—as a zero point, or point of zero potential, and represent the potentials of all the other points of the circuit by points in the diagram, such that their distances from the zero point give the intensity, their amplitude the phase of the difference of potential of the respective point with regard to the zero point; and their distance and amplitude with regard to other points of the diagram, their difference of potential from these points in intensity and phase.

Thus, for example, in an interlinked three-phase system with three voltages of equal intensity, and differing in phase by one-third of a period, we may choose the common connection of the star-connected generator as the zero point, and represent, in Fig. 28, one of the voltages, or the potential at one of the three-phase terminals, by point E_1. The potentials at the two other terminals will then be given by the points E_2 and E_3, which have the same distance from O as E_1, and are equidistant from E_1 and from each other.

The difference of potential between any pair of terminals, for instance, E_1 and E_2, is then the distance $\overline{E_2E_1}$, or $\overline{E_1E_2}$, according to the direction considered.

38. If now the three branches, $\overline{OE_1}$, $\overline{OE_2}$ and $\overline{OE_3}$, of the three-phase system are loaded equally by three currents equal in intensity and in difference of phase against their voltages,

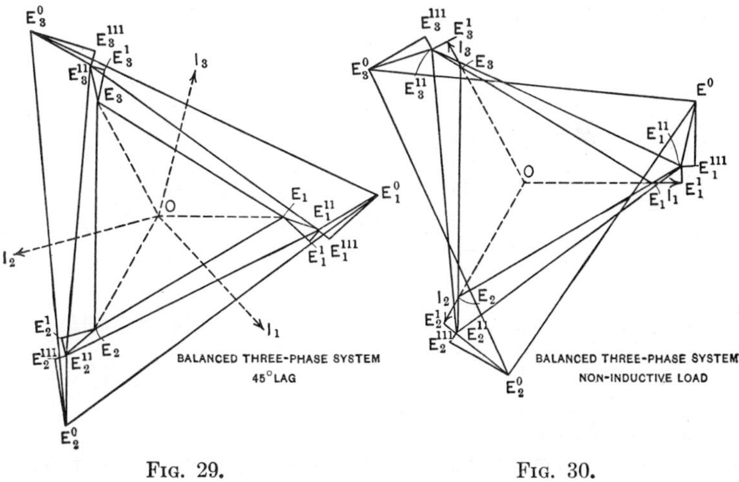

Fig. 29. Fig. 30.

these currents are represented in Fig. 29 by the vectors $\overline{OI_1} = \overline{OI_2} = \overline{OI_3} = I$, lagging behind the voltages by angles $E_1OI_1 = E_2OI_2 = E_3OI_3 = \theta$.

Let the three-phase circuit be supplied over a line of impedance, $Z_1 = r_1 + jx_1$, from a generator of internal impedance, $Z_0 = r_0 + jx_0$.

In phase $\overline{OE_1}$ the voltage consumed by resistance r_1 is represented by the distance, $\overline{E_1E_1{}^1} = Ir_1$, in phase, that is, parallel with current $\overline{OI_1}$. The voltage consumed by reactance x_1 is represented by $\overline{E_1{}^1E_1{}^{11}} = Ix_1$, 90° ahead of current $\overline{OI_1}$. The same applies to the other two phases, and it thus follows that to produce the voltage triangle, $E_1E_2E_3$, at the terminals of the consumer's circuit, the voltage triangle, $E_1{}^{11}E_2{}^{11}E_3{}^{11}$, is required at the generator terminals.

Repeating the same operation for the internal impedance of the generator, we get $\overline{E^{11}E^{111}} = Ir_0$, and parallel to $\overline{OI_1}$, $\overline{E^{111}E^0} = Ix_0$, and 90° ahead of $\overline{OI_1}$, and thus as triangle of (nominal) generated e.m.fs. of the generator, $E_1{}^0E_2{}^0E_3{}^0$.

In Fig. 29 the diagram is shown for 45° lag, in Fig. 30 for non-inductive load, and in Fig. 31 for 45° lead of the currents with regard to their voltages.

As seen, the generated e.m.f. and thus the generator excitation with lagging current must be higher, and with leading current lower, than at non-inductive load, or conversely with the same generator excitation, that is, the same internal generator e.m.f.

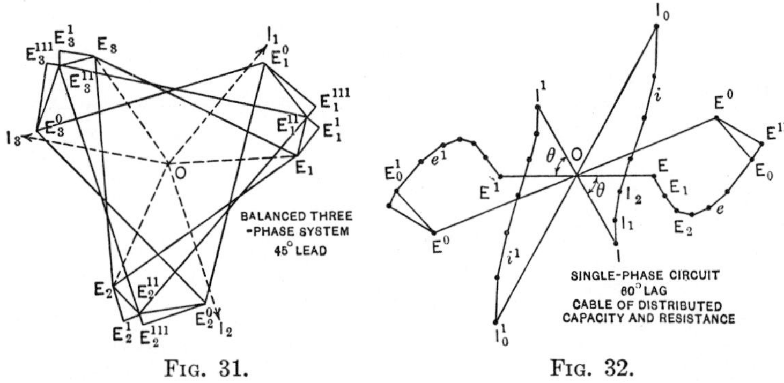

FIG. 31. FIG. 32.

triangle, $E_1{}^0E_2{}^0E_3{}^0$, the voltages at the receiver's circuit, E_1, E_2, E_3, fall off more with lagging, and less with leading current, than with non-inductive load.

39. As a further example may be considered the case of a single-phase alternating-current circuit supplied over a cable containing resistance and distributed capacity.

Let, in Fig. 32, the potential midway between the two terminals be assumed as zero point 0. The two terminal voltages at the receiver circuit are then represented by the points E and E^1, equidistant from 0 and opposite each other, and the two currents at the terminals are represented by the points I and I^1, equidistant from 0 and opposite each other, and under angle θ with E and E^1 respectively.

Considering first an element of the line or cable next to the receiver circuit. In this voltage, $\overline{EE_1}$, is consumed by the resistance of the line element, in phase with the current, $\overline{OI}$, and proportional thereto, and a current, $\overline{II_1}$, consumed by the capacity, as charging current of the line element, 90° ahead in phase of the voltage, $\overline{OE}$, and proportional thereto, so that at the generator end of this cable element current and voltage are $\overline{OI_1}$ and $\overline{OE_1}$ respectively.

Passing now to the next cable element we have again a voltage, $\overline{E_1E_2}$, proportional to and in phase with the current, $\overline{OI_1}$, and a current, $\overline{I_1I_2}$, proportional to and 90° ahead of the voltage, $\overline{OE_1}$, and thus passing from element to element along the cable to the generator, we get curves of voltages, e and e^1, and curves of currents, i and i^1, which can be called the topographical circuit characteristics, and which correspond to each other, point for point, until the generator terminal voltages, $\overline{OE_0}$ and $\overline{OE_0{}^1}$, and the generator currents, $\overline{OI_0}$ and $\overline{OI_0{}^1}$, are reached.

Again, adding $\overline{E_0E^{11}} = I_0r_0$ and parallel to $\overline{OI_1}$ and $\overline{E^{11}E^0} = I_0x_0$ and 90° ahead of $\overline{OI_0}$, gives the (nominal) generated e.m.f. of the generator $\overline{OE^0}$, where $Z_0 = r_0 + jx_0$ = internal impedance of the generator.

In Fig. 32 is shown the circuit characteristics for 60° lag of a cable containing only resistance and capacity.

Obviously by graphical construction the circuit characteristics appear more or less as broken lines, due to the necessity of using finite line elements, while in reality they are smooth curves when calculated by the differential method, as explained in "Transients in Space" [Dover, Volume III, pp. 59–221].

40. As further example may be considered a three-phase circuit supplied over a long-distance transmission line of distributed capacity, self-induction, resistance, and leakage.

Let, in Fig. 33, $\overline{OE_1}$, $\overline{OE_2}$, $\overline{OE_3}$ = three-phase voltages at receiver circuit, equidistant from each other and $= E$.

Let $\overline{OI_1}$, $\overline{OI_2}$, $\overline{OI_3}$ = three-phase currents in the receiver circuit equidistant from each other and $= I$, and making with E the phase angle, θ.

Considering again as in §3 the transmission line, element by element, we have in every element a voltage, $\overline{E_1E_1{}^1}$, consumed by the resistance in phase with the current, $\overline{OI_1}$, and proportional thereto, and a voltage, $\overline{E_1{}^1, E_1{}^{11}}$, consumed by the reactance of the line element, 90° ahead of the current, $\overline{OI_1}$, and proportional thereto.

In the same line element we have a current, $\overline{I_1I_1{}^1}$, in phase with the voltage, $\overline{OE_1}$, and proportional thereto, representing the loss of current by leakage, dielectric hysteresis, etc., and a current, $\overline{I_1^1 I_1^{11}}$, 90° ahead of the voltage, $\overline{OE_1}$, and proportional thereto, the charging current of the line element as condenser; and in this manner passing along the line, element by element, we ultimately reach the generator terminal voltages, E_1^0, E_2^0, E_3^0,

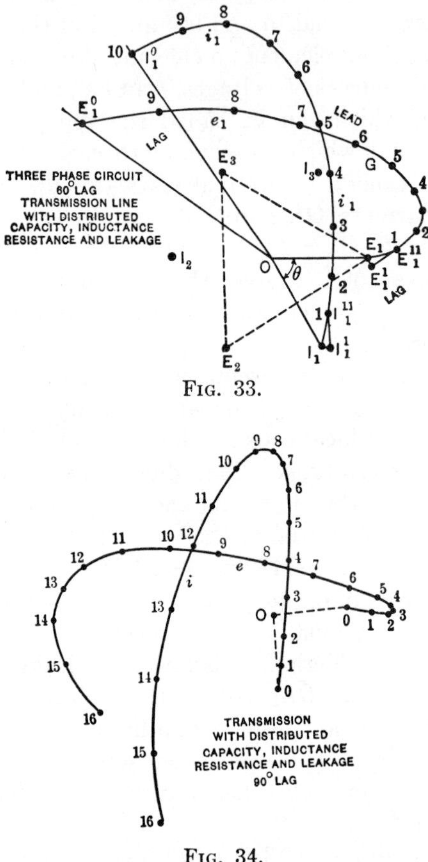

FIG. 33.

FIG. 34.

and generator currents, I_1^0, I_2^0, I_3^0, over the topographical characteristics of voltage, e_1, e_2, e_3, and of current, i_1, i_2, i_3, as shown in Fig. 33.

The circuit characteristics of current, i, and of voltage, e, correspond to each other, point for point, the one giving the current and the other the voltage in the line element.

Only the circuit characteristics of the first phase are shown, as e_1 and i_1. As seen, passing from the receiving end toward the generator end of the line, potential and current alternately rise and fall, while their phase angle changes periodically between lag and lead.

41. More markedly this is shown in Fig. 34, the topographic circuit characteristic of one of the lines with 90° lag in the receiver circuit. Corresponding points of the two characteristics, e and i, are marked by corresponding figures 0 to 16, representing equidistant points of the line. The values of voltage, current and

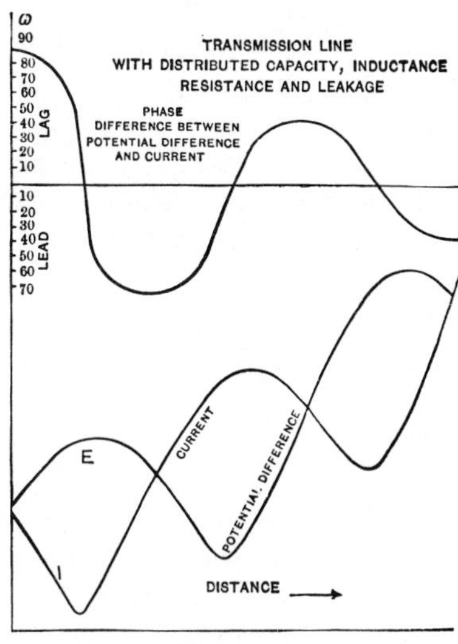

FIG. 35.

their difference of phase are plotted in Fig. 35 in rectangular coördinates with the distance as abscissas, counting from the receiving circuit toward the generator. As seen from Fig. 35, voltage and current periodically but alternately rise and fall, a maximum of one approximately coinciding with a minimum of the other, and with a point of zero phase displacement. The phase angle between current and e.m.f. changes from 90° lag to 72° lead, 44° lag, 34° lead, etc., gradually decreasing in the amplitude of its variation.

CHAPTER VII

POLAR COÖRDINATES AND POLAR DIAGRAMS

42. The graphic representation of alternating waves in rectangular coördinates, with the time as abscissæ and the instantaneous values as ordinates, gives a picture of their wave structure, as shown in Figs. 1 to 5. It does not, however, show their periodic character as well as the representation in polar coördinates, with the time as the angle or the amplitude—one complete period being represented by one revolution—and the instantaneous values as radius vectors; the polar coördinate system, in which the independent variable, the angle, is periodic, obviously lends itself better to the representation of periodic functions, as alternating waves.

Thus the two waves of Figs. 2 and 3 are represented in polar coördinates in Figs. 36 and 37 as closed characteristic curves, which, by their intersection with the radius vector, give the instantaneous value of the wave, corresponding to the time represented by the amplitude or angle of the radius vector.

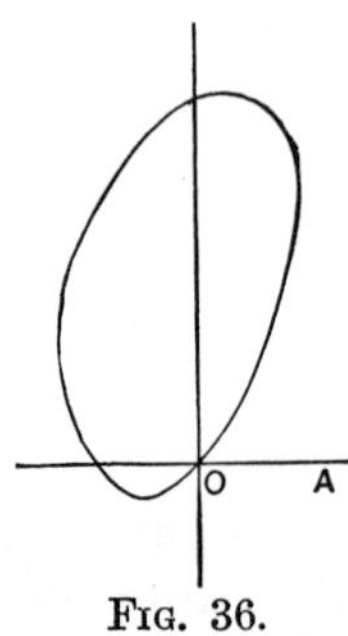

FIG. 36.

These instantaneous values are positive if in the direction of the radius vector, and negative if in opposition. Hence the two half-waves in Fig. 2 are represented by the same polar characteristic curve, which is traversed by the point of intersection of the radius vector twice per period—once in the direction of the vector, giving the positive half-wave, and once in opposition to the vector, giving the negative half-wave. In Figs. 3 and 37 where the two half-waves are different, they give different polar characteristics.

43. The sine wave, Fig. 1, is represented in polar coördinates by one circle, as shown in Fig. 38. The diameter of the characteristic curve of the sine wave, $I = \overline{OC}$, represents the intensity of the wave; and the amplitude of the diameter $\overline{OC}$, $\measuredangle\, \theta_0 = AOC$, is the phase of the wave, which, therefore, is represented analytically by the function

$$i = I \cos (\theta - \theta_0),$$

where $\theta = 2\pi\frac{t}{t_0}$ is the instantaneous value of the amplitude corresponding to the instantaneous value, i, of the wave.

The instantaneous values are cut out on the movable radius vector by its intersection with the characteristic circle. Thus, for instance, at the amplitude, $AOB_1 = \theta_1 = 2\pi\frac{t_1}{t_0}$ (Fig. 38), the instantaneous value is OB'; at the amplitude, $AOB_2 = \theta_2 = 2\pi\frac{t_2}{t_0}$, the instantaneous value is $\overline{OB''}$, and negative, since in opposition to the radius vector, OB_2.

The angle, θ, so represents the time, and increasing time is represented by an increase of angle θ in counter-clockwise rota-

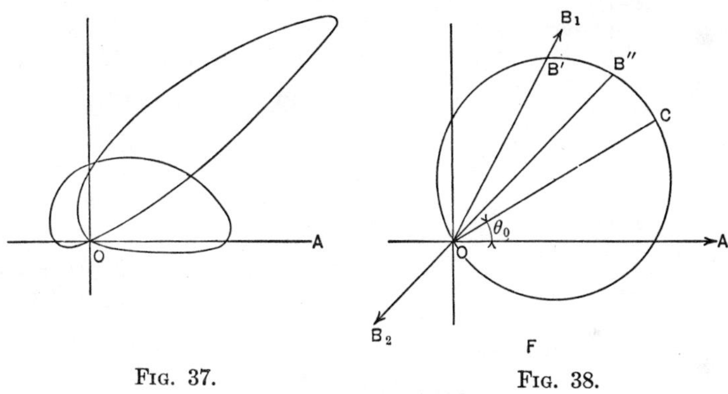

FIG. 37. FIG. 38.

tion. That is, the positive direction, or increase of time, is chosen as counter-clockwise rotation, in conformity with general custom.

The characteristic circle of the alternating sine wave is determined by the length of its diameter—the intensity of the wave; and by the amplitude of the diameter—the phase of the wave.

Hence wherever the integral value of the wave is considered alone, and not the instantaneous values, the characteristic circle may be omitted altogether, and the wave represented in intensity and in phase by the diameter of the characteristic circle.

Thus, in polar coördinates, the alternating wave may be represented in intensity and phase by the length and direction of a vector, $\overline{OC}$, Fig. 38, and its analytical expression would then be $c = \overline{OC}\cos(\theta - \theta_0)$.

This leads to a second vector representation of alternating

waves, differing from the crank diagram discussed in Chapter IV. It may be called the time diagram or polar diagram, and is used to a considerable extent in the literature, thus must be familiar to the engineer, though in the following we shall in graphic representation and in the symbolic representation based thereon, use the crank diagram of Chapters IV and V (pages 139 and 150).

In the time diagram as well as in the crank diagram, instead of the maximum value of the wave, the effective value, or square root of mean square, may be used as the vector, which is more convenient; and the maximum value is then $\sqrt{2}$ times the vector $\overline{OC}$, so that the instantaneous values, when taken from the diagram, have to be increased by the factor $\sqrt{2}$.

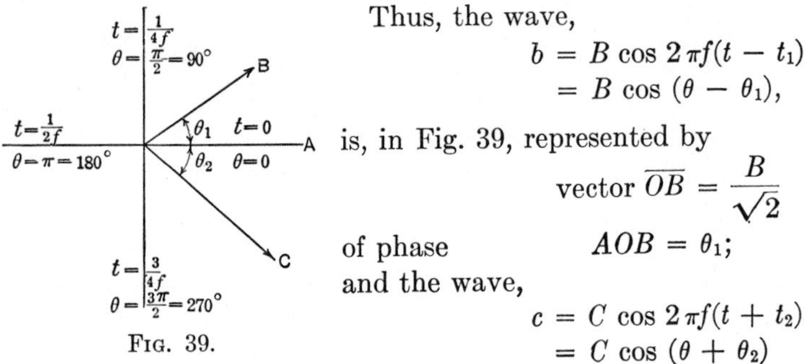

FIG. 39.

Thus, the wave,

$$b = B \cos 2\pi f(t - t_1)$$
$$= B \cos (\theta - \theta_1),$$

is, in Fig. 39, represented by

$$\text{vector } \overline{OB} = \frac{B}{\sqrt{2}}$$

of phase $\qquad AOB = \theta_1;$

and the wave,

$$c = C \cos 2\pi f(t + t_2)$$
$$= C \cos (\theta + \theta_2)$$

is, in Fig. 39, represented by

$$\text{vector } \overline{OC} = \frac{C}{\sqrt{2}}$$

of phase $\qquad AOC = -\theta_2.$

The former is said to lag by angle θ_1, the latter to lead by angle θ_2, with regard to the zero position.

The wave b lags by angle $(\theta_1 + \theta_2)$ behind wave c, or c leads b by angle $(\theta_1 + \theta_2)$.

44. To combine different sine waves, their graphical representations, or vectors, are combined by the parallelogram law.

From the foregoing considerations we have the conclusions:

The sine wave is represented graphically in polar coördinates by a vector, which by its length $\overline{OC}$, denotes the intensity, and by its amplitude, AOC, the phase, of the sine wave.

Sine waves are combined or resolved graphically, in polar coördinates, by the law of the parallelogram or the polygon of sine waves. (Fig. 40.)

Kirchhoff's laws now assume, for alternating sine waves, the form:

(*a*) The resultant of all the e.m.fs. in a closed circuit, as found by the parallelogram of sine waves, is zero if the counter e.m.fs. of resistance and of reactance are included.

(*b*) The resultant of all the currents toward a distributing point, as found by the parallelogram of sine waves, is zero.

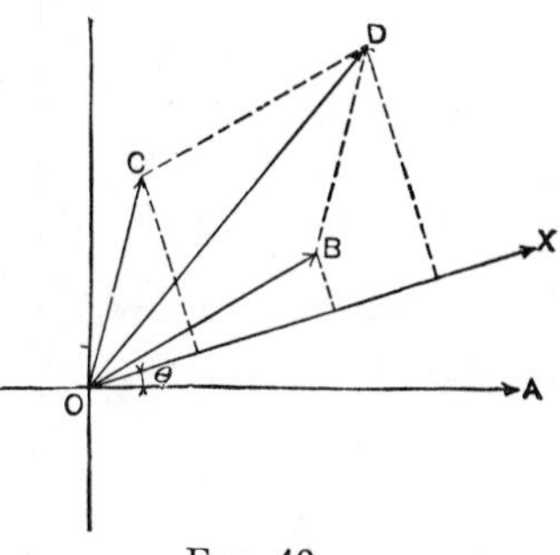

Fig. 40.

The power equation expressed graphically is as follows:

The power of an alternating-current circuit is represented in polar coördinates by the product of the current, I, into the projection of the e.m.f., E, upon the current, or by the e.m.f., E, into the projection of the current, I, upon the e.m.f., or by $IE \cos \theta$, where θ = angle of time-phase displacement.

45. The instances represented by the vector representation of the crank diagram in Chapter IV as Figs. 16, 17, 18, 19, 20,

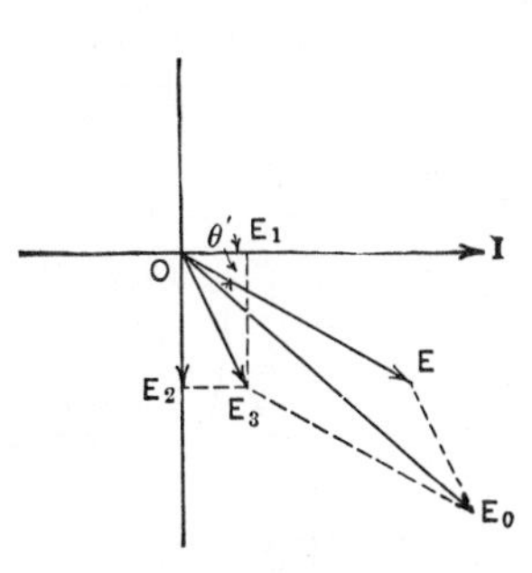

Fig. 41.

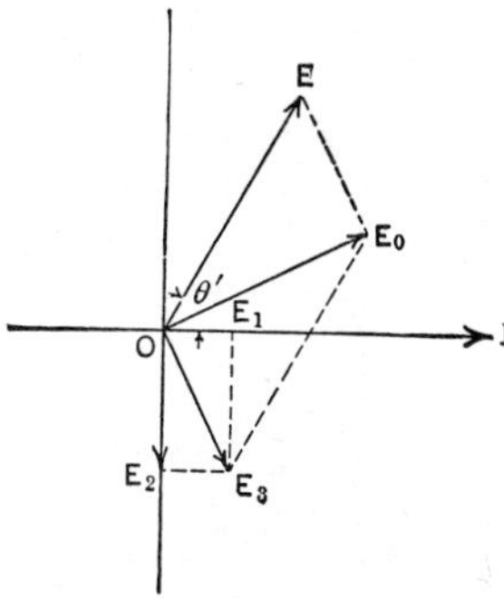

Fig. 42.

then appear in the vector representation of the time diagram or polar coordinate diagram, in the form of Figs. 41, 42, 43, 44, 45.

These figures are the reverse, or mirror image of each other. That is, the crank diagrams, turned around the horizontal (or any other axis), so as they would be seen in a mirror, are the time diagrams, and inversely.

The polar diagram, Fig. 46, of a current:

$$i = I \cos (\vartheta - \beta)$$

represented by vector $\overline{OI}$,

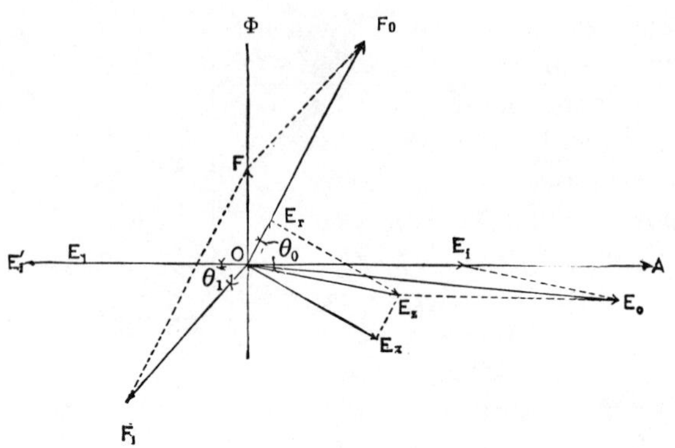

FIG. 43.

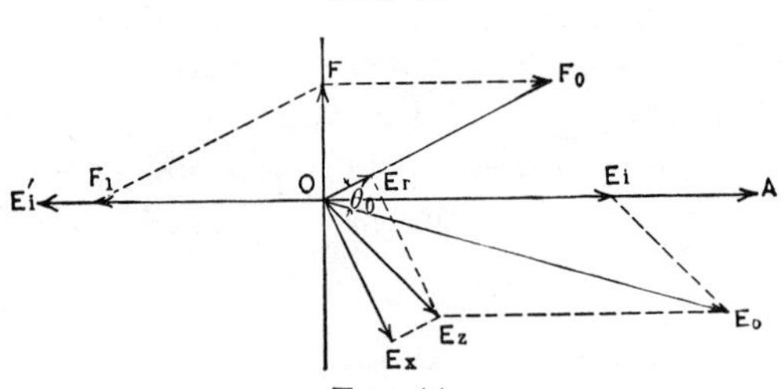

FIG. 44.

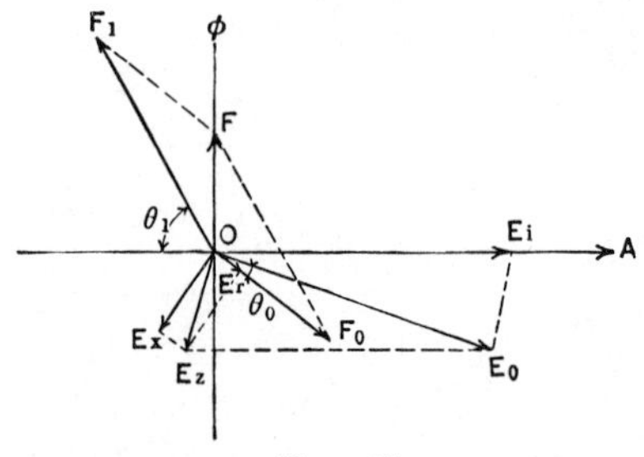

FIG. 45.

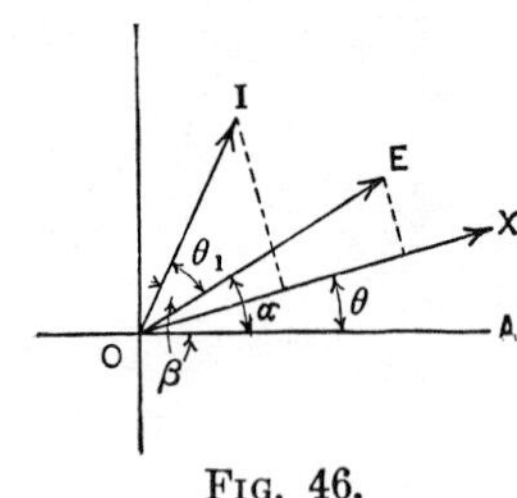

FIG. 46.

lagging behind the voltage:

$$e = E \cos (\vartheta - \alpha)$$

represented by vector $\overline{OE}$,
by angle

$$\theta_1 = \beta - \alpha$$

then means:

The voltage e reaches its maximum at the time t_1, which is represented by angle $\alpha = 2\pi\frac{t_1}{t_0}$, where t_0 = period, and the current, i, reaches its maximum at the time t_2, which is represented by angle $\beta = 2\pi\frac{t_2}{t_0}$, and since $\beta > \alpha$, the current reaches its maximum at a later time than the voltage, that is, lags behind the voltage, and the lag of the current behind the voltage is the difference between the times of their maxima, β and α, in angular measure, that is, is

$$\theta_1 = \beta - \alpha = 2\pi\frac{t_2 - t_1}{t_0}.$$

At any moment of time t, represented by angle $\theta = 2\pi\frac{t}{t_0}$, the instantaneous values of current and voltage, i and e, are the projections of $\overline{OI}$ and $\overline{OE}$ on the time radius OX drawn under angle $AOX = \theta$.

The crank diagram corresponding to the time diagram Fig. 46 is shown in Fig. 47. It means: The vectors $\overline{OI}$ and $\overline{OE}$, representing the current and the voltage respectively, rotate synchronously, and by their projections on the horizontal $\overline{OA}$ represent the instantaneous values of current and voltage. Angle $IOA = \beta$ being larger than angle $EOA = \alpha$, the current vector $\overline{OI}$ passes its maximum, in position OA, later than the voltage vector OE, that is, the current lags behind the voltage, by the difference of time corresponding to the passage of the current and voltage vectors through their maxima, in the direction OA, that is, by the time angle $\theta_1 = \beta - \alpha$.

A polar diagram, Fig. 46, with the current, $\overline{OI}$, lagging behind the voltage, $\overline{OE}$, by the angle, θ_1, thus considered as crank diagram would represent the current leading the voltage by the angle, θ_1, and a crank diagram, Fig. 47, with the current lagging behind the voltage by the angle, θ_1, would as polar diagram represent a current leading the voltage by the angle, θ_1.

46. The main difference in appearance between the crank diagram and the polar diagram therefore is that, with the same direction of rotation, lag in the one diagram is represented in the same manner as lead in the other diagram, and inversely. Or, a representation by the crank diagram looks like a representation by the polar diagram, with reversed direction of rotation, and *vice versa*. Or, the one diagram is the image of the other and can be transformed into it by reversing right and left, or top and bottom. So the crank diagram, Fig. 47, is the image of the polar diagram, Fig. 46.

In symbolic representation, based upon the crank diagram, the impedance was denoted by

$$Z = r + jx,$$

where x = inductive reactance.

In the polar diagram, the impedance thus is denoted by:

$$Z = r - jx$$

since the latter is the mirror image of the crank diagram, that is, differs from it symbolically by the interchange of $+j$ and $-j$.

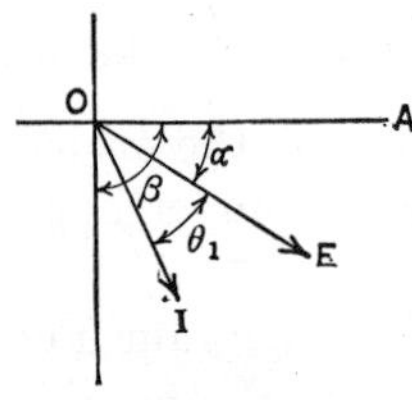

FIG. 47.

A treatise written in the symbolic representation by the polar diagram, thus can be translated to the representation by the crank diagram, and inversely, by simply reversing the signs of all imaginary quantities, that is, considering the signs of all terms with j changed.

A graphical representation in the polar diagram can be considered as a graphic representation in the crank diagram, with clockwise or right-handed rotation, and inversely.

Thus, for the engineer familiar with one representation only, but less familiar with the other, the most convenient way when meeting with a treatise in the, to him, unfamiliar representation is to consider all the diagrams as clockwise and all the signs of j reversed.

In conformity with the recommendation of the Turin Congress —however ill considered this may appear to many engineers—in the following the crank diagram will be used, and wherever conditions require the time diagram, the latter be translated to the crank diagram. It is not possible to entirely avoid the time diagram, since the crank diagram is more limited in its application.

47. The crank diagram offers the disadvantage, that it can be applied to sine waves only, while the polar diagram permits the construction of the curve of waves of any shapes, as those in Figs. 36 and 37.

In most cases, this objection is not serious, and in the diagrammatic and symbolic representation, the alternating quantities can be assumed as sine waves, that is, the general wave represented by the equivalent sine wave, that is, the sine wave of the same effective value as the general wave.

The transformation of the general wave into the equivalent sine wave, however, has to be carried out algebraically in the crank diagram, while the polar diagram permits a graphical transformation of the general wave into the equivalent sine wave.

Let Fig. 48 represent a general alternating wave. An element B_1OB_2 of this wave then has the area

$$dA = \frac{r^2 d\theta}{2},$$

and the total area of the polar curve is

$$A = \int_0^{2\pi} \frac{r^2}{2} d\theta.$$

The effective value of the wave is

$$R = \sqrt{\text{mean square}}$$
$$= \sqrt{\frac{1}{2\pi}\int_0^{2\pi} r^2 d\theta};$$

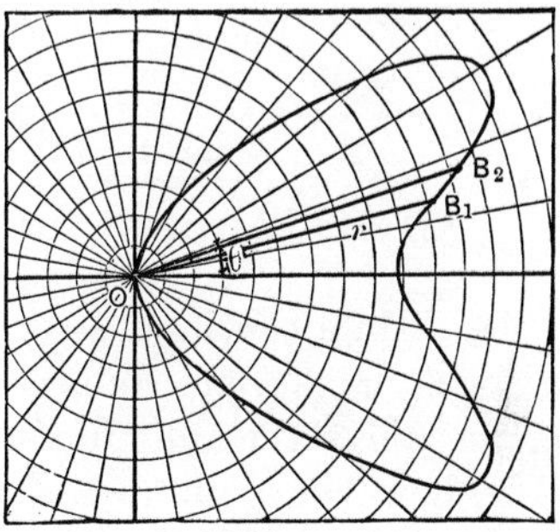

FIG. 48.

hence,

$$R^2\pi = \frac{1}{2}\int_0^{2\pi} r^2 d\theta = A.$$

The area of the polar curve of the general periodic wave, as measured by planimeter, therefore equals the area of a circle with the effective value of the wave as radius.

The effective value of the equivalent sine wave therefore is the radius of a circle having the same area as the general wave, in polar coördinates:

$$R = \sqrt{\frac{A}{\pi}}$$

The diameter of the general polar circle, therefore, is

$$R\sqrt{2} = \sqrt{\frac{2A}{\pi}}.$$

And the phase of the equivalent sine wave, or the direction of the diameter of its polar circle, is the vector bisecting the area of the general wave, in polar coordinates.

The transformation of the general alternating wave into the equivalent sine wave, therefore, is carried out by measuring the area of the general wave in polar coördinates, and drawing the sine wave circle of half this area.

CIRCUITS

CHAPTER VIII

ADMITTANCE, CONDUCTANCE, SUSCEPTANCE

48. If in a continuous-current circuit, a number of resistances, $r_1, r_2, r_3, \ldots$, are connected in series, their joint resistance, R, is the sum of the individual resistances, $R = r_1 + r_2 + r_3 + \ldots$

If, however, a number of resistances are connected in multiple or in parallel, their joint resistance, R, cannot be expressed in a simple form, but is represented by the expression

$$R = \frac{1}{\frac{1}{r_1} + \frac{1}{r_2} + \frac{1}{r_3} + \ldots}.$$

Hence, in the latter case it is preferable to introduce, instead of the term *resistance*, its reciprocal, or inverse value, the term *conductance*, $g = \frac{1}{r}$. If, then, a number of conductances, $g_1, g_2, g_3, \ldots$ are connected in parallel, their joint conductance is the sum of the individual conductances, or $G = g_1 + g_2 + g_3 + \ldots$ When using the term conductance, the joint conductance of a number of series-connected conductances becomes similarly a complicated expression

$$G = \frac{1}{\frac{1}{g_1} + \frac{1}{g_2} + \frac{1}{g_3} + \ldots}.$$

Hence the term *resistance* is preferable in case of series connection, and the use of the reciprocal term *conductance* in parallel connections; therefore,

The joint resistance of a number of series-connected resistances is equal to the sum of the individual resistances; the joint conductance of a number of parallel-connected conductances is equal to the sum of the individual conductances.

49. In alternating-current circuits, instead of the term *resistance* we have the term *impedance*, $Z = r + jx$, with its two components, the *resistance*, r, and the *reactance*, x, in the formula of Ohm's law, $\underset{\cdot}{E} = \underset{\cdot}{I}Z$. The resistance, r, gives the component of e.m.f. in phase with the current, or the power component of the e.m.f., Ir; the reactance, x, gives the component of the e.m.f. in quadrature with the current, or the wattless component of e.m.f., Ix; both combined give the total e.m.f.,

$$Iz = I\sqrt{r^2 + x^2}.$$

Since e.m.fs. are combined by adding their complex expressions, we have:

The joint impedance of a number of series-connected impedances is the sum of the individual impedances, when expressed in complex quantities.

In graphical representation impedances have not to be added, but are combined in their proper phase by the law of parallelogram in the same manner as the e.m.fs. corresponding to them.

The term impedance becomes inconvenient, however, when dealing with parallel-connected circuits; or, in other words, when several currents are produced by the same e.m.f., such as in cases where Ohm's law is expressed in the form,

$$\underset{\cdot}{I} = \frac{\dot{E}}{Z}.$$

It is preferable, then, to introduce the reciprocal of impedance, which may be called the *admittance* of the circuit, or

$$Y = \frac{1}{Z}.$$

As the reciprocal of the complex quantity, $Z = r + jx$, the admittance is a complex quantity also, or $Y = g - jb$; it consists of the component, g, which respresents the coefficient of current in phase with the e.m.f., or the power or active component, $g\underset{\cdot}{E}$, of the current, in the equation of Ohm's law,

$$\underset{\cdot}{I} = Y\underset{\cdot}{E} = (g - jb)\underset{\cdot}{E},$$

and the component, b, which represents the coefficient of current in quadrature with the e.m.f., or wattless or reactive component, $b\underset{\cdot}{E}$, of the current.

g is called the *conductance*, and b the *susceptance*, of the circuit. Hence the conductance, g, is the power component, and the susceptance, b, the wattless component, of the admittance, $Y = g - jb$, while the numerical value of admittance is

$$y = \sqrt{g^2 + b^2};$$

the resistance, r, is the power component, and the reactance, x, the wattless component, of the impedance, $Z = r + jx$, the numerical value of impedance being

$$z = \sqrt{r^2 + x^2}.$$

50. As shown, the term *admittance* implies resolving the current into two components, in phase and in quadrature with the e.m.f., or the power or active component and the wattless or reactive component; while the term *impedance* implies resolving the e.m.f. into two components, in phase and in quadrature with the current, or the power component and the wattless or reactive component.

It must be understood, however, that the conductance is not the reciprocal of the resistance, but depends upon the reactance as well as upon the resistance. Only when the reactance $x = 0$, or in continuous-current circuits, is the conductance the reciprocal of resistance.

Again, only in circuits with zero resistance ($r = 0$) is the susceptance the reciprocal of reactance; otherwise, the susceptance depends upon reactance and upon resistance.

The conductance is zero for two values of the resistance:

1. If $r = \infty$, or $x = \infty$, since in this case there is no current, and either component of the current $= 0$.

2. If $r = 0$, since in this case the current in the circuit is in quadrature with the e.m.f., and thus has no power component.

Similarly, the susceptance, b, is zero for two values of the reactance:

1. If $x = \infty$, or $r = \infty$.

2. If $x = 0$.

From the definition of admittance, $Y = g - jb$, as the reciprocal of the impedance, $Z = r + jx$,
we have

$$Y = \frac{1}{Z}, \text{ or } g - jb = \frac{1}{r + jx};$$

or, multiplying numerator and denominator on the right side by $(r - jx)$,

$$g - jb = \frac{r - jx}{(r + jx)\,(r - jx)};$$

hence, since

$$(r + jx)(r - jx) = r^2 + x^2 = z^2,$$

$$g - jb = \frac{r}{r^2 + x^2} - j\frac{x}{r^2 + x^2} = \frac{r}{z^2} - j\frac{x}{z^2}$$

or

$$g = \frac{r}{r^2 + x^2} = \frac{r}{z^2},$$

$$b = \frac{x}{r^2 + x^2} = \frac{x}{z^2};$$

and conversely

$$r = \frac{g}{g^2 + b^2} = \frac{g}{y^2},$$

$$x = \frac{b}{g^2 + b^2} = \frac{b}{y^2}.$$

By these equations, the conductance and susceptance can be calculated from resistance and reactance, and conversely.

Multiplying the equations for g and r, we get

$$gr = \frac{rg}{z^2y^2};$$

hence,

$$z^2y^2 = (r^2 + x^2)(g^2 + b^2) = 1;$$

and

$$z = \frac{1}{y} = \frac{1}{\sqrt{g^2 + b^2}}, \quad \text{the absolute value of impedance;}$$

$$y = \frac{1}{z} = \frac{1}{\sqrt{r^2 + x^2}}, \quad \text{the absolute value of admittance.}$$

51. If, in a circuit, the reactance, x, is constant, and the resistance, r, is varied from $r = 0$ to $r = \infty$, the susceptance, b, decreases from $b = \frac{1}{x}$ at $r = 0$, to $b = 0$ at $r = \infty$; while the conductance, $g = 0$ at $r = 0$, increases, reaches a maximum for $r = x$, where $g = \frac{1}{2r}$, is equal to the susceptance or $g = b$, and then decreases again, reaching $g = 0$ at $r = \infty$.

In Fig. 49, for constant reactance $x = 0.5$ ohm, the variation of the conductance, g, and of the susceptance, b, are shown as functions of the varying resistance, r. As shown, the absolute value of admittance, susceptance, and conductance are plotted in full lines, and in dotted line the absolute value of impedance,

$$z = \sqrt{r^2 + x^2} = \frac{1}{y}.$$

Obviously, if the resistance, r, is constant, and the reactance, x, is varied, the values of conductance and susceptance are merely exchanged, the conductance decreasing steadily from $g = \frac{1}{r}$ to 0, and the susceptance passing from 0 at $x = 0$ to the maximum, $b = \frac{1}{2r} = g = \frac{1}{2x}$ at $x = r$, and to $b = 0$ at $x = \infty$.

The resistance, r, and the reactance, x, vary as functions of the conductance, g, and the susceptance, b, in the same manner as g and b vary as functions of r and x.

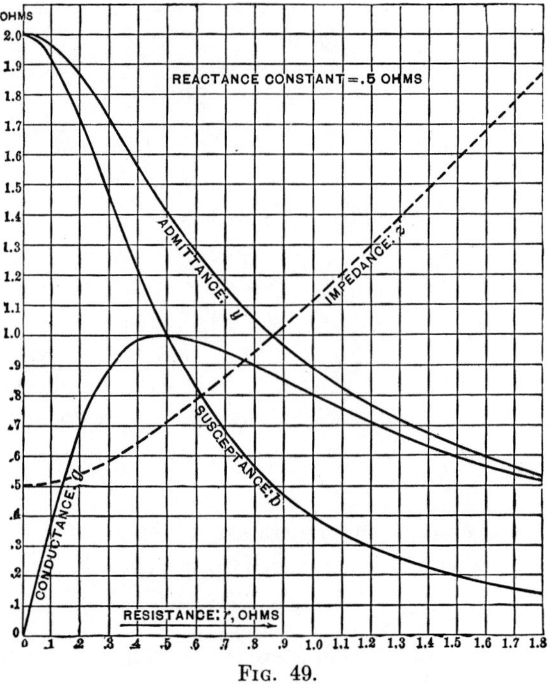

FIG. 49.

The sign in the complex expression of admittance is always opposite to that of impedance; this is obvious, since if the current lags behind the e.m.f., the e.m.f. leads the current, and conversely.

We can thus express Ohm's law in the two forms,

$$\dot{E} = \dot{I}Z,$$
$$\dot{I} = \dot{E}Y,$$

and therefore,

The joint impedance of a number of series-connected impedances is equal to the sum of the individual impedances; the joint admittance of a number of parallel-connected admittances is equal to the sum of the individual admittances, if expressed in complex quantities. In diagrammatic representation, combination by the parallelogram law takes the place of addition of the complex quantities.

52. Experimentally, impedances and admittances are most conveniently determined by establishing an alternating current in the circuit, and measuring by voltmeter, ammeter and wattmeter, the volts, e, the amperes, i, and the watts, p.

It is then,

$$\text{Impedance: } z = \frac{e}{i}.$$

$$\text{Resistance (effective): } r = \frac{p}{i^2}.$$

$$\text{Reactance: } x = \sqrt{z^2 - r^2}.$$

$$\text{Admittance: } y = \frac{i}{e}.$$

$$\text{Conductance: } g = \frac{p}{e^2}.$$

$$\text{Susceptance: } b = \sqrt{y^2 - g^2}.$$

Regarding their calculation, see Dover edition, Volume I, Part I.

CHAPTER IX

CIRCUITS CONTAINING RESISTANCE, INDUCTIVE REACTANCE, AND CONDENSIVE REACTANCE

53. Having, in the foregoing, re-established Ohm's law and Kirchhoff's laws as being also the fundamental laws of alternating-current circuits, when expressed in their complex form,

$$\dot{E} = Z\dot{I}, \qquad \text{or, } \dot{I} = Y\dot{E},$$

and

$$\Sigma \dot{E} = 0 \text{ in a closed circuit,}$$
$$\Sigma \dot{I} = 0 \text{ at a distributing point,}$$

where $\dot{E}$, $\dot{I}$, Z, Y, are the expressions of e.m.f., current, impedance, and admittance in complex quantities—these values representing not only the intensity, but also the phase, of the alternating wave—we can now—by application of these laws, and in the same manner as with continuous-current circuits, keeping in mind, however, that $\dot{E}$, $\dot{I}$, Z, Y, are complex quantities—calculate alternating-current circuits and networks of circuits containing resistance, inductive reactance, and condensive reactance in any combination, without meeting with greater difficulties than when dealing with continuous-current circuits.

It is obviously not possible to discuss with any completeness all the infinite varieties of combinations of resistance, inductive reactance, and condensive reactance which can be imagined, and which may exist, in a system of network of circuits; therefore only some of the more common or more interesting combinations will here be considered.

1. Resistance in Series with a Circuit

54. In a constant-potential system with impressed e.m.f.,

$$\dot{E}_0 = e_0 + je'_0, \qquad E_0 = \sqrt{e_0^2 + e_0'^2},$$

let the receiving circuit of impedance,

$$Z = r + jx, \qquad z = \sqrt{r^2 + x^2},$$

be connected in series with a resistance, r_0.

The total impedance of the circuit is then

$$Z + r_0 = r + r_0 + jx;$$

hence the current is

$$\dot{I} = \frac{\dot{E}_0}{Z + r_0} = \frac{\dot{E}_0}{r + r_0 + jx} = \frac{E_0(r + r_0 - jx)}{(r + r_0)^2 + x^2};$$

and the e.m.f. of the receiving circuit becomes

$$\dot{E} = \dot{I}Z = \frac{\dot{E}_0(r + jx)}{r + r_0 + jx} = \frac{\dot{E}_0\{r(r + r_0) + x^2 + jr_0x\}}{(r + r_0)^2 + x^2}$$

$$= \frac{\dot{E}_0\{z^2 + rr_0 + jr_0x\}}{z^2 + 2\,rr_0 + r_0^2};$$

or, in absolute values we have the following:

Impressed e.m.f.,

$$E_0 = \sqrt{e_0^2 + e_0'^2};$$

current,

$$I = \frac{E_0}{\sqrt{(r + r_0)^2 + x^2}} = \frac{E_0}{\sqrt{z^2 + 2\,rr_0 + r_0^2}};$$

e.m.f. at terminals of receiver circuit,

$$E = E_0\sqrt{\frac{r^2 + x^2}{(r + r_0)^2 + x^2}} = \frac{E_0z}{\sqrt{z^2 + 2rr_0 + r_0^2}};$$

difference of phase in receiver circuit, $\tan\theta = \frac{x}{r};$

difference of phase in supply circuit, $\tan\theta_0 = \frac{x}{r + r_0}$

since in general,

$$\tan(\text{phase}) = \frac{\text{imaginary component}}{\text{real component}}.$$

(*a*) If x is negligible with respect to r, as in a non-inductive receiving circuit,

$$I = \frac{E_0}{r + r_0}, \qquad E = E_0\frac{r}{r + r_0},$$

and the current and e.m.f. at receiver terminals decrease steadily with increasing r_0.

(*b*) If r is negligible compared with x, as in a wattless receiver circuit,

$$I = \frac{E_0}{\sqrt{r_0^2 + x^2}}, \quad E = E_0\frac{x}{\sqrt{r_0^2 + x^2}};$$

or, for small values of r_0,

$$I = \frac{E_0}{x}, \qquad E = E_0;$$

that is, the current and e.m.f. at receiver terminals remain approximately constant for small values of r_0, and then decrease with increasing rapidity.

In the general equations, x appears in the expressions for I and E only as x^2, so that I and E assume the same value when x is negative as when x is positive; or, in other words, series resistance acts upon a circuit with leading current, or in a condenser circuit, in the same way as upon a circuit with lagging current, or an inductive circuit.

For a given impedance, z, of the receiver circuit, the current, I, and e.m.f., E, are smaller the larger the value of r; that is, the less the difference of phase in the receiver circuit.

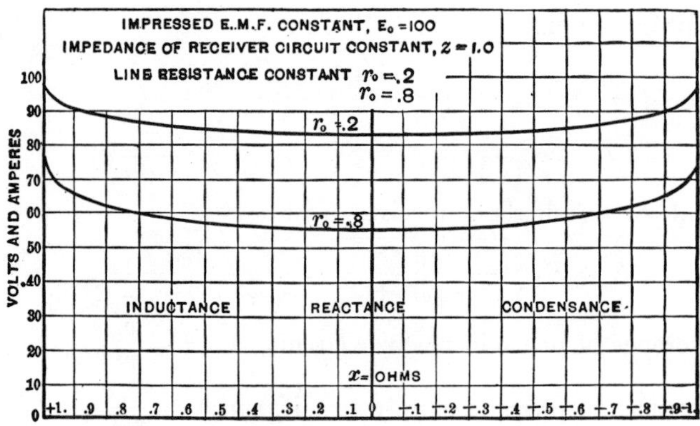

FIG. 50.—Variation of voltage at constant series resistance with phase relation of receiver circuit.

As an instance, in Fig. 50 is shown the e.m.f., E, at the receiver circuit, for E_0 = const. = 100 volts, $z = 1$ ohm; hence $I = E$, and

(a) $r_0 = 0.2$ ohm (Curve I)
(b) $r_0 = 0.8$ ohm (Curve II)

with values of reactance, $x = \sqrt{z^2 - r^2}$, for abscissæ, from $x = +1.0$ to $x = -1.0$ ohm.

As shown, I and E are smallest for $x = .0$, $r = 1.0$, or for the non-inductive receiver circuit, and largest for $x = \pm 1.0$, $r = 0$, or for the wattless circuit, in which latter a series resistance causes but a very small drop of potential.

Hence the control of a circuit by series resistance depends upon the difference of phase in the circuit.

For $r_0 = 0.8$ and $x = 0$, $x = +0.8$, $x = -0.8$, the vector diagrams are shown in Figs. 51 to 53.

In these Figs. $\overline{OE_0}$ is the supply voltage, $\overline{OE_3}$ the voltage consumed by the line resistance, and $\overline{OE}$ the receiver voltage, with its two components, $\overline{OE_1}$ in phase and $\overline{OE_2}$ in quadrature with the current.

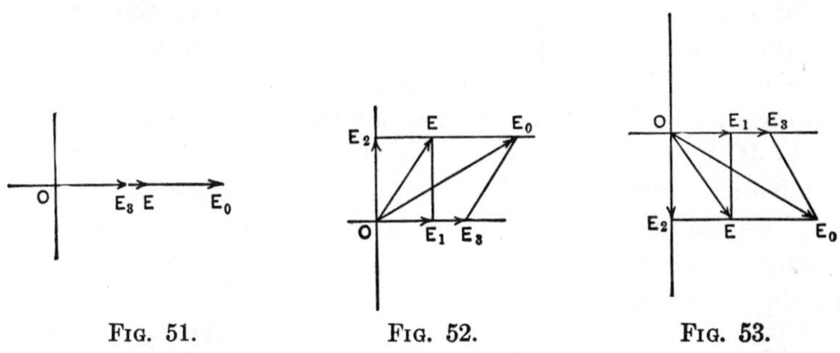

FIG. 51. FIG. 52. FIG. 53.

2. Reactance in Series with a Circuit

52. In a constant potential system of impressed e.m.f.,

$$\dot{E}_0 = e_0 + je'_0, \quad E_0 = \sqrt{e_0^2 + e_0'^2}$$

let a reactance, x_0, be connected in series in a receiver circuit of impedance,

$$\dot{Z} = r + jx, \qquad z = \sqrt{r^2 + x^2}.$$

Then, the total impedance of the circuit is

$$Z + jx_0 = r + j(x + x_0),$$

and the current is

$$\dot{I} = \frac{\dot{E}_0}{Z + jx_0} = \frac{\dot{E}_0}{r + j(x + x_0)},$$

while the difference of potential at the receiver terminals is

$$\dot{E} = \dot{I}Z = \dot{E}_0 \frac{r + jx}{r + j(x + x_0)}.$$

Or, in absolute quantities,

current,

$$I = \frac{E_0}{\sqrt{r^2 + (x + x_0)^2}} = \frac{E_0}{\sqrt{z^2 + 2\,xx_0 + x_0^2}};$$

e.m.f. at receiver terminals,

$$E = E_0\sqrt{\frac{r^2 + x^2}{r^2 + (x + x_0)^2}} = \frac{E_0 z}{\sqrt{r^2 + 2\,xx_0 + x_0^2}};$$

difference of phase in receiver circuit,

$$\tan \theta = \frac{x}{r};$$

difference of phase in supply circuit,

$$\tan \theta_0 = \frac{x + x_0}{r}.$$

(*a*) If x is small compared with r, that is, if the receiver circuit is non-inductive, I and E change very little for small values of x_0; but if x is large, that is, if the receiver circuit is of large reactance, I and E change considerably with a change of x_0.

(*b*) If x is negative, that is, if the receiver circuit contains condensers, synchronous motors, or other apparatus which produce leading currents, below a certain value of x_0 the denominator in the expression of E becomes $< z$, or $E > E_0$; that is, the reactance, x_0, raises the voltage.

(*c*) $E = E_0$, or the insertion of a series reactance, x_0, does not affect the potential difference at the receiver terminals, if

$$\sqrt{z^2 + 2x\, x_0 + x_0^2} = z;$$

or,

$$x_0 = -2x.$$

That is, if the reactance which is connected in series in the circuit is of opposite sign, but twice as large as the reactance of the receiver circuit, the voltage is not affected, but $E = E_0$, $I = \frac{E_0}{z}$. If $x_0 < -2x$, it raises, if $x_0 > -2x$, it lowers, the voltage.

We see, then, that a reactance inserted in series in an alternating-current circuit will always lower the voltage at the receiver terminals, when of the same sign as the reactance of the receiver circuit; when of opposite sign, it will lower the voltage if larger, raise the voltage if less, than twice the numerical value of the reactance of the receiver circuit.

(*d*) If $x = 0$, that is, if the receiver circuit is non-inductive, the e.m.f. at receiver terminals is

$$E = \frac{E_0 r}{\sqrt{r^2 + x_0^2}} = \frac{E_0}{\sqrt{1 + \left(\frac{x_0}{r}\right)^2}}$$

$$= E_0 \left\{ 1 - \frac{1}{2}\left(\frac{x_0}{r}\right)^2 + \frac{3}{8}\left(\frac{x_0}{r}\right)^4 - + \dots \right\}$$

$\left(\frac{1}{\sqrt{1+x}} = (1 + x)^{-1/2}\right.$ expanded by the binomial theorem

$$\left.(1 + x)^n = 1 + nx + \frac{n(n-1)}{1.2} x^2 + \dots\right).$$

Therefore, if x_0 is small compared with r,

$$E = E_0\left(1 - \frac{1}{2}\left(\frac{x_0}{r}\right)^2\right),$$

$$\frac{E_0 - E}{E_0} = \frac{1}{2}\left(\frac{x_0}{r}\right)^2.$$

That is, the percentage drop of potential by the insertion of reactance in series in a non-inductive circuit is, for small values of reactance, independent of the sign, but proportional to the square of the reactance, or the same whether it be inductive reactance or condensive reactance.

56. As an example, in Fig. 54 the changes of current, I, and of e.m.f. at receiver terminals, E, at constant impressed e.m.f.,

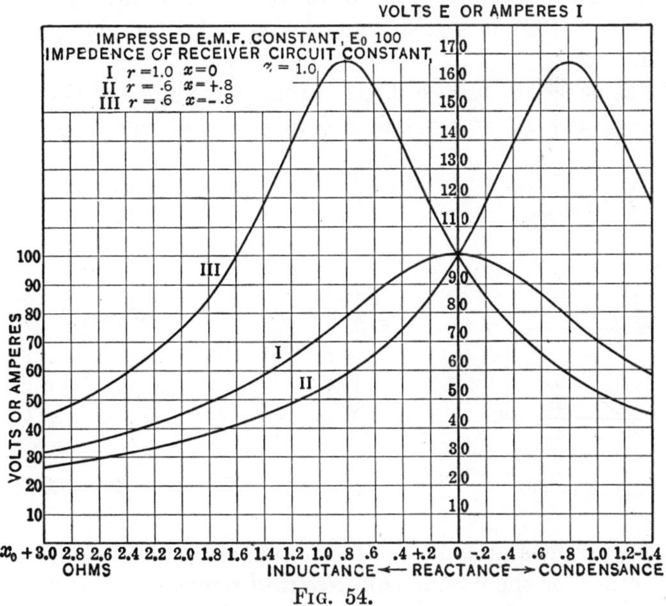

FIG. 54.

E_0, are shown for various conditions of a receiver circuit and amounts of reactance inserted in series.

Fig. 54 gives for various values of reactance, x_0 (if positive, inductive; if negative, condensive), the e.m.fs., E, at receiver terminals, for constant impressed e.m.f., $E_0 = 100$ volts, and the following conditions of receiver circuit:

$$z = 1.0,\ r = 1.0,\ x = 0 \text{ (Curve I)}$$
$$z = 1.0,\ r = 0.6,\ x = 0.8 \text{ (Curve II)}$$
$$z = 1.0,\ r = 0.6,\ x = -0.8 \text{ (Curve III)}.$$

As seen, curve I is symmetrical, and with increasing x_0 the voltage E remains first almost constant, and then drops off with increasing rapidity.

In the inductive circuit series inductive reactance, or in a condenser circuit series condensive reactance, causes the voltage to drop off very much faster than in a non-inductive circuit.

Series inductive reactance in a condenser circuit, and series condensive reactance in an inductive circuit, cause a rise of potential. This rise is a maximum for $x_0 = \pm 0.8$, or $x_0 = - x$ (the condition of resonance), and the e.m.f. reaches the value $E = 167$ volts, or $E = E_0 \frac{z}{r}$. This rise of potential by series reactance continues up to $x_0 = \pm 1.6$, or, $x_0 = - 2x$, where $E = 100$ volts again; and for $x_0 > 1.6$ the voltage drops again.

At $x_0 = \pm 0.8$, $x = \mp 0.8$, the total impedance of the circuit is $r - j\ (x + x_0) = r = 0.6$, $x + x_0 = 0$, and $\tan \theta_0 = 0$; that

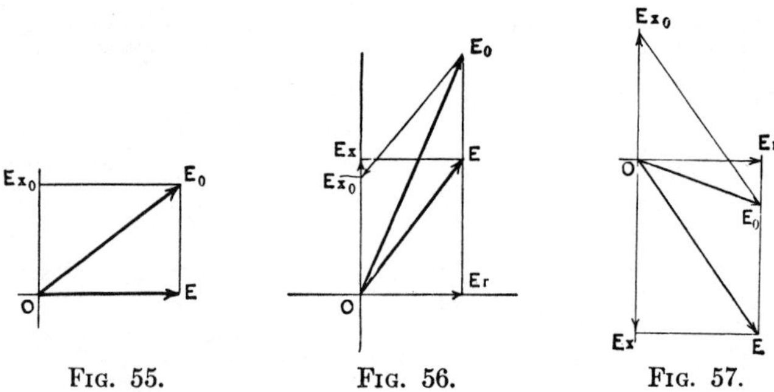

FIG. 55. FIG. 56. FIG. 57.

is, the current and e.m.f. in the supply circuit are in phase with each other, or the circuit is in *electrical resonance.*

Since a synchronous motor in the condition of efficient working acts as a condensive reactance, we get the remarkable result that, in synchronous motor circuits, choking coils, or reactive coils, can be used for raising the voltage.

In Figs. 55 to 57, the vector diagrams are shown for the conditions

$$
\begin{aligned}
E_0 = 100,\ x_0 = 0.6,\ x &= 0 \quad &\text{(Fig. 48) } E &= 85.7 \\
x &= +\,0.8 \quad &\text{(Fig. 49) } E &= 65.7 \\
x &= -\,0.8 \quad &\text{(Fig. 50) } E &= 158.1.
\end{aligned}
$$

57. In Fig. 58 the dependence of the potential, E, upon the difference of phase, θ, in the receiver circuit is shown for the constant impressed e.m.f., $E_0 = 100$; for the constant receiver impedance, $z = 1.0$ (but of various phase differences θ), and for various series reactances, as follows:

$x_0 = 0.2$ (Curve I)
$x_0 = 0.6$ (Curve II)
$x_0 = 0.8$ (Curve III)
$x_0 = 1.0$ (Curve IV)
$x_0 = 1.6$ (Curve V)
$x_0 = 3.2$ (Curve VI).

Since $z = 1.0$, the current, I, in all these diagrams has the same value as E.

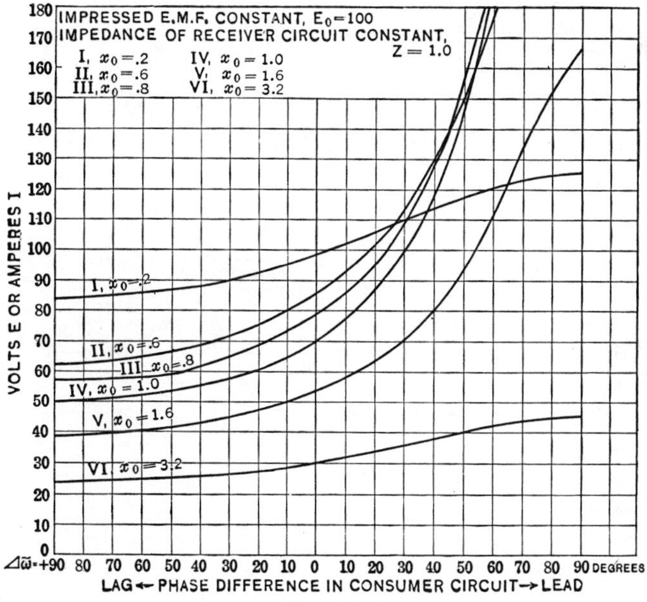

FIG. 58.

In Figs. 59 and 60, the same curves are plotted as in Fig. 58, but in Fig. 59 with the reactance, x, of the receiver circuit as abscissas; and in Fig. 60 with the resistance, r, of the receiver circuit as abscissas.

As shown, the receiver voltage, E, is always lowest when x_0 and x are of the same sign, and highest when they are of opposite sign.

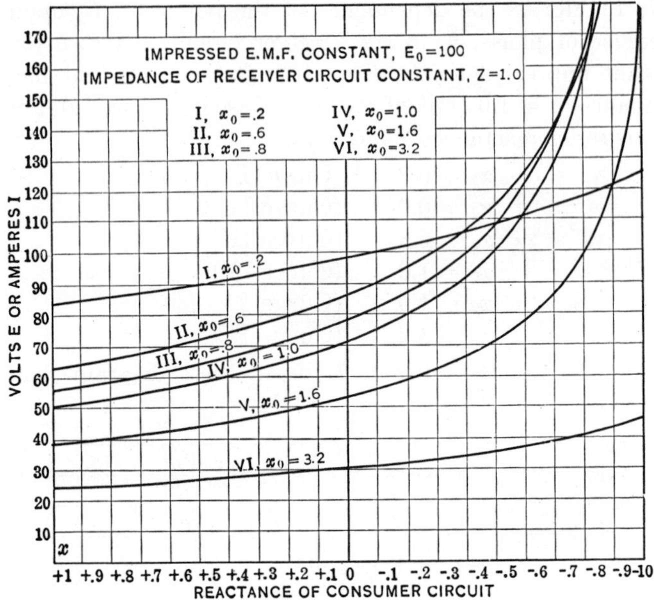
IMPRESSED E.M.F. CONSTANT, $E_0=100$
IMPEDANCE OF RECEIVER CIRCUIT CONSTANT, Z=1.0
I, $x_0=.2$
II, $x_0=.6$
III, $x_0=.8$
IV, $x_0=1.0$
V, $x_0=1.6$
VI, $x_0=3.2$
VOLTS E OR AMPERES I
REACTANCE OF CONSUMER CIRCUIT

FIG. 59.

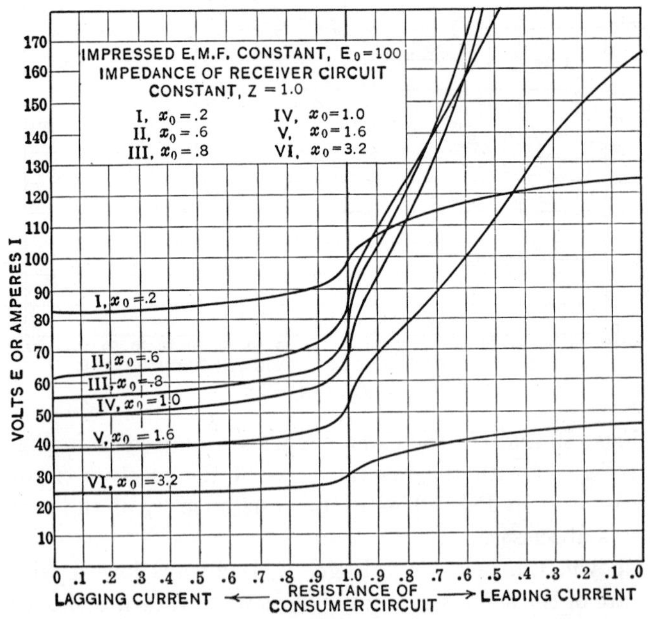
IMPRESSED E.M.F. CONSTANT, $E_0=100$
IMPEDANCE OF RECEIVER CIRCUIT
CONSTANT, Z = 1.0
I, $x_0=.2$
II, $x_0=.6$
III, $x_0=.8$
IV, $x_0=1.0$
V, $x_0=1.6$
VI, $x_0=3.2$
VOLTS E OR AMPERES I
LAGGING CURRENT
RESISTANCE OF CONSUMER CIRCUIT
LEADING CURRENT

FIG. 60.

The rise of voltage due to the balance of x_0 and x is a maximum for $x_0 = +1.0$, $x = -1.0$, and $r = 0$, where $E = \infty$; that is, absolute resonance takes place. Obviously, this condition cannot be completely reached in practice.

It is interesting to note, from Fig. 60, that the largest part of the drop of potential due to inductive reactance, and rise to condensive reactance—or conversely—takes place between $r = 1.0$ and $r = 0.9$; or, in other words, a circuit having a power-factor $\cos\theta = 0.9$ gives a drop several times larger than a non-inductive circuit, and hence must be considered as an inductive circuit.

3. Impedance in Series with a Circuit

58. By the use of reactance for controlling electric circuits, a certain amount of resistance is also introduced, due to the ohmic resistance of the conductor and the hysteretic loss, which, as will be seen hereafter, can be represented as an effective resistance.

Hence the impedance of a reactive coil (choking coil) may be written thus:

$$Z_0 = r_0 + jx_0, \qquad z_0 = \sqrt{r_0^2 + x_0^2},$$

where r_0 is in general small compared with x_0.

From this, if the impressed e.m.f. is

$$E_0 = e_0 + je'_0, \qquad E_0 = \sqrt{e_0^2 + e_0'^2},$$

and the impedance of the consumer circuit is

$$Z = r + jx, \qquad z = \sqrt{r^2 + x^2},$$

we get the current

$$I = \frac{\dot{E}_0}{Z + Z_0} = \frac{\dot{E}_0}{(r + r_0) + j(x + x_0)}$$

and the e.m.f. at receiver terminals,

$$\underset{\cdot}{E} = \underset{\cdot}{E_0} \frac{Z}{Z + Z_0} = \underset{\cdot}{E_0} \frac{r + jx}{(r + r_0) + j(x + x_0)}.$$

Or, in absolute quantities,

the current is,

$$I = \frac{E_0}{\sqrt{(r + r_0)^2 + (x + x_0)^2}} = \frac{E_0}{\sqrt{z^2 + z_0^2 + 2(rr_0 + xx_0)}},$$

the e.m.f. at receiver terminals is

$$E = \frac{E_0 z}{\sqrt{(r + r_0)^2 + (x + x_0)^2}} = \frac{E_0 z}{\sqrt{z^2 + z_0^2 + 2(rr_0 + xx_0)}};$$

the difference of phase in receiver circuit is

$$\tan\theta = \frac{x}{r};$$

and the difference of phase in the supply circuit is

$$\tan\theta = \frac{x + x_0}{r + r_0}.$$

59. In this case, the maximum drop of potential will not take place for either $x = 0$, as for resistance in series, or for $r = 0$, as for reactance in series, but at an intermediate point. The drop of voltage is a maximum; that is, E is a minimum if the denominator of E is a maximum; or, since z, z_0, r_0, x_0 are constant, if $rr_0 + xx_0$ is a maximum, that is, since $x = \sqrt{z^2 - r^2}$, if $rr_0 + x_0\sqrt{z^2 - r^2}$ is a maximum. A function, $f = rr_0 + x_0 \sqrt{z^2 - r^2}$, is a maximum when its differential coefficient equals zero. For, plotting f as curve with values of r as abscissas, at the point where f is a maximum or a minimum, this curve is for a short distance horizontal, hence the tangens-function of its tangent equals zero. The tangens-function of the tangent of a curve, however, is the ratio of the change of ordinates to the change of abscissas, or is the differential coefficient of the function represented by the curve.

Thus we have

$$f = rr_0 + x_0\sqrt{z^2 - r^2}$$

is a maximum or minimum, if

$$\frac{df}{dr} = 0$$

Differentiating, we get

$$r_0 + \frac{1}{2}\frac{x_0}{\sqrt{z^2 - r^2}}(-2r) = 0;$$

or, expanded,

$$r_0\sqrt{z^2 - r^2} - x_0 r = r_0 x - x_0 r = 0,$$

or,

$$r \div x = r_0 \div x_0.$$

That is, the drop of potential is a maximum, if the reactance factor, $\frac{x}{r}$, of the receiver circuit equals the reactance factor, $\frac{x_0}{r_0}$, of the series impedance.

60. As an example, Fig. 61 shows the e.m.f., E, at the receiver terminals, at a constant impressed e.m.f., $E_0 = 100$, a constant impedance of the receiver circuit, $z = 1.0$, and constant series impedances,

$$Z_0 = 0.3 + j\,0.4 \qquad \text{(Curve I)}$$
$$Z_0 = 1.2 + j\,1.6 \qquad \text{(Curve II)}$$

as functions of the reactance, x, of the receiver circuit.

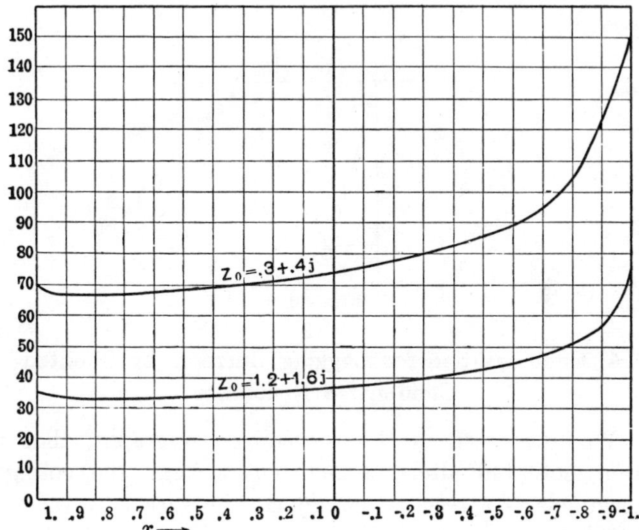

Fig. 61.

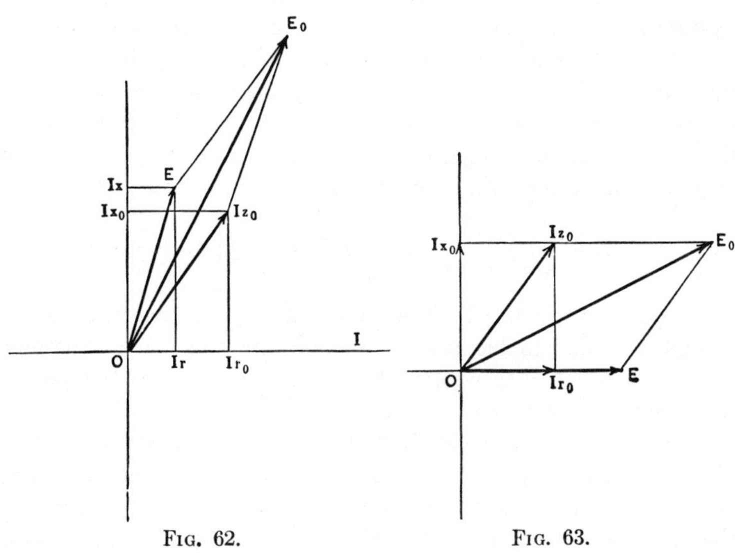

Fig. 62. Fig. 63.

Figs. 62 to 64, give the vector diagram for $E_0 = 100$, $x = 0.95$, $x = 0$, $x = -0.95$, and $Z_0 = 0.3 + 0.4j$.

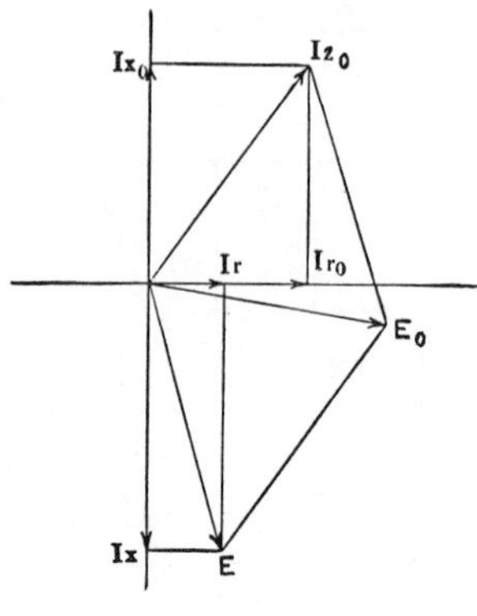

FIG. 64.

4. Compensation for Lagging Currents by Shunted Condensive Reactance

61. We have seen in the preceding paragraphs, that in a constant potential alternating-current system, the voltage at the terminals of a receiver circuit can be varied by the use of a variable reactance in series with the circuit, without loss of energy except the unavoidable loss due to the resistance and hysteresis of the reactance; and that, if the series reactance is very large compared with the resistance of the receiver circuit, the current in the receiver circuit becomes more or less independent of the resistance—that is, of the power consumed in the receiver circuit, which in this case approaches the conditions of a constant alternating-current circuit, whose current is

$$I = \frac{E_0}{\sqrt{r^2 + x_0^2}}, \text{ or, approximately, } I = \frac{E_0}{x_0}.$$

This potential control, however, causes the current taken from the mains to lag greatly behind the e.m.f., and thereby requires a much larger current than corresponds to the power consumed in the receiver circuit.

Since a condenser draws from the mains a current in leading phase, a condenser shunted across such a circuit carrying current in lagging phase compensates for the lag, the leading and the lagging current combining to form a resultant current more or less in phase with the e.m.f., and therefore proportional to the power expended.

In a circuit shown diagrammatically in Fig. 65, let the non-inductive receiver circuit of resistance, r, be connected in series with the inductive reactance, x_0, and the whole shunted by a condenser C of condensive reactance, x_c, entailing but a negligible loss of power.

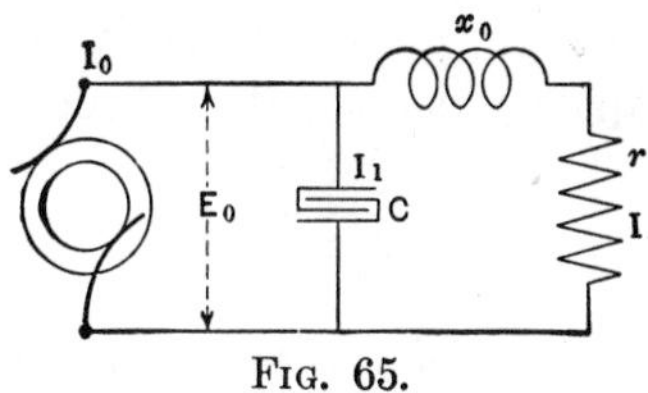

FIG. 65.

Then, if $\dot{E}_0$ = impressed e.m.f., the current in receiver circuit is

$$\dot{I} = \frac{\dot{E}_0}{r + jx_0}, \qquad I = \frac{E_0}{\sqrt{r^2 + x_0^2}};$$

the current in condenser circuit is

$$\dot{I}_1 = -\frac{\dot{E}_0}{jx_c}, \qquad I_1 = \frac{E_0}{x_c};$$

and the total current is

$$\dot{I}_0 = \dot{I} + \dot{I}_1 = \dot{E}_0 \left\{ \frac{1}{r + jx_0} - \frac{1}{jx_c} \right\}$$

$$= \dot{E}_0 \left\{ \frac{r}{r^2 + x_0^2} - j \left(\frac{x_0}{r^2 + x_0^2} - \frac{1}{x_c} \right) \right\},$$

or, in absolute terms,

$$I_0 = E_0 \sqrt{\left(\frac{r}{r^2 + x_0^2} \right)^2 + \left(\frac{x_0}{r^2 + x_0^2} - \frac{1}{x_c} \right)^2};$$

while the e.m.f. at receiver terminals is

$$\dot{E} = \dot{I}r = \dot{E}_0 \frac{r}{r + jx_0}, \qquad E = \frac{E_0 r}{\sqrt{r^2 + x_0^2}}.$$

62. The main current, $\dot{I}_0$, is in phase with the impressed e.m.f., $\dot{E}_0$, or the lagging current is completely balanced, or supplied by, the condensive reactance, if the imaginary term in the expression of $\dot{I}_0$ disappears; that is, if

$$\frac{x_0}{r^2 + x_0^2} - \frac{1}{x_c} = 0.$$

This gives, expanded,

$$x_c = \frac{r^2 + x_0^2}{x_0}.$$

Hence the capacity required to compensate for the lagging current produced by the insertion of inductive reactance in series with a non-inductive circuit depends upon the resistance and the inductive reactance of the circuit. x_0 being constant, with increasing resistance, r, the condensive reactance has to be increased, or the capacity decreased, to keep the balance.

Substituting

$$x_c = \frac{r^2 + x_0^2}{x_0},$$

we get, as the equations of the inductive circuit balanced by condensive reactance,

$$\dot{I} = \frac{\dot{E}_0}{r + jx_0} = \frac{\dot{E}_0(r - jx_0)}{r^2 + x_0^2}, \qquad I = \frac{E_0}{\sqrt{r^2 + x_0^2}};$$

$$\dot{I}_1 = \frac{j\dot{E}_0 x_0}{r^2 + x_0^2}, \qquad I_1 = \frac{E_0 x_0}{r^2 + x_0^2};$$

$$\dot{I}_0 = \frac{\dot{E}_0 r}{r^2 + x_0^2}, \qquad I_0 = \frac{E_0 r}{r^2 + x_0^2};$$

$$\dot{E} = \frac{\dot{E}_0 r}{r + jx_0}, \qquad E = \frac{E_0 r}{\sqrt{r^2 + x_0^2}};$$

and for the power expended in the receiver circuit,

$$I^2 r = \frac{E_0^2 r}{r^2 + x_0^2} = I_0 E_0;$$

that is, the main current is proportional to the expenditure of power.

For $r = 0$, we have $x_c = x_0$, as the condition of balance.

Complete balance of the lagging component of current by shunted capacity thus requires that the condensive reactance x_c be varied with the resistance, r; that is, with the varying load on the receiver circuit.

In Fig. 66 are shown, for a constant impressed e.m.f., $E_0 =$ 1000 volts, and a constant series reactance, $x_0 = 100$ ohms, values for the balanced circuit of

current in receiver circuit (Curve I),
current in condenser circuit (Curve II),
current in main circuit (Curve III),
e.m.f. at receiver terminals (Curve IV),

with the values the resistance, r, of the receiver circuit as abscissas.

63. If, however, the condensive reactance is left unchanged, $x_c = x_0$ at the no-load value, the circuit is balanced for $r = 0$, but will be overbalanced for $r > 0$, and the main current will become leading.

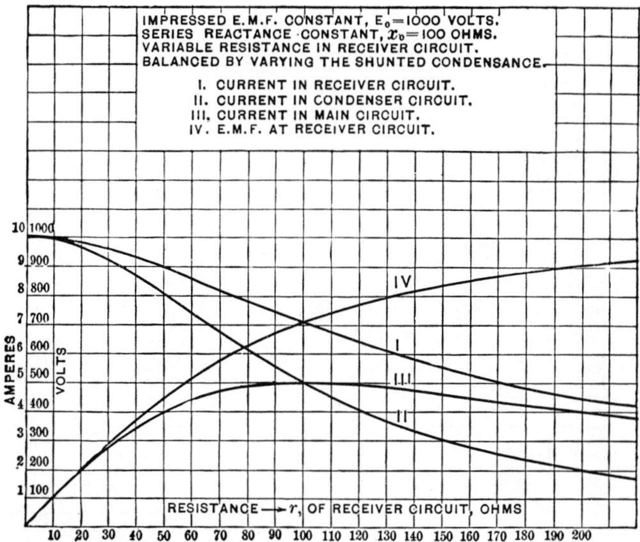

FIG. 66.—Compensation of lagging currents in receiving circuit by variable shunted condensance.

We get in this case,

$$x_c = x_0;$$

$$\dot{I} = \frac{\dot{E}_0}{r + jx_0}, \qquad I = \frac{E_0}{\sqrt{r^2 + x_0^2}};$$

$$\dot{I}_1 = \frac{j\dot{E}_0}{x_0}, \qquad I_1 = \frac{E_0}{x_0};$$

$$\dot{I}_0 = \dot{I} + \dot{I}_1 = \frac{\dot{E}_0 r}{x_0(x_0 - jr)}, \qquad I_0 = \frac{E_0 r}{x_0\sqrt{r^2 + x_0^2}};$$

$$\dot{E} = \dot{I}r = \frac{\dot{E}_0 r}{r + jx_0}, \qquad E = \frac{E_0 r}{\sqrt{r^2 + x_0^2}}.$$

The difference of phase in the main circuit is

$$\tan \theta_0 = -\frac{r}{x_0},$$

which is = 0, when $r = 0$ or at no-load, and increases with increasing resistance, as the lead of the current. At the same time, the current in the receiver circuit, I, is approximately constant for small values of r, and then gradually decreases.

In Fig. 67 are shown the values of I, I_1, I_0, E, in Curves I, II, III, IV, similarly as in Fig. 60, for $E_0 = 1000$ volts, $x_c = x_0 = 100$ ohms, and r as abscissas.

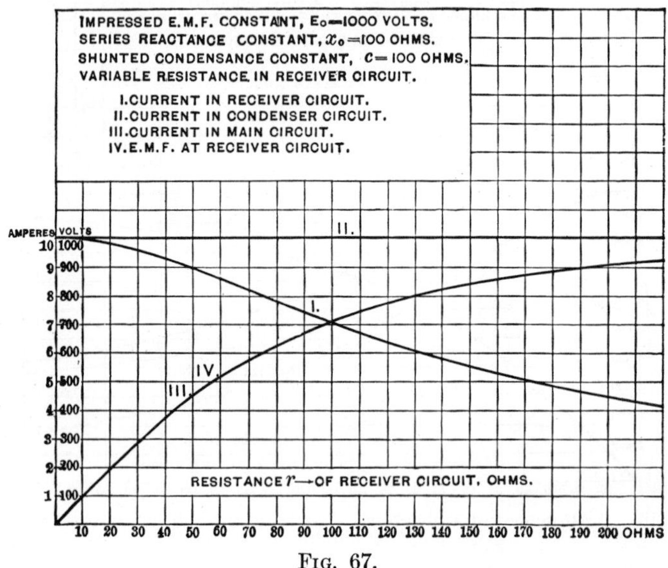

Fig. 67.

5. Constant Potential—Constant-current Transformation

64. In a constant potential circuit containing a large and constant reactance, x_0, and a varying resistance, r, the current is approximately constant, and only gradually drops off with increasing resistance, r—that is, with increasing load—but the current lags greatly behind the voltage. This lagging current in the receiver circuit can be supplied by a shunted condensance. Leaving, however, the condensance constant, $x_c = x_0$, so as to balance the lagging current at no-load, that is, at $r = 0$, it will overbalance with increasing load, that is, with increasing r, and thus the main current will become leading, while the receiver current decreases if the impressed voltage, E_0, is kept constant. Hence, to keep the current in the receiver circuit entirely constant, the impressed voltage, E_0, has to be increased with increasing resistance, r; that is, with increasing lead of the main current. Since, as explained before, in a circuit with leading current, a series inductive reactance raises the potential, to maintain the current in the receiver circuit constant under all loads, an inductive reactance, x_2, inserted in the main circuit, as shown in the diagram, Fig. 68, can be used for raising the voltage, E_0, with increasing load, and by properly choosing the inductive and the condensive reactances, practically constant current at varying load can be produced from constant voltage supply, and inversely.

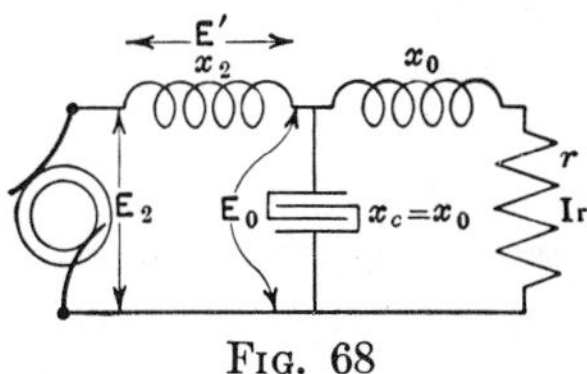

Fig. 68

The generation of alternating-current electric power almost always takes place at constant potential. For some purposes, however, as for operating series arc circuits, and to a limited extent also for electric furnaces, a constant, or approximately constant, alternating current is required.

Such constant alternating currents can be produced from constant potential circuits by means of inductive reactances, or combinations of inductive and condensive reactances; and the investigation of different methods of producing constant alternating current from constant alternating potential, or inversely, constitutes a good illustration of the application of the terms "impedance," "reactance," etc., and offers a large number of problems or examples for the application of the method of complex quantities. A number of such are given in "Theory and Calculation of Electric Circuits" [Dover Volume I, Part III].

CHAPTER X

RESISTANCE AND REACTANCE OF TRANSMISSION LINES

65. In alternating-current circuits, voltage is consumed in the feeders of distributing networks, and in the lines of long-distance transmissions, not only by the resistance, but also by the reactance, of the line. The voltage consumed by the resistance is in phase, while the voltage consumed by the reactance is in quadrature, with the current. Hence their influence upon the voltage at the receiver circuit depends upon the difference of phase between the current and the voltage in that circuit. As discussed before, the drop of potential due to the resistance is a maximum when the receiver current is in phase, a minimum when it is in quadrature, with the voltage. The change of voltage due to line reactance is small if the current is in phase with the voltage, while a drop of potential is produced with a lagging, and a rise of potential with a leading, current in the receiver circuit.

Thus the change of voltage due to a line of given resistance and reactance depends upon the phase difference in the receiver circuit, and can be varied and controlled by varying this phase difference; that is, by varying the admittance, $Y = g - jb$, of the receiver circuit.

The conductance, g, of the receiver circuit depends upon the consumption of power—that is, upon the load on the circuit—and thus cannot be varied for the purpose of regulation. Its susceptance, b, however, can be changed by shunting the circuit with a reactance, and will be increased by a shunted inductive reactance, and decreased by a shunted condensive reactance. Hence, for the purpose of investigation, the receiver circuit can be assumed to consist of two branches, a conductance, g,—the non-inductive part of the circuit—shunted by a susceptance, b, which can be varied without expenditure of energy. The two components of current can thus be considered separately, the energy component as determined by the load on the circuit, and the wattless component, which can be varied for the purpose of regulation.

Obviously, in the same way, the voltage at the receiver circuit may be considered as consisting of two components, the power component, in phase with the current, and the wattless component, in quadrature with the current. This will correspond to the case of a reactance connected in series to the non-inductive part of the circuit. Since the effect of either resolution into components is the same so far as the line is concerned, we need not make any assumption as to whether the wattless part of the receiver circuit is in shunt, or in series, to the power part.

Let

$Z_0 = r_0 + jx_0$ = impedance of the line;

$$z_0 = \sqrt{r_0^2 + x_0^2};$$

$Y = g - jb$ = admittance of receiver circuit;

$$y = \sqrt{g^2 + b^2};$$

$\dot{E}_0 = e_0 + je'_0$ = impressed voltage at generator end of line;

$$E_0 = \sqrt{e_0^2 + e_0'^2};$$

$\dot{E} = e + je'$ = voltage at receiver end of line;

$$E = \sqrt{e^2 + e'^2};$$

$\dot{I}_0 = i_0 + ji'_0$ = current in the line;

$$I_0 = \sqrt{i_0^2 + i_0'^2}.$$

The simplest condition is the non-inductive circuit.

1. Non-inductive Receiver Circuit Supplied over an Inductive Line

66. In this case, the admittance of the receiver circuit is $Y = g$, since $b = 0$.

We have then

current, $\dot{I}_0 = \dot{E}g;$

impressed voltage: $\dot{E}_0 = \dot{E} + Z_0\dot{I}_0 = \dot{E}(1 + Z_0 g).$

Hence—voltage at receiver circuit,

$$\dot{E} = \frac{\dot{E}_0}{1 + Z_0 g} = \frac{\dot{E}_0}{1 + gr_0 + jgx_0};$$

current,

$$\dot{I}_0 = \frac{\dot{E}_0 g}{1 + Z_0 g} = \frac{\dot{E}_0 g}{1 + gr_0 + jgx_0}.$$

Hence, in absolute values—voltage at receiver circuit,

$$\dot{E} = \frac{\dot{E}_0}{\sqrt{(1 + gr_0)^2 + g^2x_0^2}};$$

current,

$$I_0 = \frac{\dot{E}_0 g}{\sqrt{(1 + gr_0)^2 + g^2x_0^2}}.$$

The ratio of e.m.fs. at receiver circuit and at generator, or supply circuit, is

$$a = \frac{E}{E_0} = \frac{1}{\sqrt{(1 + gr_0)^2 + g^2x_0^2}};$$

and the power delivered in the non-inductive receiver circuit, or output,

$$P = I_0 E = \frac{E_0^2 g}{(1 + gr_0)^2 + g^2x_0^2}.$$

As a function of g, and with a given E_0, r_0, and x_0, this power is a maximum, if

$$\frac{dP}{dg} = 0;$$

that is,

$$-1 + g^2r_0^2 + g^2x_0^2 = 0;$$

hence,
conductance of receiver circuit for maximum output,

$$g_m = \frac{1}{\sqrt{r_0^2 + x_0^2}} = \frac{1}{z_0}.$$

Resistance of receiver circuit, $r_m = \dfrac{1}{g_m} = z_0$;

and, substituting this in P,

Maximum output, $P_m = \dfrac{E_0^2}{2(r_0 + z_0)} = \dfrac{E_0^2}{2\left\{r_0 + \sqrt{r_0^2 + x_0^2}\right\}}$;

and ratio of e.m.f. at receiver and at generator end of line,

$$a_m = \frac{E}{E_0} = \frac{1}{\sqrt{2\left(1 + \frac{r_0}{z_0}\right)}};$$

efficiency, $\dfrac{r_m}{r_m + r_0} = \dfrac{z_0}{r_0 + z_0}$.

That is:

The output which can be transmitted over an inductive line of resistance, r_0, and reactance, x_0—that is, of impedance, z_0—into a *non-inductive receiver circuit, is a maximum if the resistance of the receiver circuit equals the impedance of the line,* $r = z_0$, and is

$$P_m = \frac{E_0^2}{2\,(r_0 + z_0)}.$$

The output is transmitted at the efficiency of

$$\frac{z_0}{r_0 + z_0},$$

and with a ratio of e.m.fs. of

$$a_m = \frac{1}{\sqrt{2\left(1 + \frac{r_0}{z_0}\right)}}.$$

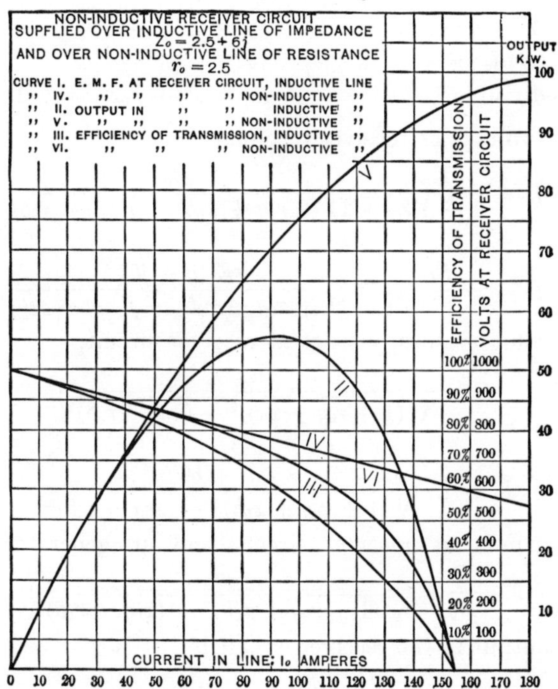

FIG. 69.—Non-inductive receiver-circuit supplied over inductive line.

67. We see from this that the maximum output which can be delivered over an inductive line is less than the output delivered over a non-inductive line of the same resistance—that is, which can be delivered by continuous currents with the same generator potential.

In Fig. 69 are shown, for the constants,

$E_0 = 1000$ volts, $Z_0 = 2.5 + 6j$; that is, $r_0 = 2.5$ ohms, $x_0 = 6$ ohms, $z_0 = 6.5$ ohms, with the current I_0 as abscissas, the values.

e.m.f. at receiver circuit, E, (Curve I);
output of transmission, P, (Curve II);
efficiency of transmission, (Curve III).

The same quantities for a non-inductive line of resistance, $r_0 = 2.5$ ohms, $x_0 = 0$, are shown in Curves IV, V, and VI.

2. Maximum Power Supplied over an Inductive Line

68. If the receiver circuit contains the susceptance, b, in addition to the conductance, g, its admittance can be written thus:

$$Y = g - jb, \qquad y = \sqrt{g^2 + b^2}.$$

Then, current, $\qquad \dot{I}_0 = \dot{E}Y;$

Impressed voltage, $\qquad \dot{E}_0 = \dot{E} + \dot{I}_0 Z_0 = \dot{E}(1 + YZ_0).$

Hence, voltage at receiver terminals,

$$\dot{E} = \frac{\dot{E}_0}{1 + YZ_0} = \frac{\dot{E}_0}{(1 + r_0 g + x_0 b) + j(x_0 g - r_0 b)};$$

current,

$$\dot{I}_0 = \frac{\dot{E}_0 Y}{1 + YZ_0} = \frac{\dot{E}_0(g - jb)}{(1 + r_0 g + x_0 b) + j(x_0 g - r_0 b)};$$

or, in absolute values, voltage at receiver circuit,

$$E = \frac{E_0}{\sqrt{(1 + r_0 g + x_0 b)^2 + (x_0 g - r_0 b)^2}};$$

current,

$$I_0 = E_0 \sqrt{\frac{g^2 + b^2}{(1 + r_0 g + x_0 b)^2 + (x_0 g - r_0 b)^2}};$$

ratio of e.m.fs. at receiver circuit and at generator circuit,

$$a = \frac{E}{E_0} = \frac{1}{\sqrt{(1 + r_0 g + x_0 b)^2 + (x_0 g - r_0 b)^2}};$$

and the output in the receiver circuit is

$$P = E^2 g = E_0{}^2 a^2 g.$$

69. (*a*) *Dependence of the output upon the susceptance of the receiver circuit.*

At a given conductance, g, of the receiver circuit, its output, $P = E_0{}^2 a^2 g$, is a maximum if a^2 is a maximum; that is, when

$$f = \frac{1}{a^2} = (1 + r_0 g + x_0 b)^2 + (x_0 g - r_0 b)^2$$

is a minimum.

The condition necessary is

$$\frac{df}{db} = 0,$$

or, expanding,

$$x_0(1 + r_0 g + x_0 b) - r_0(x_0 g - r_0 b) = 0.$$

Hence

Susceptance of receiver circuit,

$$b = -\frac{x_0^2}{r_0^2 + x_0^2} = -\frac{x_0}{z_0^2} = -b_0;$$

or

$$b + b_0 = 0,$$

that is, if the sum of the susceptances of line and of receiver circuit equals zero.

Substituting this value, we get

ratio of e.m.fs. at maximum output,

$$a_1 = \frac{E}{E_0} = \frac{1}{z_0(g + g_0)};$$

maximum output,

$$P_1 = \frac{E_0^2 g}{z_0^2(g + g_0)^2};$$

current,

$$\dot{I}_0 = \frac{\dot{E}_0 Y}{1 + Z_0 Y} = \frac{\dot{E}_0(g + jb_0)}{1 + (r_0 + jx_0)(g + jb_0)}$$

$$= \frac{\dot{E}_0(g + jb_0)}{(1 + r_0 g - x_0 b_0) - j(r_0 b_0 + x_0 g)};$$

$$I_0 = E_0\sqrt{\frac{g^2 + b_0^2}{(1 + r_0 g - x_0 b_0)^2 + (r_0 b_0 + x_0 g)^2}};$$

and, since,

$$b_0 = \frac{x_0}{r_0^2 + x_0^2}, \qquad g_0 = \frac{r_0}{r_0^2 + x_0^2},$$

it is,

$$(1 + r_0 g - x_0 b_0)^2 + (r_0 b_0 + x_0 g)^2 = \left(r_0 g + 1 - \frac{x_0^2}{r_0^2 + b^2}\right)^2$$

$$+ \left(\frac{r_0 x_0}{r_0^2 + x_0^2} + x_0 g\right)^2$$

$$= \left(r_0 g + \frac{r_0^2}{r_0^2 + x_0^2}\right)^2 + \left(x_0 g + \frac{r_0 x_0}{r_0^2 + x_0^2}\right)^2$$

$$= r_0^2(g + g_0)^2 + x_0^2(g + g_0)^2$$

$$= z_0^2(g + g_0)^2,$$

Thus, it is, current,

$$I_0 = \frac{E_0\sqrt{g^2 + b_0^2}}{z_0(g + g_0)};$$

phase difference in receiver circuit,

$$\tan \theta = \frac{b}{g} = -\frac{b_0}{g};$$

phase difference in generator circuit,

$$\tan \theta_0 = \frac{x + x_0}{r + r_0} = \frac{b_0(y^2 - y_0^2)}{g_0 y^2 + g y_0^2}.$$

70. (*b*) *Dependence of the output upon the conductance of the receiver circuit.*

At a given susceptance, b, of the receiver circuit, its output, $P = E_0^2 \alpha^2 g$, is a maximum if

$$\frac{dP}{dg} = 0, \quad \text{or} \quad \frac{d}{dg}\left(\frac{1}{P}\right) = 0,$$

or

$$\frac{d}{dg}\left(\frac{1}{\alpha^2 g}\right) = \frac{d}{dg}\left(\frac{(1 + r_0 g + x_0 b)^2 + (x_0 g - r_0 b)^2}{g}\right) = 0;$$

that is, expanding,

$$(1 + r_0 g + x_0 b)^2 + (x_0 g - r_0 b)^2 - 2\, g(r_0 + r_0^2 g + x_0^2 g) = 0;$$

or, expanding,

$$(b + b_0)^2 = g^2 - g_0^2; \quad g = \sqrt{g_0^2 + (b + b_0)^2}.$$

Substituting this value in the equation for α, §68, we get—ratio of e.m.fs.,

$$\alpha_2 = \frac{1}{z_0 \sqrt{2\{g_0^2 + (b + b_0)^2 + g_0 \sqrt{g_0^2 + (b + b_0)^2}\}}}$$

$$= \frac{1}{z_0 \sqrt{2\, g(g + g_0)}} = \frac{y_0}{\sqrt{2\, g(g + g_0)}};$$

power,

$$P_2 = \frac{E_0^2 y_0^2}{2\,(g + g_0)} = \frac{E_0^2 y_0^2}{2^{?}\{g_0 + \sqrt{g_0^2 + (b + b_0)^2}\}}$$

$$= \frac{E_0^2}{2\left\{r_0 + \sqrt{r_0^2 + \left(x_0 + x\frac{z_0^2}{z^2}\right)^2}\right\}}.$$

As a function of the susceptance, b, this power becomes a maximum for $\frac{dP_2}{db} = 0$, that is, according to §69 if

$$b = -b_0.$$

Substituting this value, we get

$$b = -b_0,\ g = g_0,\ y = y_0, \text{ hence: } Y = g - jb = g_0 + jb_0;$$

$$x = -x_0,\ r = r_0,\ z = z_0, \qquad Z = r + jx = r_0 - jx_0;$$

substituting this value, we get—

ratio of e.m.fs., $\alpha_m = \frac{y_0}{2\, g_0} = \frac{z_0}{2\, r_0};$

power, $P_m = \frac{E_0^2}{4\, r_0};$

that is, the same as with a continuous-current circuit; or, in other words, the inductive reactance of the line and of the receiver circuit can be perfectly balanced in its effect upon the output.

71. As a summary, we thus have:

The output delivered over an inductive line of impedance, $Z_0 = r_0 + jx_0$, into a non-inductive receiver circuit, is a maximum for the resistance, $r = z_0$, or conductance, $g = y_0$, of the receiver circuit, and this maximum is

$$P = \frac{E_0^2}{2\,(r_0 + z_0)},$$

at the ratio of voltages,

$$\alpha = \frac{1}{\sqrt{2\left(1 + \frac{r_0}{z_0}\right)}}.$$

With a receiver circuit of constant susceptance, b, the output, as a function of the conductance, g, is a maximum for the conductance,

$$g = \sqrt{g_0^2 + (b + b_0)^2},$$

and is

$$P = \frac{E_0^2 y_0^2}{2\,(g + g_0)},$$

at the ratio of voltages,

$$\alpha = \frac{y_0}{\sqrt{2\, g\,(g + g_0)}}.$$

With a receiver circuit of constant conductance, g, the output, as a function of the susceptance, b, is a maximum for the susceptance $b = -\,b_0$, and is

$$P = \frac{E_0^2 g}{z_0^2\,(g + g_0)^2},$$

at the ratio of voltages,

$$\alpha = \frac{1}{z_0\,(g + g_0)}.$$

The maximum output which can be delivered over an inductive line, as a function of the admittance or impedance of the receiver circuit, takes place when $Z = r_0 - jx_0$, or $Y = g_0 + jb_0$; that is, when the resistance or conductance of receiver circuit and line are equal, the reactance or susceptance of the receiver circuit and line are equal but of opposite sign, and is $P = \frac{E_0^2}{4 r_0}$, or independent of the reactances, but equal to the output of a

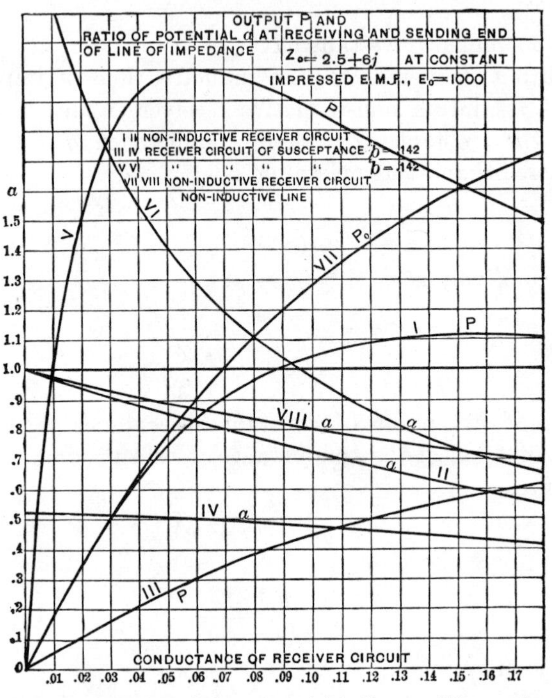

FIG. 70.—Variation of the potential in line at different loads.

continuous-current circuit of equal line resistance. The ratio of voltages is, in this case, $a = \frac{z_0}{2 r_0}$, while in a continuous-current circuit it is equal to 0.5. The efficiency is equal to 50 per cent.

72. As an example, in Fig. 70 are shown for the constants

$E_0 = 1000$ volts, and $\quad Z_0 = 2.5 + 6j$; that is, for

$r_0 = 2.5$ ohms, $\quad x_0 = 6$ ohms, $\quad z_0 = 6.5$ ohms, and with the variable conductances as abscissas, the values of the

output, in Curve I, Curve III, and Curve V;
ratio of voltages, in Curve II, Curve IV, and Curve VI;
Curves I and II refer to a non-inductive receiver circuit;
Curves III and IV refer to a receiver circuit of constant susceptance . . . . $b = 0.142$

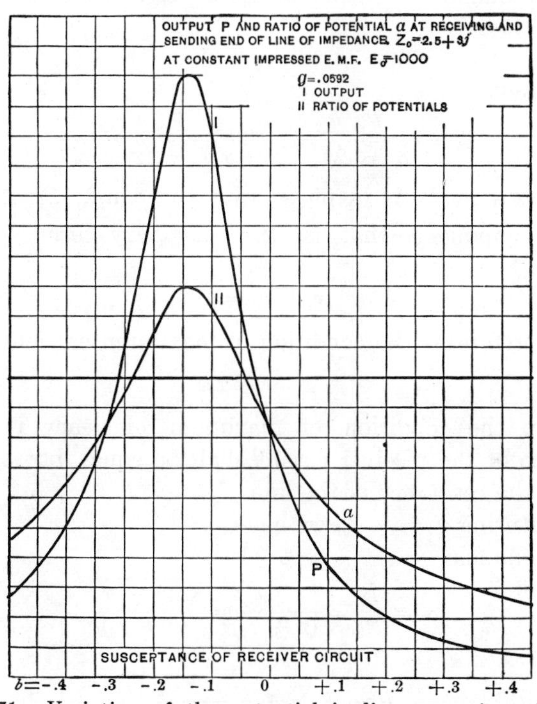

FIG. 71.—Variation of the potential in line at various loads.

Curves V and VI refer to a receiver circuit of constant susceptance . . . . $b = -0.142$
Curves VII and VIII refer to a non-inductive receiver circuit and non-inductive line.

In Fig. 71 the output is shown as Curve I, and the ratio of voltages as Curve II, for the same line constants, for a constant conductance, $g = 0.0592$ ohm, and for variable susceptances, b, of the receiver circuit.

3. Maximum Efficiency

73. The output for a given conductance, g, of a receiver circuit is a maximum if $b = -b_0$. This, however, is generally not the condition of maximum efficiency.

The loss of power in the line is constant if the current is constant; the output of the generator for a given current and given generator voltage is a maximum if the current is in phase with the voltage at the generator terminals. Hence the condition of maximum output at given loss, or of maximum efficiency is

$$\tan \theta_0 = 0.$$

The current is

$$\dot{I}_0 = \frac{\dot{E}_0}{Z + Z_0} = \frac{\dot{E}_0}{(r + r_0) + j(x + x_0)};$$

The current, $\dot{I}_0$, is in phase with the e.m.f., $\dot{E}_0$, if its quadrature component—that is, the imaginary term—disappears, or

$$x + x_0 = 0.$$

This, therefore, is the condition of maximum efficiency,

$$x = -x_0.$$

Hence, the condition of maximum efficiency is that the reactance of the receiver circuit shall be equal, but of opposite sign, to the reactance of the line.

Substituting $x = -x_0$, we have:
ratio of e.m.fs.,

$$a = \frac{E}{E_0} = \frac{z}{(r + r_0)} = \frac{\sqrt{r^2 + x_0^2}}{(r + r_0)};$$

power,

$$P = E_0^2 g a^2 = \frac{E_0^2 r}{(r + r_0)^2},$$

and depending upon the resistance only, and not upon the reactance.

This power is a maximum if $g = g_0$, as shown before; hence, substituting $g = g_0$, $r = r_0$,

maximum power at maximum efficiency, $P_m = \frac{E_0^2}{4 r_0}$,

at a ratio of potentials, $a_m = \frac{z_0}{2 r_0}$,

or the same result as in §70.

In Fig. 72 are shown, for the constants,

$E_0 = 100$ volts,

$Z_0 = 2.5 + 6j$; $r_0 = 2.5$ ohms, $x_0 = 6$ ohms, $z_0 = 6.5$ ohms,

and with the variable conductances, g, of the receiver circuit as abscissas, the

Output at maximum efficiency, (Curve I);

Volts at receiving end of line, (Curve II);

Efficiency $= \dfrac{r}{r + r_0}$, (Curve III).

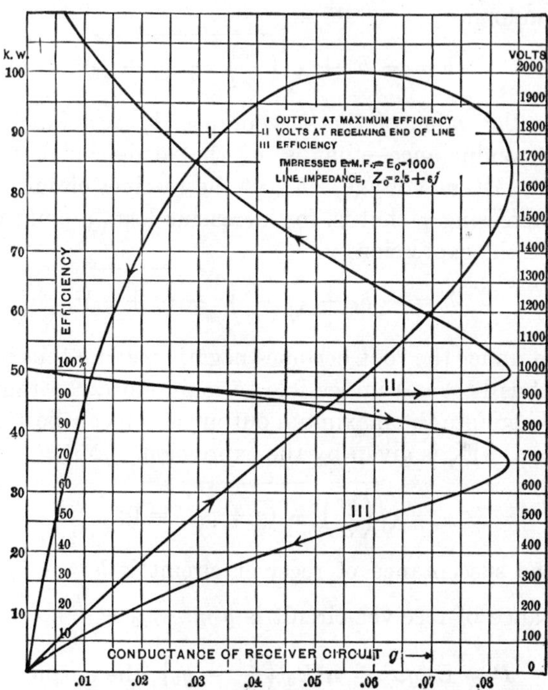

Fig. 72.—Load characteristics of transmission lines.

4. Control of Receiver Voltage by Shunted Susceptance

74. By varying the susceptance of the receiver circuit, the voltage at the receiver terminals is varied greatly. Therefore, since the susceptance of the receiver circuit can be varied at will, it is possible, at a constant generator voltage, to adjust the receiver susceptance so as to keep the voltage constant at the receiver end of the line, or to vary it in any desired manner, and independently of the generator voltage, within certain limits.

The ratio of voltages is

$$\alpha = \frac{E}{E_0} = \frac{1}{\sqrt{(1 + r_0 g + x_0 b)^2 + (x_0 g - r_0 b)^2}}.$$

If at constant generator voltage E_0 the receiver voltage E shall be constant,

$$\alpha = \text{constant};$$

hence,

$$(1 + r_0 g + x_0 b)^2 + (x_0 g - r_0 b)^2 = \frac{1}{\alpha^2};$$

or, expanding,

$$b = - b_0 + \sqrt{\left(\frac{y_0}{\alpha}\right)^2 - (g + g_0)^2},$$

which is the value of the susceptance, b, as a function of the receiver conductance—that is, of the load—which is required to yield constant voltage, αE_0, at the receiver circuit.

For increasing g, that is, for increasing load, a point is reached where, in the expression

$$b = - b_0 + \sqrt{\left(\frac{y_0}{\alpha}\right)^2 - (g + g_0)^2},$$

the term under the root becomes negative, and b thus imaginary, and it thus becomes impossible to maintain a constant voltage, αE_0. Therefore the maximum output which can be transmitted at voltage, αE_0, is given by the expression

$$\sqrt{\left(\frac{y_0}{\alpha}\right)^2 - (g + g_0)^2} = 0;$$

hence the susceptance of receiver circuit is $b = - b_0$, and the conductance of receiver circuit is $g = - g_0 + \frac{y_0}{\alpha}$,

$$P = E_0^2 g \alpha^2 = \alpha^2 E_0^2 \left(\frac{y_0}{\alpha} - g_0\right), \text{ the output.}$$

75. If $\alpha = 1$, that is, if the voltage at the receiver circuit equals the generator voltage,

$$g = y_0 - g_0; \; P = E_0^2(y_0 - g_0).$$

If $\alpha = 1$, when $g = 0$, $b = 0$

when $g > 0$, $b < 0$;

if $\alpha > 1$, when $g = 0$, or $g > 0$, $b < 0$,

that is, condensive reactance;

if $\alpha < 1$, when $g = 0$, $b > 0$,

$$\text{when } g = -g_0 + \sqrt{\left(\frac{y_0}{\alpha}\right)^2 - b_0^2}, \quad b = 0;$$

$$\text{when } g > -g_0 + \sqrt{\left(\frac{y_0}{\alpha}\right)^2 - b_0^2}, \quad b < 0,$$

or, in other words, if $\alpha < 1$, the phase difference in the main line must change from lag to lead with increasing load.

76. The value of α giving the maximum possible output in a receiver circuit is determined by $\frac{dP}{d\alpha} = 0$;

expanding
$$2\,\alpha\left(\frac{y_0}{\alpha} - g_0\right) - \frac{\alpha^2 y^0}{\alpha^2} = 0;$$

hence
$$y_0 = 2\,\alpha g_0,$$

and
$$\alpha = \frac{y_0}{2\,g_0} = \frac{1}{2\sqrt{g_0 r_0}} = \frac{z_0}{2\,r_0};$$

the maximum output is determined by

$$g = -g_0 + \frac{y_0}{\alpha} = g_0;$$

and is,
$$P = \frac{E_0^2}{4\,r_0}.$$

From
$$\alpha = \frac{y_0}{2\,g_0} = \frac{z_0}{2\,r_0},$$

the line reactance, x_0, can be found, which delivers a maximum output into the receiver circuit at the ratio of voltages, α, as

$$z_0 = 2\,r_0\alpha,$$
$$x_0 = r_0\sqrt{4\,\alpha^2 - 1};$$

for $\alpha = 1$,

$$z_0 = 2\,r_0;$$
$$x_0 = r_0\sqrt{3}.$$

If, therefore, the line impedance equals $2\,\alpha$ times the line resistance, the maximum output, $P = \frac{E_0^2}{4\,r_0}$, is transmitted into the receiver circuit at the ratio of voltages, α.

If $z_0 = 2\,r_0$, or $x_0 = r_0\sqrt{3}$, the maximum output, $P = \frac{E_0^2}{4\,r_0}$, can be supplied to the receiver circuit, without change of voltage at the receiver terminals.

Obviously, in an analogous manner, the law of variation of the susceptance of the receiver circuit can be found which is required to increase the receiver voltage proportionally to the load; or, still more generally, to cause any desired variation of the voltage at the receiver circuit independently of any variation of the generator voltage, as, for instance, to keep the voltage of a receiver circuit constant, even if the generator voltage fluctuates widely.

77. In Figs. 73, 74, and 75 are shown, with the output, $P = E_0^2 g \alpha^2$, as abscissas, and a constant impressed voltage,

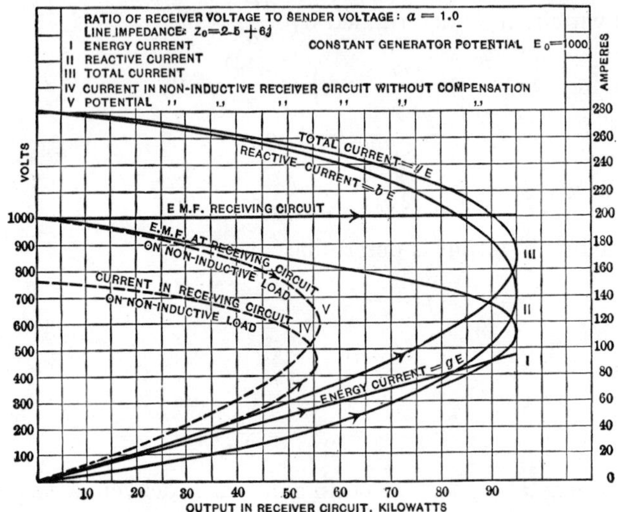

FIG. 73.—Variation of voltage of transmission lines.

$E_0 = 1000$ volts, and a constant line impedance, $Z_0 = 2.5 + 6j$, or $r_0 = 2.5$ ohms, $x_0 = 6$ ohms, $z = 6.5$ ohms, the following values:

power component of current, gE, (Curve I);
reactive, or wattless component of current, bE, (Curve II);
total current, yE, (Curve III),
and power factor at generator for the following conditions:

$$\alpha = 1.0 \text{ (Fig. 73)}; \quad \alpha = 0.7 \text{ (Fig. 74)}; \quad \alpha = 1.3 \text{ (Fig. 75)}.$$

For the non-inductive receiver circuit (in dotted lines), the curve of e.m.f., E, and of the current, $I = gE$, are added in the three diagrams for comparison, as Curves IV and V.

As shown, the output can be increased greatly, and the voltage at the same time maintained constant, by the judicious

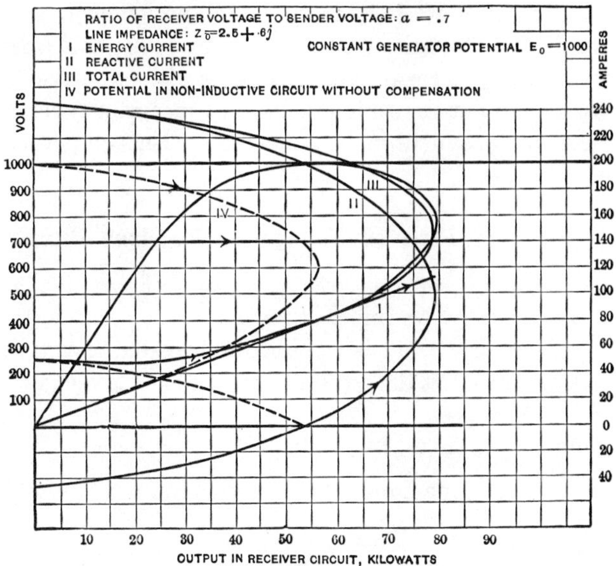

FIG. 74.—Variation of voltage of transmission lines.

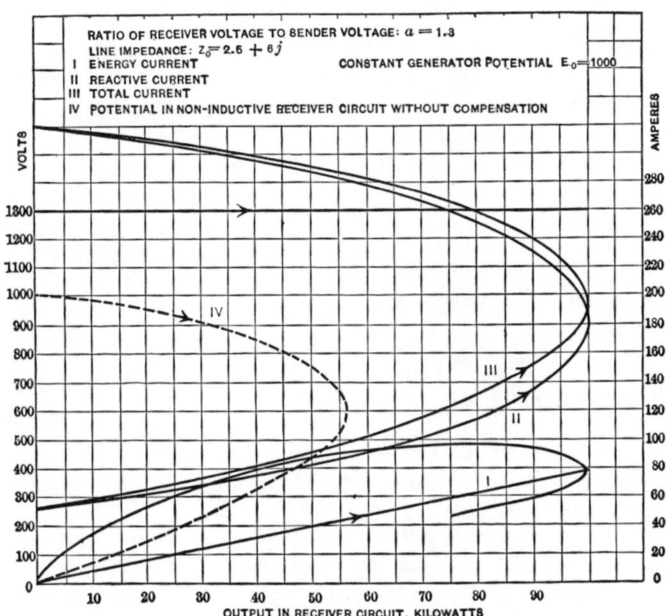

FIG. 75.—Variation of voltage of transmission lines.

use of shunted reactance, so that a much larger output can be transmitted over the line with no drop, or even with a rise, of voltage. Shunted susceptance, therefore, is extensively used for voltage control of transmission lines, by means of synchronous condensers, or by synchronous converters with compound field winding.

5. Maximum Rise of Voltage at Receiver Circuit

78. Since, under certain circumstances, the voltage at the receiver circuit may be higher than at the generator, it is of interest to determine what is the maximum value of voltage, E, that can be produced at the receiver circuit with a given generator voltage, E_0.

The condition is that

$$\alpha = \text{maximum or } \frac{1}{\alpha^2} = \text{minimum};$$

that is,

$$\frac{d\left(\frac{1}{\alpha^2}\right)}{dg} = 0, \quad \frac{d\left(\frac{1}{\alpha^2}\right)}{db} = 0;$$

substituting,

$$\frac{1}{\alpha^2} = (1 + r_0 g + x_0 b)^2 + (x_0 g - r_0 b)^2,$$

and expanding, we get,

$$\frac{d\left(\frac{1}{\alpha^2}\right)}{dg} = 0; \quad g = -\frac{r_0}{z_0^2}$$

—a value which is impossible, since neither r_0 nor g can be negative. The next possible value is $g = 0$—a wattless circuit.

Substituting this value, we get,

$$\frac{1}{\alpha^2} = (1 + x_0 b)^2 + r_0^2 b^2;$$

and by substituting, in

$$\frac{d\left(\frac{1}{\alpha^2}\right)}{db} = 0, \quad b = -\frac{x_0}{z_0^2} = -b_0,$$

$$b + b_0 = 0;$$

that is, the sum of the susceptances = 0, or the inductive susceptance of the line is balanced by the capacity susceptance of the load.

Substituting

$$b = -b_0,$$

we have

$$\alpha = \frac{1}{\sqrt{r_0 g_0}} = \frac{z_0}{r_0} = \frac{y_0}{g_0}.$$

The current in this case is

$$I = \alpha b E_0 = \frac{x_0 E_0}{z_0 r_0},$$

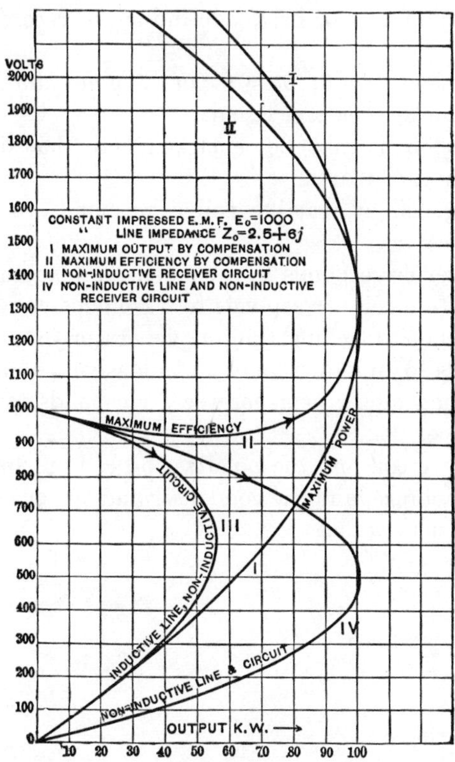

FIG. 76.—Efficiency and output of transmission lines.

or somewhat less than the current at complete resonance, that is, when the line inductive reactance, x_0, is balanced by the capacity reactance, x, of the load, $x = -x_0$; in which latter case the current is assuming wattless receiver circuit, and is in phase with the voltage, E_0.

79. As summary to this chapter, in Fig. 76 are plotted, for a constant generator e.m.f., $E_0 = 1000$ volts, and a line impedance, $Z_0 = 2.5 + 6j$, or $r_0 = 2.5$ ohms, $x_0 = 6$ ohms, $z_0 = 6.5$ ohms, and with the receiver output as abscissas and the receiver voltages as ordinates, curves representing

the condition of maximum output, (Curve I);

the condition of maximum efficiency, (Curve II);

the condition $b = 0$, or a non-inductive receiver circuit, (Curve III);

the condition $b = 0$, $b_0 = 0$, or a non-inductive line and non-inductive receiver circuit.

In conclusion, it may be remarked here that of the sources of susceptance, or reactance,

a choking coil or reactive coil corresponds to an inductive reactance;

a condenser corresponds to a condensive reactance;

a polarization cell corresponds to a condensive reactance;

a synchronous machine (motor, generator or converter) corresponds to an inductive or a condensive reactance at will;

an induction motor or generator corresponds to an inductive reactance.

The reactive coil and the polarization cell are specially suited for series reactance, and the condenser and synchronous machine for shunted susceptance.

CHAPTER XI

PHASE CONTROL

80. At constant voltage, e_0, impressed upon a circuit, as a transmission line, resistance, r, inserted in series with the receiving circuit, causes the voltage, e, at the receiver circuit to decrease with increasing current, I, through the resistance. The decrease of the voltage, e, is greatest if the current, I, is in phase with the voltage, e—less if the current is not in phase. Inductive reactance in series with the receiving circuit, e, at constant impressed e.m.f., e_0, causes the voltage, e, to drop less with a unity power-factor current, I, but far more with a lagging current, and causes the voltage, e, to rise with a leading current.

While series resistance always causes a drop of voltage, series inductive reactance, x, may cause a drop of voltage or a rise of voltage, depending on whether the current is lagging or leading. If the supply line contains resistance, r, as well as reactance, x, and the phase of the current, I, can be varied at will, by producing in the receiver circuit lagging or leading currents, the change of voltage, e, with a change of load in the circuit can be controlled. For instance, by changing the current from lagging at no-load to lead at heavy load the reactance, x, can be made to lower the voltage at light load and raise it at overload, and so make up for the increasing drop of voltage with increasing load, caused by the resistance, r, that is, to maintain constant voltage, or even a voltage, e, which rises with the load on the receiving circuit, at constant voltage, e_0, at the generator side of the line. Or the wattless component of the current can be varied so as to maintain unity power-factor at the generator end of the line, e_0, etc.

This method of controlling a circuit supplied over an inductive line, by varying the phase relation of the current in the circuit, has been called "phase control," and is used to a great extent, especially in the transmission of three-phase power for conversion to direct current by synchronous converters for railroading, and in the voltage control at the receiving end of very long high voltage transmission lines.

It requires a receiving circuit in which, independent of the load, a lagging or leading component of current can be produced at will. Such is the case in synchronous motors or converters: in a synchronous motor a lagging current can be produced by decreasing, a leading current by increasing, the field excitation.

81. If in a direct-current motor, at constant impressed voltage, the field excitation and therefore the field magnetism is decreased, the motor speed increases, as the armature has to revolve faster to consume the impressed e.m.f., and if the field excitation is increased, the motor slows down. A synchronous motor, however, cannot vary in speed, since it must keep in step with the impressed frequency, and if, therefore, at constant impressed voltage the field excitation is decreased below that which gives a field magnetism, that at the synchronous speed consumes the impressed voltage, the field magnetism still must remain the same, and the armature current thus changes in phase in such a manner as to magnetize the field and make up for the deficiency in the field excitation. That is, the armature current becomes lagging. Inversely, if the field excitation of the synchronous motor is increased, the magnetic flux still must remain the same as to correspond to the impressed voltage at synchronous speed, and the armature current so becomes demagnetizing—that is, leading.

By varying the field excitation of a synchronous motor or converter, quadrature components of current can be produced at will, proportional to the variation of the field excitation from the value that gives a magnetic flux, which at synchronous speed just consumes the impressed voltage (after allowing for the impedance of the motor).

Phase control of transmission lines is especially suited for circuits supplying synchronous motors or converters; since such machines, in addition to their mechanical or electrical load, can with a moderate increase of capacity carry or produce considerable values of wattless current. For instance, a quadrature component of current equal to 50 per cent. of the power component of current consumed by a synchronous motor would increase the total current only to $\sqrt{1 + 0.5^2} = 1.118$, or 11.8 per cent., while a quadrature component of current equal to 30 per cent. of the power component of the current would give an increase of 4.4 per cent. only, that is, could be carried by the motor armature without any appreciable increase of the motor heating.

Phase control depends upon the inductive reactance of the line or circuit between generating and receiving voltage, e_0 and e, and where the inductive reactance of the transmission line is not sufficient, additional reactance may be inserted in the form of reactive coils or high internal reactance transformers. This is usually the case in railway transmissions to synchronous converters. Phase control is extensively used for voltage control in railway power transmission by compounded synchronous converters. It is also used to a considerable extent in very long distance transmission, for controlling the voltage and the power-factor; in a distribution system for controlling the power-factor of the system.

While, therefore, the resistance, r, of the line is fixed, as it would not be economical to increase it, the reactance, x, can be increased beyond that given by line and transformer, by the insertion of reactive coils, and therefore can be adjusted so as to give best results in phase control, which are usually obtained when the quadrature component of the current is a minimum.

82. Let, then,

e = voltage at receiving circuit, chosen as zero vector.

$I = i - ji'$ = current in receiving circuit, comprising a power component, i, which depends upon the load in the receiving circuit, and a quadrature component, i', which can be varied to suit the requirements of regulation, and is considered positive when lagging, negative when leading.

$\dot{E}_0 = e'_0 - je_0''$ = voltage impressed upon the system at the generator end, or supply voltage, and the absolute value is

$$e_0 = \sqrt{e'_0{}^2 + e''_0{}^2}.$$

$Z = r + jx$ = impedance of the circuit between voltage e and voltage e_0, and the absolute value is $z = \sqrt{r^2 + x^2}$.

If e = terminal voltage of receiving station, e_0 = terminal voltage of generating station, Z = impedance of transmission line; if e = nominal induced e.m.f. of receiving synchronous machine, that is, voltage corresponding to its field excitation, and e_0 = nominal induced e.m.f. of generator; Z also includes the synchronous impedance of both machines, and of step-up and step-down transformers, where used,

It is

$$\dot{E}_0 = e + Z\dot{I},$$

or,

$$E_0 = (e + ri + xi') - j(ri' - xi), \tag{1}$$

and in absolute value we have

$$e_0^2 = (e + ri + xi')^2 + (ri' - xi)^2. \tag{2}$$

This is the fundamental equation of phase control, giving the relation of the two voltages, e and e_0, with the two components of current, i and i', and the circuit constants r and x.

From equation (2), follows:

$$e = \sqrt{e_0^2 - (ri' - xi)^2} - (ri + xi'), \tag{3}$$

expressing the receiver voltage, e, as a function of e_0 and $\dot{I}$.

And:

$$i' = \pm\sqrt{\frac{e_0^2}{z^2} - \left(\frac{er}{z^2} + i\right)^2} - \frac{ex}{z^2}. \tag{4}$$

Denoting

$$\tan\theta = \frac{x}{r}, \tag{5}$$

where θ is the phase angle of the line impedance, we have

$$r = z\cos\theta \text{ and } x = z\sin\theta \tag{6}$$

and

$$i' = \pm\sqrt{\frac{e_0^2}{z^2} - \left(\frac{e\cos\theta}{z} + i\right)^2} - \frac{e\sin\theta}{z}, \tag{7}$$

gives the reactive component of the current, i', required by the power component of the current, i, at the voltages, e and e_0.

83. The phase angle of the impressed e.m.f., $\dot{E}_0$, is, from (1),

$$\tan\theta_0 = \frac{ri' - xi}{e + ri + xi'}, \tag{8}$$

the phase angle of the current

$$\tan\theta_1 = \frac{i'}{i}, \tag{9}$$

hence, to bring the current, I, into phase with the impressed e.m.f., $\dot{E}_0$, or produce unity power-factor at the generator terminal, e_0, it must be

$$\theta_0 = \theta_1;$$

hence,

$$\frac{ri' - xi'}{e + ri + xi'} = \frac{i'}{i},$$

and herefrom follows:

$$i' = \frac{\pm \sqrt{e^2 - 4\,x^2 i^2} - e}{2\,x}, \tag{10}$$

hence always negative, or leading, but $i' = 0$ for $i = 0$, or at no-load.

From equation (10) follows that i' becomes imaginary, if the term under the square root, $(e^2 - 4\,x^2 i^2)$, becomes negative, that is, if

$$i > \frac{e}{2\,x},$$

that is, the maximum load, or power component of current, at which unity power-factor can still be maintained at the supply voltage, e_0, is given by

$$i_m = \frac{e}{2\,x}, \tag{11}$$

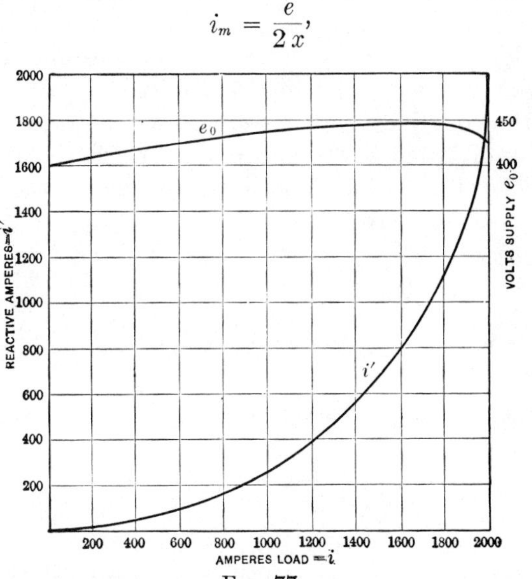

FIG. 77.

and the leading quadrature component of current required to compensate for the line reactance x at maximum current, i_m, is from equation (10),

$$i_m' = \frac{e}{2\,x}, \tag{12}$$

that is, in this case of the maximum load which can be delivered at e, with unity power-factor at e_0, the total current, $\dot{I}$, leads the receiver voltage, e, by 45°.

Substituting the value, i', of equation (10), which compensates for the line reactance, x, and so gives unity power-factor at e_0, into equation (2), gives as required supply voltage e_0.

$$e_0 = \frac{e^2 z^2}{2\,x^2} + \frac{(x - r)\,(e - 2\,ix)\sqrt{e^2 - 4\,i^2 x^2}}{2\,x}. \tag{13}$$

As illustration are shown, in Fig. 77, with the load current, i, as abscissas, the values of leading quadrature component of current, i', and of generator voltage, e_0, for the constants

$$e = 400 \text{ volts};\ r = 0.05 \text{ ohm, and } x = 0.10 \text{ ohm.}$$

84. More frequently than for controlling the power-factor, phase control is used for controlling the voltage, that is, to maintain the receiver voltage, e, constant, or raise it with increasing load, i, at constant generator voltage, e_0.

In this case, equation (4) gives the quadrature component of current, i', required by current, i, at constant receiver voltage, e, and constant generator voltage, e_0.

Since the equation (4) of i' contains a square root, the maximum value of i, that is, the maximum load which can be carried at constant voltage, e and e_0, is given by equating the term under the square root to zero

$$\frac{e_0{}^2}{z^2} - \left(\frac{er}{z^2} + i\right)^2 = 0;$$

as

$$i_m = \frac{e_0 z - er}{z^2} = \frac{e_0 - e \cos\theta}{z}, \tag{14}$$

and the corresponding quadrature component of current, by (4), is

$$i'_m = -\frac{ex}{z^2} = -\frac{e \sin\theta}{z}, \tag{15}$$

that is, leading.

From equation (14) follows as the impedance, z, which, at constant line-resistance, r, gives the maximum value of i_m

$$\frac{di_m}{dz} = 0;$$

hence,

$$z_m = 2\,r\frac{e}{e_0}, \tag{16}$$

and for this value of impedance, z_m, substituting in (14) and (15)

$$i_{mm} = \frac{e_0{}^2}{4\,r}, \qquad \text{and} \qquad i'_{mm} = \frac{e_0{}^2}{4\,r}. \tag{17}$$

The maximum load, i, which can be delivered at constant voltage, e, therefore depends upon the line impedance, and the voltages, e and e_0.

Since e_0 and e are not very different from each other, the ratio $\frac{e}{e_0}$ in equation (16) is approximately unity, and the impedance, z, which permits maximum load to be transmitted, is approximately twice the line resistance, r, or rather slightly less.

$$z \leqq 2r,$$

gives

$$x \leqq r\sqrt{3}.$$

A relatively low line-reactance, x, so gives maximum output. In practice, a far higher reactance, x, is used, since it gives sufficient output and a lesser quadrature component of current.

By substituting $i = 0$ in equation (4), the value of the quadrature component of current at no-load is found as

$$\left.\begin{aligned} i'_0 &= \frac{\sqrt{e_0{}^2 z^2 - e^2 r^2} - ex}{z^2} \\ &= \frac{\sqrt{e_0{}^2 - e^2 \cos^2\theta} - e\sin\theta}{z} \end{aligned}\right\} \quad (18)$$

This can be written in the form

$$i'_0 = \frac{\sqrt{(e_0{}^2 - e^2) + e^2 \sin^2\theta} - e\sin\theta}{z},$$

and then shows that for $e = e_0$, $i'_0 = 0$, or no quadrature component of current exists at no-load; for $e > e_0$, $i'_0 < 0$ or negative, that is, the quadrature component of current is already leading at no-load. For: $e < e_0$, $i'_0 > 0$ or lagging, that is, the quadrature component of current i'_0 is lagging at no-load, becomes zero at some load, and leading at still higher loads.

The latter arrangement, $e < e_0$, is generally used, as the quadrature component of current passes through zero at some intermediate load, and so is less over the range of required load than it would be if i'_0 were 0 or negative.

From (18) follows that the larger z, and at constant resistance r, also x, the smaller the quadrature component of current. That is, increase of the line reactance, x, reduces the quadrature current at no-load, i'_0, and in the same way at load, that is, improves the power-factor of the circuit, and so is desirable, and the insertion of reactive coils in the line for this reason customary.

Increase of reactance, however, reduces the maximum output i_m, and too large a reactance is for this reason objectionable.

Let

$$i = i_1$$

be the load at which the quadrature component of current vanishes, $i' = 0$, that is, the receiver circuit has unity power-factor.

Substituting $\quad i = i_1,\ i' = 0$ into equation (2) gives

$$e_0^2 = (e + ri_1)^2 + x^2 i_1^2 \tag{19}$$

and, substituting (19) in (4), (18), (14), gives reactive component of current

$$i' = \sqrt{\frac{e^2 \sin^2 \theta}{z^2} + \frac{2\,e \cos \theta}{z}(i_1 - i) + (i_1^2 - i^2)} - \frac{e \sin \theta}{z}, \tag{20}$$

and at no-load

$$i'_0 = \sqrt{\frac{e^2 \sin^2 \theta}{z^2} + \frac{2\,ei_1 \cos \theta}{z} + i_1^2} - \frac{e \sin \theta}{z}, \tag{21}$$

Maximum output current

$$i_m = \sqrt{\frac{e^2}{z^2} + \frac{2\,ei_1 \cos \theta}{z} + i_1^2} - \frac{e \cos \theta}{z}. \tag{22}$$

85. Of importance in phase control for constant voltage, e, at constant e_0, are the three currents

i_1, the power component of current at which the quadrature component of current vanishes: $i' = 0$.

i_m, the maximum load which can be transmitted at constant voltage, e.

i'_0, the reactive component of current at no-load.

The equation of phase control, (2), however, contains only two quantities which can be chosen: The reactance, x, which can be increased by inserting reactive coils, and the generator voltage, e_0, which can be made anything desired, even with an existing generating station, since between e and e_0 practically always transformers are interposed, and their ratio can be chosen so as to correspond to any desired generator voltage, e_0, as they usually are supplied with several voltage steps.

Of the three quantities, i_1, i_m and i'_0, only two can be chosen, and the constants, x and e_0, derived therefrom. The third current then also follows, and if the value found for it does not suit the requirements of the problem, other values have to be tried. For instance, choosing i_1 as corresponding to three-fourths load, and i'_0 fairly small, gives very good power-factors over the whole range of load, but a relatively low value of i_m, and where very great overload capacities are required, i_m may not be sufficient, and i_1 may have to be chosen corresponding to full-load and a higher value of i'_0 permitted, that is, some sacrifice made in the power-factor, in favor of overload capacity.

So, for instance, the values may be chosen

i_1, corresponding to full-load,

and required that i'_0 does not exceed half of full-load current;

$$i'_0 < 0.5 i_1,$$

and that the synchronons converter or motor can carry at least 100 per cent. overload, that is,

$$i_m > 2\, i_1.$$

We then can put, $i_m = 2\, i_1\, c$ and $i'_0 = \dfrac{0.5 i_1}{c}$, (23)

and substitute (23) in (19), (22) and determine x, e_o, c.

86. The variation of the reactive current, i' with the load, i, equation (4), is brought about by varying the field excitation of the receiving synchronous machine. Where the load on the synchronous machine is direct-current output, as in a motor generator and especially a converter, the most convenient way of varying the field excitation with the load is automatically, by a series field-coil traversed by the direct-current output. The field windings of converters intended for phase control—as for the supply of power to electric railways, from substations fed by a high-potential alternating-current transmission line—are compound-wound, and the shunt field is adjusted for under-excitation, so as to produce at no-load the lagging current, i'_0, and the series field adjusted so as to make the reactive component of current, i', disappear at the desired load, i_1.

In this case, however, the variation of the field excitation by the series field is directly proportional to the load, as is also the variation of i', that is, it varies from $i' = i'_0$ for $i = 0$, to $i' = 0$ for $i = i_1$, and can be expressed by the equation

$$\left.\begin{aligned} i' &= i'_0\left(1 - \frac{i}{i_1}\right) \\ &= q(i_1 - i) \end{aligned}\right\} \qquad (24)$$

where

$$q = \frac{i'_0}{i_1} \qquad (25)$$

is the ratio of (reactive) no-load current, i'_0, to (effective) non-inductive load current, i_1.

To maintain constant voltage, e, at constant, e_0, the required variation of i' is not quite linear, and with a linear variation of i', as given by a compound field-winding on the synchronous machine, the receiver-voltage, e, at constant impressed voltage does not remain perfectly constant, but when adjusted for the same value at no-load and at full-load, e is slightly high at intermediate loads, low at higher loads. It is, however, sufficiently constant for all practical purposes.

Choosing then the full-load current, i_1, and the no-load current, $i'_0 = qi_1$, and let the reactive component of current, i', by a compound field-winding vary as a linear function of the load, i:

$$i' = q(i_1 - i).$$

Then, substituting i_1 and $i'_0 = qi_1$ in the equations (2) for phase control:

No-load: $\quad i = 0, \quad i' = qi_1;$

$$e_0^2 = (e + qxi_1)^2 + qri_1^2. \tag{26}$$

Full load: $\quad i_1 = i_1, \quad i' = 0;$

$$e_0^2 = (e + ri_1)^2 + xi_1^2. \tag{27}$$

From these equations (26) and (27) then calculate the required reactance, x, and the generator voltage, e_0, as:

$$x = \frac{\dfrac{qe}{i_1} \pm \sqrt{\dfrac{e^2}{i_1^2}(1 + q^2) - \left[\dfrac{e}{i_1} + r(1 - q^2)\right]^2}}{1 - q^2}, \tag{28}$$

and from (27) or (26) the voltage, e_0.

The terminal voltage at the receiving circuit then is, by equation (3):

$$e = \sqrt{e_0^2 - [qri_1 - (qr + x)i]^2} - ((r - qx)i + qxi_1). \tag{29}$$

As an example is shown, in Fig. 78, the curve of receiving voltage, e, with the load, i, as abscissas, for the values:

e = 400 volts at no-load and at full-load,
i_1 = 500 amp. at full-load, power component of current,
i'_0 = 200 amp., lagging reactive or quadrature component of current at no-load,

hence $q = 0.4$,
$i' = 200 - 0.4\, i$,
and $r = 0.05$ ohm.

From equation (28) then follows:

$$x = 0.381 \pm 0.165 \text{ ohm}.$$

Choosing the lower value:

$$x = 0.216 \text{ ohm}.$$

gives, from equation (27):

$$e_0 = 443.4 \text{ volts};$$

hence

$$e = \sqrt{196{,}420 + 5.76\,i - 0.0576\,i^2} - (43.2 - 0.0264\,i).$$

For comparison is shown, in Fig. 78, the receiving voltage, e', at the same supply voltage, $e_0 = 443.4$ volts, but without phase control, that is, with a non-inductive receiver-circuit.

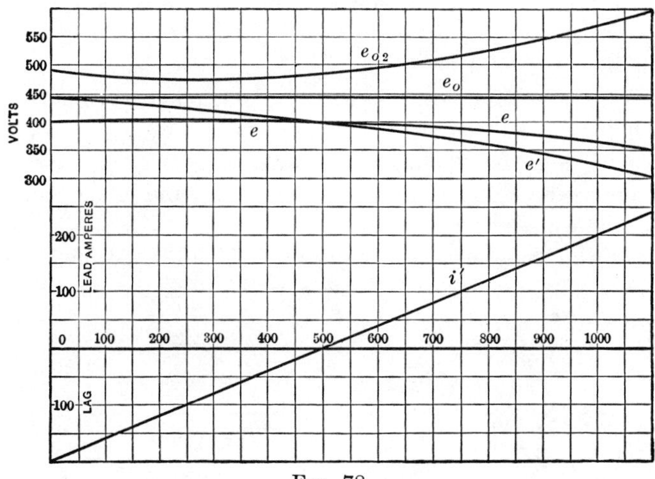

Fig. 78.

87. Equation (28) shows that there are two values of x: x_1 and x_2; and corresponding thereto two values of e_0: e_{01} and e_{02}, which as constant-supply voltage give the same receiver-voltage, e, at no-load and at full-load, and so approximately constant receiver-voltage throughout.

One of the two reactances, x_2, is much larger than the other, x_1, and the corresponding voltage, e_{02}, accordingly larger than e_{01}.

In addition to the terminal voltage, e, at the receiver-circuit, there are therefore two further points of constant voltage in the system: e_{01}, distant from e by the resistance, r, and reactance, x_1, and : e_{02}, distant from e_{01} by the reactance $x_0 = x_2 - x_1$.

That is, by the proper choice of the reactances, x_1 and x_0, three points of the system can be maintained automatically at approximately constant voltage, by phase control: e, e_{01} and e_{02}.

Such *multiple-phase control* can advantageously be employed by using:

e as the terminal voltage of the receiving circuit,

e_{01} as the generator terminal voltage e_0, and

e_{02} as the nominal induced e.m.f. of the generator, that is, the voltage corresponding to the field-excitation. Constancy of e_{02} accordingly means constant field-excitation.

That is, with constant field-excitation of the generator, the voltage remains approximately constant, by multiple-phase control, at the generator busbars as well as at the terminals of the receiving circuit, at the end of the transmission line of resistance, r.

In this case:

x_1 = reactance of transmission line plus reactive coils inserted in the line (usually at the receiving station).

$x_0 = x_2 - x_1$ = synchronous reactance of the generator plus reactive coils inserted between generator and generator busbars, where necessary.

Since the generator also contains a small resistance, r_0, the two values of reactance, x_1 and $x_2 = x_1 + x_0$, are given by the equation (28) as:

$$x_1 = \frac{\frac{qe}{i_1} - \sqrt{\frac{e^2}{i_1^2}(1 + q^2) - \left[\frac{e}{i_1} + r(1 - q^2)\right]^2}}{1 - q^2},$$

and

$$x_2 = \frac{\frac{qe}{i_1} + \sqrt{\frac{e_2}{i_1^2}(1 + q^2) - \left[\frac{e}{i_1} + (r + r_0)\ (1 - q)^2\right]^2}}{1 - q^2}$$

Assuming in above example:

$$r_0 = 0.01 \text{ ohm}$$

gives

$$x_2 = 0.440 \text{ ohm};$$

hence,

$$x_0 = 0.224 \text{ ohm}.$$

The curve of nominal generated e.m.f., e_{02}, of the generator is shown in Fig. 78 as e_{02}.

That is, at constant field-excitation, corresponding to a nominal generated e.m.f.,

$$e_{02} = 488.2 \text{ volts.}$$

The generator of synchronous impedance,

$$Z_0 = r_0 + jx_0 = 0.01 + 0.224\,j \text{ ohms,}$$

maintains approximately constant voltage at its own terminals, or at the generator busbars,

$$e_0 = 443.4 \text{ volts,}$$

and at the same time maintains constant voltage,

$$e = 400 \text{ volts,}$$

at the end of a transmission line of impedance,

$$Z = r + jx_1 = 0.06 + 0.216\,j \text{ ohms,}$$

if by phase control in the receiving circuit, by compounded converter, the reactive or quadrature component of current, i', is varied with the load or power component of current, i, and proportional thereto, that is:

$$\begin{aligned} i' &= q(i_1 - i) \\ &= 200 - 0.4\,i. \end{aligned}$$

88. To adjust a circuit experimentally for phase control for constant voltage, by overcompounded synchronous converter: at constant-supply voltage and no-load on the converter—with the transmission line with its transformers, reactances, etc., or an impedance equal thereto, in the circuit between converter and supply voltage—the shunt field of the converter is adjusted by the field rheostat so as to give the desired direct-current voltage at the converter brushes. Then load is put on the converter, and, without changing the supply voltage or the adjustment of the shunt field, the rheostat or shunt across the series field of the converter is adjusted so as to give the desired direct-current voltage.

If the supply voltage can be varied, as is usually provided for by different voltage taps on the transformer, then, before adjusting the converter fields as described above, first the proper supply voltage is found. This is done by loading the converter with the current, at which unity power-factor at the converter is desired—for instant full-load—and then varying the converter shunt field so as to get minimum alternating-current input, and varying the supply voltage so as to get—at minimum alternating-current input—the desired direct-current voltage. Where the supply voltage can only be varied in definite steps: at some voltage step, the converter field—at the desired non-inductive load—is adjusted for minimum alternating-current input; if then the direct-current voltage is too low, the transformer connections are changed to the next higher supply voltage step; if the direct-current voltage is too high, the change is made to the next lower supply voltage step, until that supply voltage step is found, which, at the adjustment of the converter field for minimum alternating-current input, brings the direct-current voltage nearest to that desired. Then for this supply voltage step, the converter field circuits are adjusted for phase control, as above described.

POWER AND EFFECTIVE CONSTANTS

CHAPTER XII

EFFECTIVE RESISTANCE AND REACTANCE

89. The resistance of an electric circuit is determined:

1. By direct comparison with a known resistance (Wheatstone bridge method, etc.).

This method gives what may be called the true ohmic resistance of the circuit.

2. By the ratio:

$$\frac{\text{Volts consumed in circuit}}{\text{Amperes in circuit}}.$$

In an alternating-current circuit, this method gives, not the resistance of the circuit, but the impedance,

$$z = \sqrt{r^2 + x^2}.$$

3. By the ratio:

$$r = \frac{\text{Power consumed}}{(\text{Current})^2};$$

where, however, the "power" does not include the work done by the circuit, and the counter e.m.fs. representing it, as, for instance, in the case of the counter e.m.f. of a motor.

In alternating-current circuits, this value of resistance is the power coefficient of the e.m.f.,

$$r = \frac{\text{Power component of e.m.f.}}{\text{Total current}}.$$

It is called the *effective resistance* of the circuit, since it represents the effect, or power, expended by the circuit. The power coefficient of current,

$$g = \frac{\text{Power component of current}}{\text{Total e.m.f.}},$$

is called the *effective conductance* of the circuit.

In the same way, the value,

$$x = \frac{\text{Wattless component of e.m.f.}}{\text{Total current}},$$

is the *effective reactance,* and

$$b = \frac{\text{Wattless component of current}}{\text{Total e.m.f.}},$$

is the *effective susceptance* of the circuit.

While the true ohmic resistance represents the expenditure of power as heat inside of the electric conductor by a current of uniform density, the effective resistance represents the total expenditure of power.

Since in an alternating-current circuit, in general power is expended not only in the conductor, but also outside of it, through hysteresis, secondary currents, etc., the effective resistance frequently differs from the true ohmic resistance in such way as to represent a larger expenditure of power.

In dealing with alternating-current circuits, it is necessary, therefore, to substitute everywhere the values "effective resistance," "effective reactance," "effective conductance," and "effective susceptance," to make the calculation applicable to general alternating-current circuits, such as inductive reactances containing iron, etc.

While the true ohmic resistance is a constant of the circuit, depending only upon the temperature, but not upon the e.m.f., etc., the effective resistance and effective reactance are, in general, not constants, but depend upon the e.m.f., current, etc. This dependence is the cause of most of the difficulties met in dealing analytically with alternating-current circuits containing iron.

90. The foremost sources of energy loss in alternating-current circuits, outside of the true ohmic resistance loss, are as follows:

1. Molecular friction, as,
   (*a*) Magnetic hysteresis;
   (*b*) Dielectric hysteresis.
2. Primary electric currents, as,
   (*a*) Leakage or escape of current through the insulation, brush discharge, corona.
   (*b*) Eddy currents in the conductor or unequal current distribution.

3. Secondary or induced currents, as,
   (*a*) Eddy or Foucault currents in surrounding magnetic materials;
   (*b*) Eddy or Foucault currents in surrounding conducting materials;
   (*c*) Secondary currents of mutual inductance in neighboring circuits.
4. Induced electric charges, electrostatic induction or influence.

While all these losses can be included in the terms effective resistance, etc., the magnetic hysteresis and the eddy currents are the most frequent and important sources of energy loss.

Magnetic Hysteresis

91. In an alternating-current circuit surrounded by iron or other magnetic material, energy is expended outside of the conductor in the iron, by a kind of molecular friction, which, when the energy is supplied electrically, appears as magnetic hysteresis, and is caused by the cyclic reversals of magnetic flux in the iron in the alternating magnetic field.

To examine this phenomenon, first a circuit may be considered, of very high inductive reactance, but negligible true ohmic resistance; that is, a circuit entirely surrounded by iron, as, for instance, the primary circuit of an alternating-current transformer with open secondary circuit.

The wave of current produces in the iron an alternating magnetic flux which generates in the electric circuit an e.m.f.—the counter e.m.f. of self-induction. If the ohmic resistance is negligible, that is, practically no e.m.f. consumed by the resistance, all the impressed e.m.f. must be consumed by the counter e.m.f. of self-induction, that is, the counter e.m.f. equals the impressed e.m.f.; hence, if the impressed e.m.f. is a sine wave, the counter e.m.f., and, therefore, the magnetic flux which generates the counter e.m.f., must follow a sine wave also. The alternating wave of current is not a sine wave in this case, but is distorted by hysteresis. It is possible, however, to plot the current wave in this case from the hysteretic cycle of magnetic flux.

From the number of turns, n, of the electric circuit, the effective counter e.m.f., E, and the frequency, f, of the current, the maximum magnetic flux, Φ, is found by the formula:

$$E = \sqrt{2}\,\pi n f \Phi\, 10^{-8};$$

hence,

$$\Phi = \frac{E\, 10^8}{\sqrt{2}\, \pi n f}.$$

A maximum flux, Φ, and magnetic cross-section, A, give the maximum magnetic induction, $B = \frac{\Phi}{A}$.

If the magnetic induction varies periodically between $+ B$ and $- B$, the magnetizing force varies between the corresponding values $+ f$ and $- f$, and describes a looped curve, the cycle of hysteresis.

If the ordinates are given in lines of magnetic force, the abscissas in tens of ampere-turns, then the area of the loop equals the energy consumed by hysteresis in ergs per cycle.

From the hysteretic loop the instantaneous value of magnetizing force is found, corresponding to an instantaneous value of magnetic flux, that is, of generated e.m.f.; and from the magnetizing force, f, in ampere-turns per units length of magnetic circuit, the length, l, of the magnetic circuit, and the number of turns, n, of the electric circuit, are found the instantaneous values of current, i, corresponding to a magnetizing force, f, that is, magnetic induction, B, and thus generated e.m.f., e, as:

$$i = \frac{fl}{n}.$$

92. In Fig. 79, four magnetic cycles are plotted, with maximum values of magnetic induction, B = 2,000, 6,000, 10,000, and 16,000, and corresponding maximum magnetizing forces, f = 1.8, 2.8, 4.3, 20.0. They show the well-known hysteretic loop, which becomes pointed when magnetic saturation is approached.

These magnetic cycles correspond to sheet iron or sheet steel, of a hysteretic coefficient, $\eta = 0.0033$, and are given with ampere-turns per centimeter as abscissas, and kilolines of magnetic force as ordinates.

In Figs. 80 and 81, the curve of magnetic induction as derived from the generated e.m.f. is a sine wave. For the different values of magnetic induction of this sine curve, the corresponding values of magnetizing force f, hence of current, are taken from Fig. 79, and plotted, giving thus the exciting current required to produce the sine wave of magnetism; that is, the wave of current which a sine wave of impressed e.m.f. will establish in the circuit.

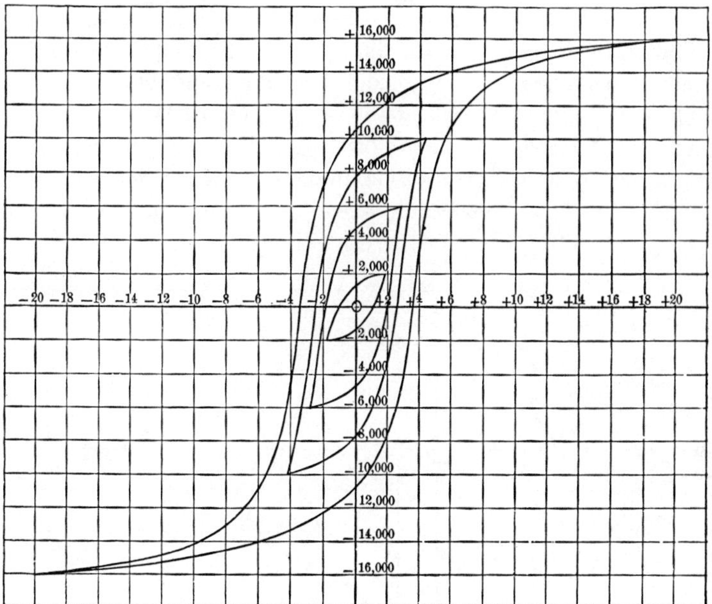

FIG. 79.—Hysteretic cycle of sheet iron.

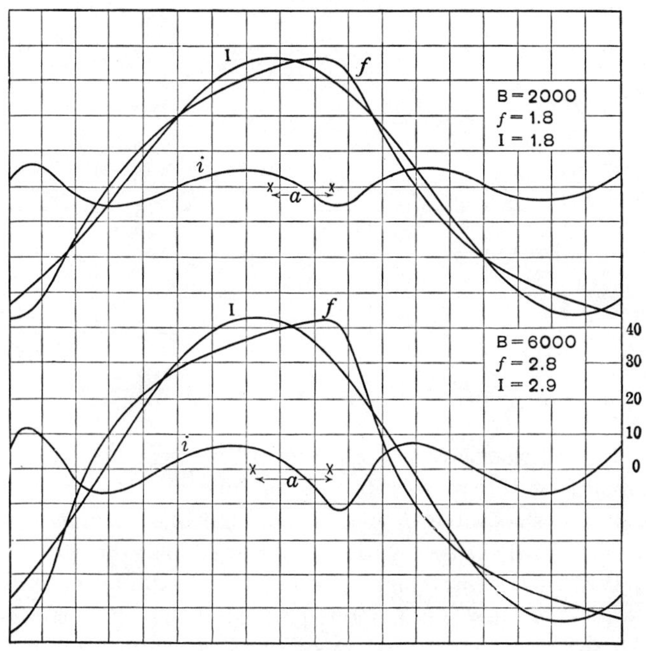

FIG. 80.

As shown in Figs. 80 and 81, these waves of alternating current are not sine waves, but are distorted by the super-position of higher harmonics, and are complex harmonic waves. They reach their maximum value at the same time with the maximum of magnetism, that is, 90° ahead of the maximum generated e.m.f., and hence about 90° behind the maximum impressed e.m.f., but pass the zero line considerably ahead of the zero value of magnetism of 42°, 52°, 50° and 41°, respectively.

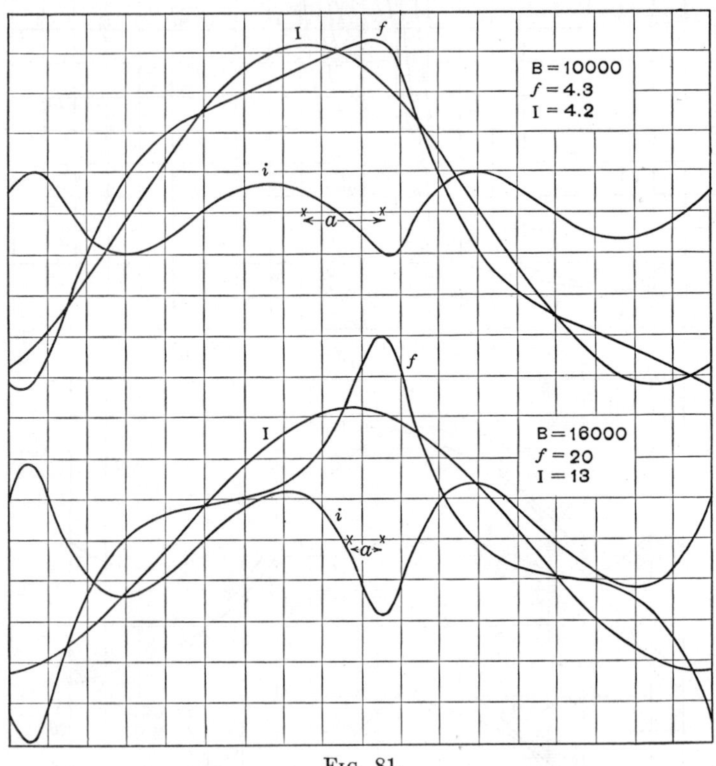

Fig. 81.

The general character of these current waves is, that the maximum point of the wave coincides in time with the maximum point of the sine wave of magnetism; but the current wave is bulged out greatly at the rising, and hollowed in at the decreasing, side. With increasing magnetization, the maximum of the current wave becomes more pointed, as shown by the curves of Fig. 81, for $B = 10,000$; and at still higher saturation a peak is formed at the maximum point, as in the curve for $B = 16{,}000$. This is the case when the curve of magnetization reaches within the range of magnetic saturation, since in the proximity of saturation the current near the maximum point of magnetization has to rise abnormally to cause even a small increase of magnetization. The four curves, Figs. 80 and 81 are not drawn to the same scale. The maximum values of magnetizing force, corresponding to the maximum values of magnetic induction, $B =$ 2,000, 6,000, 10,000, and 16,000 lines of force per square centimeter, are $f = 1.8, 2.8, 4.3$, and 20.0 ampere-turns per centimeter. In the different diagrams these are represented in the ratio of 8:6:4:1, in order to bring the current curves to approximately the same height. The magnetizing force, in c.g.s. units, is

$$H = 4\pi f/10 = 1.257f.$$

93. The distortion of the current waves, f, in Figs. 80 and 81, is almost entirely due to the magnetizing current, and is caused by the disproportionality between magnetic induction, B, and magnetizing force, f, as exhibited by the magnetic characteristic or saturation curve, and is very little due to hysteresis.

Resolving these curves, f, of Figs. 80 and 81 into two components, one in phase with the magnetic induction, B, or symmetrical thereto, hence in quadrature with the induced e.m.f., and therefore wattless: the magnetizing current, i_m; and the other, in time quadrature with the magnetic induction, B, hence in phase, or symmetrical, with the generated e.m.f., that is, representing power: the hysteresis power-current, i_h. Then we see that the hysteresis power-current, i_h, is practically a sine wave, while the magnetizing current, i_m, differs considerably from a sine wave, and tends toward peakedness—the more the higher the magnetic induction, B, that is, the more magnetic saturation is approached, so that for $B = 16{,}000$ a very high peak is shown, and the wave of magnetizing current, i_m, does not resemble a sine wave at all, but at the maximum value is nearly four times higher than a sine wave of the same instantaneous values near zero induction would have.

These curves of hysteresis power-current, i_h, and magnetizing current, i_m, derived by resolving the distorted current curves, f, of Figs. 80 and 81, are plotted in Fig. 82, the last one, corresponding to $B = 16{,}000$, with one-quarter the ordinates of the first three.

As curves, symmetrical with regard to the maximum value of $B - i_m -$, and to the zero value of $B - i_h -$, these curves are constructed thus:

Let

$$b = B \sin \phi = \text{sine wave of magnetic induction,}$$

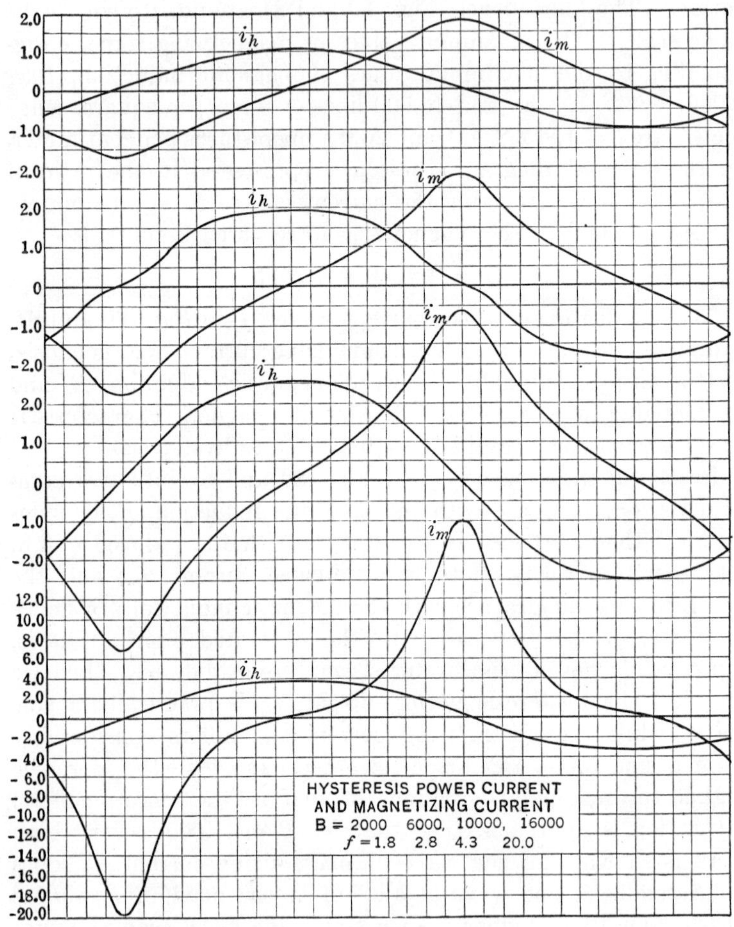

FIG. 82.

then

$$i_m = \tfrac{1}{2}(f_\phi + f_{180-\phi}),$$
$$i_h = \tfrac{1}{2}(f_\phi + f_{-\phi}).$$

That is, i_m is the average value of f for an angle ϕ, and its supplementary angle $180 - \phi$, i_h the average value of f for an angle ϕ and its negative angle $- \phi$.

94. The distortion of the wave of magnetizing current is as large as shown here only in an iron-closed magnetic circuit expending power by hysteresis only, as in an ironclad transformer on open secondary circuit. As soon as the circuit expends power in any other way, as in resistance or by mutual

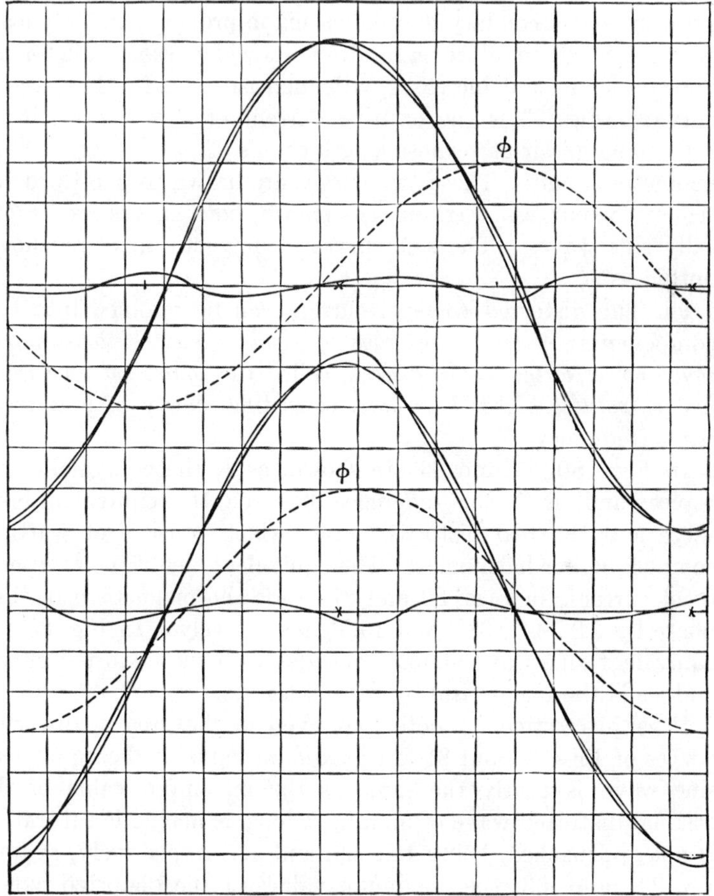

FIG. 83.—Distortion of current wave by hysteresis.

inductance, or if an air-gap is introduced in the magnetic circuit, the distortion of the current wave rapidly decreases and practically disappears, and the current becomes more sinusoidal. That is, while the distorting component remains the same, the sinusoidal component of the current greatly increases, and obscures the distortion. For example, in Fig. 83, two waves are shown corresponding in magnetization to the last curve of Fig. 80, as the one most distorted. The first curve in Fig. 83 is the current wave of a transformer at 0.1 load. At higher loads the distortion is correspondingly still less, except where the magnetic flux of self-induction, that is, flux passing between primary and secondary and increasing in proportion to the load, is so large as to reach saturation, in which case a distortion appears again and increases with increasing load. The second curve of Fig. 83 is the exciting current of a magnetic circuit containing an air-gap whose length equals $\frac{1}{400}$ the length of the magnetic circuit. These two curves are drawn to one-third the size of the curve in Fig. 80. As shown, both curves are practically sine waves. The sine curves of magnetic flux are shown dotted as ϕ.

95. The distorted wave of current can be resolved into two components: *A true sine wave of equal effective intensity and equal power to the distorted wave*, called the *equivalent sine wave*, and a *wattless higher harmonic*, consisting chiefly of a term of triple frequency.

In Figs. 80, 81 and 83 are shown, as I, the equivalent sine waves, and as i, the difference between the equivalent sine wave and the real distorted wave, which consists of wattless complex higher harmonics. The equivalent sine wave of m.m.f. or of current, in Figs. 80 and 81, leads the magnetism in time phase by 34°, 44°, 38°, and 15.5°, respectively. In Fig. 83 the equivalent sine wave almost coincides with the distorted curve, and leads the magnetism by only 9 degrees.

It is interesting to note that even in the greatly distorted curves of Figs. 80 and 81 the maximum value of the equivalent sine wave is nearly the same as the maximum value of the original distorted wave of m.m.f., so long as magnetic saturation is not approached, being 1.8, 2.9, and 4.2, respectively, against 1.8, 2.8, and 4.3, the maximum values of the distorted curve. Since, by the definition, the effective value of the equivalent sine wave is the same as that of the distorted wave, it follows that this distorted wave of exciting current shares with the sine wave the feature, that the maximum value and the effective value have the ratio of $\sqrt{2} \div 1$. Hence, below saturation, the maximum value of the distorted curve can be calculated from the effective value—which is given by the reading of an electrodynamometer—by using the same ratio that applies to a true sine wave, and the magnetic characteristic can thus be determined by means of alternating currents, with sufficient exactness, by the electrodynamometer method, in the range below saturation, that is, by alternating-current voltmeter and ammeter.

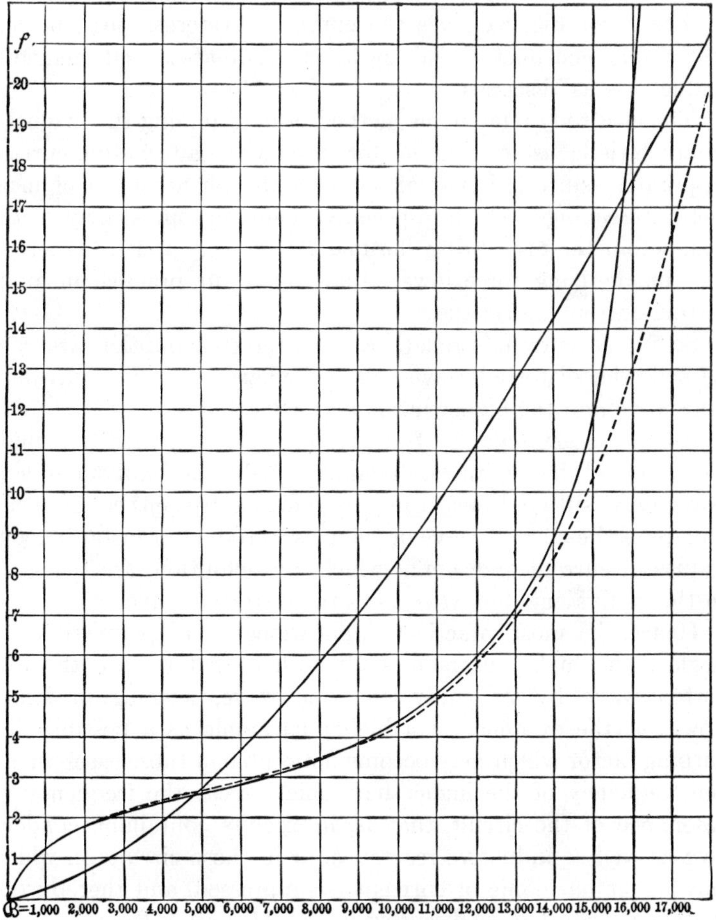

FIG. 84.—Magnetization and hysteresis curve.

96. In Fig. 84 is shown the true magnetic characteristic of a sample of average sheet iron, as found by the method of slow reversals with the magnetometer; for comparison there is shown in dotted lines the same characteristic, as determined with alternating currents by the electrodynamometer, with ampere-turns per centimeter as ordinates and magnetic inductions as abscissas. As represented, the two curves practically coincide up to a value of $B = 13{,}000$; that is, up to fairly high inductions. For higher saturations, the curves rapidly diverge, and the electrodynamometer curve shows comparatively small magnetizing forces producing apparently very high magnetizations.

The same Fig. 84 gives the curve of hysteretic loss, in ergs per cubic centimeter and cycle, as ordinates, and magnetic inductions as abscissas.

The electrodynamometer method of determining the magnetic characteristic is preferable for use with alternating-current apparatus, since it is not affected by the phenomenon of magnetic "creeping," which, especially at low densities, may in the magnetometer tests bring the magnetism very much higher, or the magnetizing force lower, than found in practice in alternating-current apparatus.

So far as current strength and power consumption are concerned, the distorted wave can be replaced by the equivalent sine wave and the higher harmonics neglected.

All the measurements of alternating currents, with the single exception of instantaneous readings, yield the equivalent sine wave only, since all measuring instruments give either the mean square of the current wave or the mean product of instantaneous values of current and e.m.f., which, by definition, are the same in the equivalent sine wave as in the distorted wave.

Hence, in most practical applications it is permissible to neglect the higher harmonics altogether, and replace the distorted wave by its equivalent sine wave, keeping in mind, however, the existence of a higher harmonic as a possible disturbing factor which may become noticeable in those cases where the frequency of the higher harmonic is near the frequency of resonance of the circuit, that is, in circuits containing condensive as well as inductive reactance, or in those circuits in which the higher harmonic of currrent is suppressed, and thereby the voltage is distorted, as discussed in Vol. II, Part I, Chapter 1.

97. The equivalent sine wave of exciting current leads the sine wave of magnetism by an angle α, which is called the *angle of hysteretic advance of phase.* Hence the current lags behind the e.m.f. by the time angle $(90° - \alpha)$, and the power is, therefore,

$$P = IE \cos (90° - \alpha) = IE \sin \alpha.$$

Thus the exciting current, I, consists of a power component, $I \sin \alpha$, called the *hysteretic or magnetic power current*, and a wattless component, $I \cos \alpha$, which is called the *magnetizing current*. Or, conversely, the e.m.f. consists of a power component, $E \sin \alpha$, the *hysteretic power component*, and a wattless component, $E \cos \alpha$, the e.m.f. consumed by *self-induction*.

Denoting the absolute value of the impedance of the circuit, $\frac{E}{I}$, by z—where z is determined by the magnetic characteristic of the iron and the shape of the magnetic and electric circuits —the impedance is represented, in phase and intensity, by the symbolic expression,

$$Z = r + jx = z \sin \alpha + jz \cos \alpha;$$

and the admittance by,

$$Y = g - jb = \frac{1}{z} \sin \alpha - j\frac{1}{z} \cos \alpha = y \sin \alpha - jy \cos \alpha.$$

The quantities z, r, x, and y, g, b are, however, not constants as in the case of the circuit without iron, but depend upon the intensity of magnetization, B—that is, upon the e.m.f. This dependence complicates the investigation of circuits containing iron.

In a circuit entirely inclosed by iron, α is quite considerable, ranging from 30° to 50° for values below saturation. Hence, even with negligible true ohmic resistance, no great lag can be produced in ironclad alternating-current circuits.

98. The loss of energy by hysteresis due to molecular magnetic friction is, with sufficient exactness, proportional to the 1.6th power of magnetic induction, B. Hence it can be expressed by the formula:

$$W_H = \eta B^{1.6}$$

where—

W_H = loss of energy per cycle, in ergs or (c.g.s.) units ($= 10^{-7}$ joules) per cubic centimeter,

B = maximum magnetic induction, in lines of force per sq. cm., and η = the *coefficient of hysteresis*.

This I found to vary in iron from 0.001 to 0.0055. As a safe mean, 0.0033 can be accepted for common annealed sheet iron or sheet steel, 0.002 for silicon steel and 0.0010 to 0.0015 for specially selected low hysteresis steel. In gray cast iron, η averages

0.013; it varies from 0.0032 to 0.028 in cast steel, according to the chemical or physical constitution; and reaches values as high as 0.08 in hardened steel (tungsten and manganese steel). Soft nickel and cobalt have about the same coefficient of hysteresis as gray cast iron; in magnetite I found $\eta = 0.023$.

In the curves of Figs. 79 to 84, $\eta = 0.0033$.

At the frequency, f, the loss of power in the volume, V, is, by this formula,

$$P = \eta f V B^{1.6}\, 10^{-7} \text{ watts}$$
$$= \eta f V \left(\frac{\Phi}{A}\right)^{1.6} 10^{-7} \text{ watts},$$

where A is the cross-section of the total magnetic flux, Φ.

The maximum magnetic flux, Φ, depends upon the counter e.m.f. of self-induction,

$$E = \sqrt{2}\pi f n \Phi\, 10^{-8},$$

or

$$\Phi = \frac{E\, 10^8}{2\, \pi f n},$$

where n = number of turns of the electric circuit, f = frequency.

Substituting this in the value of the power, P, and canceling, we get,

$$P = \eta \frac{E^{1.6}}{f^{0.6}} \frac{V\, 10^{5.8}}{2^{0.8}\, \pi^{1.6} A^{1.6} n^{1.6}} = 58\, \eta \frac{E^{1.6}}{f^{0.6}} \frac{V\, 10^3}{A^{1.6} n^{1.6}};$$

or

$$P = \frac{KE^{1.6}}{f^{0.6}}, \text{ where } K = \eta \frac{V\, 10^{5.8}}{2^{0.8}\, \pi^{1.6} A^{1.6} n^{1.6}} = 58\, \eta \frac{V\, 10^3}{A^{1.6} n^{1.6}};$$

or, substituting $\eta = 0.0033$, we have $K = 191.4 \dfrac{V}{A^{1.6} n^{1.6}}$;

or, substituting $V = Al$, where l = length of magnetic circuit,

$$K = \frac{\eta l\, 10^{5.8}}{2^{0.8}\, \pi^{1.6} A^{0.6} n^{1.6}} = \frac{58\, \eta l\, 10^3}{A^{0.6} n^{1.6}} = 191.4 \frac{l}{A^{0.6} n^{1.6}};$$

and

$$P = \frac{58 \eta E^{1.6} l\, 10^3}{f^{0.6} A^{0.6} n^{1.6}} = \frac{191.4\, E^{1.6} l}{f^{0.6} A^{0.6} n^{1.6}}.$$

In Figs. 85, 86, and 87 is shown a curve of hysteretic loss, with the loss of power as ordinates, and
in curve 85, with the e.m.f., E, as abscissas,

for $l = 6$, $A = 20$, $f = 100$, and $n = 100$;

in curve 86, with the number of turns as abscissas, for

$l = 6$, $A = 20$, $f = 100$, and $E = 100$;

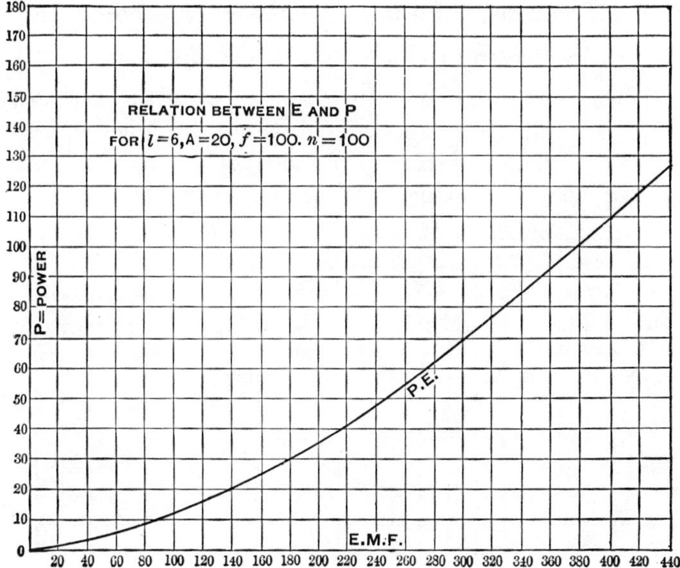

FIG. 85.—Hysteresis loss as function of E.M.F.

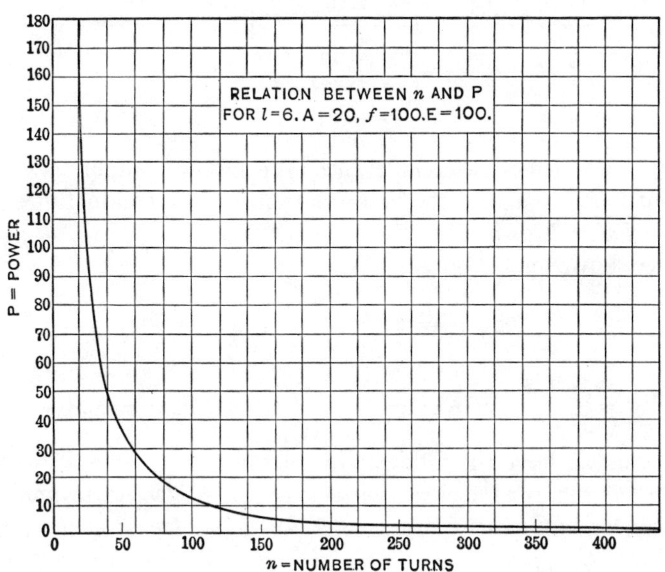

FIG. 86.

in curve 87, with the frequency, f, or the cross-section, A, as abscissas, for $l = 6$, $n = 100$, and $E = 100$.

As shown, the hysteretic loss is proportional to the 1.6^{th} power of the e.m.f., inversely proportional to the 1.6^{th} power of the number of turns, and inversely proportional to the 0.6^{th} power of the frequency and of the cross-section.

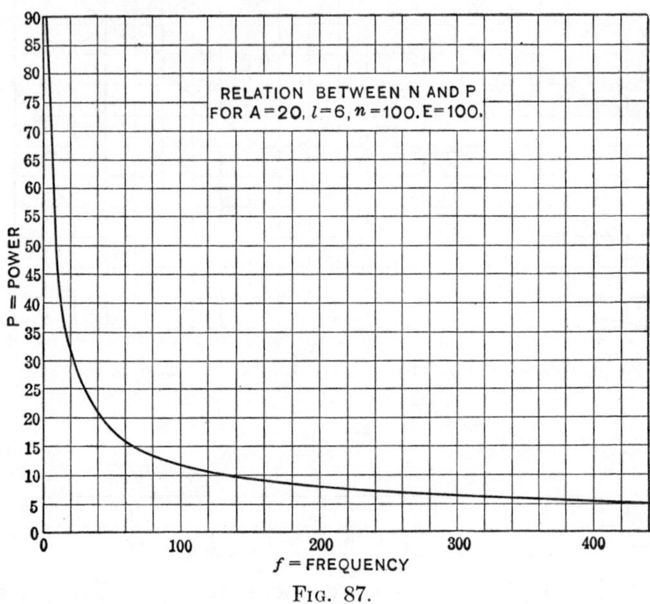

FIG. 87.

99. If g = effective conductance, the power component of a current is $I = Eg$, and the power consumed in a conductance, g, is $P = IE = E^2g$.

Since, however,

$$P = K\frac{E^{1.6}}{f^{0.6}}, \text{ we have } K\frac{E^{1.6}}{f^{0.6}} = E^2g;$$

it is:

$$g = \frac{K}{f^{0.6}E^{0.4}} = \frac{58\,\eta l\,10^3}{E^{0.4}f^{0.6}A^{0.6}n^{1.6}} = 191.4\frac{l}{E^{0.4}f^{0.6}A^{0.6}n^{1.6}}.$$

From this we have the following deduction:

The effective conductance due to magnetic hysteresis is proportional to the coefficient of hysteresis, η, and to the length of the magnetic circuit, l, and inversely proportional to the 0.4^{th} power of the e.m.f., to the 0.6^{th} power of the frequency, f, and of the cross-section *of the magnetic circuit, A, and to the 1.6^{th} power of the number of turns, n.*

Hence, the effective hysteretic conductance increases with decreasing e.m.f., and decreases with increasing e.m.f.; it varies, however, much slower than the e.m.f., so that, if the hysteretic conductance represents only a part of the total power consumption, it can, within a limited range of variation—as, for instance, in constant-potential transformers—be assumed as constant without serious error.

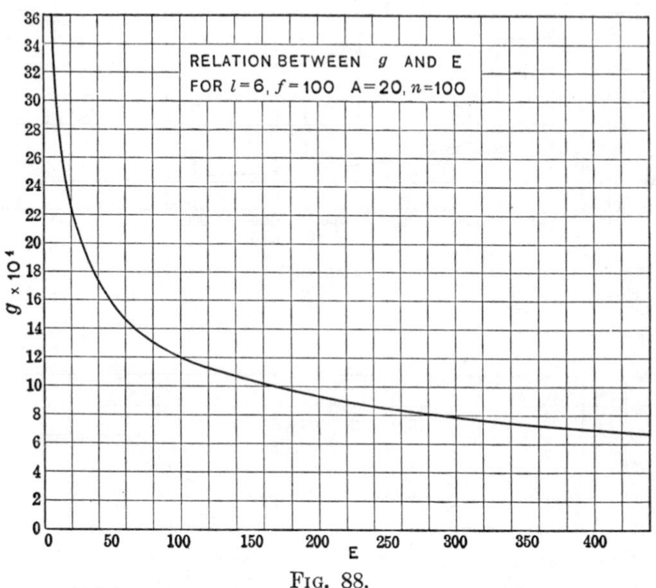

FIG. 88.

In Figs. 88, 89, and 90, the hysteretic conductance, g, is plotted, for $l = 6$, $E = 100$, $f = 100$, $A = 20$ and $n = 100$, respectively, with the conductance, g, as ordinates, and with

E as abscissas in Curve 88.
f as abscissas in Curve 89.
n as abscissas in Curve 90.

As shown, a variation in the e.m.f. of 50 per cent. causes a variation in g of only 14 per cent., while a variation in f or A by 50 per cent. causes a variation in g of 21 per cent.

If $\mathcal{R}$ = magnetic reluctance of a circuit, F_A = maximum

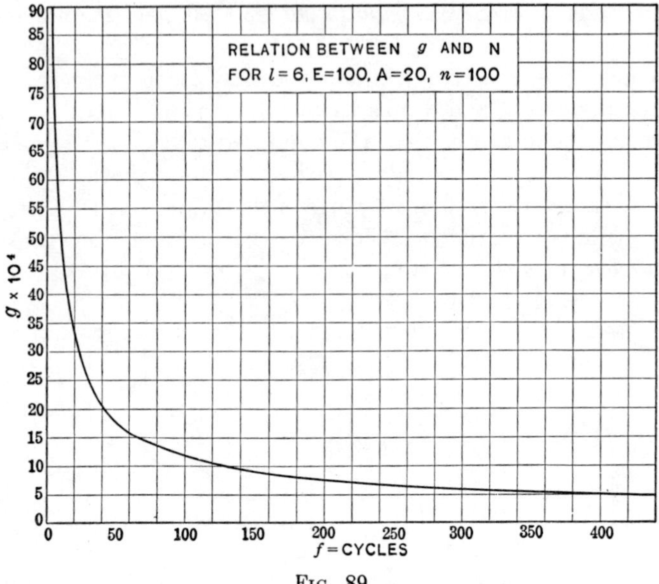
RELATION BETWEEN g AND N
FOR l = 6, E=100, A=20, n=100
g × 10⁴
0
5
10
15
20
25
30
35
40
45
50
55
60
65
70
75
80
85
90
0
50
100
150
200
250
300
350
400
f = CYCLES

FIG. 89.

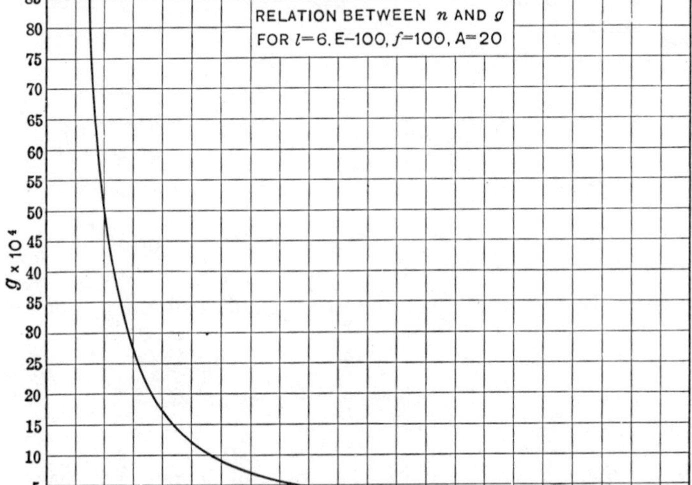
RELATION BETWEEN n AND g
FOR l=6, E–100, f=100, A=20
g × 10⁴
0
5
10
15
20
25
30
35
40
45
50
55
60
65
70
75
80
85
90
0
50
100
150
200
250
300
350
400
n= NUMBER OF TURNS

FIG. 90.

m.m.f., I = effective current, since $I\sqrt{2}$ = maximum current, the magnetic flux,

$$\Phi = \frac{F_A}{\mathcal{R}} = \frac{nI\sqrt{2}}{\mathcal{R}}.$$

Substituting this in the equation of the counter e.m.f. of self-induction,

$$E = \sqrt{2}\,\pi f n \Phi\, 10^{-8},$$

we have

$$E = \frac{2\,\pi n^2 f I\, 10^{-8}}{\mathcal{R}};$$

hence, the absolute admittance of the circuit is

$$y = \sqrt{g^2 + b^2} = \frac{I}{E} = \frac{\mathcal{R}\, 10^8}{2\,\pi n^2 f} = \frac{a\mathcal{R}}{f},$$

where

$$a = \frac{10^8}{2\,\pi n^2}, \text{ a constant.}$$

Therefore, the absolute admittance, y, of a circuit of negligible resistance is proportional to the magnetic reluctance, $\mathcal{R}$, and inversely proportional to the frequency, f, and to the square of the number of turns, n.

100. In a circuit containing iron, the reluctance, $\mathcal{R}$, varies with the magnetization; that is, with the e.m.f. Hence the admittance of such a circuit is not a constant, but is also variable.

In an ironclad electric circuit—that is, a circuit whose magnetic field exists entirely within iron, such as the magnetic circuit of a well-designed alternating-current transformer—$\mathcal{R}$ is the reluctance of the iron circuit. Hence, if μ = permeability since

$$\mathcal{R} = \frac{F_A}{\Phi},$$

and

$$F_A = lF = \frac{10}{4\,\pi} lH = \text{m.m.f.},$$

$$\Phi = A\mathcal{B} = \mu A H = \text{magnetic flux},$$

and

$$\mathcal{R} = \frac{10\,l}{4\,\pi\mu A};$$

substituting this value in the equation of the admittance,

$$y = \frac{\mathcal{R}\, 10^8}{2\pi n^2 f},$$

we have

$$y = \frac{l\,10^9}{8\,\pi^2 n^2 \mu A f} = \frac{c}{f\mu},$$

where

$$c = \frac{l\,10^9}{8\,\pi^2 n^2 A} = \frac{127\,l\,10^5}{n^2 A}.$$

Therefore, in an ironclad circuit, the absolute admittance, y, is inversely proportional to the frequency, f, to the permeability, μ, to the cross-section, A, and to the square of the number of turns, n; and directly proportional to the length of the magnetic circuit, l.

The conductance is

$$g = \frac{K}{f^{0.6} E^{0.4}};$$

and the admittance,

$$y = \frac{c}{f\mu};$$

hence, the angle of hysteretic advance is

$$\sin\alpha = \frac{g}{y} = \frac{K\mu f^4}{cE^{0.4}};$$

or, substituting for A and c (§119),

$$\sin\alpha = \mu\frac{f^{0.4}}{E^{0.4}}\,\frac{\eta l\,10^{5.8}}{2^{0.8}\pi^{1.6}A^{0.6}n^{1.6}}\,\frac{8\,\pi^2 n^2 A}{l\,10^9}$$
$$= \frac{\mu\eta f^{0.4} n^{0.4} A^{0.4} \pi^{0.4}\,2^{2.2}}{E^{0.4}\,10^{3.2}};$$

or, substituting

$$E = 2^{0.5}\pi f n A \mathfrak{B} 10^{-8},$$

we have

$$\sin\alpha = \frac{4\,\mu\eta}{\mathfrak{B}^{0.4}},$$

which is independent of frequency, number of turns, and shape and size of the magnetic and electric circuit.

Therefore, in an ironclad inductance, the angle of hysteretic advance, α, depends upon the magnetic constants, permeability and coefficient of hysteresis, and upon the maximum magnetic induction, but is entirely independent of the frequency, of the shape and other conditions of the magnetic and electric circuit; and, therefore, all ironclad magnetic circuits constructed of the same quality of iron and using the same magnetic density, give the same angle of hysteretic advance, and the same power factor of their electric energizing circuit.

The angle of hysteretic advance, α, in a closed circuit transformer and the no-load power factor, depend upon the quality of the iron, and upon the magnetic density only.

The sine of the angle of hysteretic advance equals 4 times the product of the permeability and coefficient of hysteresis, divided by the 0.4^{th} power of the magnetic density.

101. If the magnetic circuit is not entirely ironclad, and the magnetic structure contains air-gaps, the total reluctance is the sum of the iron reluctance and of the air reluctance, or

$$\mathcal{R} = \mathcal{R}_i + \mathcal{R}_a;$$

hence the admittance is

$$y = \sqrt{g^2 + b^2} = \frac{a}{f}(\mathcal{R}_i + \mathcal{R}_a).$$

Therefore, in a circuit containing iron, the admittance is the sum of the admittance due to the iron part of the circuit, $y_i = \frac{a\mathcal{R}_i}{f}$, and of the admittance due to the air part of the circuit, $y_a = \frac{a\mathcal{R}_a}{f}$, if the iron and the air are in series in the magnetic circuit.

The conductance, g, represents the loss of power in the iron, and, since air has no magnetic hysteresis, is not changed by the introduction of an air-gap. Hence the angle of hysteretic advance of phase is

$$\sin \alpha' = \frac{g}{y} = \frac{g}{y_i + y_a} = \frac{g}{y_i} \frac{\mathcal{R}_i}{\mathcal{R}_i + \mathcal{R}_a},$$

and a maximum, $\frac{g}{y_i}$, for the ironclad circuit, but decreases with increasing width of the air-gap. The introduction of the air-gap of reluctance, $\mathcal{R}_a$, decreases $\sin \alpha$ in the ratio,

$$\frac{\mathcal{R}_i}{\mathcal{R}_i + \mathcal{R}_a}.$$

In the range of practical application, from $B = 2{,}000$ to $B = 14{,}000$, the permeability of iron usually exceeds 1,000, while $\sin \alpha$ in an ironclad circuit varies in this range from 0.51 to 0.69. In air, $\mu = 1$.

If, consequently, 1 per cent. of the length of the iron consists of an air-gap, the total reluctance only varies by a few per cent., that is, remains practically constant; while the angle of hysteretic advance varies from $\sin \alpha = 0.035$ to $\sin \alpha = 0.064$. Thus g is negligible compared with b, and b is practically equal to y.

Therefore, in an electric circuit containing iron, but forming an open magnetic circuit whose air-gap is not less than 1/100 the length of the iron, the susceptance is practically constant and equal to the admittance, so long as saturation is not yet approached, or,

$$b = \frac{\mathcal{R}_a}{f}, \text{ or: } x = \frac{f}{\mathcal{R}_a}.$$

The angle of hysteretic advance is small, and the hysteretic conductance is

$$g = \frac{K}{E^{0.4} f^{0.6}}.$$

The current wave is practically a sine wave.

As an example, in Fig. 83, Curve II, the current curve of a circuit is shown, containing an air-gap of only 1/400 of the length of the iron, giving a current wave much resembling the sine shape, with an hysteretic advance of 9°.

102. To determine the electric constants of a circuit containing iron, we shall proceed in the following way:

Let

$$E = \text{counter e.m.f. of self-induction}$$

then from the equation,

$$E = \sqrt{2}\,\pi n f \Phi 10^{-8},$$

where f = frequency, n = number of turns,
we get the magnetism, Φ, and by means of the magnetic cross-section, A, the maximum magnetic induction: $B = \frac{\Phi}{A}.$

From B, we get, by means of the magnetic characteristic of the iron, the magnetizing force, $= f$ ampere-turns per centimeter length where

$$f = \frac{10}{4\pi} H$$

if H = magnetizing force in c.g.s. units.

Hence,
if l_i = length of iron circuit, $F_i = l_i f$ = ampere-turns required in the iron;
if l_a = length of air circuit, $F_a = \frac{10\, l_a B}{4\pi}$ = ampere-turns required in the air;

hence, $F = F_i + F_a$ = total ampere-turns, maximum value, and $\frac{F}{\sqrt{2}}$ = effective value. The exciting current is

$$I = \frac{F}{n\sqrt{2}},$$

and the absolute admittance,

$$y = \sqrt{g^2 + b^2} = \frac{I}{E}.$$

If F_i is not negligible as compared with F_a, this admittance, y, is variable with the e.m.f., E.

If V = volume of iron, η = coefficient of hysteresis, the loss of power by hysteresis due to molecular magnetic friction is

$$P = \eta f V B^{1.6};$$

hence the hysteretic conductance is $g = \frac{P}{E^2}$, and variable with the e.m.f., E.

The angle of hysteretic advance is

$$\sin \alpha = \frac{g}{y};$$

the susceptance,

$$b = \sqrt{y^2 - g^2};$$

the effective resistance,

$$r = \frac{g}{y^2};$$

and the reactance,

$$x = \frac{b}{y^2}.$$

103. As conclusions, we derive from this chapter the following:

1. In an alternating-current circuit surrounded by iron, the current produced by a sine wave of e.m.f. is not a true sine wave, but is distorted by hysteresis, and inversely, a sine wave of current requires waves of magnetism and e.m.f. differing from sine shape.

2. This distortion is excessive only with a closed magnetic circuit transferring no energy into a secondary circuit by mutual inductance.

3. The distorted wave of current can be replaced by the equivalent sine wave—that is, a sine wave of equal effective intensity and equal power—and the superposed higher harmonic, con-

sisting mainly of a term of triple frequency, may be neglected except in resonating circuits.

4. Below saturation, the distorted curve of current and its equivalent sine wave have approximately the same maximum value.

5. The angle of hysteretic advance—that is, the phase difference between the magnetic flux and equivalent sine wave of m.m.f.—is a maximum for the closed magnetic circuit, and depends there only upon the magnetic constants of the iron, upon the permeability, μ, the coefficient of hysteresis, η, and the maximum magnetic induction, as shown in the equation,

$$\sin \alpha = \frac{4\, \mu \eta}{B^{0.4}}.$$

6. The effect of hysteresis can be represented by an admittance $Y = g - jb$, or an impedance, $Z = r + jx$.

7. The hysteretic admittance, or impedance, varies with the magnetic induction; that is, with the e.m.f., etc.

8. The hysteretic conductance, g, is proportional to the coefficient of hysteresis, η, and to the length of the magnetic circuit, l, inversely proportional to the 0.4th power of the e.m.f. E, to the 0.6th power of frequency, f, and of the cross-section of the magnetic circuit, A, and to the 1.6th power of the number of turns of the electric circuit, n, as expressed in the equation,

$$g = \frac{58\, \eta l\, 10^3}{E^{0.4} f^{0.6} A^{0.6} n^{1.6}}.$$

9. The absolute value of hysteretic admittance,

$$y = \sqrt{g^2 + b^2},$$

is proportional to the magnetic reluctance: $\mathcal{R} = \mathcal{R}_i + \mathcal{R}_a$, and inversely proportional to the frequency, f, and to the square of the number of turns, n, as expressed in the equation,

$$y = \frac{(\mathcal{R}_i + \mathcal{R}_a)\, 10^8}{2\, \pi f n^2}.$$

10. In an ironclad circuit, the absolute value of admittance is proportional to the length of the magnetic circuit, and inversely proportional to cross-section, A, frequency, f, permeability, μ and square of the number of turns, n, or

$$y_i = \frac{127\, l\, 10^5}{n^2 A f \mu}.$$

11. In an open magnetic circuit, the conductance, g, is the same as in a closed magnetic circuit of the same iron part.

12. In an open magnetic circuit, the admittance, y, is practically constant, if the length of the air-gap is at least $1/100$ of the length of the magnetic circuit, and saturation be not approached.

13. In a closed magnetic circuit, conductance, susceptance, and admittance can be assumed as constant through a limited range only.

14. From the shape and the dimensions of the circuits, and the magnetic constants of the iron, all the electric constants, g, b, y; r, x, z, can be calculated.

104. The preceding applies to the alternating magnetic circuit, that is, circuit in which the magnetic induction varies between equal but opposite limits: $B_1 = + B_0$ and $B_2 = - B_0$.

In a pulsating magnetic circuit, in which the induction B varies between two values B_1 and B_2, which are not equal numerically, and which may be of the same sign or of opposite sign, that is in which the hysteresis cycle is unsymmetrical, the law of the 1.6^{th} power still applies, and the loss of energy per cycle is proportional to the 1.6^{th} power of the amplitude of the magnetic variation:

$$W_H = \eta \left(\frac{B_1 - B_2}{2}\right)^{1.6}$$

but the hysteresis coefficient η is not the same as for alternating magnetic circuits, but increases with increasing average value $\frac{B_1 + B_2}{2}$ of the magnetic induction.

Such unsymmetrical magnetic cycles occur in some types of inductor alternators, in which the magnetic induction does not reverse, but pulsates between a high and a low value in the same direction.

Unsymmetrical magnetic cycles occasionally occur—and give trouble—in transformers by the entrance of a stray direct current (railway return) over the ground connection, or when an unsuitable transformer connection is used on a synchronous converter feeding a three-wire system.

Very unsymmetrical cycles may give very much higher losses than symmetrical cycles of the same amplitude.

For more complete discussion of unsymmetrical cycles see "Theory and Calculation of Electric Circuits" [Dover Volume I; Part III, pp. 401–404].

CHAPTER XIII

FOUCAULT OR EDDY CURRENTS

105. While magnetic hysteresis due to molecular friction is a magnetic phenomenon, eddy currents are rather an electrical phenomenon. When iron passes through a magnetic field, a loss of energy is caused by hysteresis, which loss, however, does not react magnetically upon the field. When cutting an electric conductor, the magnetic field produces a current therein. The m.m.f. of this current reacts upon and affects the magnetic field, more or less; consequently, an alternating magnetic field cannot penetrate deeply into a solid conductor, but a kind of screening effect is produced, which makes solid masses of iron unsuitable for alternating fields, and necessitates the use of laminated iron or iron wire as the carrier of magnetic flux.

Eddy currents are true electric currents, though existing in minute circuits; and they follow all the laws of electric circuits.

Their e.m.f. is proportional to the intensity of magnetization, B, and to the frequency, f.

Eddy currents are thus proportional to the magnetization, B, the frequency, f, and to the electric conductivity, λ, of the iron; hence, can be expressed by

$$i = b\lambda Bf.$$

The power consumed by eddy currents is proportional to their square, and inversely proportional to the electric conductivity, and can be expressed by

$$P = b^2\lambda B^2 f^2;$$

or, since Bf is proportional to the generated e.m.f., E, in the equation

$$E = \sqrt{2}\,\pi A n f B\, 10^{-8},$$

it follows that, *The loss of power by eddy currents is proportional to the square of the e.m.f., and proportional to the electric conductivity of the iron;* or,

$$P = aE^2\lambda.$$

Hence, that component of the effective conductance which is due to eddy currents is

$$g = \frac{P}{E^2} = a\lambda;$$

that is, *The equivalent conductance due to eddy currents in the iron is a constant of the magnetic circuit; it is independent of e.m.f., frequency, etc., but proportional to the electric conductivity of the iron,* λ.

Eddy currents, like magnetic hysteresis, cause an advance of phase of the current by an *angle of advance*, β; but unlike hysteresis, eddy currents in general do not distort the current wave.

The angle of advance of phase due to eddy currents is

$$\sin \beta = \frac{g}{y},$$

where y = absolute admittance of the circuit, g = eddy current conductance.

While the equivalent conductance, g, due to eddy currents, is a constant of the circuit, and independent of e.m.f., frequency, etc., the loss of power by eddy currents is proportional to the square of the e.m.f. of self-induction, and therefore proportional to the square of the frequency and to the square of the magnetization.

Only the power component, gE, of eddy currents, is of interest, since the wattless component is identical with the wattless component of hysteresis, discussed in the preceding chapter.

106. To calculate the loss of power by eddy currents,

Let V = volume of iron;
B = maximum magnetic induction;
f = frequency;
λ = electric conductivity of iron;
ϵ = coefficient of eddy currents.

The loss of energy per cubic centimeter, in ergs per cycle, is

$$w = \epsilon\lambda f B^2;$$

hence, the total loss of power by eddy currents is

$$P = \epsilon\lambda V f^2 B^2\, 10^{-7} \text{ watts},$$

and the equivalent conductance due to eddy currents is

$$g = \frac{P}{E^2} = \frac{10\,\epsilon\lambda l}{2\,\pi^2 A n^2} = \frac{0.507\,\epsilon\lambda l}{A n^2},$$

where

l = length of magnetic circuit,
A = section of magnetic circuit,
n = number of turns of electric circuit.

The coefficient of eddy currents, ϵ, depends merely upon the shape of the constituent parts of the magnetic circuit; that is, whether of iron plates or wire, and the thickness of plates or the diameter of wire, etc.

The two most important cases are:

(*a*) Laminated iron.
(*b*) Iron wire.

107. (*a*) *Laminated Iron.*

Let, in Fig. 91,

d = thickness of the iron plates;
B = maximum magnetic induction;
f = frequency;
λ = electric conductivity of the iron.

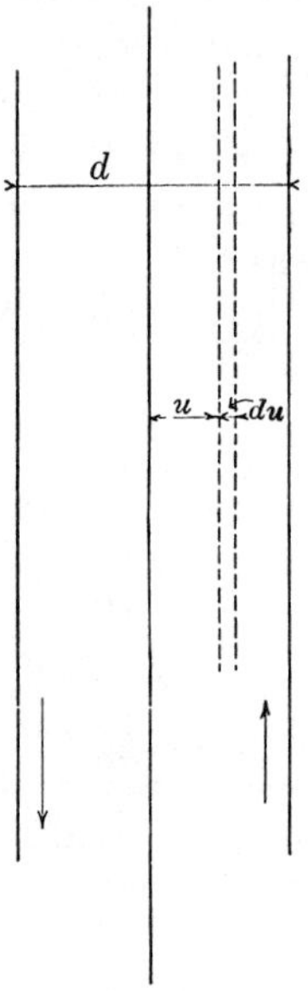

Then, if u is the distance of a zone, du, from the center of the sheet, the conductance of a zone of thickness, du, and of one centimeter length and width is λdu; and the magnetic flux cut by this zone is Bu. Hence, the e.m.f. induced in this zone is

$$\delta E = \sqrt{2}\,\pi f B u, \text{ in c.g.s. units.}$$

FIG. 91.

This e.m.f. produces the current, $dI = \delta E\,\lambda du = \sqrt{2}\,\pi f B\, u du$, in c.g.s. units, provided the thickness of the plate is negligible as compared with the length, in order that the current may be assumed as parallel to the sheet, and in opposite directions on opposite sides of the sheet.

The power consumed by the current in this zone, du, is

$$dP = \delta E dI = 2\,\pi^2 f^2 B^2 \lambda u^2 du,$$

in c.g.s. units or ergs per second, and, consequently, the total power consumed in one square centimeter of the sheet of thickness, d, is

$$\delta P = \int_{-\frac{d}{2}}^{+\frac{d}{2}} dP = 2\,\pi^2 f^2 B^2 \lambda \int_{-\frac{d}{2}}^{+\frac{d}{2}} u^2 du$$

$$= \frac{\pi^2 f^2\ {}^2\lambda d^3}{6}, \text{ in c.g.s. units;}$$

the power consumed per cubic centimeter of iron is, therefore,

$$p = \frac{\delta P}{d} = \frac{\pi^2 f^2 B^2 \lambda\, d^{2}}{6}, \text{ in c.g.s. units or erg-seconds,}$$

and the energy consumed per cycle and per cubic centimeter of iron is

$$w = \frac{p}{f} = \frac{\pi^2 \lambda\, d^2 f B^2}{6} \text{ ergs.}$$

The coefficient of eddy currents for laminated iron is, therefore,

$$\epsilon = \frac{\pi^2 d^2}{6} = 1.645\, d^2,$$

where λ is expressed in c.g.s. units. Hence, if λ is expressed in practical units or 10^{-9} c.g.s. units,

$$\epsilon = \frac{\pi^2\, d^2\, 10^{-9}}{6} = 1.645\, d^2\, 10^{-9}.$$

Substituting for the conductivity of sheet iron the approximate value.

$$\lambda = 10^5,\text{[1]}$$

we get as the coefficient of eddy currents for laminated iron,

$$\epsilon = \frac{\pi^2}{6}\, d^2\, 10^{-9} = 1.645\, d^2\, 10^{-9};$$

loss of energy per cubic centimeter and cycle,

$$W = \epsilon \lambda f B^2 = \frac{\pi^2}{6}\, d^2 \lambda f B^2\, 10^{-9} = 1.645\, d^2 \lambda f B^2\, 10^{-9} \text{ ergs}$$

$$= 1.645\, d^2 f B^2\, 10^{-4} \text{ ergs;}$$

or, $W = \epsilon \lambda f B^2 10^{-7} = 1.645\, d^2 f B^2 10^{-11}$ joules.

The loss of power per cubic centimeter at frequency, f, is

$$p = fW = \epsilon \lambda f^2 B^2 10^{-7} = 1.645\, d^2 f^2 B^2 10^{-11} \text{ watts;}$$

the total loss of power in volume, V, is

$$P = Vp = 1.645\, V d^2 f^2 B^2 10^{-11} \text{ watts.}$$

As an example,

$d = 1$ mm. $= 0.1$ cm.; $f = 100$; $B = 5{,}000$; $V = 1{,}000$ c.c.;
$\epsilon = 1{,}645 \times 10^{-11}$;
$W = 4{,}110$ ergs
$= 0.000411$ joules;
$p = 0.0411$ watts;
$P = 41.4$ watts.

[1] In some of the modern silicon steels used for transformer iron, λ reaches values as low as 2×10^4, and even lower; and the eddy current losses are reduced in the same proportion (1915).

108. (*b*) *Iron Wire.*

Let, in Fig. 92, d = diameter of a piece of iron wire; then if u is the radius of a circular zone of thickness, du, and one centimeter in length, the conductance of this zone is $\frac{\lambda\, du}{2\,\pi u}$, and the magnetic flux inclosed by the zone is $Bu^2\pi$.

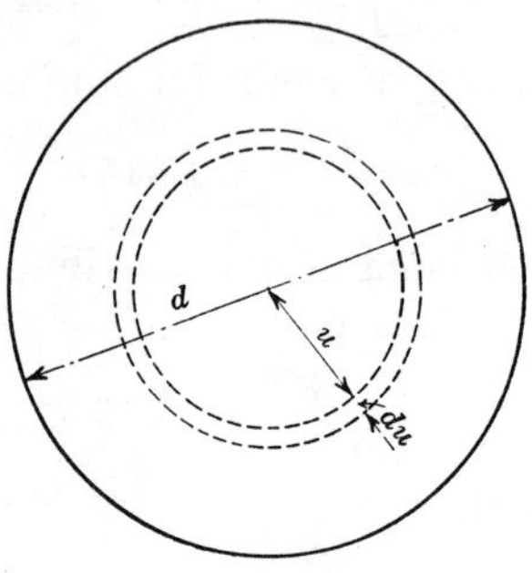

FIG. 92.

Hence, the e.m.f. generated in this zone is

$$\delta E = \sqrt{2}\,\pi^2 fBu^2 \text{ in c.g.s. units,}$$

and the current produced thereby is

$$dI = \frac{\lambda\, du}{2\pi u} \times \sqrt{2}\,\pi^2 fBu^2$$

$$= \frac{\sqrt{2}\pi}{2} \lambda fBu\, du, \text{ in c.g.s. units.}$$

The power consumed in this zone is, therefore,

$$dP = \delta E dI = \pi^3 \lambda f^2 B^2 u^3 du, \text{ in c.g.s. units;}$$

consequently, the total power consumed in one centimeter length of wire is

$$\delta P = \int_0^{\frac{d}{2}} dW = \pi^3\, \lambda f^2 B^2 \int_0^{\frac{d}{2}} u^3 du$$

$$= \frac{\pi^3}{64} \lambda f^2 B^2 d^4, \text{ in c.g.s. units.}$$

Since the volume of one centimeter length of wire is

$$v = \frac{d^2\pi}{4},$$

the power consumed in one cubic centimeter of iron is

$$p = \frac{\delta P}{v} = \frac{\pi^2}{16} \lambda f^2 B^2 d^2, \text{ in c.g.s. units or erg-seconds,}$$

and the energy consumed per cycle and cubic centimeter of iron is

$$W = \frac{p}{f} = \frac{\pi^2}{16} \lambda f B^2 d^2 \text{ ergs.}$$

Therefore, the coefficient of eddy currents for iron wire is

$$\epsilon = \frac{\pi^2}{16} d^2 = 0.617\, d^2;$$

or, if λ is expressed in practical units, or 10^{-9} c.g.s. units,

$$\epsilon = \frac{\pi^2}{16} d^2\, 10^{-9} = 0.617\, d^2\, 10^{-9}.$$

Substituting

$$\lambda = 10^5,$$

we get as the coefficient of eddy currents for iron wire,

$$\epsilon = \frac{\pi^2}{16} d^2\, 10^{-9} = 0.617\, d^2\, 10^{-9}.$$

The loss of energy per cubic centimeter of iron, and per cycle, becomes

$$\begin{aligned} W &= \epsilon \lambda f B^2 = \frac{\pi^2}{16} d^2 \lambda f B^2\, 10^{-9} = 0.617\, d^2 \lambda f B^2\, 10^{-9} \\ &= 0.617\, d^2 f B^2\, 10^{-4} \text{ ergs,} \\ &= \epsilon \lambda f B^2\, 10^{-7} = 0.617\, d^2 f B^2\, 10^{-11} \text{ joules;} \end{aligned}$$

loss of power per cubic centimeter at frequency, f,

$$p = fW = \epsilon \lambda N^2 B^2\, 10^{-7} = 0.617\, d^2 N^2 B^2\, 10^{-11} \text{ watts;}$$

total loss of power in volume, V,

$$P = Vp = 0.617\, V d^2 f^2 B^2\, 10^{-11} \text{ watts.}$$

As an example,

$d = 1$ mm., $= 0.1$ cm.; $f = 100$; $B^2 = 5{,}000$; $V = 1{,}000$ cu. cm.

Then,

$$\begin{aligned} \epsilon &= 0.617 \times 10^{-11}, \\ W &= 1{,}540 \text{ ergs} = 0.000154 \text{ joules,} \\ p &= 0.0154 \text{ watts,} \\ P &= 1.54 \text{ watts,} \end{aligned}$$

hence very much less than in sheet iron of equal thickness.

109. *Comparison of sheet iron and iron wire.*

If

d_1 = thickness of lamination of sheet iron, and
d_2 = diameter of iron wire,

the eddy current coefficient of sheet iron being

$$\epsilon_1 = \frac{\pi^2}{6} d_1^2 10^{-9},$$

and the eddy current coefficient of iron wire

$$\epsilon_2 = \frac{\pi^2}{16} d_2^2 10^{-9},$$

the loss of power is equal in both—other things being equal—if $\epsilon_1 = \epsilon_2$; that is, if

$$d_2^2 = 8/3\, d_1^2, \text{ or } d_2 = 1.63\, d_1.$$

It follows that the diameter of iron wire can be 1.63 times or, roughly, $1\frac{2}{3}$ as large as the thickness of laminated iron, to give the same loss of power through eddy currents, as shown in Fig. 93.

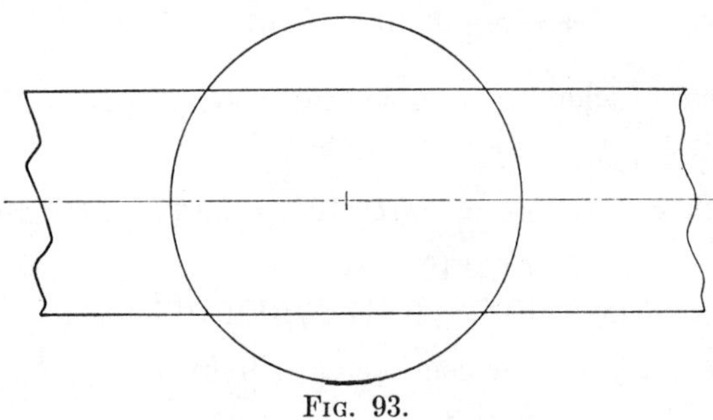

FIG. 93.

110. *Demagnetizing, or screening effect of eddy currents.* The formulas derived for the coefficient of eddy currents in laminated iron and in iron wire hold only when the eddy currents are small enough to neglect their magnetizing force. Otherwise the phenomenon becomes more complicated; the magnetic flux in the interior of the lamina, or the wire, is not in phase with the flux at the surface, but lags behind it. The magnetic flux at the surface is due to the impressed m.m.f., while the flux in the interior is due to the resultant of the impressed m.m.f. and to the m.m.f. of eddy currents; since the eddy currents lag 90 degrees behind the flux producing them, their resultant with the impressed m.m.f., and therefore the magnetism in the interior, is made lagging. Thus, progressing from the surface toward the interior, the magnetic flux gradually lags more and more in phase, and at the same time decreases in intensity. While the complete analytical solution of this phenomenon is beyond the scope of this book, a determination of the magnitude of this demagnetization, or screening effect, sufficient to determine whether it is negligible, or whether the subdivision of the iron has to be increased to make it negligible, can be made by calculating the maximum magnetizing effect, which cannot be exceeded by the eddys.

Assuming the magnetic density as uniform over the whole cross-section, and therefore all the eddy currents in phase with each other, their total m.m.f. represents the maximum possible value, since by the phase difference and the lesser magnetic density in the center the resultant m.m.f. is reduced.

In laminated iron of thickness d, the current in a zone of thickness du, at distance u from center of sheet, is

$$\begin{aligned} dI &= \sqrt{2}\,\pi f B\lambda u\, du \text{ units (c.g.s.)} \\ &= \sqrt{2}\,\pi f B\lambda u\, du\, 10^{-8} \text{ amp.;} \end{aligned}$$

hence the total current in the sheet is

$$\begin{aligned} I &= \int_0^{\frac{d}{2}} dI = \sqrt{2}\,\pi f B\lambda\, 10^{-8} \int_0^{\frac{d}{2}} u\, du \\ &= \frac{\sqrt{2}\,\pi}{8} f B\lambda d^2\, 10^{-8} \text{ amp.} \end{aligned}$$

Hence, the maximum possible demagnetizing ampere-turns, acting upon the center of the lamina, are

$$\begin{aligned} I &= \frac{\sqrt{2}\,\pi}{8} f B\lambda d^2\, 10^{-8} = 0.555 f B\lambda d^2\, 10^{-8}, \\ &= 0.555\, f B\lambda d^2\, 10^{-8} \text{ ampere-turns per cm.} \end{aligned}$$

Example: $d = 0.1$ cm., $f = 100$, $B = 5000$, $\lambda = 10^5$, or $I = 2.775$ ampere-turns per cm.

111. In iron wire of diameter d, the current in a tubular zone of du thickness and u radius is

$$dI = \frac{\sqrt{2}}{2} \pi f B\lambda u\, du\, 10^{-8} \text{ amp.;}$$

hence, the total current is

$$\begin{aligned} I &= \int_0^{\frac{d}{2}} dI = \frac{\sqrt{2}}{2} \pi f B\lambda\, 10^{-8} \int_0^{\frac{d}{2}} u\, dx \\ &= \frac{\sqrt{2}}{16} \pi f B\lambda d^2\, 10^{-8} \text{ amp.} \end{aligned}$$

Hence, the maximum possible demagnetizing ampere-turns, acting upon the center of the wire, are

$$I = \frac{\sqrt{2}\,\pi}{16} fB\lambda d^2\, 10^{-8} = 0.2775\, fB\lambda d^2\, 10^{-8}$$
$$= 0.2775\, fB\lambda d^2\, 10^{-8} \text{ ampere-turns per cm.}$$

For example, if $d = 0.1$ cm., $f = 100$, $B = 5000$, $\lambda = 10^5$, then $I = 1.338$ ampere-turns per cm.; that is, half as much as in a lamina of the thickness d.

For a more complete investigation of the screening effect of eddy currents in laminated iron, see "Transients in Space" [Dover Volume III, pp. 59–221].

112. Besides the eddy, or Foucault, currents proper, which exist as parasitic currents in the interior of the iron lamina or wire, under certain circumstances eddy currents also exist in larger orbits from lamina to lamina through the whole magnetic structure. Obviously a calculation of these eddy currents is possible only in a particular structure. They are mostly surface currents, due to short circuits existing between the laminæ at the surface of the magnetic structure.

Furthermore, eddy currents are produced outside of the magnetic iron circuit proper, by the magnetic stray field cutting electric conductors in the neighborhood, especially when drawn toward them by iron masses behind, in electric conductors passing through the iron of an alternating field, etc. All these phenomena can be calculated only in particular cases, and are of less interest, since they can and should be avoided.

The power consumed by such large eddy currents frequently increases more than proportional to the square of the voltage, when approaching magnetic saturation, by the magnetic stray field reaching unlaminated conductors, and so, while negligible at normal voltage, this power may become large at over-normal voltage.

Eddy Currents in Conductor, and Unequal Current Distribution

113. If the electric conductor has a considerable size, the alternating magnetic field, in cutting the conductor, may set up differences of potential between the different parts thereof, thus giving rise to local or eddy currents in the copper. This phenomenon can obviously be studied only with reference to a particular case, where the shape of the conductor and the distribution of the magnetic field are known.

Only in the case where the magnetic field is produced by the current in the conductor can a general solution be given. The alternating current in the conductor produces a magnetic field, not only outside of the conductor, but inside of it also; and the lines of magnetic force which close themselves inside of the conductor generate e.m.fs. in their interior only. Thus the counter e.m.f. of self-induction is largest at the axis of the conductor, and least at its surface; consequently, the current density at the surface will be larger than at the axis, or, in extreme cases, the current may not penetrate at all to the center, or a reversed current may exist there. Hence it follows that only the exterior part of the conductor may be used for the conduction of electricity, thereby causing an increase of the ohmic resistance due to unequal current distribution.

The general discussion of this problem, as applicable to the distribution of alternating current in very large conductors, as the iron rails of the return circuit of alternating-current railways, is given in "Transients in Space" [Dover Volume III, pp. 59–221].

In practice, this phenomenon is observed mainly with very high frequency currents, as lightning discharges, wireless telegraph and lightning arrester circuits; in power-distribution circuits it has to be avoided by either keeping the frequency sufficiently low or having a shape of conductor such that unequal current-distribution does not take place, as by using a tubular or a flat conductor, or several conductors in parallel.

114. It will, therefore, here be sufficient to determine the largest size of round conductor, or the highest frequency, where this phenomenon is still negligible.

In the interior of the conductor, the current density is not only less than at the surface, but the current lags in phase behind the current at the surface, due to the increased effect of self-induction. This time-lag of the current causes the magnetic fluxes in the conductor to be out of phase with each other, making their resultant less than their sum, while the lesser current density in the center reduces the total flux inside of the conductor. Thus, by assuming, as a basis for calculation, a uniform current density and no difference of phase between the currents in the different layers of the conductor, the unequal distribution is found larger than it is in reality. Hence this assumption brings us on the safe side, and at the same time greatly simplifies the calculation; however, it is permissible only where the current density is still fairly uniform.

Let Fig. 94 represent a cross-section of a conductor of radius, R, and a uniform current density,

$$i = \frac{I}{R^2 \pi},$$

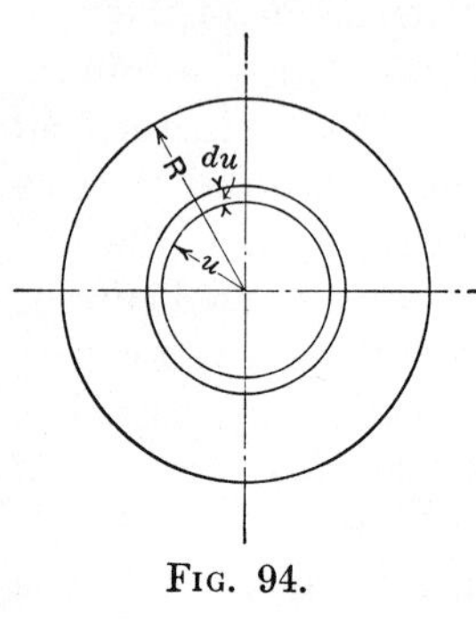

FIG. 94.

where I = total current in conductor.

The magnetic reluctance of a tubular zone of unit length and thickness du, of radius u, is

$$\mathcal{R}_u = \frac{2\, u\pi}{du}.$$

The current inclosed by this zone is $I_u = iu^2\pi$, and therefore, the m.m.f. acting upon this zone is

$$F_u = 0.4\,\pi I_u = 0.4\,\pi^2 iu^2,$$

and the magnetic flux in this zone is

$$d\Phi = \frac{F_u}{\mathcal{R}_u} = 0.2\,\pi iu\,du.$$

Hence, the total magnetic flux inside the conductor is

$$\Phi = \int_0^R d\Phi = \frac{2\,n}{10}\,i\int_0^R u\,du = \frac{\pi i R^2}{10} = \frac{I}{10}.$$

From this we get, as the excess of counter e.m.f. at the axis of the conductor over that at the surface,

$$\Delta E = \sqrt{2}\,\pi f\Phi\, 10^{-8} = \sqrt{2}\,\pi f I\, 10^{-9}, \text{ per unit length,}$$
$$= \sqrt{2}\,\pi^2 fiR^2\, 10^{-9};$$

and the reactivity, or specific reactance at the center of the conductor, becomes $k = \frac{\Delta E}{i} = \sqrt{2}\,\pi^2 fR^2\, 10^{-9}$.

Let ρ = resistivity, or specific resistance, of the material of the conductor.

We have then,

$$\frac{k}{\rho} = \frac{\sqrt{2}\,\pi^2 fR^2\, 10^{-9}}{\rho};$$

and

$$\frac{\rho}{\sqrt{k^2 + \rho^2}},$$

the ratio of current densities at center and at periphery.

For example, if, in copper, $\rho = 1.7 \times 10^{-6}$, and the percentage decrease of current density at center shall not exceed 5 per cent., that is,

$$\rho \div \sqrt{k^2 + \rho^2} = 0.95 \div 1,$$

we have

$$k = 0.51 \times 10^{-6};$$

hence

$$0.51 \times 10^{-6} = \sqrt{2}\, \pi^2 f R^2\, 10^{-9},$$

or

$$fR^2 = 36.3;$$

hence, when

$f =$	125	100	60	25
$R =$	0.541	0.605	0.781	1.21 cm.
$D = 2R =$	1.08	1.21	1.56	2.42 cm.

Hence, even at a frequency of 125 cycles, the effect of unequal current distribution is still negligible at one centimeter diameter of the conductor. Conductors of this size are, however, excluded from use at this frequency by the external self-induction, which is several times larger than the resistance. We thus see that unequal current distribution is usually negligible in practice.

The above calculation was made under the assumption that the conductor consists of unmagnetic material. If this is not the case, but the conductor of iron of permeability μ, then $d\Phi = \frac{\mu F_u}{\mathcal{R}_u}$; and thus ultimately, $k = \sqrt{2}\, \pi^2 f \mu R^2\, 10^{-9}$, and $\frac{k}{\rho} = \sqrt{2}\, \pi^2 \frac{f \mu R^2\, 10^{-9}}{\rho}$. Thus, for instance, for iron wire at $\rho = 10 \times 10^{-6}$, $\mu = 500$, it is, permitting 5 per cent. difference between center and outside of wire, $k = 3.2 \times 10^{-6}$, and $fR^2 = 0.46$;

hence, when

$f =$	125	100	60	25
$R =$	0.061	0.068	0.088	0.136 cm.;

thus the effect is noticeable even with relatively small iron wire.

Mutual Induction

115. When an alternating magnetic field of force includes a secondary electric conductor, it generates therein an e.m.f. which produces a current, and thereby consumes energy if the circuit of the secondary conductor is closed.

Particular cases of such secondary currents are the eddy or Foucault currents previously discussed.

Another important case is the generation of secondary e.m.fs. in neighboring circuits; that is, the interference of circuits running parallel with each other.

In general, it is preferable to consider this phenomenon of mutual induction as not merely producing a power component and a wattless component of e.m.f. in the primary conductor, but to consider explicitly both the secondary and the primary circuit, as will be done in the chapter on the alternating-current transformer.

Only in cases where the energy transferred into the secondary circuit constitutes a small part of the total primary energy, as in the discussion of the disturbance caused by one circuit upon a parallel circuit, may the effect on the primary circuit be considered analogously as in the chapter on eddy currents by the introduction of a power component, representing the loss of power, and a wattless component, representing the decrease of self-induction.

Let

$x = 2\pi fL$ = reactance of main circuit; that is, L = total number of interlinkages with the main conductor, of the lines of magnetic force produced by unit current in that conductor;

$x_1 = 2\pi fL_1$ = reactance of secondary circuit; that is, L_1 = total number of interlinkages with the secondary conductor, of the lines of magnetic force produced by unit current in that conductor;

$x_m = 2\pi fL_1$ = mutual inductive reactance of the circuits; that is, L_m = total number of interlinkages with the secondary conductor, of the lines of magnetic force produced by unit current in the main conductor, or total number of interlinkages with the main conductor of the lines of magnetic force produced by unit current in the secondary conductor.

Obviously: $x_m^2 \geqq xx_1$.[1]

[1] As self-inductance L, L_1, the total flux surrounding the conductor is here meant. Usually in the discussion of inductive apparatus, especially of transformers, as the self-inductance of circuit is denoted that part of the magnetic flux which surrounds one circuit but not the other circuit; and as mutual inductance flux which passes between both circuits. Hence, the total self-inductance, L, is in this case equal to the sum of the self-inductance, L_l, and mutual inductance, L_m.

The object of this distinction is to separate the wattless part, L_1, of the

Let r_1 = resistance of secondary circuit. Then the impedance of secondary circuit is

$$Z_1 = r_1 + jx_1,\quad z_1 = \sqrt{r_1^2 + x_1^2};$$

e.m.f. generated in the secondary circuit, $E_1 = -jx_m I$, where I = primary current. Hence, the secondary current is

$$I_1 = \frac{E_1}{z_1} = \frac{-jx_m}{r_1 + jx_1} I;$$

and the e.m.f. generated in the primary circuit by the secondary current, I_1, is

$$E = -jx_m I_1 = \frac{-x_m^2}{r_1 + jx_1} I;$$

or, expanded,

$$E = \left\{\frac{-x_m^2 r_1}{r_1^2 + x_1^2} + \frac{jx_m^2 x_1}{r_1^2 + x_2^1}\right\} I.$$

Hence, the e.m.f. consumed thereby,

$$E' = \left\{\frac{x_m^2 r_1}{r_1^2 + x_1^2} - \frac{jx_m^2 x_1}{r_1^2 + x_1^2}\right\} I = (r + jx)I.$$

$$r = \frac{x_m^2 r_1}{r_1^2 + x_1^2} = \text{effective resistance of mutual inductance;}$$

$$x = \frac{-x_m^2 x_1}{r_1^2 + x_1^2} = \text{effective reactance of mutual inductance.}$$

The susceptance of mutual inductance is negative, or of opposite sign from the reactance of self-inductance. Or,

Mutual inductance consumes energy and decreases the self-inductance.

For the calculation of the mutual inductance between circuits L_m, see pp. 21–24.

---

total self-inductance, L, from that part, L_m, which represents the transfer of e.m.f. into the secondary circuit, since the action of these two components is essentially different.

Thus, in alternating-current transformers it is customary—and will be done later in this book—to denote as the self-inductance, L, of each circuit only that part of the magnetic flux produced by the circuit which passes between both circuits, and thus acts in "choking" only, but not in transforming; while the flux surrounding both circuits is called the mutual inductance, or useful magnetic flux.

With this denotation, in transformers the mutual inductance, L_m, is usually very much greater than the self-inductance, L', and L_1', while, if the self-inductance, L and L_1, represent the total flux, their product is larger than the square of the mutual inductance, L_m; or

$$LL_1 \geq L_m^2; \qquad (L' + L_m)(L_1' + L_m) \geq L_m^2.$$

CHAPTER XIV

DIELECTRIC LOSSES

Dielectric Hysteresis

116. Just as magnetic hysteresis and eddy currents give a power component in the inductive reactance, as "effective resistance," so the energy losses in the dielectric lead to a power component in the condensive reactance, which may be represented by an "effective resistance of dielectric losses" or an "effective conductance of dielectric losses."

In the alternating magnetic field, power is consumed by magnetic hysteresis. This is proportional to the frequency, and to the 1.6^{th} power of the magnetic density, and is considerable, amounting in a closed magnetic circuit to 40 to 60 per cent. of the total volt-amperes.

In the dielectric field, the energy losses usually are very much smaller, rarely amounting to more than a few per cent., though they may at high temperature in cables rise as high as 40 to 60 per cent. The foremost such losses are: leakage, that is, i^2r loss of the current passing by conduction (as "dynamic current") through the resistance of the dielectric; corona, that is, losses due to a partial or local breakdown of the electrostatic field, and dielectric hysteresis or phenomena of similar nature.

It is doubtful whether a true dielectric hysteresis, that is, a molecular dielectric friction, exists. A dielectric loss, proportional to the frequency and to the 1.6^{th} power of the dielectric field:

$$P = nfD^{1.6}$$

has been observed in rotating dielectric fields, but is so small, that it usually is overshadowed by the other losses.

In alternating dielectric fields in solid materials, such as in condensers, coil insulation, etc., a loss is commonly observed which gives an approximately constant power-factor of the electric energizing circuit, over a wide range of voltage and of frequency, from less than a fraction of 1 per cent. up to a few per cent.

Constancy of the power-factor with the frequency, means that the loss is proportional to the frequency, as the current i, and thus the volt-ampere input, ei, are proportional to the frequency. Constancy of the power-factor with the voltage, means that the loss is proportional to the square of the voltage, as the current i is proportional to the voltage, and the volt-ampere input ei thus proportional to the square of the voltage. This loss thus would be approximated by the expression:

$$P = \eta f D^2$$

and thus seems to be akin to magnetic hysteresis, except that at least a part of this dielectric loss is possibly consumed in chemical and mechanical disintegration of the insulating material, while the magnetic hysteresis loss is entirely converted to heat.

Leakage

117. The eddy current losses in the magnetic field are the i^2r loss of the currents flowing in the magnetic material, and as such are proportional to the square of the frequency and of the magnetic density:

$$P = \epsilon \gamma f^2 B^2$$

where γ = conductivity of the magnetic material.

This expression obviously holds only as long as the m.m.f. of the eddy currents is not sufficient to appreciably affect the magnetic flux distribution.

As corresponding hereto in the dielectric field may be considered the conduction loss through the resistance of the dielectric.

In a homogeneous dielectric of electric conductivity γ (usually very low) and specific capacity or permittivity k, if:

l = thickness of the dielectric,
A = area or cross-section,
e = impressed alternating-current voltage, effective value,

the dielectric capacity of the material is:

$$C = \frac{kA}{l}$$

and the capacity susceptance:

$$b = 2\,\pi f C = \frac{2\,\pi f k A}{l}$$

hence the current passing through the dielectric as capacity current or "displacement current," is:

$$i_0 = eb = 2\,\pi f C e = \frac{2\,\pi f k A}{l}\,e$$

The conductance of the dielectric is:

$$g = \frac{\gamma A}{l}$$

hence, the current, conducted through the dielectric, or leakage current:

$$i_1 = eg = \frac{\gamma A}{l}\,e$$

thus, the total current:

$$\dot{I} = i_0 + j i_1 = \frac{eA}{l}\{\gamma + 2\pi f k j\}$$

here the j denotes, that the current component i_0 is in quadrature ahead of the voltage e.

The absolute value of the current thus is:

$$i = \sqrt{i_0^2 + i_1^2} = \frac{eA}{l}\sqrt{\gamma^2 + (2\,\pi f k)^2}$$

and the power consumption:

$$P = e i_1 = \frac{e^2 \gamma A}{l}$$

or, since the dielectric density D is proportional to the voltage gradient $\frac{e}{l}$ and the permittivity:

$$D = \frac{ek}{4\,\pi v^2 l}$$

(where $v = 3 \times 10^{10}$ = velocity of light, see Dover edition, Volume I, Part I).

Thus:

$$P = \frac{(4\,\pi v^2)^2 \gamma V D^2}{k^2}$$

where

$$V = Al = \text{volume}$$

The power-factor then is:

$$p = \frac{P}{ei} = \frac{\gamma}{\sqrt{\gamma^2 + (2\,\pi f k)^2}}$$

Or, if, as usually the case, the conductivity γ is small compared with the susceptivity $2\pi fk$:

$$p = \frac{\gamma}{2\pi fk}$$

that is, the power-factor is inverse proportional to the frequency.

The observation of leakage losses and leakage resistance thus is best made at low frequencies or at direct-current voltage.

While, however, in magnetic materials the conductivity γ is fairly constant, varying only with the temperature, like that of all metals, the very low conductivity of the dielectric is often not even approximately constant, but may vary with the temperature, the voltage, etc., sometimes by many thousand per cent.

118. While in a homogeneous dielectric field, the leakage current power losses are independent of the frequency and herein differ from the magnetic eddy current losses, which latter are proportional to the square of the frequency, in non-homogeneous dielectric fields, leakage current losses may depend on the frequency.

As an instance, let us consider a dielectric consisting of two layers of different constants, for instance, a layer of mica and a layer of varnished cloth, such as is sometimes used in high-voltage armature insulation.

Let γ_1 = electric conductivity,
k_1 = permittivity or specific capacity,
l_1 = thickness and,
A_1 = area or section

of the first layer of the dielectric, and

$$\gamma_2,\ k_2,\ l_2,\ A_2$$

the corresponding values of the second layer.

It is then:

$$\left.\begin{aligned} g &= \frac{\gamma A}{l} = \text{electric conductance} \\ C &= \frac{kA}{l} = \text{electrostatic capacity of the layer of dielectric, hence:} \\ b &= 2\pi fC = \frac{2\pi fkA}{l} = \text{capacity susceptance, and} \end{aligned}\right\} \quad (1)$$

$$\left.\begin{aligned}
Y &= g + jb = \text{admittance, thus:} \\
Z &= \frac{1}{Y} = r - jx = \text{impedance, where:} \\
r &= \frac{g}{y^2} = \text{vector resistance (not ohmic resistance,} \\
&\qquad \text{but energy component of impedance,} \\
&\qquad \text{see paragraph 89.)} \\
x &= \frac{b}{y^2} = \text{vector reactance, and} \\
y &= \sqrt{g^2 + b^2} = \text{absolute admittance,} \\
(z &= \sqrt{r^2 + x^2} = \text{absolute impedance.)}
\end{aligned}\right\} \quad (2)$$

If then, $\dot{E}_1$ = potential drop across the first, $\dot{E}_2$ = potential drop across the second layer of dielectric,

$$\dot{E} = \dot{E}_1 + \dot{E}_2 = \text{voltage impressed upon the dielectric.} \quad (3)$$

The current i, which traverses the dielectric, partly by conduction through its resistance, partly by capacity as displacement current, then is the same in both layers, as they are in series in the dielectric field, and it is:

$$\left.\begin{aligned}
\dot{E}_1 &= i(r_1 - jx_1) \\
\dot{E}_2 &= i(r_2 - jx_2)
\end{aligned}\right\} \quad (4)$$

and, by (3):

$$\dot{E} = i\ \{(r_1 + r_2) - j(x_1 + x_2)\} \quad (6)$$

or, absolute:

$$e = i\sqrt{(r_1 + r_2)^2 + (x_1 + x_2)^2} \quad (6)$$

Thus, the current:

$$i = \frac{e}{\sqrt{(r_1 + r_2)^2 + (x_1 + x_2)^2}} \quad (7)$$

the apparent power, or volt-ampere input:

$$Q = ei = \frac{e^2}{\sqrt{(r_1 + r_2)^2 + (x_1 + x_2)^2}} \quad (8)$$

the power consumed in the dielectric is:

$$\begin{aligned}
P &= i^2(r_1 + r_2) \\
&= \frac{e^2(r_1 + r_2)}{(r_1 + r_2)^2 + (x_1 + x_2)^2}
\end{aligned} \quad (9)$$

and the power-factor:

$$P = \frac{P}{Q} = \frac{r_1 + r_2}{\sqrt{(r_1 + r_2)^2 + (x_1 + x_2)^2}} \quad (10)$$

119. Let us consider some special cases:

(*a*) If the conductivity, γ_1 and γ_2, of the two layers of dielectric is so small that the conduction current, ge, is negligible compared with the capacity current, $2\pi fCe$.

In this case, r_1 and r_2 are negligible compared with x_1 and x_2, and it is:

$$\left.\begin{aligned} i &= \frac{e}{x_1 + x_2} \\ P &= \frac{e^2(r_1 + r_2)}{(x_1 + x_2)^2} \\ p &= \frac{r_1 + r_2}{x_1 + x_2} \end{aligned}\right\} \tag{11}$$

Substituting now for the impedance quantities $Z = r - jx$, which have no direct physical meaning in the dielectric field, the admittance quantities $Y = g + jb$, which have the physical meaning that g is the effective ohmic conductance, b the capacity susceptance, it is:

g negligible compared with b and y, and $b = y$.

Thus, by (2):

$$i = \frac{eb_1b_2}{b_1 + b_2} = \frac{2\pi fC_1C_2e}{C_1 + C_2} \tag{12}$$

hence proportional to the frequency f:

$$P = \frac{e^2(g_1b_2^2 + g_2b_1^2)}{(b_1 + b_2)^2} = \frac{e^2(g_1C_2^2 + g_2C_1^2)}{(C_1 + C_2)^2} \tag{13}$$

hence, the loss of power by current leakage in the dielectric in this case is independent of the frequency.

$$p = \frac{g_1\dfrac{b_2}{b_1} + g_2\dfrac{b_1}{b_2}}{b_1 + b_2} = \frac{g_1\dfrac{C_2}{C_1} + g_2\dfrac{C_1}{C_2}}{2\pi f(C_1 + C_2)} \tag{14}$$

hence, in this case the power-factor is inverse proportional to the frequency.

(*b*) If in both layers the leakage current is large compared with the capacity current, that is, $2\pi fCe$ negligible compared with ge.

In this case, x_1 and x_2 are negligible compared with r_1 and r_2, and:

$$\left.\begin{aligned} i &= \frac{e}{r_1 + r_2} \\ Q &= \frac{e^2}{r_1 + r_2} \\ P &= \frac{e^2}{r_1 + r_2} \\ p &= 1 \end{aligned}\right\} \tag{15}$$

and as in this case r_1 and r_2 are the effective ohmic resistance of the dielectric, all the quantities are independent of the frequency; that is, the case is one of simple conduction.

120 (*c*) If in the first layer the leakage is negligible compared with the capacity current, but is not negligible in the second layer. That is, in a two-layer insulation, one layer leaks badly.

Assuming for simplicity that the two layers have the same capacity, $C = C_1 = C_2$. If the two capacities are unequal, the treatment is analogous, but merely the equations somewhat more complicated.

Let the conductance of the second layer $= g$, the capacity susceptance $2\pi fC = b$.

It is then:

r_1 negligible compared with the other quantities.

$$\left.\begin{aligned} r_2 &= \frac{g}{g^2+b^2} \\ x_1 &= \frac{1}{b} \\ x_2 &= \frac{b}{g^2+b^2} \end{aligned}\right\} \qquad (16)$$

Substituting these values in equations (7) (8) (9) (10) gives:

$$i = \frac{e(g^2+b^2)}{g\sqrt{1+\left(\frac{g}{b}+\frac{2b}{g}\right)^2}} = \frac{e(g^2+(2\pi fC)^2)}{g\sqrt{1+\left(\frac{g}{2\pi fC}+\frac{4\pi fC}{g}\right)^2}} \qquad (17)$$

$$P = \frac{e^2(g^2+b^2)}{g\left\{1+\left(\frac{g}{b}+\frac{2b}{g}\right)\right\}} = \frac{e^2(g^2+(2\pi fC)^2)}{g\left\{1+\left(\frac{g}{2\pi fC}+\frac{4\pi fC}{g}\right)^2\right\}} \qquad (18)$$

$$p = \frac{1}{\sqrt{1+\left(\frac{g}{b}+2\frac{b}{g}\right)^2}} = \frac{1}{\sqrt{1+\left(\frac{g}{2\pi fC}+\frac{4\pi fC}{g}\right)^2}} \qquad (19)$$

As seen, in this case current, power loss and power-factor depend on the frequency, but in a more complex manner.

With changing values of the conductance from low values, where g is negligible compared with the other terms, but the other terms negligible compared with $\frac{1}{g}$, up to high conductivity, where $\frac{1}{g}$ is negligible, but the terms with g predominate, the *current* changes from:

low g:

$$i = \pi f C e,$$

proportional to the frequency, to:
high g:

$$i = 2\pi f C e.$$

Again proportional to the frequency, but twice as large, and at intermediate values of g, the current changes more rapidly than proportional to the frequency. The *loss of power* changes from:
low g:

$$P = \frac{ge}{4},$$

or independent of the frequency, to:
high g:

$$P = \frac{(2\pi f C)^2 e}{g},$$

or proportional to the square of the frequency. The *power-factor* changes from:
low g:

$$p = \frac{g}{4\pi f C},$$

or inverse proportional to the frequency, to:
high g:

$$p = \frac{2\pi f C}{g},$$

or proportional to the frequency.
And over a considerable range of intermediate values of conductance, g, the power-factor, therefore, remains approximately constant; or, inversely, with changing frequency and constant g and b, the power-factor changes from proportionality with the frequency at low frequencies, up to inverse proportionality at high frequencies, and thereby passes through a maximum.

The value of g, for which the power-factor in equation (19) is a maximum, is found by differentiating: $\frac{dp}{dg} = 0$, as:

$$g = 2\sqrt{2}\,\pi f C \tag{20}$$

and this maximum power-factor is $p_0 = \frac{1}{3}$.

For $C_2 > C_1$, higher, for $C_2 < C_1$, lower values of power-factor maximum result, where C_2 is the leaky dielectric.

As illustration, Fig. 95 gives the values of power-factor, p, from equation (19), as function of $\frac{g}{b} = \frac{g}{2\pi fC} = \frac{g}{2\pi fk}$ as abscissæ.

A dielectric circuit, in which the power-factor decreases with increasing frequency, for instance, is that of the capacity of the transmission line; a dielectric circuit, in which the power-factor increases with the frequency, is that of the aluminum-cell lightning arrester.

121. As seen, in the dielectric circuit, that is, in insulators in which the current is essentially a displacement current, the

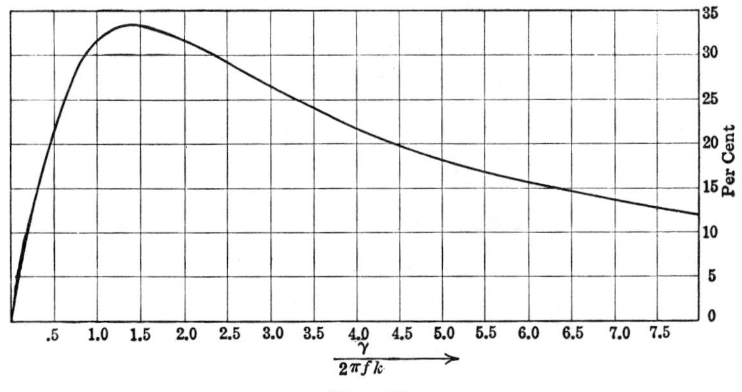

Fig. 95.

relations between voltage, current, power, phase angle and power-factor can be represented by the same symbolic equations as the relations between voltage, current, power and power-factor in metallic conductors, in which the current flow is dynamic, by the introduction of the effective admittance of the dielectric circuit, or part of circuit:

$$Y = g + jb,$$

where g is the effective conductance of the dielectric circuit, or the energy component of the admittance, representing the energy consumption by leakage, dielectric hysteresis, corona, etc., and $b = 2\pi fC$ is the capacity susceptance. Instead of the admittance Y, its reciprocal, the impedance $Z = r - jx$, may be used.

The main differences between the dielectric and the electrodynamic circuit are:

In the dielectric circuit, the susceptance, b, is positive, the reactance, x, negative; the current normally leads the voltage, that is, capacity effects predominate and inductive effects are usually absent.

In the dynamic circuit, the reactance, x, usually is positive, the susceptance, b, negative; the current usually lags, that is, inductive effects predominate and capacity effects are usually absent.

In the dielectric circuit, the admittance terms, $Y = g + jb$, have a physical meaning as the effective conductance and the capacity susceptance, $2\pi fC$, but the impedance terms, $Z = r - jx$, are only derived quantities, without direct physical meaning: the vector resistance, r, is not the effective ohmic resistance of the dielectric, $\frac{1}{g}$, but is also depending on the capacity, $r = \frac{g}{g^2 + b^2}$, and the vector reactance, x, is not the condensive reactance, $\frac{1}{b} = \frac{1}{2\pi fC}$, but also depends on the conductance, $x = \frac{b}{g^2 + b^2}$.

In the dynamic circuit, the impedance terms, $Z = r + jx$, have a direct physical meaning, as effective ohmic resistance, r, and as self-inductive reactance, $2\pi fL$, while the admittance terms, $Y = g - jb$, are derived quantities, and the vector conductance, g, is not the reciprocal of the resistance, r, the vector susceptance, b, not the reciprocal of the reactance, x, as discussed in preceding chapters.

Physically, the most prominent difference between the dielectric circuit and the dynamic circuit is that for the displacement current of the dielectric circuit, that is, for the electrostatic flux, all space is conducting, while for the dynamic current, most materials are practically non-conductors, and the dynamic circuit thus is sharply defined in the extent of the flow of the current, while the dielectric circuit is not. The dielectric circuit thus is similar to the magnetic circuit; for the magnetic circuit all space is conducting also, that is, can carry magnetic flux. An uninsulated submarine electric circuit would be more nearly similar, in the distribution of current flow, to the dielectric and the magnetic circuit.

In the electric circuit, the conductor through which the current flows is generally sharply defined and of a uniform section, which is small compared with the length, and the conductor thus can be approximated as a linear conductor, that is, the current distribution throughout the conductor section assumed as uniform. With the dielectric and the magnetic circuit this is rarely the case, and such circuits thus have to be investigated from place to place across the section of the current flow. This brings in the consideration of dielectric current density or dielectric flux density, and corresponding thereto magnetic flux density, as commonly used terms, while dynamic current density, that is, current per unit section of conductor, is far less frequently considered.

Thus, in the dielectric circuit, instead of admittance $Y = g + jb$, commonly the admittance per unit section and unit length of the dielectric circuit, or the *admittivity*, $v = \gamma - j\beta$, has to be considered, where γ = conductivity of the dielectric (or effective conductivity, including all other energy losses), and $\beta = 2\,\pi fk$ = susceptivity, where k = permittivity or specific capacity of the material.

We then have:

$$\left.\begin{aligned} \underset{\cdot}{I} &= \int (\gamma + 2\,\pi fkj)\,\frac{d\dot{E}}{dl}\,dA \\ \underset{\cdot}{E} &= \int \frac{\gamma - 2\,\pi fkj}{\gamma^2 + (2\,\pi fk)^2}\,\frac{d\dot{I}}{dA}\,dl \end{aligned}\right\} \qquad (20)$$

122. With the extended industrial use of very high voltage, the explicit study of the dielectric field has become of importance, and it is not safe merely to consider the thickness of the insulation in relation to the voltage impressed upon it.

In an ununiform electric conductor, the relation of the voltage to the length of the conductor does not determine whether the conductor is safe or whether locally, due to small cross-section or high resistivity, unsafe current densities may cause destructive heating, but the adaptability of the conductor to the current carried by it must be considered throughout its entire length. So in the dielectric field, the thickness of the dielectric may be such that the voltage impressed upon it may give a very safe average voltage gradient or average dielectric flux density, and the dielectric nevertheless may break down, due to local concentration of the dielectric flux density in the insulating material. Thus, for instance, in the dielectric field between parallel conductors, at a voltage far below that which would jump from conductor to conductor, locally at the conductor surface the concentration of electrostatic stress exceeds the dielectric strength of air, and causes it to break down as corona. In solid dielectrics, under similar conditions, the breakdown due to local over-stress often may change the flux distribution so as to gradually extend throughout the entire dielectric, until puncture results.

Corona

123.—In the magnetic field, with increasing magnetizing force, f, or magnetic field intensity, H, the magnetic flux density, B, increases, but for high field intensities the flux density ceases to be even approximately proportional to the field intensity, and finally, at very high field intensities, H, the "metallic magnetic induction," $B_0 = B - H$, reaches a finite limiting value, which with iron is not far from $B_0 = 20{,}000$, the so-called "saturation value."

In the dielectric field, with increasing voltage gradient, g, or dielectric field intensity, K, the dielectric flux density, D, increases proportional thereto, until a finite limiting field intensity, K_0, or voltage gradient, g_0, is reached, beyond which the dielectric cannot be stressed, but breaks down and becomes dynamically conducting, that is, punctures, and thereby short-circuits the dielectric field.

The voltage gradient, g_0, at which disruption of the dielectric occurs is called the "disruptive strength" or "dielectric strength" of the dielectric. With air at atmospheric pressure and temperature, it is $g_0 = 30$ kv. per centimeter. Thus under alternating electric stress, air punctures at 21 kv. effective per centimeter $\left(\frac{30}{\sqrt{2}}\right)$. The dielectric strength of air is over a very wide range proportional to the air density, and thus proportional to the barometric pressure and inverse proportional to the absolute temperature. Air is one of the weakest dielectrics, and liquids and still more solids show far higher values of dielectric strength, up to and beyond a million volts per centimeter.

124. If then in a uniform dielectric field, such as that between parallel plates A and B as shown in Fig. 96, the voltage is gradually increased, as soon as the voltage maximum reaches a gradient of $g_0 = 30$ kv. in the gap between the metal plates, the air in this gap ceases to sustain the voltage, a spark passes, usually followed by the arc, and the potential difference across this gap drops from $g_0 l$—where l is the distance between the metal plates A and B—to practically nothing, and the electric circuit thereby ceases to include a dielectric field.

Assuming now that the gap between the metal plates does not contain a homogeneous dielectric, but one consisting of several layers of different dielectric strength and different permittivity. For instance, we put two glass plates, a and b, of thickness l_0 into the gap, as shown in Fig. 97, thereby leaving an air space, c, of $l - 2\,l_0$. The dielectric flux density in the field is still uniform

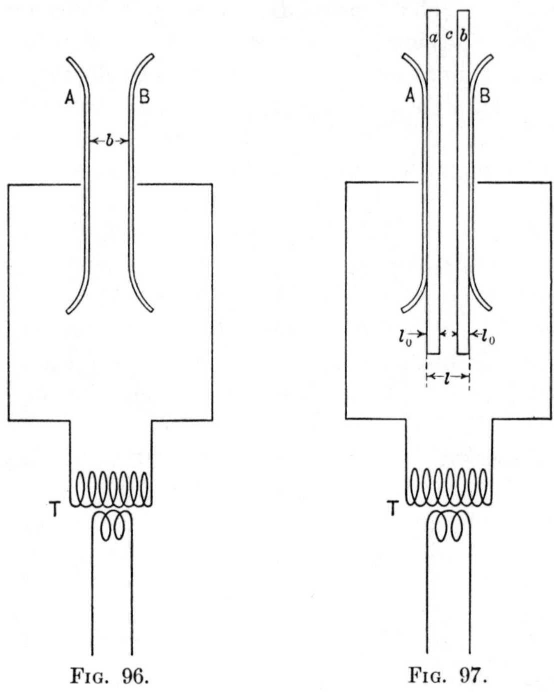

FIG. 96. FIG. 97.

throughout the field section, but the voltage gradient in the different layers, a, b and c, is not the same, is not the average gradient, $g = \frac{e}{l}$, of the gap, but is inverse proportional to the permittivities:

$$g_0 \div g_1 = \frac{1}{k_0} \div \frac{1}{k_1}$$

where k_0 is the permittivity of the layers, a and b, k_1 the permittivity of the layer c (=1, if this layer is air). The potential drop across a and b thus is $l_0 g_0$, across c is $(l - 2\,l_0)g_1$, and the total voltage thus:

$$e = 2\,l_0\,g_0 + (l - 2\,l_0)g_1,$$

or, substituting $g_0 = \frac{g_1 k_1}{k_0}$ gives:

$$e = g_1\left\{2l_0\left(\frac{k_1}{k_0} - 1\right) + l\right\}$$

hence:

$$g_1 = \frac{e}{2l_0\left(\frac{k_1}{k_0} - 1\right) + l} = \frac{ek_0}{2l_0(k_1 - k_0) + lk_0}$$

and

$$g_0 = \frac{e\frac{k_1}{k_0}}{2l_0\left(\frac{k_1}{k_0} - 1\right) + l} = \frac{ek_1}{2l_0(k_1 - k_0) + lk_0}$$

Depending on the values of k_1 and k_0, either g_0 or g_1 may be higher than the average gradient

$$g = \frac{e}{l}.$$

To illustrate on a numerical instance:

Let the distance between the metal plates A and B be $l = 1$ cm. With nothing but air at atmospheric pressure and temperature between the plates, the gap would break down by a spark discharge, and short-circuit the circuit of Fig. 96; at $e = 30$ kv. maximum, and at $e = 25$ kv., no discharge would occur.

Assuming now two glass plates, a and b, each of 0.3 cm. thickness and permittivity $k_0 = 4$, were inserted, leaving an air-gap of 0.4 cm. of permittivity $k_1 = 1$. At $e = 25$ kv. the gradients thus would be, by above equation:

In the glass plates:

$$g_1 = 8.4 \text{ kv. per cm.}$$

In the air-gap:

$$g_0 = 35.7 \text{ kv. per cm.}$$

The air would thus be stressed beyond its dielectric strength, and would break down by spark discharge. This would drop the gradient in the air down to practically $g'_0 = 0$, and the gradient in the glass plates thus would become:

$$g'_1 = \frac{25}{0.6} = 41.7 \text{ kv. per cm.}$$

Thus the insertion of the glass plates would cause the air-gap to break down. The dynamic current which flows through the air-gap in this case would not be the short-circuit current of the electric circuit, as would be the case in the absence of the glass plates but it would merely be the capacity current of the glass plates; and it would not be followed by the arc, but passes as a uniform bluish glow discharge, or as pink streamers—corona.

125. If the dielectric field is not uniform, but varying in density as, for instance, the field between two spheres or the field between two parallel wires, then with increasing voltage the breakdown gradient will not be reached simultaneously throughout the entire field, as in a uniform field, but it is first reached in the denser portion of the field—at the surface of the spheres or parallel wires, where the lines of dielectric force converge. Thus the dielectric will first break down at the denser portion of the field, and short-circuit these portions by the flow of dynamic current. This, however, changes the voltage gradient in the rest of the field, and may raise it so as to break down the entire field, or it may not do so.

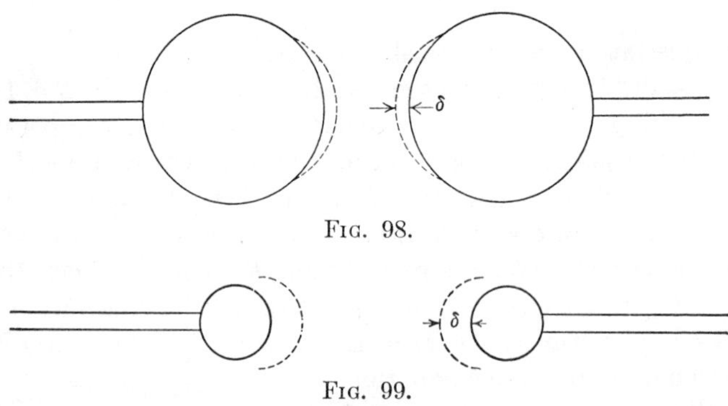

FIG. 98.

FIG. 99.

For instance, in the dielectric field between two spheres at distance l from each other, as shown in Figs. 98 and 99, with increasing potential difference, e, finally the breakdown gradient of the air, $g_0 = 30$ kv. = cm., is reached at the surface of the spheres, and up to a certain distance δ beyond it, and in this space δ the air breaks down, becomes conducting, and the space up to the distance δ is filled with corona. As the result, the conducting terminals of the dielectric field are not the original spheres, but the entire space filled by the corona, that is, the terminals are increased in size, and the convergency of the dielectric flux lines, that is, the voltage gradient at the effective terminals, is reduced. At the same time the gap between the effective terminals is reduced by 2δ, and the average voltage gradient thereby increased.

If the latter effect is greater—as is the case with large spheres at short distance from each other—the air becomes over-stressed at the edge δ of the corona formed by the original field, the corona spreads farther, and so on, until the entire field breaks down, that is, no stable corona forms, but immediate disruptive discharge. Inversely, with small spheres at considerable distance from each other, the formation of corona very soon increases the size of the effective terminals so as to bring the voltage gradient at the edge of the corona down to the disruptive gradient, g_0, and the corona spreads no farther. In this case then, with increasing voltage, at a certain voltage, e_0, corona begins to form at the terminals, first as bluish glow, then as violet streamers, which spread farther and farther with increasing voltage, until finally the disruptive spark passes between the terminals. In this case, corona precedes the disruptive discharge.

Experience shows that the voltage, e_v, at which corona begins at the surface is not the voltage at which the breakdown gradient of air, $g_0 = 30$, is reached at the sphere surface, but e_v is the voltage at which the breakdown gradient, g_0, has extended up to a certain small but definite distance the "energy distance" from the spheres. That is, dielectric breakdown of the air requires a finite volume of over-stressed air, that is, a finite amount of dielectric energy. As the result, when corona begins, the gradient at the terminal surface, g_v, is higher than the breakdown gradient, g_0, the more so the more the flux lines converge, that is, the smaller the spheres (or parallel wires) are.

126. With the development of high-voltage transmission at 100 kv. and over, the electrical industry has entered the range of voltage, where corona appears on parallel wires of sizes such as are industrially used. Such corona consumes power, and thereby introduces an energy component into the expression of the line capacity, a corona conductance.

The power consumption by the corona is approximately proportional to the frequency, its power factor therefore independent of the frequency.

The power consumption by the corona is proportional to the square of the excess voltage over that voltage, e_0, which brings the dielectric field at the conductor surface up to the breakdown gradient, g_0.

However, corona does not yet appear at the voltage, e_0, which produces the breakdown gradient, g_0, at the conductor surface, but at the higher voltage, e_v, which has extended the breakdown gradient by the energy distance from the conductor surface. Then the corona power begins with a finite value, and in the range between e_0 and e_v it is indefinite, depending on the surface condition of the conductor.

The equations of the power consumption by corona in parallel conductors are:

$$P = a(f + c)(e - e_0)^2$$

where:

P = power loss in kilowatts per kilometer length of single-line conductor;
e = effective value of the voltage between the line conductor and neutral in kilovolts;[1]
f = frequency;
c = 25;

and a is given by the equation:

$$a = \frac{A}{\delta}\sqrt{\frac{r}{s}}$$

where:

r = radius of conductor in centimeters;
s = distance between conductor and return conductor in centimeters;
δ = density of the air, referred to 25°C. and 76 cm. barometer;
A = 241;

and:

e_0 = effective disruptive critical voltage to neutral, given in kilovolts by the equation (natural logarithm)

$$e_0 = m_0 g_0 \, \delta r \log \frac{s}{r}.$$

where:

g_0 = 21.1 kv. per centimeter effective = breakdown gradient of air;
m_0 = surface constant of the conductor.

It is:

m_0 = 1 for perfectly smooth polished wire;
m_0 = 0.98 to 0.93 for roughened or weathered wire;

[1] = ½ the voltage between the conductors in a single-phase circuit, $1/\sqrt{3}$ times the voltage between the conductors in a three-phase circuit.

decreasing to:

m_0 = 0.87 to 0.83 for 7-strand cable (r being the outer radius of the cable).[1]

Materially higher losses occur in snow storms and rain.

For further discussion of the dielectric field and the power losses in it, see F. W. Peek's "Dielectric Phenomena in High-voltage Engineering."

[1] "Dielectric Phenomena in High-voltage Engineering," by F. W. Peek, Jr., page 200.

CHAPTER XV

DISTRIBUTED CAPACITY, INDUCTANCE, RESISTANCE, AND LEAKAGE

127. In the foregoing, the phenomena causing loss of energy in an alternating-current circuit have been discussed; and it has been shown that the mutual relation between current and e.m.f. can be expressed by two of the four constants:

power component of e.m.f., in phase with current, and = current × effective resistance, or r;

reactive component of e.m.f., in quadrature with current, and = current × effective reactance, or x;

power component of current, in phase with e.m.f., and = e.m.f. × effective conductance, or g;

reactive component of current, in quadrature with e.m.f., and = e.m.f. × effective susceptance, or b.

In many cases the exact calculation of the quantities, r, x, g, b, is not possible in the present state of the art.

In general, r, x, g, b, are not constants of the circuit, but depend—besides upon the frequency—more or less upon e.m.f., current, etc. Thus, in each particular case it becomes necessary to discuss the variation of r, x, g, b, or to determine whether, and through what range, they can be assumed as constant.

In what follows, the quantities r, x, g, b, will always be considered as the coefficients of the power and reactive components of current and e.m.f.—that is, as the *effective* quantities—so that the results are directly applicable to the general electric circuit containing iron and dielectric losses.

Introducing now, in Chapters VIII, to XI, instead of "ohmic resistance," the term "effective resistance," etc., as discussed in the preceding chapter, the results apply also—within the range discussed in the preceding chapter—to circuits containing iron and other materials producing energy losses outside of the electric conductor.

128. As far as capacity has been considered in the foregoing chapters, the assumption has been made that the condenser or other source of negative reactance is shunted across the circuit at a definite point. In many cases, however, the condensive reactance is distributed over the whole length of the conductor, so that the circuit can be considered as shunted by an infinite number of infinitely small condensers infinitely near together, as diagrammatically shown in Fig. 100.

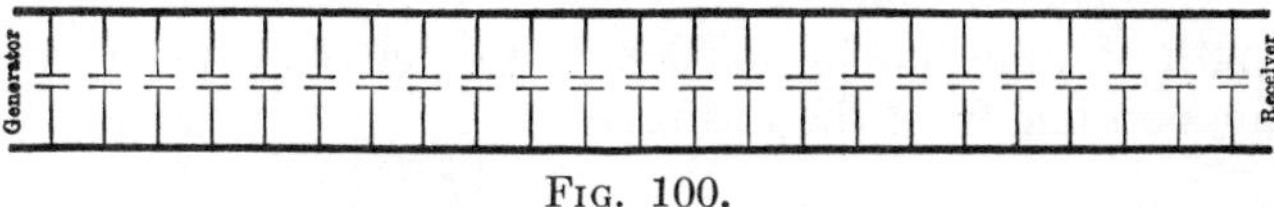

FIG. 100.

In this case the intensity as well as phase of the current, and consequently of the counter e.m.f. of inductive reactance and resistance, vary from point to point; and it is no longer possible to treat the circuit in the usual manner by the vector diagram.

This phenomenon is especially noticeable in long-distance lines, in underground cables, and to a certain degree in the high-potential coils of alternating-current transformers for very high voltage and also in high frequency circuits. It has the effect that not only the e.m.fs., but also the currents, at the beginning, end, and different points of the conductor, are different in intensity and in phase.

Where the capacity effect of the line is small, it may with sufficient approximation be represented by one condenser of the same capacity as the line, shunted across the line at its middle. Frequently it makes no difference either, whether this condenser is considered as connected across the line at the generator end, or at the receiver end, or at the middle.

A better approximation is to consider the line as shunted at the generator and at the motor end, by two condensers of one-sixth the line capacity each, and in the middle by a condenser of two-thirds the line capacity. This approximation, based on Simpson's rule, assumes the variation of the electric quantities in the line as parabolic. If, however, the capacity of the line is considerable, and the condenser current is of the same magnitude as the main current, such an approximation is not permissible, but each line element has to be considered as an infinitely small condenser, and the differential equations based thereon integrated. Or the phenomena occurring in the circuit can be investigated graphically by the method given in Chap. VI, §39, page 162, by dividing the circuit into a sufficiently large number of sections or line elements, and then passing from line element to line element, to construct the topographic circuit characteristics.

129. It is thus desirable to first investigate the limits of applicability of the approximate representation of the line by one or by three condensers.

Assuming, for instance, that the line conductors are of 1 cm. diameter, and at a distance from each other of 50 cm., and that the length of transmission is 50 km., we get the capacity of the transmission line from the formula—

$$C = 1.11 \times 10^{-6}\, kl \div 4 \log_e 2\frac{d}{\delta} \text{ microfarads,}$$

where

k = dielectric constant of the surrounding medium = 1 in air;
l = length of conductor = 5×10^6 cm.;
d = distance of conductors from each other = 50 cm.;
δ = diameter of conductor = 1 cm.

Hence $C = 0.3$ microfarad,

the condensive reactance is $x = \frac{10_6}{2\,\pi f C}$ ohms,

where f = frequency; hence at f = 60 cycles,

$$x = 8{,}900 \text{ ohms;}$$

and the charging current of the line, at E = 20,000 volts, becomes,

$$i_0 = \frac{E}{x} = 2.25 \text{ amp.}$$

The resistance of 100 km. of wire of 1 cm. diameter is 22 ohms; therefore, at 10 per cent. = 2,000 volts loss in the line, the main current transmitted over the line is

$$I = \frac{2{,}000}{22} = 91 \text{ amp.}$$

representing about 1,800 kw.

In this case, the condenser current thus amounts to less than 2.5 per cent., and hence can still be represented by the approximation of one condenser shunted across the line.

If the length of transmission is 150 km., and the voltage, 30,000,

condensive reactance at 60 cycles,	x = 2,970 ohms;
charging current,	i_0 = 10.1 amp.;
line resistance,	r = 66 ohms;
main current at 10 per cent. loss,	I = 45.5 amp.

The condenser current is thus about 22 per cent. of the main current, and the approximate calculation of the effect of line capacity still fairly accurate.

At 300 km. length of transmission it will, at 10 per cent. loss and with the same size of conductor, rise to nearly 90 per cent. of the main current, thus making a more explicit investigation of the phenomena in the line necessary.

In many cases of practical engineering, however, the capacity effect is small enough to be represented by the approximation of one; or, three condensers shunted across the line.

130. (*A*) *Line capacity represented by one condenser shunted across middle of line.*

Let

$Y = g - jb$ = admittance of receiving circuit;
$Z = r + jx$ = impedance of line;
b_c = condenser susceptance of line.

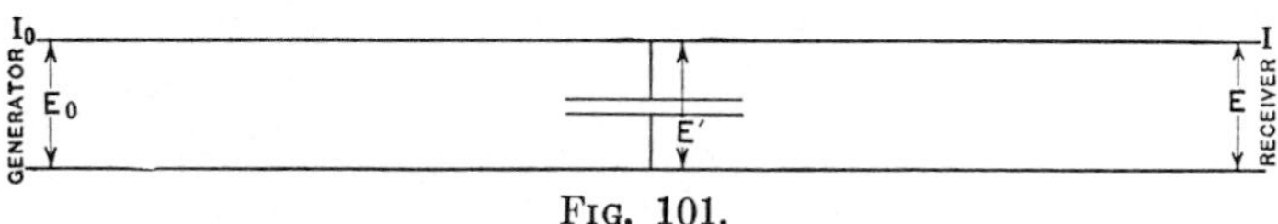

FIG. 101.

Denoting in Fig. 101.

the e.m.f., and current in receiving circuit by $\dot{E}$, $\dot{I}$,
the e.m.f. at middle of line by $\dot{E}'$,
the e.m.f., and current at generator by $\dot{E}_0$, $\dot{I}_0$;

we have,

$$\dot{I} = \dot{E}\,(g - jb);$$

$$\dot{E}' = \dot{E} + \frac{r + jx}{2}\dot{I}$$

$$= \dot{E}\left\{1 + \frac{(r + jx)\,(g - jb)}{2}\right\};$$

$$\dot{I}_0 = \dot{I} + jb_c\dot{E}'$$

$$= \dot{E}\left\{g - jb + jb_c\left[1 + \frac{(r + jx)\,(g - jb)}{2}\right]\right\};$$

$$\dot{E}_0 = \dot{E}' + \frac{r + jx}{2}\dot{I}_0$$

$$= \dot{E}\left\{1 + \frac{(r + jx)\,(g - jb)}{2} + \frac{(r + jx)\,(g - jb)}{2} + \frac{jb_c(r + jx)}{2} + jb_c\,\frac{(r + jx)^2\,(g - jb)}{4}\right\};$$

or, expanding,

$$\dot{I}_0 = \dot{E}\left[\left\{g + \frac{b_c}{2}(rb - xg)\right] - j\left[(b - b_c) - \frac{b_c}{2}(rg + xb)\right]\right\};$$

$$\dot{E}_0 = \dot{E}\left\{1 + (r + jx)(g - jb) + \frac{jb_c}{2}(r + jx)\left[1 + \frac{(r + jx)(g - jb)}{2}\right]\right\}$$

$$= \dot{E}\left\{1 + (r + jx)\left(g - jb + \frac{jb_c}{2}\right) + \frac{jb_c}{4}(r + jx)^2(g - jb)\right\}.$$

131. *Distributed condensive reactance, inductive reactance, leakage, and resistance.*

In some cases, especially in very long circuits, as in lines conveying alternating-power currents at high potential over extremely long distances by overhead conductors or underground cables, or with very feeble currents at extremely high frequency, such as telephone currents, the consideration of the *line resistance*—which consumes e.m.fs. in phase with the current —and of the *line reactance*—which consumes e.m.fs. in quadrature with the current—is not sufficient for the explanation of the phenomena taking place in the line, but several other factors have to be taken into account.

In long lines, especially at high potentials, the *electrostatic capacity* of the line is sufficient to consume noticeable currents. The charging current of the line condenser is proportional to the difference of potential, and is one-fourth period ahead of the e.m.f. Hence, it will either increase or decrease the main current, according to the relative phase of the main current and the e.m.f.

As a consequence, the current changes in intensity as well as in phase, in the line from point to point; and the e.m.f. consumed by the resistance and inductive reactance therefore also changes in phase and intensity from point to point, being dependent upon the current.

Since no insulator has an infinite resistance, and as at high potentials not only leakage, but even direct *escape of electricity* into the air, takes place by corona, we have to recognize the existence of a current approximately proportional and in phase with the e.m.f. of the line. This current represents consumption of power, and is, therefore, analogous to the e.m.f. consumed by resistance, while the condenser current and the e.m.f. of self-induction are wattless or reactive.

Furthermore, the alternating current in the line produces in all neighboring conductors secondary currents, which react upon the primary current, and thereby introduce e.m.fs. of *mutual inductance* into the primary circuit. Mutual inductance is neither in phase nor in quadrature with the current, and can therefore be resolved into a *power component* of mutual inductance in phase with the current, which acts as an increase of resistance, and into a *reactive component* in quadrature with the current, which decreases the self-inductance.

This mutual inductance is not always negligible, as, for instance, its disturbing influence in telephone circuits shows.

The alternating voltage of the line induces, by *electrostatic influence*, electric charges in neighboring conductors outside of the circuit, which retain corresponding opposite charges on the line wires. This electrostatic influence requires a current proportional to the e.m.f. and consisting of a *power component*, in phase with the e.m.f., and a *reactive component*, in quadrature thereto.

The alternating electromagnetic field of force set up by the line current produces in some materials a loss of energy by magnetic hysteresis, or an expenditure of e.m.f. in phase with the current, which acts as an increase of resistance. This electromagnetic hysteretic loss may take place in the conductor proper if iron wires are used, and will then be very serious at high frequencies, such as those of telephone currents.

The effect of *eddy currents* has already been referred to under "mutual inductive reactance," of which it is a power component.

The alternating electrostatic field of force expends energy in dielectrics by *corona* and *dielectric hysteresis*. In concentric cables, where the electrostatic gradient in the dielectric is comparatively large, the dielectric losses may at high potentials consume appreciable amounts of energy. The dielectric loss appears in the circuit as consumption of a current, whose component in phase with the e m.f. is the *dielectric power current*, which may be considered as the power component of the capacity current.

Besides this, there is the increase of ohmic resistance due to *unequal distribution of current*, which, however, is usually not large enough to be noticeable.

Furthermore, the electric field of the conductor progresses with a finite velocity, the velocity of light, hence lags behind the flow of power in the conductor, and so also introduces power components, depending on current as well as on potential difference.

132. This gives, as the most general case, and per unit length of line:

e.m.fs. *consumed in phase with the current*, I, and $= rI$, representing consumption of power, and due to:

Resistance, and its increase by unequal current distribution; to the power component of *mutual inductive reactance* or to *induced currents;* to the power component of *self-inductive reactance* or to *electromagnetic hysteresis*, and to *radiation*.

e.m.fs. *consumed in quadrature with the current*, I, and $= xI$, wattless, and due to:

Self-inductance, and *mutual inductance*.

Currents consumed in phase with the e.m.f., E, and $= g\,E$, representing consumption of power, and due to:

Leakage through the insulating material, including silent discharge and *corona;* power component of *electrostatic influence;* power component of *capacity* or *dielectric hysteresis*, and to *radiation*.

Currents consumed in quadrature to the e.m.f., E, and $= bE$, being wattless, and due to:

Capacity and *electrostatic influence*.

Hence we get four constants:

Effective resistance, r,
Effective reactance, x,
Effective conductance, g,
Effective susceptance, $- b$,

per unit length of line, which represents the coefficients, per unit lenght of line, of

e.m.f. consumed in phase with current;
e.m.f. consumed in quadrature with current;
current consumed in phase with e.m.f.;
current consumed in quadrature with e.m.f.;

or,

$$Z = r + jx,$$
$$Y = g + jb,$$

and, absolute,

$$z = \sqrt{r^2 + x^2},$$
$$y = \sqrt{g^2 + b^2}.$$

The complete investigation of a circuit or line containing distributed capacity, inductive reactance, resistance, etc., leads to functions which are products of exponential and of trigonometric functions. That is, the current and potential difference along the line, l, are given by expressions of the form:

$$\epsilon^{+al}(A \cos \beta l + B \sin \beta l).$$

Such functions of the distance, l, or position on the line, while alternating in time, differ from the true alternating waves in that the intensities of successive half-waves progressively increase or decrease with the distance. Such functions are called oscillating waves, and, as such, are beyond the scope of this book, but are more fully treated in "Transients in Space," [Dover, Volume III, pp. 59–221]. There also will be found the discussion of the phenomena of distributed capacity in high-potential transformer windings, the effect of the finite velocity of propagation of the electric field, etc.

For most purposes, however, in calculating long-distance transmission lines and other circuits of distributed constants, the following approximate solutions of the general differential equation of the circuit offers sufficient exactness.

133. The impedance of an element, dl, of the line is:

$$Zdl$$

and the voltage, $d\dot{E}$, consumed by the current, $\dot{I}$, in this line element dl:

$$d\dot{E} = Z\dot{I}dl$$

The admittance of the line element, dl, is:

$$Ydl$$

hence the current, $d\dot{I}$, consumed by the voltage, $d\dot{E}$, of this line element dl:

$$d\dot{I} = Y\dot{E}dl$$

This gives the two equations of the transmission line:

$$\frac{d\dot{E}}{dl} = Z\dot{I}$$

$$\frac{d\dot{I}}{dl} = Y\dot{E}$$

Differentiating the first equation, and substituting therein the second, gives:

$$\frac{d^2\dot{E}}{dl^2} = ZY\dot{E} \tag{1}$$

and from the first equation follows:

$$\dot{I} = \frac{1}{Z}\frac{d\dot{E}}{dl} \tag{2}$$

Equation (1) is integrated by:

$$\dot{E} = A\epsilon^{Bl} \tag{3}$$

and, substituting (3) in (1), gives:

$$B^2 = ZY$$

hence:

$$B = +\sqrt{ZY} \text{ and } -\sqrt{ZY}$$

There exist thus two values of B, which make (3) a solution of (1), and the most general solution, therefore, is:

$$\dot{E} = A_1\epsilon^{+\sqrt{ZY}l} + A_2\epsilon^{-\sqrt{ZY}l} \tag{4}$$

Substituting (4) in (2) gives:

$$\dot{I} = \sqrt{\frac{Y}{Z}}\left\{A_1\epsilon^{+\sqrt{ZY}l} - A_2\epsilon^{-\sqrt{ZY}l}\right\} \tag{5}$$

where l is counted from some point of the line as starting point, for instance, from the step-down end as $l = 0$.

If then:

$\dot{E}_0$ = voltage at step-down end of the line,

$\dot{I}_0$ = current at step-down end,

it is, for:

$$l = 0;$$

$$\dot{E}_0 = A_1 + A_2$$

$$\dot{I}_0 = \sqrt{\frac{Y}{Z}}\left\{A_1 - A_2\right\}$$

hence:

$$\left.\begin{aligned} A_1 &= \frac{1}{2}\left\{\dot{E}_0 + \sqrt{\frac{Z}{Y}}\dot{I}_0\right\} \\ A_2 &= \frac{1}{2}\left\{\dot{E}_0 - \sqrt{\frac{Z}{Y}}\dot{I}_0\right\} \end{aligned}\right\} \tag{6}$$

and, substituting (6) into (5):

$$\left.\begin{aligned} \dot{E} &= \dot{E}_0\frac{\epsilon^{+\sqrt{ZY}l} + \epsilon^{-\sqrt{ZY}l}}{2} + \sqrt{\frac{Z}{Y}}\dot{I}_0\frac{\epsilon^{+\sqrt{ZY}l} - \epsilon^{-\sqrt{ZY}l}}{2} \\ \dot{I} &= \dot{I}_0\frac{\epsilon^{+\sqrt{ZY}l} + \epsilon^{-\sqrt{ZY}l}}{2} + \sqrt{\frac{Y}{Z}}\dot{E}_0\frac{\epsilon^{+\sqrt{ZY}l} - \epsilon^{-\sqrt{ZY}l}}{2} \end{aligned}\right\} \tag{7}$$

Substituting in (7) for the exponential function the infinite series:

$$\epsilon^{\pm\sqrt{ZY}l} = 1 \pm \sqrt{ZY}l + \frac{ZYl^2}{|\underline{2}} \pm \frac{ZY\sqrt{ZY}l^3}{|\underline{3}} + \frac{Z^2Y^2l^4}{|\underline{4}} + \ldots$$

gives:

$$\left.\begin{aligned} \dot{E}_1 &= \dot{E}_0\left\{1+\frac{ZYl^2}{2}+\ldots\right\} + Zl\dot{I}_0\left\{1+\frac{ZYl^2}{6}+\ldots\right\} \\ \dot{I} &= \dot{I}_0\left\{1+\frac{ZYl^2}{2}+\ldots\right\} + Yl\dot{E}_0\left\{1+\frac{ZYl^2}{6}+\ldots\right\} \end{aligned}\right\} \quad (8)$$

134. If then: $l = l_0$ is the total length of line, and
$Z_0 = l_0 Z$ = total line impedance,
$Y_0 = l_0 Y$ = total line admittance,

the equations of voltage $\dot{E}_1$ and current $\dot{I}_1$ at the end l_0 of the line are given by substituting $l = l_0$ into equations (8), as:

$$\left.\begin{aligned} \dot{E}_1 &= \dot{E}_0\left\{1+\frac{Z_0Y_0}{2}+\ldots\right\} + Z_0\dot{I}_0\left\{1+\frac{Z_0Y_0}{6}+\ldots\right\} \\ \dot{I}_1 &= \dot{I}_0\left\{1+\frac{Z_0Y_0}{2}+\ldots\right\} + Y_0\dot{E}_0\left\{1+\frac{Z_0I_0}{6}+\ldots\right\} \end{aligned}\right\} \quad (9)$$

Since Z_0 is the line impedance, and thus $Z_0\dot{I}$ the impedance voltage, $\frac{Z_0\dot{I}}{\dot{E}}$ is the impedance voltage, as fraction of the total voltage. Since Y_0 is the line admittance, $Y_0\dot{E}$ is the charging current, and $\frac{Y_0\dot{E}}{\dot{I}}$ the charging current as fraction of the total current. The product of these two fractions is:

$$\frac{Z_0\dot{I}}{\dot{E}} \times \frac{Y_0\dot{E}}{\dot{I}} = Z_0Y_0.$$

Z_0Y_0 thus is the product of impedance voltage and charging current of the line, expressed as fraction of total voltage and total current, respectively, hence is a small quantity, and its higher powers can therefore almost always be neglected even in very long transmission lines, and the equation (9) approximated to:

$$\left.\begin{aligned} \dot{E}_1 &= \dot{E}_0\left\{1+\frac{Z_0Y_0}{2}\right\} + Z_0\dot{I}_0\left\{1+\frac{Z_0Y_0}{6}\right\} \\ \dot{I}_1 &= \dot{I}_0\left\{1+\frac{Z_0Y_0}{2}\right\} + Y_0\dot{E}_0\left\{1+\frac{Z_0Y_0}{6}\right\} \end{aligned}\right\} \quad (10)$$

These equations are simpler than those often given by representing the line capacity by a condenser shunted across the middle of the line, and are far more exact. They give the generator voltage and current, $\dot{E}_1$ repectively $\dot{I}_1$, by the step-down voltage and current, $\dot{E}_0$ and $\dot{I}_0$ respectively.

Inversely, if $\dot{E}_0$ and $\dot{I}_0$ are chosen as the values at the generator end, the values at the step-down end are given by substituting $l = -l_0$ in equations (8), as:

$$\left.\begin{aligned} \dot{E}_1 &= \dot{E}_0\left\{1 + \frac{Z_0Y_0}{2}\right\} - Z_0\dot{I}_0\left\{1 + \frac{Z_0Y_0}{6}\right\} \\ \dot{I}_1 &= \dot{I}_0\left\{1 + \frac{Z_0Y_0}{2}\right\} - Y_0\dot{E}_0\left\{1 + \frac{Z_0Y_0}{6}\right\} \end{aligned}\right\} \qquad (11)$$

Neglecting the line conductance: $g_0 = 0$, gives:

$$Y_0 = +jb_0$$

and:

$$Z_0 = r_0 + jx_0$$

hence, substituted in equations (10) and (11), and expanded, gives

$$\left.\begin{aligned} \dot{E}_1 &= \dot{E}_0\left\{1 - \frac{b_0x_0}{2} + j\frac{b_0r_0}{2}\right\} \pm \dot{I}_0(r_0 + jx_0)\left\{1 - \frac{b_0x_0}{6} = j\frac{b_0r_0}{6}\right\} \\ \dot{I}_1 &= \dot{I}_0\left\{1 - \frac{b_0x_0}{2} + j\frac{b_0r_0}{2}\right\} \pm jb_0\dot{E}_0\left\{1 - \frac{b_0x_0}{6} + j\frac{b_0r_0}{6}\right\} \end{aligned}\right\} \quad (12)$$

where the upper sign holds, if $\dot{E}_0$, $\dot{I}_0$ are at the step-down end, $\dot{E}_1$, $\dot{I}_1$ at the generator end of the line, and the lower sign holds, if $\dot{E}_0$, $\dot{I}_0$ are at the generator end, $\dot{E}_1$, $\dot{I}_1$ at the step-down end of the line.

As seen, the equations (12) are just as simple as those of a circuit containing the resistance, inductance and capacity localized, and are amply exact for practically all cases. Where a still closer approximation should be required, the next term of equations (8) and (9) may be included.

In many cases, the $\frac{Z_0Y_0}{6}$ term in (10) and (11) may also be dropped, giving the still simpler equation:

$$\left.\begin{aligned} \dot{E}_1 &= \dot{E}_0\left\{1 + \frac{Z_0Y_0}{2}\right\} \pm Z_0\dot{I}_0 \\ \dot{I}_1 &= \dot{I}_0\left\{1 + \frac{Z_0Y_0}{2}\right\} \pm Y_0\dot{E}_0 \end{aligned}\right\} \qquad (13)$$

CHAPTER XVI

POWER, AND DOUBLE-FREQUENCY QUANTITIES IN GENERAL

135. Graphically, alternating currents and voltages are represented by vectors, of which the length represents the intensity, the direction the phase of the alternating wave. The vectors generally issue from the center of coördinates.

In the topographical method, however, which is more convenient for complex networks, as interlinked polyphase circuits, the alternating wave is represented by the straight line between two points, these points representing the absolute values of potential (with regard to any reference point chosen as coördinate center), and their connection the difference of potential in phase and intensity.

Algebraically these vectors are represented by complex quantities. The impedance, admittance, etc., of the circuit is a complex quantity also, in symbolic denotation.

Thus current, voltage, impedance, and admittance are related by multiplication and division of complex quantities in the same way as current, voltage, resistance, and conductance are related by Ohm's law in direct-current circuits.

In direct-current circuits, power is the product of current into voltage. In alternating-current circuits, if

$$\underset{\cdot}{E} = e^1 + je^{11},$$
$$\underset{\cdot}{I} = i^1 + ji^{11},$$

the product,

$$P_0 = \underset{\cdot}{E}\underset{\cdot}{I} = (e^1i^1 - e^{11}i^{11}) + j(e^{11}i^1 + e^1i^{11}),$$

is not the power; that is, multiplication and division, which are correct in the inter-relation of current, voltage, impedance, do not give a correct result in the inter-relation of voltage, current, power. The reason is, that $\underset{\cdot}{E}$ and $\underset{\cdot}{I}$ are vectors of the same frequency, and Z a constant numerical factor or "operator," which thus does not change the frequency.

The power, P, however, is of double frequency compared with $\dot{E}$ and $\dot{I}$, that is, makes a complete wave for every half wave of $\dot{E}$ or $\dot{I}$, and thus cannot be represented by a vector in the same diagram with $\dot{E}$ and $\dot{I}$.

$P_0 = \dot{E}\dot{I}$ is a quantity of the same frequency with $\dot{E}$ and $\dot{I}$, and thus cannot represent the power.

136. Since the power is a quantity of double frequency of $\dot{E}$ and $\dot{I}$, and thus a phase angle, θ, in $\dot{E}$ and $\dot{I}$ corresponds to a phase angle, $2\,\theta$, in the power, it is of interest to investigate the product, $\dot{E}\dot{I}$, formed by doubling the phase angle.

Algebraically it is,

$$\begin{aligned} \dot{P} = \dot{E}\dot{I} &= (e^{1} + je^{11})(i^{1} + ji^{11}) \\ &= (e^{1}i^{1} + j^{2}e^{11}i^{11}) + (ie^{11}i^{1} + e^{1}ji^{11}). \end{aligned}$$

Since $j^2 = -1$, that is, 180° rotation for $\dot{E}$ and $\dot{I}$, for the double-frequency vector, P, $j^2 = +1$, or 360° rotation, and

$$\begin{aligned} j \times 1 &= j, \\ 1 \times j &= -j. \end{aligned}$$

That is, multiplication with j reverses the sign, since it denotes a rotation by 180° for the power, corresponding to a rotation of 90° for $\dot{E}$ and $\dot{I}$.

Hence, substituting these values, we have

$$\dot{P} = [\dot{E}\dot{I}] = (e^{1}i^{1} + e^{11}i^{11}) + j(e^{11}i^{1} - e^{1}i^{11}).$$

The symbol $[\dot{E}\dot{I}]$ here denotes the transfer from the frequency of $\dot{E}$ and $\dot{I}$ to the double frequency of $\dot{P}$.

The product, $\dot{P} = [\dot{E}\dot{I}]$, consists of two components: the real component,

$$P^{1} = [\dot{E}\dot{I}]^{1} = (e^{1}i^{1} + e^{11}i^{11});$$

and the imaginary component,

$$jP^{j} = j[\dot{E}\dot{I}]^{j} = j(e^{11}i^{1} - e^{1}i^{11}).$$

The component,

$$P^{1} = [\dot{E}\dot{I}]^{1} = (e^{1}i^{1} + e^{11}i^{11}),$$

is the true or "effective" power of the circuit, $= EI \cos(\dot{E}\dot{I})$.

The component,

$$P^{j} = [\dot{E}\dot{I}]^{j} = (e^{11}i^{1} - e^{1}i^{11}),$$

is what may be called the "reactive power," or the wattless or quadrature volt-amperes of the circuit, $= EI \sin(\dot{E}\dot{I})$.

The real component will be distinguished by the index 1; the imaginary or reactive component by the index, j.

By introducing this symbolism, the power of an alternating circuit can be represented in the same way as in the direct-current circuit, as the symbolic product of current and voltage.

Just as the symbolic expression of current and voltage as complex quantity does not only give the mere intensity, but also the phase,

$$\dot{E} = e^1 + je^{11}$$

$$E = \sqrt{e^{1^2} + e^{11^2}}$$

$$\tan \theta = \frac{e^{11}}{e^1},$$

so the double-frequency vector product $P = [\dot{E}\dot{I}]$ denotes more than the mere power, by giving with its two components, $P^1 = [\dot{E}\dot{I}]^1$ and $P^j = [\dot{E}\dot{I}]^j$, the true power volt-ampere, or "effective power," and the wattless volt-amperes, or "reactive power."

If

$$\dot{E} = e^1 + je^{11},$$

$$\dot{I} = i^1 + ji^{11},$$

then

$$E = \sqrt{e^{1^2} + e^{11^2}},$$

$$I = \sqrt{i^{1^2} + i^{11^2}},$$

and

$$P^1 = [\dot{E}\dot{I}]^1 = (e^1 i^1 + e^{11} i^{11}),$$

$$P^j = [\dot{E}\dot{I}]^j = (e^{11} i^1 - e^1 i^{11}),$$

or

$$P^{1^2} + P^{j^2} = e^{1^2} i^{1^2} + e^{11^2} i^{11^2} + e^{11^2} i^{1^2} + e^{1^2} i^{11^2}$$
$$= (e^{1^2} + e^{11^2})(i^{1^2} + i^{11^2}) = (EI)^2 = P_a^2$$

where P_a = total volt-amperes of circuit. That is,

The effective power, P^1, and the reactive power, P^j, are the two rectangular components of the total apparent power, P_a, of the circuit.

Consequently,

In symbolic representation as double-frequency vector products, powers can be combined and resolved by the parallelogram of vectors just as currents and voltages in graphical or symbolic representation.

The graphical methods of treatment of alternating-current phenomena are here extended to include double-frequency quantities, as power, torque, etc.

$$p = \frac{P^1}{P_a} = \cos\theta = \text{power-factor.}$$

$$q = \frac{P^j}{P_a} = \sin\theta = \text{induction factor}$$

of the circuit, and the general expression of power is

$$\underset{\cdot}{P} = P_a\,(p + jq) = P_a\,(\cos\theta + j\sin\theta).$$

137. The introduction of the double-frequency vector product, $\underset{\cdot}{P} = [\underset{\cdot}{E}\underset{\cdot}{I}]$, brings us outside of the limits of algebra, however, and the commutative principle of algebra, $a \times b = b \times a$, does not apply any more, but we have

$$[\underset{\cdot}{E}\underset{\cdot}{I}] \text{ unlike } [\underset{\cdot}{I}\underset{\cdot}{E}]$$

since

$$[\underset{\cdot}{E}\underset{\cdot}{I}] = [\underset{\cdot}{E}\underset{\cdot}{I}]^1 + j[\underset{\cdot}{E}\underset{\cdot}{I}]^j$$

$$[\underset{\cdot}{I}\underset{\cdot}{E}] = [\underset{\cdot}{I}\underset{\cdot}{E}]^1 + j[\underset{\cdot}{I}\underset{\cdot}{E}]^j = [\underset{\cdot}{E}\underset{\cdot}{I}]^1 - j[\underset{\cdot}{E}\underset{\cdot}{I}]^j,$$

we have

$$[\underset{\cdot}{E}\underset{\cdot}{I}]^1 = [\underset{\cdot}{I}\underset{\cdot}{E}]^1$$

$$[\underset{\cdot}{E}\underset{\cdot}{I}]^j = -\,[\underset{\cdot}{I}\underset{\cdot}{E}]^j$$

that is, the imaginary component reverses its sign by the interchange of factors.

The physical meaning is, that if the reactive power, $[\underset{\cdot}{E}\underset{\cdot}{I}]^j$, is lagging with regard to $\underset{\cdot}{E}$, it is leading with regard to $\underset{\cdot}{I}$.

The reactive component of power is absent, or the total apparent power is effective power, if

$$[\underset{\cdot}{E}\underset{\cdot}{I}]^j = (e^{11}i^1 - e^1 i^{11}) = 0;$$

that is,

$$\frac{e^{11}}{e^1} = \frac{i^{11}}{i^1}$$

or,

$$\tan(\underset{\cdot}{E}) = \tan(\underset{\cdot}{I});$$

that is, $\underset{\cdot}{E}$ and $\underset{\cdot}{I}$ are in phase or in opposition.

The effective power is absent, or the total apparent power reactive, if

$$[\underset{\cdot}{E}\underset{\cdot}{I}]^1 = (e^1 i^1 + e^{11} i^{11}) = 0;$$

that is,
$$\frac{e^{11}}{e^{1}} = -\frac{i^{1}}{i^{11}}$$

or,
$$\tan \dot{E} = -\cot \dot{I};$$

that is, $\dot{E}$ and $\dot{I}$ are in quadrature.

The reactive component of power is lagging (with regard to $\dot{E}$ or leading with regard to $\dot{I}$) if
$$[\dot{E}\dot{I}]^{j} > 0,$$

and leading if
$$[\dot{E}\dot{I}]^{j} < 0.$$

The effective power is negative, that is, power returns, if
$$[\dot{E}\dot{I}]^{1} < 0.$$

We have,
$$[\dot{E}, -\dot{I}] = [-\dot{E}, \dot{I}] = -[\dot{E}\dot{I}]$$
$$[-\dot{E}, -\dot{I}] = +[\dot{E}\dot{I}]$$

that is, when representing the power of a circuit or a part of a circuit, current and voltage must be considered in their proper *relative* phases, but their phase relation with the remaining part of the circuit is immaterial.

We have further,
$$[\dot{E}, j\dot{I}] = -j[\dot{E}, \dot{I}] = [\dot{E}, \dot{I}]^{j} - j[\dot{E}, \dot{I}]^{1}$$
$$[j\dot{E}, \dot{I}] = j[\dot{E}, \dot{I}] = -[\dot{E}, \dot{I}]^{j} + j[\dot{E}, \dot{I}]^{1}$$
$$[j\dot{E}, j\dot{I}] = [\dot{E}, \dot{I}] = [\dot{E}\dot{I}]^{1} + j[\dot{E}, \dot{I}]^{j}$$

138. Expressing voltage and current in polar coördinates;
$$\dot{E} = e^{1} + je^{11} = e(\cos\alpha + j\sin\alpha)$$
$$\dot{I} = i^{1} + ji^{11} = i(\cos\beta + j\sin\beta)$$

gives the vector power:
$$\dot{P} = ei\{(\cos\alpha\cos\beta + j^{2}\sin\alpha\sin\beta) + (j\sin\alpha\cos\beta + \cos\alpha j\sin\beta)\}$$

and since, by the change to double frequency:
$$+j^{2} = +1$$
$$+aj = -ja$$

it is:
$$\dot{P} = ei\{(\cos\alpha\cos\beta + \sin\alpha\sin\beta) + j(\sin\alpha\sin\beta - \cos\alpha\cos\beta)\}$$
$$\dot{P} = ei\{\cos(\alpha - \beta) + j\sin(\alpha - \beta)\}$$

and:

the effective power:

$$P^1 = ei \cos(\alpha - \beta)$$

the reactive power:

$$P^j = ei \sin(\alpha - \beta)$$

We thus must note the distinction:

$$\underset{\cdot}{E} = Z\underset{\cdot}{I} = (r + jx)(i^1 + ji^{11}) = zi(\cos\gamma + j\sin\gamma)(\cos\beta + j\sin\beta)$$
$$= (ri^1 - xi^{11}) + j(ri^{11} + xi^1) = zi\{\cos(\gamma + \beta) + j\sin(\gamma + \beta)\}$$

and:

$$\underset{\cdot}{P} = [\underset{\cdot}{E}, \underset{\cdot}{I}] = [\underset{\cdot}{E}, \underset{\cdot}{I}]^1 + j[\underset{\cdot}{E}, \underset{\cdot}{I}]^j$$
$$= [(e^1 + je^{11}), (i^1 + ji^{11})] = ei[(\cos\alpha + j\sin\alpha), (\cos\beta + j\sin\beta)]$$
$$= (e^1i^1 + e^{11}i^{11}) + j(e^{11}i^1 - e^1i^{11}) = ei\{\cos(\alpha - \beta) + j\sin(\alpha - \beta)\}$$

139. If $\underset{\cdot}{P}_1 = [\underset{\cdot}{E}_1\underset{\cdot}{I}_1], \underset{\cdot}{P}_2 = [\underset{\cdot}{E}_2\underset{\cdot}{I}_2] \ . \ . \ . \ \underset{\cdot}{P}_n = [\underset{\cdot}{E}_n\underset{\cdot}{I}_n]$

are the symbolic expressions of the power of the different parts of a circuit or network of circuits, the total power of the whole circuit or network of circuits is

$$\underset{\cdot}{P} = \underset{\cdot}{P}_1 + \underset{\cdot}{P}_2 + \ . \ . \ . \ . \ + \underset{\cdot}{P}_n,$$
$$P^1 = P^1_1 + P^1_2 + \ . \ . \ . \ . \ + P_n^1,$$
$$P^j = P_2^j + P_2^j \ . \ . \ . \ . \ + P_n^j.$$

In other words, the total power in symbolic expression (effective as well as reactive) of a circuit or system is the sum of the powers of its individual components in symbolic expression.

The first equation is obviously directly a result from the law of conservation of energy.

One result derived herefrom is, for instance:

If in a generator supplying power to a system the current is out of phase with the e.m.f. so as to give the reactive power P^j, the current can be brought into phase with the generator e.m.f. or the load on the generator made non-inductive by inserting anywhere in the circuit an apparatus producing the reactive power—P^j; that is, compensation for wattless currents in a system takes place regardless of the location of the compensating device.

Obviously, wattless currents exist between the compensating device and the source of wattless currents to be compensated for, and for this reason it may be advisable to bring the compensator as near as possible to the circuit to be compensated.

140. Like power, torque in alternating apparatus is a double-frequency vector product also, of magnetism and m.m.f. or current, and thus can be treated in the same way.

In an induction motor, for instance, the torque is the product of the magnetic flux in one direction into the component of secondary current in phase with the magnetic flux in time, but in quadrature position therewith in space, times the number of turns of this current, or since the generated e.m.f. is in quadrature and proportional to the magnetic flux and the number of turns, the torque of the induction motor is the product of the generated e.m.f. into the component of secondary current in quadrature therewith in time and space, or the product of the secondary current into the component of generated e.m.f. in quadrature therewith in time and space.

Thus, if

$\dot{E}^1 = e^1 + je^{11}$ = generated e.m.f. in one direction in space,

$\dot{I}_2 = i^1 + ji^{11}$ = secondary current in the quadrature direction in space,

the torque is

$$D = [\dot{E}\dot{I}]^j = e^{11}i^1 - e^1i^{11}.$$

By this equation the torque is given in watts, the meaning being that $D = [EI]^j$ is the power which would be exerted by the torque at synchronous speed, or the torque in synchronous watts.

The torque proper is then

$$D_0 = \frac{D}{2\pi fp},$$

where

p = number of pairs of poles of the motor.
f = frequency.

In the polyphase induction motor, if $\dot{I}_2 = i^1 + ji^{11}$ is the secondary current in quadrature position, in space, to e.m.f. $\dot{E}_1$, the current in the same direction in space as $\dot{E}_1$ is $\dot{I}_1 = j\dot{I}_2 = -i^{11} + ji^1$; thus the torque can also be expressed as

$$D = [\dot{E}_1\dot{I}_1]^1 = e^{11}i^1 - e^1i^{11}.$$

It is interesting to note that the expression of torque,

$$D = [\dot{E}\dot{I}]^j,$$

and the expression of power,

$$P = [\dot{E}\dot{I}]^1,$$

are the same in character, but the former is the imaginary, the latter the real component. Mathematically, torque, in synchronous watts, can so be considered as imaginary power, and inversely. Physically, "imaginary" means quadrature component; torque is defined as force times leverage, that is, force times length in quadrature position with force; while energy is defined as force times length in the direction of the force. Expressing quadrature position by "imaginary," thus gives torque of the dimension of imaginary energy; and "synchronous watts," which is torque times frequency, or torque divided by time, thus becomes of the dimension of imaginary power. Thus, in its complex imaginary form, the vector product of force and length contains two quadrature components, of which the one is energy, the other is torque:

$$P = [f, l] = [f, l]^1 + j[f, l]^j$$

and

$$[f, l]^1 = \text{energy}$$
$$[f, l]^j = \text{torque.}$$

INDUCTION APPARATUS

CHAPTER XVII

THE ALTERNATING-CURRENT TRANSFORMER

141. The simplest alternating-current apparatus is the transformer. It consists of a magnetic circuit interlinked with two electric circuits, a primary and a secondary. The primary circuit is excited by an impressed e.m.f., while in the secondary circuit an e.m.f. is generated. Thus, in the primary circuit power is consumed, and in the secondary a corresponding amount of power is produced.

Since the same magnetic circuit is interlinked with both electric circuits, the e.m.f. generated per turn must be the same in the secondary as in the primary circuit; hence, the primary generated e.m.f. being approximately equal to the impressed e.m.f., the e.m.fs. at primary and at secondary terminals have approximately the ratio of their respective turns. Since the power produced in the secondary is approximately the same as that consumed in the primary, the primary and secondary currents are approximately in inverse ratio to the turns.

142. Besides the magnetic flux interlinked with both electric circuits—which flux, in a closed magnetic circuit transformer, has a circuit of low reluctance—a magnetic cross-flux passes between the primary and secondary coils, surrounding one coil only, without being interlinked with the other. This magnetic cross-flux is proportional to the current in the electric circuit, or rather, the ampere-turns or m.m.f., and so increases with the increasing load on the transformer, and constitutes what is called the self-inductive or leakage reactance of the transformer; while the flux surrounding both coils may be considered as mutual inductive reactance. This cross-flux of self-induction does not generate e.m.f. in the secondary circuit, and is thus, in general, objectionable, by causing a drop of voltage and a decrease of output. It is this cross-flux, however, or flux of self-inductive reactance, which is utilized in special transformers, to secure automatic regulation, for constant power, or for constant current, and in this case is exaggerated by separating primary and secondary coils. In the constant potential transformer, however, the primary and secondary coils are brought as near together as possible, or even interspersed, to reduce the cross-flux.

There is, however, a limit, to which it is safe to reduce the cross-flux, as at short-circuit at the secondary terminals, it is the e.m.f. of self-induction of this cross-flux which limits the current, and with very low self-induction, these currents may become destructive by their mechanical forces. Therefore experience shows that in large power transformers it is not safe to go below 4 to 6 per cent. cross-flux.

As will be seen, by the self-inductive reactance of a circuit, not the total flux produced by, and interlinked with, the circuit is understood, but only that (usually small) part of the flux which surrounds one circuit without interlinking with the other circuit.

143. The alternating magnetic flux of the magnetic circuit surrounding both electric circuits is produced by the combined magnetizing action of the primary and of the secondary current.

This magnetic flux is determined by the e.m.f. of the transformer, by the number of turns, and by the frequency.

If

Φ = maximum magnetic flux,
f = frequency,
n = number of turns of the coil,

the e.m.f. generated in this coil is

$$E = \sqrt{2}\pi f n \Phi\, 10^{-8} = 4.44\, f n \Phi\, 10^{-8} \text{ volts;}$$

hence, if the e.m.f., frequency, and number of turns are determined, the maximum magnetic flux is

$$\Phi = \frac{E\, 10^8}{\sqrt{2}\,\pi f n}.$$

To produce the magnetism, Φ, of the transformer, a m.m.f. of F ampere-turns is required, which is determined by the shape and the magnetic characteristic of the iron, in the manner discussed in Chapter XII (page 231).

144. Consider as instance, a closed magnetic circuit transformer. The maximum magnetic induction is $B = \frac{\Phi}{A}$, where A = the cross-section of magnetic circuit.

To induce a magnetic density, B, a magnetizing force of f ampere-turns maximum is required, or $\frac{f}{\sqrt{2}}$ ampere-turns effective, per unit length of the magnetic circuit; hence, for the total magnetic circuit, of length, l,

$$F = \frac{lf}{\sqrt{2}} \text{ ampere-turns;}$$

or

$$I = \frac{F}{n} = \frac{lf}{n\sqrt{2}} \text{ amp. eff.}$$

where n = number of turns.

At no-load, or open secondary circuit, this m.m.f., F, is furnished by the *exciting current*, I_{00}, improperly called the *leakage current*, of the transformer; that is, that small amount of primary current which passes through the transformer at open secondary circuit.

In a transformer with open magnetic circuit, such as the "hedgehog" transformer, the m.m.f., F, is the sum of the m.m.f. consumed in the iron and in the air part of the magnetic circuit (see Chapter XII).

The power component of the exciting current represents the power consumed by hysteresis and eddy currents and the small ohmic loss.

The exciting current is not a sine wave, but is, at least in the closed magnetic circuit transformer, greatly distorted by hysteresis, though less so in the open magnetic circuit transformer. It can, however, be represented by an equivalent sine wave, I_{00}, of equal intensity and equal power with the distorted wave, and a wattless higher harmonic, mainly of triple frequency.

Since the higher harmonic is small compared with the total exciting current, and the exciting current is only a small part of the total primary current, the higher harmonic can, for most practical cases, be neglected, and the exciting current represented by the equivalent sine wave.

This equivalent sine wave, I_{00}, leads the wave of magnetism, Φ, by an angle, α, the angle of hysteretic advance of phase, and consists of two components—the hysteretic power current in quadrature with the magnetic flux, and therefore in phase with the generated e.m.f. $= I_{00} \sin \alpha$; and the magnetizing current, in phase with the magnetic flux, and therefore in quadrature with the generated e.m.f., and so wattless, $= I_{00} \cos \alpha$.

The exciting current, I_{00}, is determined from the shape and magnetic characteristic of the iron, and the number of turns; the hysteretic power current is

$$I_{00} \sin \alpha = \frac{\text{power consumed in the iron}}{\text{generated e.m.f.}}.$$

145. Graphically, the polar diagram of m.m.fs., of a transformer is constructed thus:

Let, in Fig. 102, $\overline{O\Phi}$ = the magnetic flux in intensity and phase (for convenience, as intensities, the effective values are

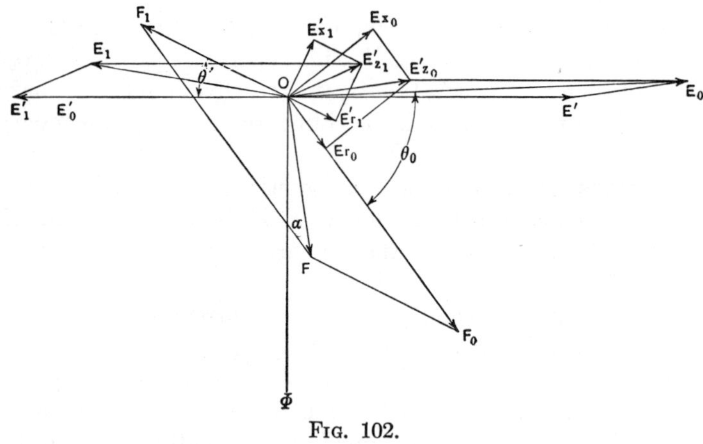

Fig. 102.

used throughout), assuming its phase as the downwards vertical; that is, counting the time from the moment where the rising magnetism passes its zero value.

Then the resultant m.m.f. is represented by the vector, $\overline{OF}$, leading $\overline{O\Phi}$ by the angle, $FO\Phi = \alpha$.

The generated e.m.fs. have the phase 180°, that is, are plotted toward the left, and represented by the vectors, $\overline{OE'_0}$ and $\overline{OE'_1}$.

If, now, θ' = angle of lag in the secondary circuit, due to the total (internal and external) secondary reactance, the secondary current, I_1, and hence the secondary m.m.f., $F_1 = n_1 \; I_1$ lag behind E'_1 by an angle θ', and have the phase, $180° + \theta'$, represented by the vector $\overline{OF_1}$. Constructing a parallelogram of m.m.fs., with $\overline{OF}$ as the diagonal and $\overline{OF_1}$ as one side, the other side or $\overline{OF_0}$ is the primary m.m.f., in intensity and phase, and hence, dividing by the number of primary turns, n_0, the primary current is $I_0 = \frac{F_0}{n_0}$.

To complete the diagram of e.m.fs., we have now,

In the primary circuit:

e.m.f. consumed by resistance is $I_0 r_0$, in phase with I_0, and represented by the vector, $\overline{OE_{r0}}$;

e.m.f. consumed by reactance is $I_0 x_0$, 90° ahead of I_0, and represented by the vector, $\overline{OE_{x0}}$;

e.m.f. consumed by induced e.m.f. is E', equal and opposite to E'_0, and represented by the vector, $\overline{OE'}$.

Hence, the total primary impressed e.m.f. by combination of $\overline{OE_{r0}}$, $\overline{OE_{x0}}$, and $\overline{OE'}$ by means of the parallelogram of e.m.fs. is

$$E_0 = \overline{OE_0},$$

and the difference of phase between the primary impressed e.m.f. and the primary current is

$$\theta_0 = E_0 O F_0.$$

In the secondary circuit:

Counter e.m.f. of resistance is $I_1 r_1$ in opposition with I_1, and represented by the vector, $\overline{OE'_{r1}}$;

Counter e.m.f. of reactance is $I_1 x_1$, 90° behind I_1, and represented by the vector, $\overline{OE'_{x1}}$.

Generated e.m.fs., E'_1, represented by the vector, $\overline{OE'_1}$.

Hence, the secondary terminal voltage, by combination of $\overline{OE'_{r1}}$, $\overline{OE'_{x1}}$ and $\overline{OE'_1}$ by means of the parallelogram of e.m.fs. is

$$E_1 = \overline{OE_1},$$

and the difference of phase between the secondary terminal voltage and the secondary current is

$$\theta_1' = E_1 O F_1.$$

As seen, in the primary circuit the "components of impressed e.m.f. required to overcome the counter e.m.fs." were used for convenience, and in the secondary circuit the "counter e.m.fs."

146. In the construction of the transformer diagram, it is usually preferable not to plot the secondary quantities, current and e.m.f., direct, but to reduce them to correspondence with the primary circuit by multiplying by the ratio of turns, $\alpha = \frac{n_0}{n_1}$, for the reason that frequently primary and secondary e.m.fs., etc., are of such different magnitude as not to be easily represented on the same scale; or the primary circuit may be reduced to the secondary in the same way. In either case, the vectors representing the two generated e.m.fs. coincide, or $\overline{OE'}_1 = \overline{OE'}_0$.

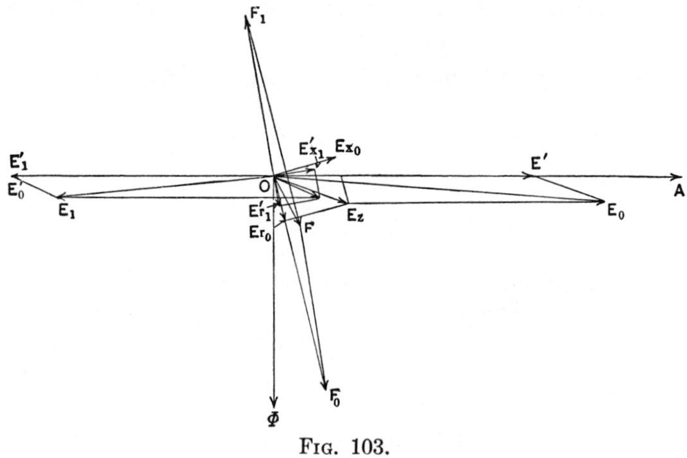

FIG. 103.

Figs. 103 to 109 give the polar diagram of a transformer having the constants, reduced to the secondary circuit:

$r_0 = 0.2$ ohm,	$b_0 = 0.173$ mhos,
$x_0 = 0.33$ ohm,	$E'_1 = 100$ volts,
$r_1 = 0.167$ ohm,	$I_1 = 60$ amp.,
$x_1 = 0.25$ ohm,	$\alpha = 30°$.
$g_0 = 0.100$ mhos,	

For the conditions of secondary circuit:

$\theta'_1 = 80°$ lag	in Fig. 103	$\theta'_1 = 20°$ lead	in Fig. 107
50° lag	" 104	50° lead	" 108
20° lag	" 105	80° lead	" 109
0, or in phase,	" 106		

As shown, with a change of θ'_1 the other quantities, E_0, I_1, I_0, etc., change in intensity and direction. The loci described

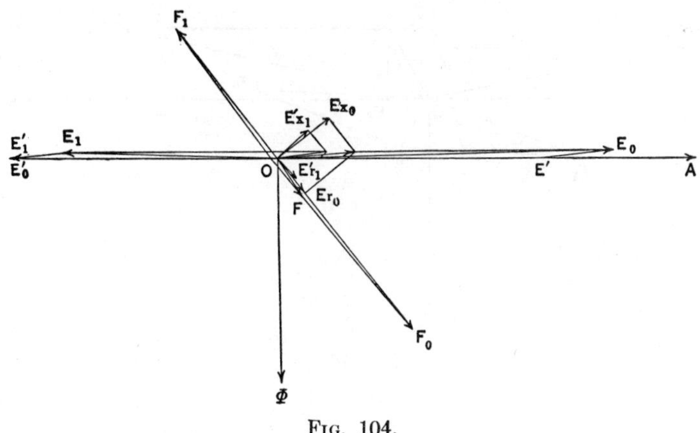

Fig. 104.

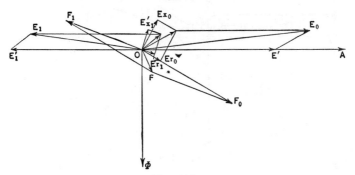

Fig. 105.

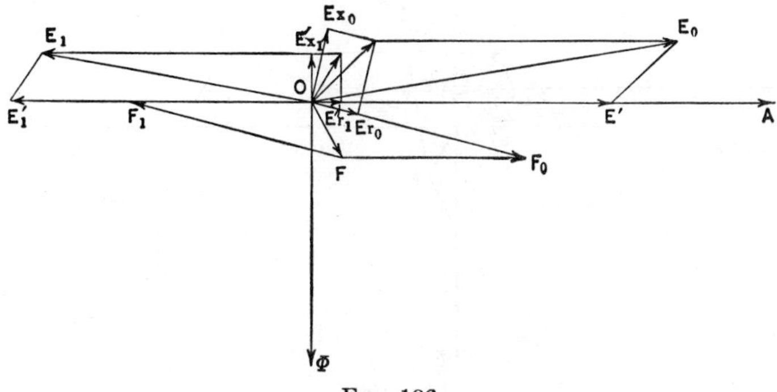

Fig. 106.

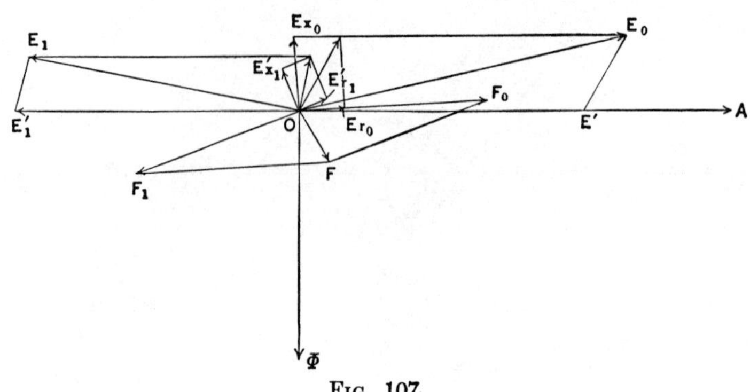

FIG. 107.

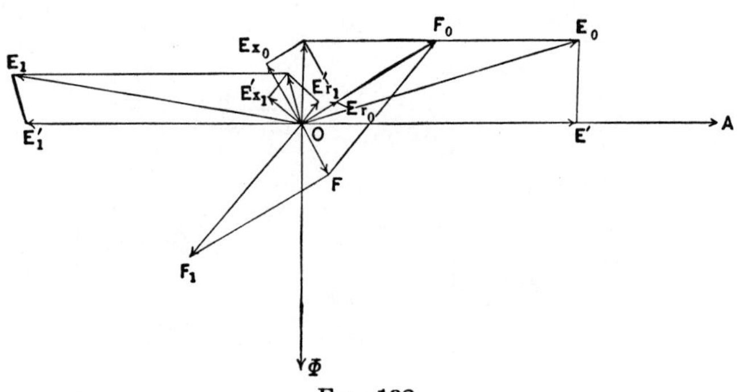

FIG. 108.

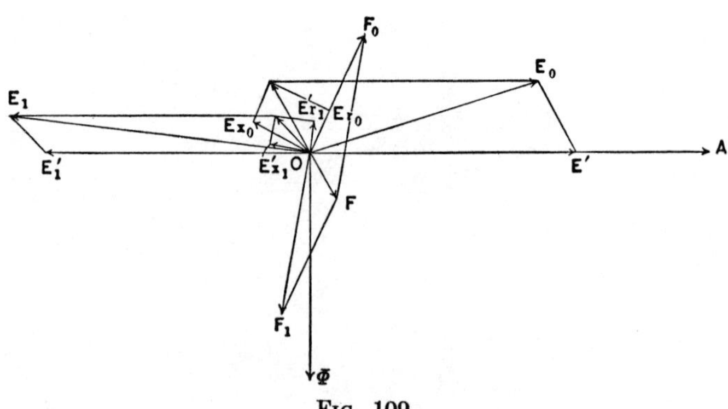

FIG. 109.

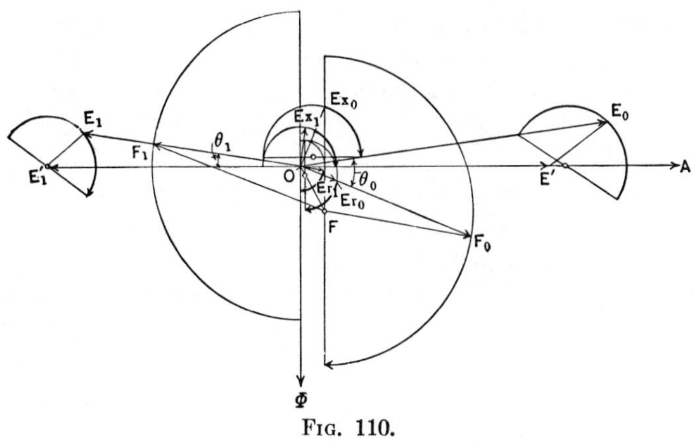

FIG. 110.

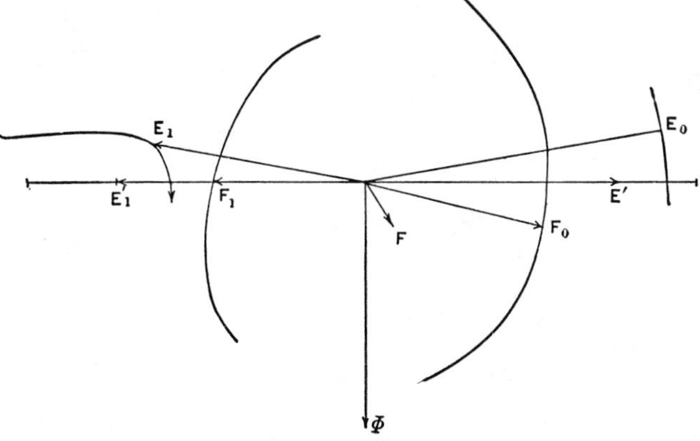

FIG. 111.

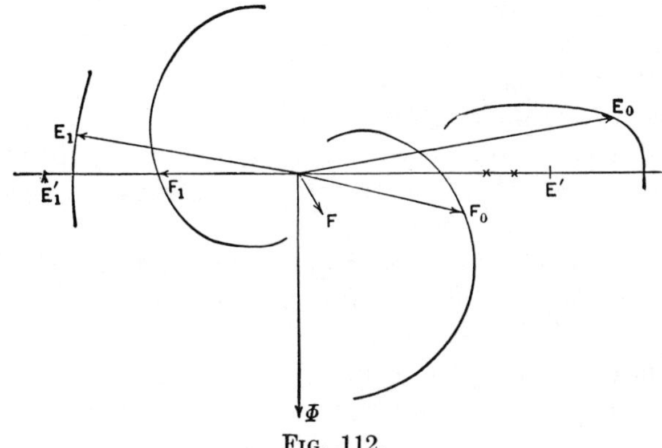

FIG. 112.

by them are circles, and are shown in Fig. 110, with the point corresponding to non-inductive load marked. The part of the locus corresponding to a lagging secondary current is shown in thick full lines, and the part corresponding to leading current in thin full lines.

147. This diagram represents the condition of constant secondary generated e.m.f., E'_1, that is, corresponding to a constant maximum magnetic flux.

By changing all the quantities proportionally from the diagram of Fig. 110, the diagrams for the constant primary im-

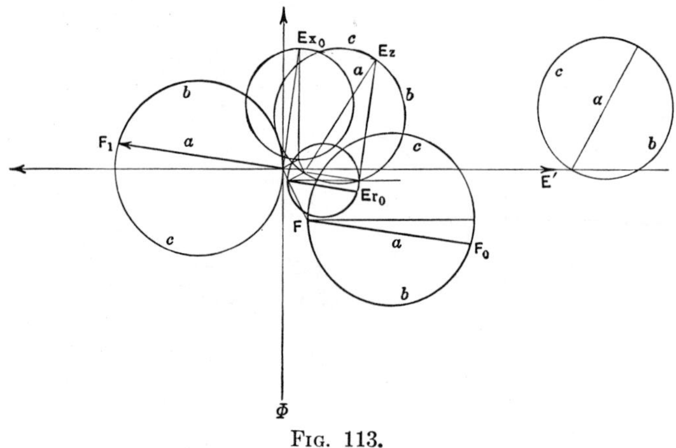

FIG. 113.

pressed e.m.f. (Fig. 111), and for constant secondary terminal voltage (Fig. 112), are derived. In these cases, the locus gives curves of higher order.

Fig. 113 gives the locus of the various quantities when the load is changed from full-load, $I_1 = 60$ amp. in a non-inductive secondary external circuit, to no-load or open-circuit:

(*a*) By increase of secondary current; (*b*) by increase of secondary inductive resistance; (*c*) by increase of secondary condensive reactance.

As shown in (*a*), the locus of the secondary terminal voltage, E_1, and thus of E_0, etc., are straight lines; and in (*b*) and (*c*), parts of one and the same circle; (*a*) is shown in full lines, (*b*) in heavy full lines, and (*c*) in light full lines. This diagram corresponds to constant maximum magnetic flux; that is, to constant secondary generated e.m.f. The diagrams representing constant primary impressed e.m.f. and constant secondary terminal voltage can be derived from the above by proportionality.

148. It must be understood, however, that for the purpose of making the diagrams plainer, by bringing the different values to somewhat nearer the same magnitude, the constants chosen for these diagrams represent not the magnitudes found in actual transformers, but refer to greatly exaggerated internal losses.

In practice, about the following magnitudes would be found:

$r_0 = 0.01$ ohm; $\quad x_1 = 0.00025$ ohm;
$x_0 = 0.033$ ohm; $\quad g_0 = 0.001$ mho;
$r_1 = 0.00008$ ohm; $\quad b_0 = 0.00173$ mho;

that is, about one-tenth as large as assumed. Thus the changes of the values of E_0, E_1, etc., under the different conditions will be very much smaller.

Symbolic Method

149. In symbolic representation by complex quantities the transformer problem appears as follows:

The exciting current, I_{00}, of the transformer depends upon the primary e.m.f., which dependence can be represented by an admittance, the "primary admittance," $Y_0 = g_0 - jb_0$, of the transformer.

The resistance and reactance of the primary and the secondary circuit are represented in the impedance by

$$Z_0 = r_0 + jx_0, \quad \text{and} \quad Z_1 = r_1 + jx_1.$$

Within the limited range of variation of the magnetic density in a constant-potential transformer, admittance and impedance can usually, and with sufficient exactness, be considered as constant.

Let

n_0 = number of primary turns in series;

n_1 = number of secondary turns in series;

$a = \frac{n_0}{n_1}$ = ratio of turns;

$Y_0 = g_0 - jb_0$ = primary admittance

$= \frac{\text{Exciting current}}{\text{Primary induced e.m.f.}}$;

$$Z_0 = r_0 + jx_0 = \text{primary impedance}$$
$$= \frac{\text{e.m.f. consumed in primary coil by resistance and reactance}}{\text{Primary current}};$$
$$Z_1 = r_1 + jx_1 = \text{secondary impedance}$$
$$= \frac{\text{e.m.f. consumed in secondary coil by resistance and reactance}}{\text{Secondary current}};$$

where the reactances, x_0 and x_1, refer to the true self-induction only, or to the cross-flux passing between primary and secondary coils; that is, interlinked with one coil only.

Let also

$Y = g - jb$ = total admittance of secondary circuit, including the internal impedance;
E_0 = primary impressed e.m.f.;
$\dot{E}'$ = e.m.f. consumed by primary counter e.m.f.;
$\dot{E}_1$ = secondary terminal voltage;
$\dot{E}'_1$ = secondary generated e.m.f.;
$\dot{I}_0$ = primary current, total;
$\dot{I}_{00}$ = primary exciting current;
$\dot{I}_1$ = secondary current.

Since the primary counter e.m.f., $\dot{E}_0'$, and the secondary generated e.m.f., E'_1, are proportional by the ratio of turns, a,

$$\dot{E}'_0 = + a\dot{E}'_1. \tag{1}$$
$$\dot{E}'_0 = - \dot{E}'.$$

The secondary current is

$$\dot{I}_1 = Y\dot{E}'_1. \tag{2}$$

consisting of a power component, $g\dot{E}_1'$, and a reactive component, $b\dot{E}'_1$.

To this secondary current corresponds the component of primary current.

$$\dot{I}'_0 = \frac{- Y\dot{E}'_1}{a} = \frac{Y\dot{E}'}{a^2}. \tag{3}$$

The primary exciting current is

$$\dot{I}_{00} = Y_0\dot{E}'. \tag{4}$$

Hence, the total primary current is

$$\dot{I}_0 = \dot{I}'_0 + \dot{I}_{00} \tag{5}$$
$$= \frac{Y\dot{E}'}{a^2} + Y_0\dot{E}',$$

or,

$$\dot{I}_0 = \frac{\dot{E}'}{a^2}\{Y + a^2Y_0\} \tag{6}$$
$$= -\frac{\dot{E}'_1}{a^2}\{Y + a^2Y_0\}.$$

The e.m.f. consumed in the secondary coil by the internal impedance is $Z_1\dot{I}_1$.

The e.m.f. generated in the secondary coil by the magnetic flux is $\dot{E}'_1$.

Therefore, the secondary terminal voltage is

$$\dot{E}_1 = \dot{E}'_1 - Z_1\dot{I}_1;$$

or, substituting (2), we have

$$\dot{E}_1 = \dot{E}'_1\{1 - Z_1Y\}. \tag{7}$$

The e.m.f. consumed in the primary coil by the internal impedance is $Z_0\dot{I}_0$.

The e.m.f. consumed in the primary coil by the counter e.m.f. is $\dot{E}'$.

Therefore, the primary impressed e.m.f. is

$$\dot{E}_0 = \dot{E}' + Z_0\dot{I}_0,$$

or, substituting (6),

$$\left.\begin{aligned}\dot{E}_0 &= \dot{E}'\left\{1 + Z_0Y_0 + \frac{Z_0Y}{a^2}\right\}\\ &= -a\dot{E}'_1\left\{1 + Z_0Y_0 + \frac{Z_0Y}{a^2}\right\}.\end{aligned}\right\} \tag{8}$$

150. We thus have,

primary e.m.f., $\dot{E}_0 = -a\dot{E}'_1\left\{1 + Z_0Y_0 + \frac{Z_0Y}{a^2}\right\}$, (8)

secondary e.m.f., $\dot{E}_1 = \dot{E}'_1\{1 - Z_1Y\}$, (7)

primary current, $\dot{I}_0 = -\frac{\dot{E}'_1}{a}\{Y + a^2Y_0\}$, (6) secondary current, $\dot{I}_1 = Y\dot{E}_1{}^1$, (2)

as functions of the secondary generated e.m.f., $\dot{E}_1'$, as parameter.

From the above we derive

Ratio of transformation of e.m.fs.:

$$\frac{\dot{E}_0}{\ddot{E}_1} = -a\frac{1 + Z_0Y_0 + \frac{Z_0Y}{a^2}}{1 - Z_1Y}. \tag{9}$$

Ratio of transformations of currents:

$$\frac{\dot{I}_0}{\dot{I}_1} = -\frac{1}{a}\left\{1 + a^2\frac{Y_0}{Y}\right\}. \tag{10}$$

From this we get, at constant primary impressed e.m.f.,

$$\dot{E}_0 = \text{constant};$$

secondary generated e.m.f.,

$$\dot{E}'_1 = -\frac{\dot{E}_0}{a}\frac{1}{1 + Z_0Y_0 + \frac{Z_0Y}{a^2}};$$

e.m.f. generated per turn,

$$\delta\dot{E} = -\frac{\dot{E}_0}{n_0}\frac{1}{1 + Z_0Y_0 + \frac{Z_0Y}{a^2}};$$

secondary terminal voltage,

$$\dot{E}_1 = -\frac{\dot{E}_0}{a}\frac{1 - Z_1Y}{1 + Z_0Y_0 + \frac{Z_0Y}{a^2}}; \tag{11}$$

primary current,

$$\dot{I}_0 = \frac{\dot{E}_0}{a^2}\frac{Y + a^2Y_0}{1 + Z_0Y_0 + \frac{Z_0Y}{a^2}} = \dot{E}_0\frac{\frac{Y}{a^2} + Y_0}{1 + Z_0Y_0 + \frac{Z_0Y}{a^2}};$$

secondary current,

$$\dot{I}_1 = -\frac{\dot{E}_0}{a}\frac{Y}{1 + Z_0Y_0 + \frac{Z_0Y}{a^2}}.$$

At constant secondary terminal voltage,

$$\dot{E}_1 = \text{const.};$$

secondary generated e.m.f.,

$$\dot{E}'_1 = \frac{\dot{E}_1}{1 - Z_1 Y};$$

e.m.f. generated per turn,

$$\dot{\delta E} = \frac{\dot{E}_1}{n_1} \frac{1}{1 - Z_1 Y};$$

primary impressed e.m.f.,

$$\dot{E}_0 = -a\dot{E}_1 \frac{1 + Z_0 Y_0 + \frac{Z_0 Y}{a^2}}{1 - Z_1 Y}; \qquad (12)$$

primary current,

$$\dot{I}_0 = -\frac{\dot{E}_1}{a} \frac{Y + a^2 Y_0}{1 - Z_1 Y};$$

secondary current,

$$\dot{I}_1 = \dot{E}_1 \frac{Y}{1 - Z_1 Y}.$$

151. Some interesting conclusions can be drawn from these equations.

The apparent impedance of the total transformer is

$$Z_t = \frac{\dot{E}_0}{\dot{I}_0} = a^2 \frac{1 + Z_0 Y_0 + \frac{Z_0 Y}{a^2}}{Y + a^2 Y_0} \qquad (13)$$

$$= \frac{1 + Z_0\left(Y_0 + \frac{Y}{a^2}\right)}{Y_0 + \frac{Y}{a^2}};$$

$$Z_t = \frac{1}{Y_0 + \frac{Y}{a^2}} + Z_0. \qquad (14)$$

Substituting now, $\frac{Y}{a^2} = Y'$, the total secondary admittance, reduced to the primary circuit by the ratio of turns, it is

$$Z_t = \frac{1}{Y_0 + Y'} + Z_0. \qquad (15)$$

$Y_0 + Y'$ is the total admittance of a divided circuit with the exciting current of admittance, Y_0, and the secondary current of admittance, Y' (reduced to primary), as branches. Thus,

$$\frac{1}{Y_0 - Y'} = Z'_0 \qquad (16)$$

is the impedance of this divided circuit, and

$$Z_t = Z'_0 + Z_0. \tag{17}$$

That is,

The alternate-current transformer, of primary admittance Y_0, total secondary admittance Y, and primary impedance Z_0, is equivalent to, and can be replaced by, a divided circuit with the branches of admittance Y_0, the exciting current, and admittance $Y' = \frac{Y}{a^2}$, the secondary current, fed over mains of the impedance Z_0, the internal primary impedance.

This is shown diagrammatically in Fig. 114.

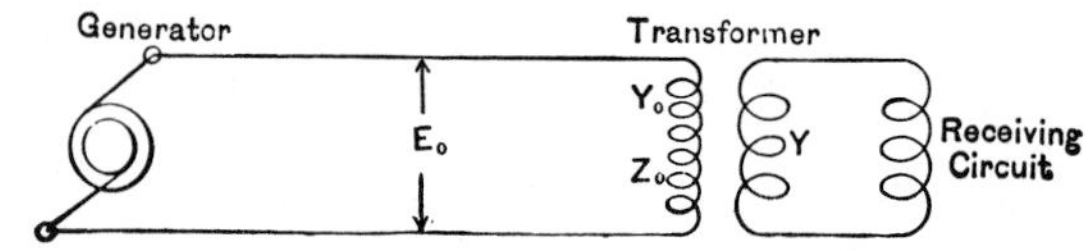

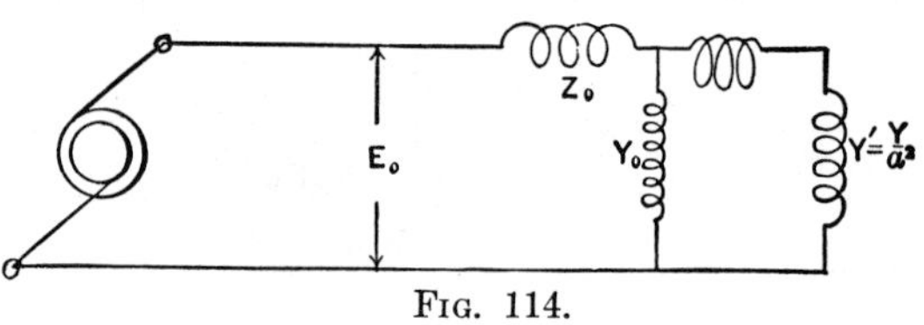

FIG. 114.

152. Separating now the internal secondary impedance from the external secondary impedance, or the impedance of the consumer circuit, it is

$$\frac{1}{Y} = Z_1 + Z; \tag{18}$$

where Z = external secondary impedance,

$$Z = \frac{\dot{E}_1}{I_1}. \tag{19}$$

Reduced to primary circuit, it is

$$\frac{1}{Y'} = \frac{a^2}{Y} = a^2 Z_1 + a^2 Z$$
$$= Z'_1 + Z'. \tag{20}$$

That is,

An alternate-current transformer, of primary admittance Y_0, *primary impedance* Z_0, *secondary impedance* Z_1, *and ratio of turns* a, *can, when the secondary circuit is closed by an impedance,* Z *(the impedance of the receiver circuit), be replaced, and is equivalent to a circuit of impedance,* $Z' = a^2Z$, *fed over mains of the impedance,* $Z_0 + Z'_1$, *where* $Z'_1 = a^2Z_1$, *shunted by a circuit of admittance,* Y_0, *which latter circuit branches off at the points,* a, b, *between the impedances,* Z_0 *and* Z'_1.

This is represented diagrammatically in Fig. 115.

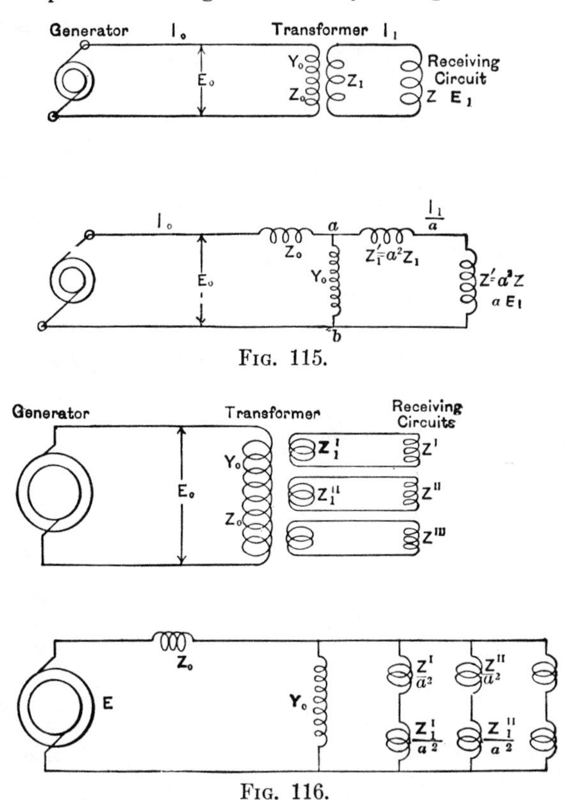

FIG. 115.

FIG. 116.

It is obvious, therefore, that if the transformer contains several independent secondary circuits, they are to be considered as branched off at the points a, b, in diagram, Fig. 115, as shown in diagram, Fig. 116.

It therefore follows:

An alternate-current transformer, of s *secondary coils, of the internal impedances,* Z_1^I, Z_1^{II},... Z_1^s, *closed by external secondary circuits of the impedances,* Z^I, Z^{II},... Z^s, *is equivalent to a divided circuit of* $s + 1$ *branches, one branch of admittance,* Y_0, *the exciting current, the other branches of the impedances,* $Z_1^I + Z^I$, $Z_1^{II} + Z^{II}$,... $Z_1^s + Z^s$, *the latter impedances being reduced to the primary circuit by the ratio of turns, and the whole divided circuit being fed by the primary impressed e.m.f.,* E_0, *over mains of the impedance,* Z_0.

Consequently, transformation of a circuit merely changes all the quantities proportionally, introduces in the mains the impedance, $Z_0 + Z'_1$, and a branch circuit between Z_0 and Z'_1, of admittance Y_0.

Thus, double transformation will be represented by diagram, Fig. 117.

With this the discussion of the alternate-current transformer ends, by becoming identical with that of a divided circuit containing resistances and reactances.

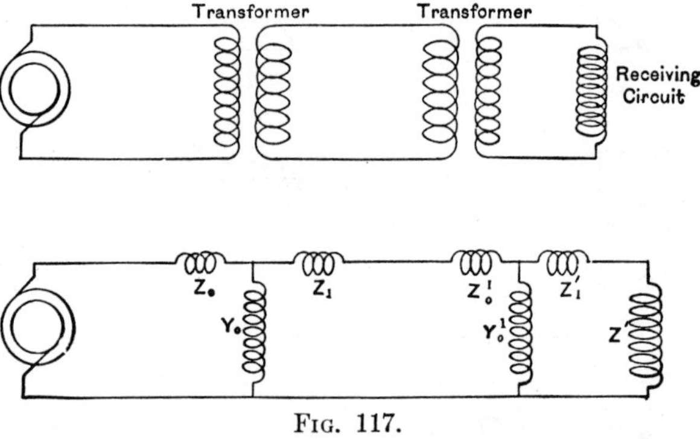

Fig. 117.

Such circuits have been discussed in detail in Chapter IX, and the results derived there are now directly applicable to the transformer, giving the variation and the control of secondary terminal voltage, resonance phenomena, etc.

Thus, for instance, if $Z'_1 = Z_0$, and the transformer contains an additional secondary coil, constantly closed by a condensive reactance of such size that this auxiliary circuit, together with the exciting circuit, gives the reactance, $-x_0$, with a non-inductive secondary circuit, $Z_1 = r_1$, we get the condition of transformation from constant primary potential to constant secondary current, and inversely.

153. As seen, the alternating-current transformer is characterized by the constants:

Ratio of turns: $a = \frac{n_0}{n_1}$.

Exciting admittance: $Y_0 = g_0 - jb_0$.

Self-inductive impedances: $Z_0 = r_0 + jx_0$.

$Z_1 = r_1 + jx_1$.

Since the effect of the secondary impedance is essentially the same as that of the primary impedance (the only difference being, that no voltage is consumed by the exciting current in the secondary impedance, but voltage is consumed in the primary impedance, though very small in a constant-potential transformer), the individual values of the two impedances, Z_0 and Z_1, are of less importance than the resultant or total impedance of the transformer, that is, the sum of the primary impedance plus the secondary impedance reduced to the primary circuit;

$$Z' = Z_0 + a^2 Z_1,$$

and the transformer accordingly is characterized by the two constants:

Exciting admittance, $Y_0 = g_0 - jb_0$.
Total self-inductive impedance, $Z' = r' + jx'$.

Especially in constant-potential transformers with closed magnetic circuit—as usually built—the combination of both impedances into one, Z', is permissible as well within the errors of observation.

Experimentally, the exciting admittance, $Y_0 = g_0 - jb_0$, and the total self-inductive impedance, $Z' = r' + jx'$, are determined by operating the transformer at its normal frequency:

1. With open secondary circuit, and measuring volts e_0, amperes i_0, and watts w_0, input—excitation test.

2. With the secondary short-circuited, and measuring volts e_1, amperes i_1, and watts p_1, input. (In this case, usually a far lower impressed voltage is required—impedance test.)

It is then:

$$y_0 = \frac{i_0}{e_0}, \qquad z' = \frac{e_1}{i_1},$$

$$g_0 = \frac{p_0}{e_0^2}, \qquad r' = \frac{p_1}{i_1^2},$$

$$b_0 = \sqrt{y_0^2 + g_0^2}. \qquad x' = \sqrt{z'^2 - r'^2}.$$

If a separation of the total impedance Z' into the primary impedance and the secondary impedance is desired, as a rule the secondary reactance reduced to the primary can be assumed as equal to the primary reactance:

$$a^2 x_1 = x_0,$$

except if from the construction of the transformer it can be seen that one of the circuits has far more reactance than the other, and then judgment or approximate calculation must guide in the division of the total reactance between the two circuits.

If the total effective resistance, r', as derived by wattmeter, equals the sum of the ohmic resistances of primary and of secondary reduced to the primary:

$$r' = r_0 + a^2 r,$$

the ohmic resistances, r_0 and r_1, as measured by Wheatstone bridge or by direct current, are used.

If the effective resistance is greater than the resultant of the ohmic resistances:

$$r' > r_0 + a^2 r_1,$$

the difference:

$$r'' = r' - (r_0 + a^2 r_1)$$

may be divided between the two circuits in proportion to the ohmic resistances, that is, the effective resistance distributed between the two circuits in the proportion of their ohmic resistances, so giving the effective resistances of the two circuits, r'_0 and r'_1, by:

$$r'_0 \div r'_1 = r_0 \div r_1;$$

or, if from the construction of the transformer as the use of large solid conductors, it can be seen that the one circuit is entirely or mainly the seat of the power loss by hysteresis, eddies, etc., which is represented by the additional effective resistance, r'', this resistance, r'', is entirely or mainly assigned to this circuit.

In general, it therefore may be assumed:

$$\left.\begin{aligned} x_0 &= \frac{x'}{2}, \\ x_1 &= \frac{x'}{2a^2}, \end{aligned}\right\} \qquad \left.\begin{aligned} r'_0 &= r_0 \frac{r'}{r_0 + a^2 r_1}, \\ r'_1 &= r_1 \frac{r'}{r_0 + a^2 r_1}. \end{aligned}\right\}$$

Usually, the excitation test is made on the low-voltage coil, the impedance test on the high-voltage coil, and then reduced to the same coil as primary. Hereby the currents and voltages are more nearly of the same magnitude in both tests.

154. In the calculation of the transformer:

The exciting admittance, Y_0, is derived by calculating the total exciting current from the ampere-turns excitation, the magnetic characteristic of the iron and the dimensions of the main magnetic circuit, that is the magnetic circuit interlinked with primary and secondary coils. The conductance, g_0, is derived from the hysteresis loss in the iron, as given by magnetic density, hysteresis coefficient and dimensions of magnetic circuit, allowance being made for eddy currents in the iron.

The ohmic resistances, r_0 and r_1, are found from the dimensions of the electric circuit, and, where required, allowance made for the additional effective resistance, r''.

The reactances, x_0 and x_1, are calculated by calculating the leakage flux, that is the magnetic flux produced by the total primary respectively secondary ampere-turns, and passing between primary and secondary coils, and within the primary respectively secondary coil, in a magnetic circuit consisting largely of air. In this case, the iron part of the magnetic leakage circuit can as a rule be neglected.

PART III

THEORY OF ELECTRIC CIRCUITS

CHAPTER I

ELECTRIC CONDUCTION. SOLID AND LIQUID CONDUCTORS

1. When electric power flows through a circuit, we find phenomena taking place outside of the conductor which directs the flow of power, and also inside thereof. The phenomena outside of the conductor are conditions of stress in space which are called the electric field, the two main components of the electric field being the electromagnetic component, characterized by the circuit constant inductance, L, and the electrostatic component, characterized by the electric circuit constant capacity, C. Inside of the conductor we find a conversion of energy into heat; that is, electric power is consumed in the conductor by what may be considered as a kind of resistance of the conductor to the flow of electric power, and so we speak of resistance of the conductor as an electric quantity, representing the power consumption in the conductor.

Electric conductors have been classified and divided into distinct groups. We must realize, however, that there are no distinct classes in nature, but a gradual transition from type to type.

Metallic Conductors

2. The first class of conductors are the metallic conductors. They can best be characterized by a negative statement—that is, metallic conductors are those conductors in which the conduction of the electric current converts energy into no other form but heat. That is, a consumption of power takes place in the metallic conductors by conversion into heat, and into heat only. Indirectly, we may get light, if the heat produced raises the temperature high enough to get visible radiation as in the incandescent lamp filament, but this radiation is produced from heat, and directly the conversion of electric energy takes place into heat. Most of the metallic conductors cover, as regards their specific resistance, a rather narrow range, between about 1.6 microhm-cm. (1.6×10^{-6}) for copper, to about 100 microhm-cm. for cast iron, mercury, high-resistance alloys, etc. They, therefore, cover a range of less than 1 to 100.

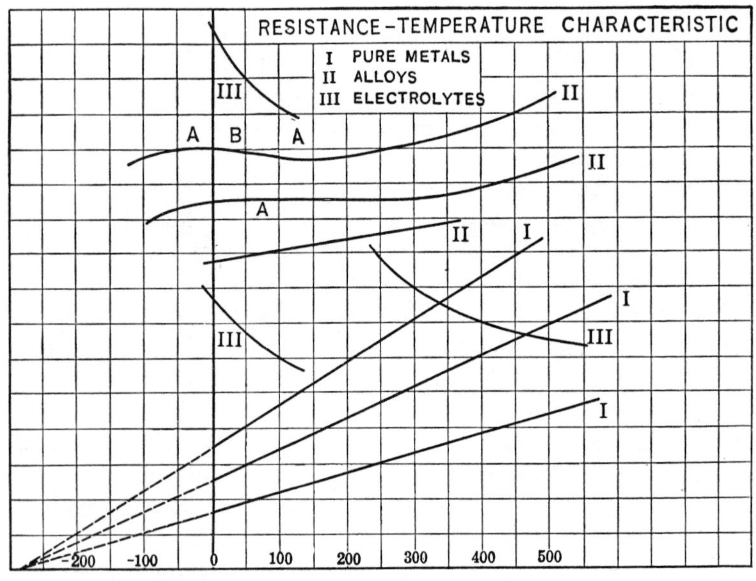

FIG. 1.

A characteristic of metallic conductors is that the resistance is approximately constant, varying only slightly with the temperature, and this variation is a rise of resistance with increase of temperature—that is, they have a positive temperature coefficient. In the pure metals, the resistance apparently is approximately proportional to the absolute temperature—that is, the temperature coefficient of resistance is constant and such that the resistance plotted as function of the temperature is a straight line which points toward the absolute zero of temperature, or, in other words, which, prolonged backward toward falling temperature, would reach zero at $-273°$C., as illustrated by curves I on Fig. 1. Thus, the resistance may be expressed by

$$r = r_0 T \tag{1}$$

where T is the absolute temperature.

In alloys of metals we generally find a much lower temperature coefficient, and find that the resistance curve is no longer a straight line, but curved more or less, as illustrated by curves II, Fig. 1, so that ranges of zero temperature coefficient, as at A in curve II, and even ranges of negative temperature coefficient, as at B in curve II, Fig. 1, may be found in metallic conductors which are alloys, but the general trend is upward. That is, if we extend the investigation over a very wide range of temperature, we find that even in those alloys which have a negative temperature coefficient for a limited temperature range, the average temperature coefficient is positive for a very wide range of temperature—that is, the resistance is higher at very high and lower at very low temperature, and the zero or negative coefficient occurs at a local flexure in the resistance curve.

3. The metallic conductors are the most important ones in industrial electrical engineering, so much so, that when speaking of a "conductor," practically always a metallic conductor is understood. The foremost reason is, that the resistivity or specific resistance of all other classes of conductors is so very much higher than that of metallic conductors that for directing the flow of current only metallic conductors can usually come into consideration.

As, even with pure metals, the change of resistance of metallic conductors with change of temperature is small—about ⅓ per cent. per degree centigrade—and the temperature of most apparatus during their use does not vary over a wide range of temperature, the resistance of metallic conductors, r, is usually assumed as constant, and the value corresponding to the operating temperature chosen. However, for measuring temperature rise of electric currents, the increase of the conductor resistance is frequently employed.

Where the temperature range is very large, as between room temperature and operating temperature of the incandescent lamp filament, the change of resistance is very considerable; the resistance of the tungsten filament at its operating temperature is about nine times its cold resistance in the vacuum lamp, twelve times in the gas-filled lamp.

Thus the metallic conductors are the most important. They require little discussion, due to their constancy and absence of secondary energy transformation.

Iron makes an exception among the pure metals, in that it has an abnormally high temperature coefficient, about 30 per cent. higher than other pure metals, and at red heat, when approaching the temperature where the iron ceases to be magnetizable, the temperature coefficient becomes still higher, until the temperature is reached where the iron ceases to be magnetic. At this point its temperature coefficient becomes that of other pure metals. Iron wire—usually mounted in hydrogen to keep it from oxidizing —thus finds a use as series resistance for current limitation in vacuum arc circuits, etc.

Electrolytic Conductors

4. The conductors of the second class are the electrolytic conductors. Their characteristic is that the conduction is accompanied by chemical action. The specific resistance of electrolytic conductors in general is about a million times higher than that of the metallic conductors. They are either fused compounds, or solutions of compounds in solvents, ranging in resistivity from 1.3 ohm-cm., in 30 per cent. nitric acid, and still lower in fused salts, to about 10,000 ohm-cm. in pure river water, and from there up to infinity (distilled water, alcohol, oils, etc.). They are all liquids, and when frozen become insulators.

Characteristic of the electrolytic conductors is the negative temperature coefficient of resistance; the resistance decreases with increasing temperature—not in a straight, but in a curved line, as illustrated by curves III in Fig. 1.

When dealing with electrical resistances, in many cases it is more convenient and gives a better insight into the character of the conductor, by not considering the resistance as a function of the temperature, but the voltage consumed by the conductor as a function of the current under stationary condition. In this case, with increasing current, and so increasing power consumption, the temperature also rises, and the curve of voltage for increasing current so illustrates the electrical effect of increasing temperature. The advantage of this method is that in many cases we get a better view of the action of the conductor in an electric circuit by eliminating the temperature, and relating only electrical quantities with each other. Such volt-ampere characteristics of electric conductors can easily and very accurately be determined, and, if desired, by the radiation law approximate values of the temperature be derived, and therefrom the temperature-resistance curve calculated, while a direct measurement of the resistance over a very wide range of temperature is extremely difficult, and often no more accurate.

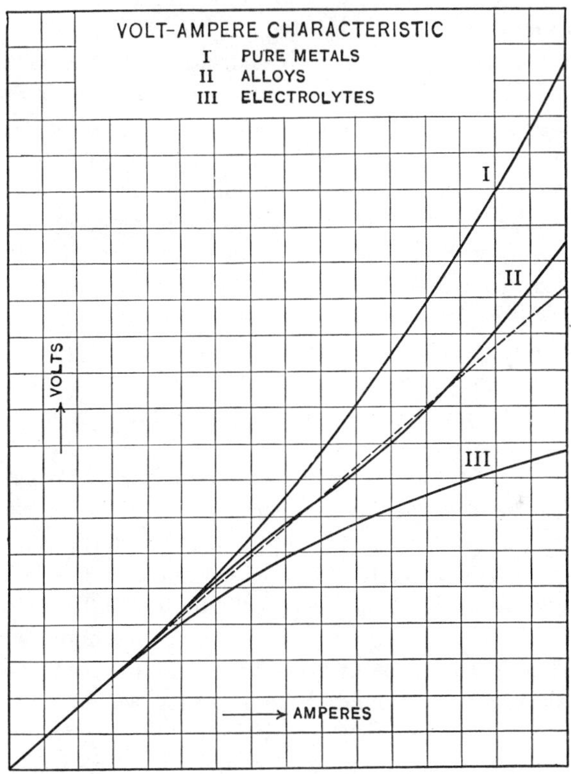

Fig. 2.

In Fig. 2, therefore, are shown such volt-ampere characteristics of conductors. The dotted straight line is the curve of absolutely constant resistance, which does not exist. Curves I and II show characteristics of metallic conductors, curve III of electrolytic conductors. As seen, for higher currents I and II rise faster, and III slower than for low currents.

It must be realized, however, that the volt-ampere characteristic depends not only on the material of the conductor, as the temperature-resistivity curve, but also on the size and shape of the conductor, and its surroundings. For a long and thin conductor in horizontal position in air, it would be materially different numerically from that of a short and thick conductor in different position at different surrounding temperature. However, qualitatively it would have the same characteristics, the same characteristic deviation from straight line, etc., merely shifted in their numerical values. Thus it characterizes the general nature of the conductor, but where comparisons between different conductor materials are required, either they have to be used in the same shape and position, when determining their volt-ampere characteristics, or the volt-ampere characteristics have to be reduced to the resistivity-temperature characteristics. The volt-ampere characteristics become of special importance with those conductors, to which the term resistivity is not physically applicable, and therefore the "effective resistivity" is of little meaning, as in gas and vapor conduction (arcs, etc.).

5. The electrolytic conductor is characterized by chemical action accompanying the conduction. This chemical action follows Faraday's law:

The amount of chemical action is proportional to the current and to the chemical equivalent of the reaction.

The product of the reaction appears at the terminals or "electrodes," between the electrolytic conductor or "electrolyte," and the metallic conductors. Approximately, 0.01 mg. of hydrogen are produced per coulomb or ampere-second. From this electrochemical equivalent of hydrogen, all other chemical reactions can easily be calculated from atomic weight and valency. For instance, copper, with atomic weight 63 and valency 2, has the equivalent $63/2 = 31.5$ and copper therefore is deposited at the negative terminal or "cathode," or dissolved at the positive terminal or "anode," at the rate of 0.315 mg. per ampere-second; aluminum, atomic weight 28 and valency 3, at the rate of 0.093 mg. per ampere-second, etc.

The chemical reaction at the electrodes represents an energy transformation between electrical and chemical energy, and as the rate of electrical energy supply is given by current times voltage, it follows that a voltage drop or potential difference occurs at the electrodes in the electrolyte. This is in opposition to the current, or a counter e.m.f., the "counter e.m.f. of electrochemical polarization," and thus consumes energy, if the chemical reaction requires energy—as the deposition of copper from a solution of a copper salt. It is in the same direction as the current, thus producing electric energy, if the chemical reaction produces energy, as the dissolution of copper from the anode.

As the chemical reaction, and therefore the energy required for it, is proportional to the current, the potential drop at the electrodes is independent of the current density, or constant for the same chemical reaction and temperature, except in so far as secondary reactions interfere. It can be calculated from the chemical energy of the reaction, and the amount of chemical reaction as given by Faraday's law. For instance: 1 amp.-sec. deposits 0.315 mg. copper. The voltage drop, e, or polarization voltage, thus must be such that e volts times 1 amp.-sec., or e watt-sec. or joules, equals the chemical reaction energy of 0.315 mg. copper in combining to the compound from which it is deposited in the electrolyte.

If the two electrodes are the same and in the same electrolyte at the same temperature, and no secondary reaction occurs, the reactions are the same but in opposite direction at the two electrodes, as deposition of copper from a copper sulphate solution at the cathode, solution of copper at the anode. In this case, the two potential differences are equal and opposite, their resultant thus zero, and it is said that "no polarization occurs."

If the two reactions at the anode and cathode are different, as the dissolution of zinc at the anode, the deposition of copper at the cathode, or the production of oxygen at the (carbon) anode, and the deposition of zinc at the cathode, then the two potential differences are unequal and a resultant remains. This may be in the same direction as the current, producing electric energy, or in the opposite direction, consuming electric energy. In the first case, copper deposition and zinc dissolution, the chemical energy set free by the dissolution of the zinc and the voltage produced by it, is greater than the chemical energy consumed in the deposition of the copper, and the voltage consumed by it, and the resultant of the two potential differences at the electrodes thus is in the same direction as the current, hence may produce this current. Such a device, then, transforms chemical energy into electrical energy, and is called a *primary cell* and a number of them, a *battery*. In the second case, zinc deposition and oxygen production at the anode, the resultant of the two potential differences at the electrodes is in opposition to the current; that is, the device consumes electric energy and converts it into chemical energy, as *electrolytic cell.*"

Both arrangements are extensively used: the battery for producing electric power, especially in small amounts, as for hand lamps, the operation of house bells, etc. The electrolytic cell is used extensively in the industries for the production of metals as aluminum, magnesium, calcium, etc., for refining of metals as copper, etc., and constitutes one of the most important industrial applications of electric power.

A device which can efficiently be used, alternately as battery and as electrolytic cell, is the *secondary cell* or *storage battery.* Thus in the lead storage battery, when discharging, the chemical reaction at the anode is conversion of lead peroxide into lead oxide, at the cathode the conversion of lead into lead oxide; in charging, the reverse reaction occurs.

6. Specifically, as "polarization cell" is understood a combination of electrolytic conductor with two electrodes, of such character that no permanent change occurs during the passage of the current. Such, for instance, consists of two platinum electrodes in diluted sulphuric acid. During the passage of the current, hydrogen is given off at the cathode and oxygen at the anode, but terminals and electrolyte remain the same (assuming that the small amount of dissociated water is replaced).

In such a polarization cell, if e_0 = counter e.m.f. of polarization (corresponding to the chemical energy of dissociation of water, and approximately 1.6 volts) at constant temperature and thus constant resistance of the electrolyte, the current, i, is proportional to the voltage, e, minus the counter e.m.f. of polarization, e_0:

$$i = \frac{e - e_0}{r} \tag{2}$$

In such a case the curve III of Fig. 2 would with decreasing current not go down to zero volts, but would reach zero amperes at a voltage $e = e_o$, and its lower part would have the shape as shown in Fig. 3. That is, the current begins at voltage, e_0, and below this voltage, only a very small "diffusion" current flows.

When dealing with electrolytic conductors, as when measuring their resistance, the counter e.m.f. of polarization thus must be considered, and with impressed voltages less than the polarization voltage, no permanent current flows through the electrolyte, or rather only a very small "leakage" current or "diffusion" current, as shown in Fig. 3. When closing the circuit, however, a transient current flows. At the moment of circuit closing, no counter e.m.f. exists, and current flows under the full impressed voltage. This current, however, electrolytically produces a hydrogen and an oxygen film at the electrodes, and with their gradual formation, the counter e.m.f. of polarization increases and decreases the current, until it finally stops it. The duration of this transient depends on the resistance of the electrolyte and on the surface of the electrodes, but usually is fairly short.

7. This transient becomes a permanent with alternating impressed voltage. Thus, when an alternating voltage, of a maxi-

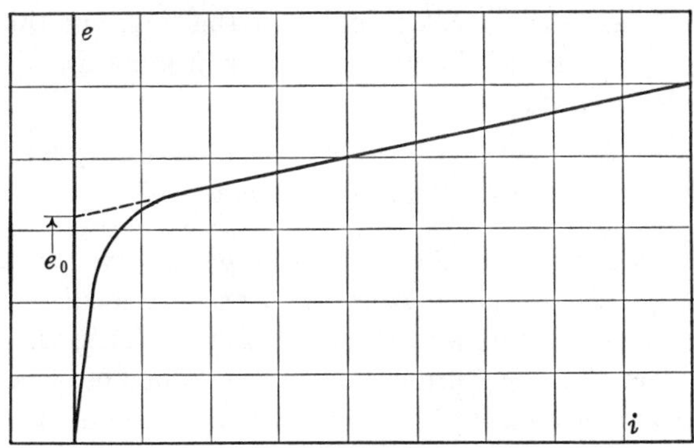

FIG. 3.

mum value lower than the polarization voltage, is impressed upon an electrolytic cell, an alternating current flows through the cell, which produces the hydrogen and oxygen films which hold back the current flow by their counter e.m.f. The current thus flows ahead of the voltage or counter e.m.f. which it produces, as a leading current, and the polarization cell thus acts like a condenser, and is called an "electrolytic condenser." It has an enormous electrostatic capacity, or "effective capacity," but can stand low voltage only—1 volt or less—and therefore is of limited industrial value. As chemical action requires appreciable time, such electrolytic condensers show at commercial frequencies high losses of power by what may be called "chemical hysteresis," and therefore low efficiences, but they are alleged to become efficient at very low frequencies. For this reason, they have been proposed in the secondaries of induction motors, for power-factor compensation. Iron plates in alkaline solution, as sodium carbonate, are often considered for this purpose.

NOTE.—The aluminum cell, consisting of two aluminum plates with an electrolyte which does not attack aluminum, often is called an electrolytic condenser, as its current is leading; that is, it acts as capacity. It is, however, not an electrolytic condenser, and the counter e.m.f., which gives the capacity effect, is not electrolytic polarization. The aluminum cell is a true electrostatic condenser, in which the film of alumina, formed on the positive aluminum plates, is the dielectric. Its characteristic is, that the condenser is self-healing; that is, a puncture of the alumina film causes a current to flow, which electrolytically produces alumina at the puncture hole, and so closes it. The capacity is very high, due to the great thinness of the film, but the energy losses are considerable, due to the continual puncture and repair of the dielectric film.

Pyroelectric Conductors

8. A third class of conductors are the *pyroelectric conductors* or *pyroelectrolytes*. In some features they are intermediate between the metallic conductors and the electrolytes, but in their essential characteristics they are outside of the range of either. The metallic conductors as well as the electrolytic conductors give a volt-ampere characteristic in which, with increase of current, the voltage rises, faster than the current in the metallic conductors, due to their positive temperature coefficient, slower than the current in the electrolytes, due to their negative temperature coefficient.

The characteristic of the pyroelectric conductors, however, is such a very high negative temperature coefficient of resistance, that is, such rapid decrease of resistance with increase of temperature, that over a wide range of current the voltage decreases with increase of current. Their volt-ampere characteristic thus has a shape as shown diagrammatically in Fig. 4—though not all such conductors may show the complete curve, or parts of the curve may be physically unattainable: for small currents, range (1), the voltage increases approximately proportional to the current, and sometimes slightly faster, showing the positive temperature coefficient of metallic conduction. At a the temperature coefficient changes from positive to negative, and the voltage begins to increase slower than the current, similar as in electrolytes, range (2). The negative temperature coefficient rapidly increases, and the voltage rise become slower, until at point *b* the negative temperature coefficient has become so large, that the voltage begins to decrease again with increasing current, range (3). The maximum voltage point *b* thus divides the range of rising characteristic (1) and (2), from that of decreasing characteristic, (3). The negative temperature coefficient reaches a maximum and then decreases again, until at point *c* the negative temperature coefficient has fallen so that beyond this minimum voltage point *c* the voltage again increases with increasing current, range (4),

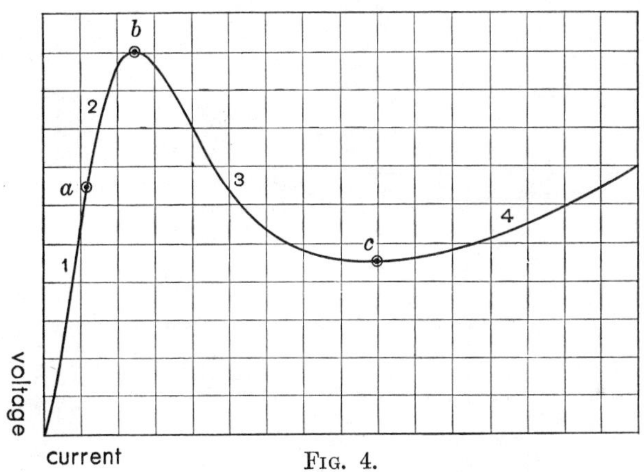

FIG. 4.

though the temperature coefficient remains negative, like in electrolytic conductors.

In range (1) the conduction is purely metallic, in range (4) becomes purely electrolytic, and is usually accompanied by chemical action.

Range (1) and point *a* often are absent and the conduction begins already with a slight negative temperature coefficient.

The complete curve, Fig. 4, can be observed only in few substances, such as magnetite. Minimum voltage point *c* and range (4) often is unattainable by the conductor material melting or being otherwise destroyed by heat before it is reached. Such, for instance, is the case with cast silicon. The maximum voltage point *b* often is unattainable, and the passage from range (2) to range (3) by increasing the current therefore not feasible, because the maximum voltage point b is so high, that disruptive discharge occurs before it is reached. Such for instance is the case in glass, the Nernst lamp conductor, etc.

9. The curve, Fig. 3, is drawn only diagrammatically, and the lower current range exaggerated, to show the characteristics. Usually the current at point b is very small compared with that at point c; rarely more than one-hundredth of it, and the actual proportions more nearly represented by Fig. 5. With pyroelectric conductors of very high value of the voltage b, the currents in the range (1) and (2) may not exceed one-millionth of that at (3). Therefore, such volt-ampere characteristics are

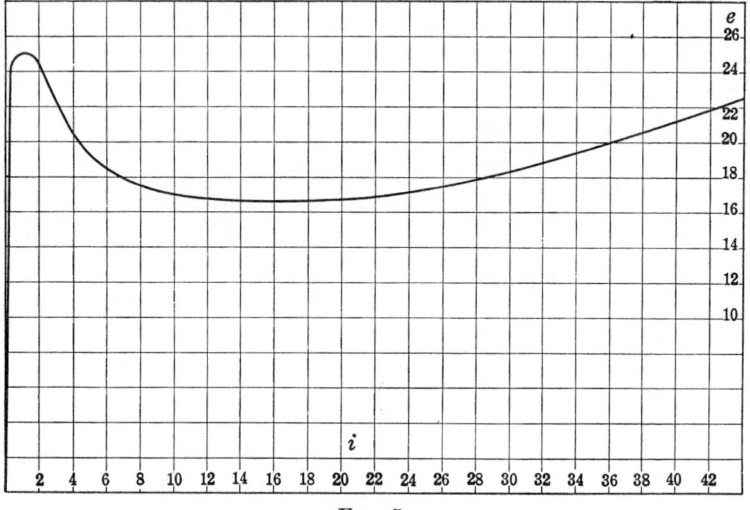

FIG. 5.

often plotted with $\sqrt{i}$ as abscissæ, to show the ranges in better proportions.

Pyroelectric conductors are metallic silicon, boron, some forms of carbon as anthracite, many metallic oxides, especially those of the formula $M^{(2)} M_2^{(3)} O_4$, where $M^{(2)}$ is a bivalent, $M^{(3)}$ a trivalent metal (magnetite, chromite), metallic sulphides, silicates such as glass, many salts, etc.

Intimate mixtures of conductors, as graphite, coke, powdered metal, with non-conductors as clay, carborundum, cement, also have pyroelectric conduction. Such are used, for instance, as "resistance rods" in lightning arresters, in some rheostats, as cement resistances for high-frequency power dissipation in reactances, etc. Many, if not all so-called "insulators" probably are in reality pyroelectric conductors, in which the maximum voltage point b is so high, that the range (3) of decreasing characteristic can be reached only by the application of external heat, as in the Nernst lamp conductor, or can not be reached at all, because chemical dissociation begins below its temperature, as in organic insulators.

Fig. 6 shows the volt-ampere characteristics of two rods of cast silicon, 10 in. long and 0.22 in. in diameter, with $\sqrt{i}$ as ab-

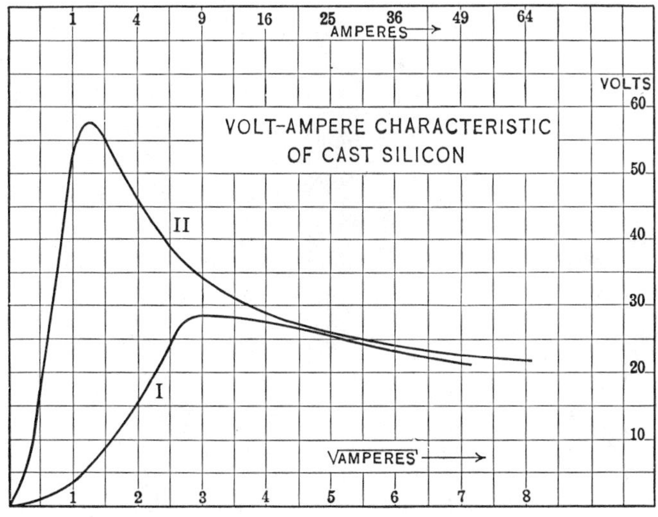

FIG. 6.

scissæ and Fig. 7 their approximate temperature-resistance characteristics. The curve II of Fig. 7 is replotted in Fig. 8, with log r as ordinates. Where the resistivity varies over a very wide range, it often is preferable to plot the logarithm of the resistivity. It is interesting to note that the range (3) of curve II, between 700° and 1400°, is within the errors of observation represented by the expression

$$r = 0.01E^{-\frac{9080}{T}}$$

where T is the absolute temperature (−273°C. as zero point).

The difference between the two silicon rods is, that the one contains 1.4 per cent., the other only 0.1 per cent. carbon; besides this, the impurities are less than 1 per cent.

As seen, in these silicon rods the range (4) is not yet reached at the melting point.

Fig. 9 shows the volt-ampere characteristic, with $\sqrt{i}$ as abscissæ, and Fig. 10 the approximate resistance temperature char-

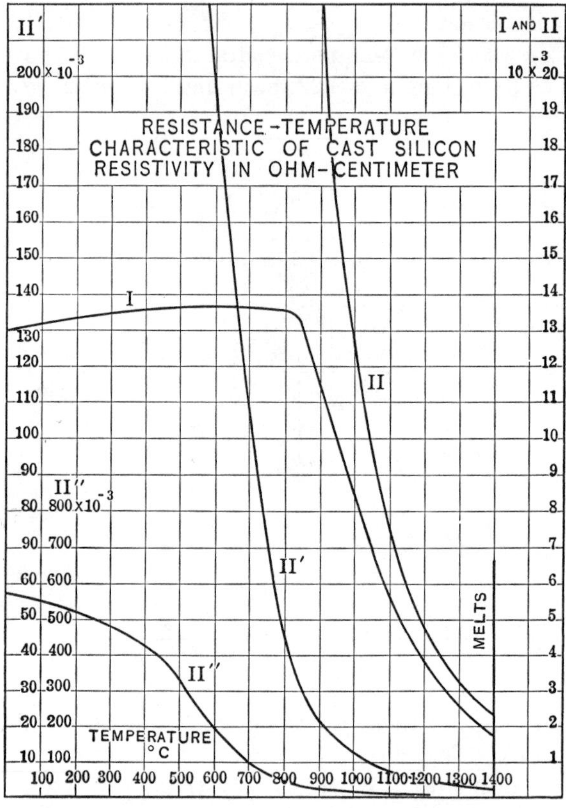

FIG. 7.

acteristic derived therefrom, with log r as ordinates, of a magnetic rod 6 in. long and ¾ in. in diameter, consisting of 90 per cent. magnetite (Fe_3O_4), 9 per cent. chromite ($FeCr_2O_4$) and 1 per cent. sodium silicate, sintered together.

10. As result of these volt-ampere characteristics, Figs. 4 to 10, pyroelectric conductors as structural elements of an electric circuit show some very interesting effects, which may be illustrated on the magnetite rod, Fig. 9. The maximum terminal voltage, which can exist across this rod in stationary conditions, is 25 volts at 1 amp. With increasing terminal voltage, the current thus gradually increases, until 25 volts is reached, and then without further increase of the impressed voltage the current rapidly rises to short-circuit values. Thus, such resistances can be used as excess-voltage cutout, or, when connected between circuit and ground, as excess-voltage grounding device: below 24 volts, it

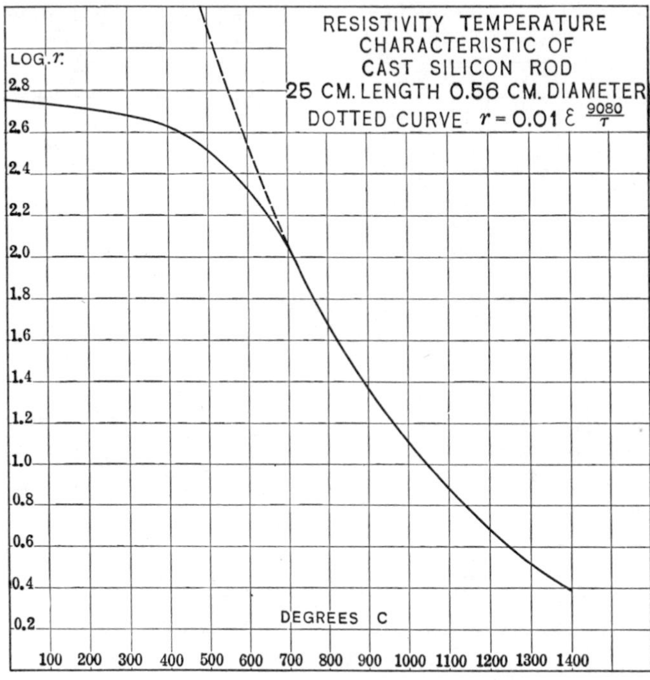

Fig. 8.

bypasses a negligible current only, but if the voltage rises above 25 volts, it short-circuits the voltage and so stops a further rise, or operates the circuit-breaker, etc. As the decrease of resistance is the result of temperature rise, it is not instantaneous; thus the rod does not react on transient voltage rises, but only on lasting ones.

Within a considerable voltage range—between 16 and 25 volts —three values of current exist for the same terminal voltage. Thus at 20 volts between the terminals of the rod in Fig. 9, the current may be 0.02 amp., or 4.1 amp., or 36 amp. That is, in series in a constant-current circuit of 4.1 amp. this rod would show the same terminal voltage as in a 0.02-amp. or a 36-amp. constant-current circuit, 20 volts. On constant-potential supply, however, only the range (1) and (2), and the range (4) is stable, but the range (3) is unstable, and here we have a conductor, which is unstable in a certain range of currents, from point *b* at 1 amp. to point *c* at 20 amp. At 20 volts impressed upon the rod, 0.02 amp. may pass through it, and the conditions are stable. That is, a tendency to increase of current would check itself by requiring an increase of voltage beyond that supplied, and a decrease of

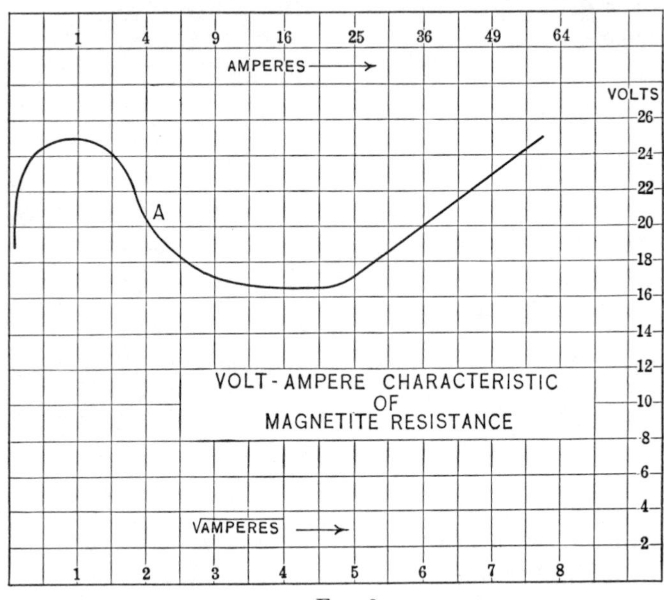

FIG. 9.

current would reduce the voltage consumption below that employed, and thus be checked. At the same impressed 20 volts, 36 amp. may pass through the rod—or 1800 times as much as before—and the conditions again are stable. A current of 4.1 amp. also would consume a terminal voltage of 20, but the condition now is unstable; if the current increases ever so little, by a momentary voltage rise, then the voltage consumed by the rod decreases, becomes less than the terminal voltage of 20, and the current thus increases by the supply voltage exceeding the consumed voltage. This, however, still further decreases the consumed voltage and thereby increases the current, and the current rapidly rises, until conditions become stable at 36 amp. Inversely, a momentary decrease of the current below 4.1 amp. increases the voltage required by the rod, and this higher voltage not being available at constant supply voltage, the current decreases.

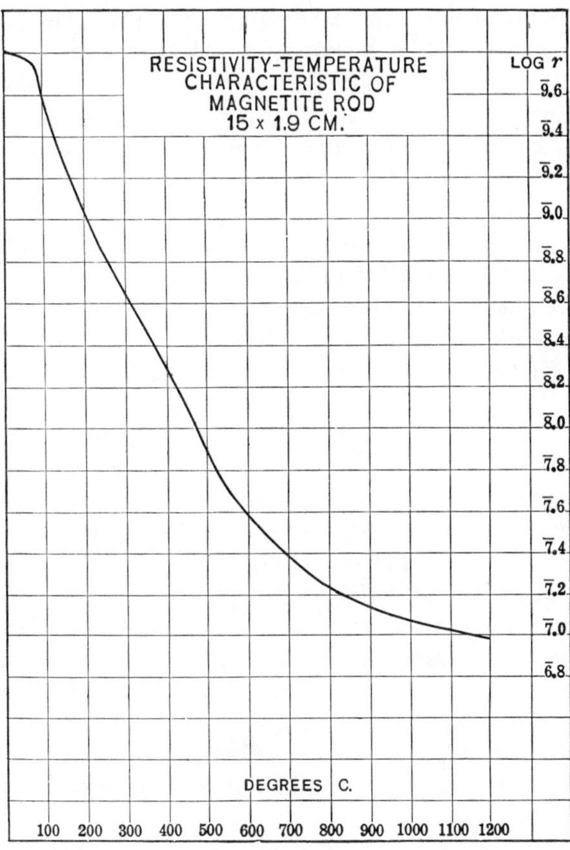

FIG. 10.

This, however, still further increases the required voltage and decreases the current, until conditions become stable at 0.02 amp.

With the silicon rod II of Fig. 6, on constant-potential supply, with increasing voltage the current and the temperature increases gradually, until 57.5 volts are reached at about 450°C.; then, without further voltage increase, current and temperature rapidly increase until the rod melts. Thus:

Condition of stability of a conductor on constant-voltage supply is, that the volt-ampere characteristic is rising, that is, an increase of current requires an increase of terminal voltage.

A conductor with falling volt-ampere characteristic, that is, a conductor in which with increase of current the terminal voltage decreases, is unstable on constant-potential supply.

11. An important application of pyroelectric conduction has been the glower of the Nernst lamp, which before the development of the tungsten lamp was extensively used for illumination.

Pyroelectrolytes cover the widest range of conductivities; the alloys of silicon with iron and other metals give, depending on their composition, resistivities from those of the pure metals up to the lower resistivities of electrolytes: 1 ohm per cm.^{3}; borides, carbides, nitrides, oxides, etc., gave values from 1 ohm per cm.^{3} or less, up to megohms per cm.^{3}, and gradually merge into the materials which usually are classed as "insulators."

The pyroelectric conductors thus are almost the only ones available in the resistivity range between the metals, 0.0001 ohm-cm. and the electrolytes, 1 ohm-cm.

Pyroelectric conductors are industrially used to a considerable extent, since they are the only solid conductors, which have resistivities much higher than metallic conductors. In most of the industrial uses, however, the dropping volt-ampere characteristic is not of advantage, is often objectionable, and the use is limited to the range (1) and (2) of Fig. 3. It, therefore, is of importance to realize their pyroelectric characteristics and the effect which they have when overlooked beyond the maximum voltage point. Thus so-called "graphite resistances" or "carborundum resistances," used in series to lightning arresters to limit the discharge, when exposed to a continual discharge for a sufficient time to reach high temperature, may practically short-circuit and thereby fail to limit the current.

12. From the dropping volt-ampere characteristic in some pyroelectric conductors, especially those of high resistance, of very high negative temperature coefficient and of considerable cross-section, results the tendency to unequal current distribution and the formation of a "luminous streak," at a sudden application of high voltage. Thus, if the current passing through a graphite-clay rod of a few hundred ohms resistance is gradually increased, the temperature rises, the voltage first increases and then decreases, while the rod passes from range (2) into the range (3) of the volt-ampere characteristic, but the temperature and thus the current density throughout the section of the rod is fairly uniform. If, however, the full voltage is suddenly applied, such as by a lightning discharge throwing line voltage on the series resistances of a lightning arrester, the rod heats up very rapidly, too rapidly for the temperature to equalize throughout the rod section, and a part of the section passes the maximum voltage point *b* of Fig. 4 into the range (3) and (4) of low resistance, high current and high temperature, while most of the section is still in the high-resistance range (2) and never passes beyond this range, as it is practically short-circuited. Thus, practically all the current passes by an irregular luminous streak through a small section of the rod, while most of the section is relatively cold and practically does not participate in the conduction. Gradually, by heat conduction the temperature and the current density may become more uniform, if before this the rod has not been destroyed by temperature stresses. Thus, tests made on such conductors by gradual application of voltage give no information on their behavior under sudden voltage application. The liability to the formation of such luminous streaks naturally increases with decreasing heat conductivity of the material, and with increasing resistance and temperature coefficient of resistance, and with conductors of extremely high temperature coefficient, such as silicates, oxides of high resistivity, etc., it is practically impossible to get current to flow through any appreciable conductor section, but the conduction is always streak conduction.

Some pyroelectric conductors have the characteristic that their resistance increases permanently, often by many hundred per cent. when the conductor is for some time exposed to high-frequency electrostatic discharges.

Coherer action, that is, an abrupt change of conductivity by an electrostatic spark, a wireless wave, etc., also is exhibited by some pyroelectric conductors.

13. Operation of pyroelectric conductors on a constant-voltage circuit, and in the unstable branch (3), is possible by the insertion of a series resistance (or reactance, in alternating-current circuits) of such value, that the resultant volt-ampere characteristic is stable, that is, rises with increase of current. Thus, the conductor in Fig. 4, shown as *I* in Fig. 11, in series with the metallic resistance giving characteristic *A*, gives the resultant characteristic *II* in Fig. 11, which is stable over the entire range. *I* in series with a smaller resistance, of characteristic B, gives the resultant characteristic III. In this, the unstable range has contracted to from b' to c'. Further discussion of the instability of such conductors, the effect of resistance in stablizing them, and the result-

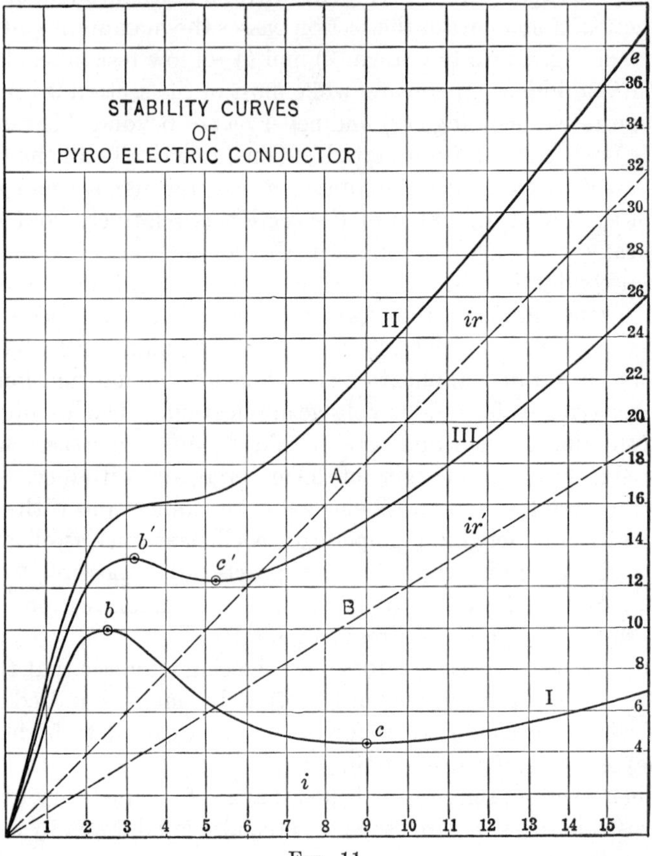

FIG. 11.

ant "stability curve" are found in the chapter on "Instability of Circuits: The Arc," pp. 489–528.

14. It is doubtful whether the pyroelectric conductors really form one class, or whether, by the physical nature of their conduction, they should not be divided into at least two classes:

1. True pyroelectric conductors, in which the very high negative temperature coefficient is a characteristic of the material.

In this class probably belong silicon and its alloys, boron, magnetite and other metallic oxides, sulphides, carbides, etc.

2. Conductors which are mixtures of materials of high conductivity, and of non-conductors, and derive their resistance from the contact resistance between the conducting particles which are separated by non-conductors. As contact resistance shares with arc conduction the dropping volt-ampere characteristic, such mixtures thereby imitate pyroelectric conduction. In this class probably belong the graphite-clay rods industrially used. Powders of metals, graphite and other good conductors also belong in this class.

The very great increase of resistance of some conductors under electrostatic discharges probably is limited to this class, and is the result of the high current density of the condenser discharge burning off the contact points.

Coherer action probably is limited also to those conductors, and is the result of the minute spark at the contact points initiating conduction.

Carbon

15. In some respects outside of the three classes of conductors thus far discussed, in others intermediate between them, is one of

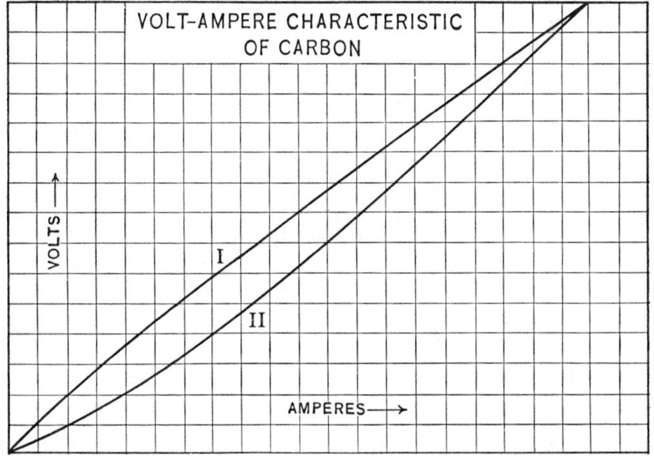

Fig. 12.

the industrially most important conductors, *carbon*. It exists in a large variety of modifications of different resistance characteristics, which all are more or less intermediate between three typical forms:

1. **Metallic Carbon.**—It is produced from carbon deposited on an incandescent filament, from hydrocarbon vapors at a partial vacuum, by exposure to the highest temperatures of the electric furnace. Physically, it has metallic characteristics: high elas-

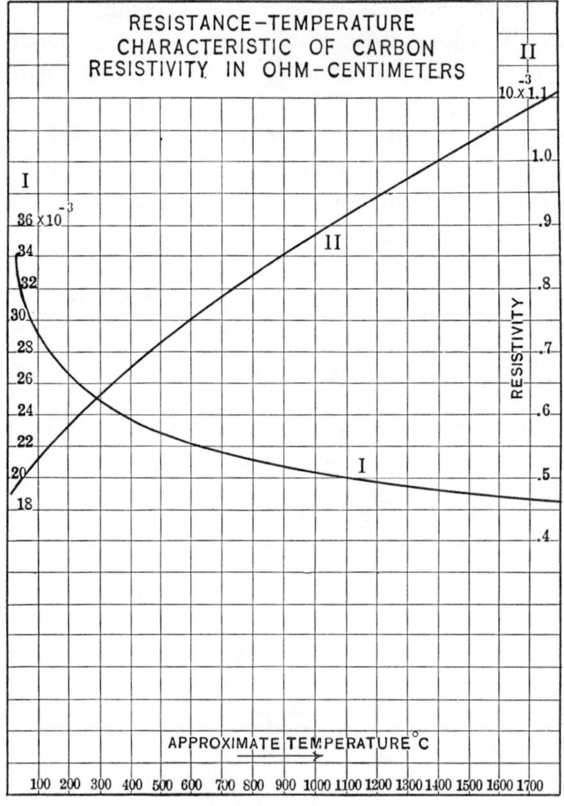

FIG. 13.

ticity, metallic luster, etc., and electrically it has a relatively low resistance approaching that of metallic conduction, and a positive temperature coefficient of resistance, of about 0.1 per cent. per degree C.—that is, of the same magnitude as mercury or cast iron.

The coating of the "Gem" filament incandescent lamp consists of this modification of carbon.

2. **Amorphous carbon,** as produced by the carbonization of cellulose. In its purest form, as produced by exposure to the highest temperatures of the electric furnace, it is characterized by a relatively high resistance, and a negative temperature coefficient of resistance, its conductivity increasing by about 0.1 per cent. per degree C.

3. **Anthracite.**—It has an extremely high resistance, is practically an insulator, but has a very high negative temperature coefficient of resistance, and thus becomes a fairly good conductor at high temperature, but its heat conductivity is so low, and the negative temperature coefficient of resistance so high, that the conduction is practically always streak conduction, and at the high temperature of the conducting luminous streak, conversion to graphite occurs, with a permanent decrease of resistance.

(1) thus shows the characteristics of metallic conduction, (2) those of electrolytic conduction, and (3) those of pyroelectric conduction.

Fig. 12 shows the volt-ampere characteristics, and Fig. 13 the resistance-temperature characteristics of amorphous carbon—curve I—and metallic carbon— curve II.

Insulators

16. As a fourth class of conductors may be considered the so-called "insulators," that is, conductors which have such a high specific resistance, that they can not industrially be used for conveying electric power, but on the contrary are used for restraining the flow of electric power to the conductor, or path, by separating the conductor from the surrounding space by such an insulator. The insulators also have a conductivity, but their specific resistance is extremely high. For instance, the specific resistance of fiber is about 10^{12}, of mica 10^{14}, of rubber 10^{16} ohm-cm., etc.

As, therefore, the distinction between conductor and insulator is only qualitative, depending on the application, and more particularly on the ratio of voltage to current given by the source of power, sometimes a material may be considered either as insulator or as conductor. Thus, when dealing with electrostatic machines, which give high voltages, but extremely small currents, wood, paper, etc., are usually considered as conductors, while for the low-voltage high-current electric lighting circuits they are insulators, and for the high-power very high-voltage transmission circuits they are on the border line, are poor conductors and poor insulators.

Insulators usually, if not always, have a high negative temperature coefficient of resistance, and the resistivity often follows approximately the exponential law,

$$r = r_0 E^{-aT} \tag{3}$$

where T = temperature. That is, the resistance decreases by the same percentage of its value, for every degree C. For instance, it decreases to one-tenth for every 25°C. rise of temperature, so that at 100°C. it is 10,000 times lower than at 0°C. Some temperature-resistance curves, with log r as ordinates, of insulating materials are given in Fig. 14.

As the result of the high negative temperature coefficient, for a sufficiently high temperature, the insulating material, if not destroyed by the temperature, as is the case with organic materials, becomes appreciably conducting, and finally becomes a fairly good conductor, usually an electrolytic conductor.

Thus the material of the Nernst lamp (rare oxides, similar to the Welsbach mantle of the gas industry), is a practically perfect insulator at ordinary temperatures, but becomes conducting at high temperature, and is then used as light-giving conductor.

Fig. 15 shows for a number of high-resistance insulating materials the temperature-resistance curve at the range where the resistivity becomes comparable with that of other conductors.

17. Many insulators, however, more particularly the organic materials, are chemically or physically changed or destroyed, before the temperature of appreciable conduction is reached, though even these show the high negative temperature coefficient. With some, as varnishes, etc., the conductivity becomes sufficient, at high temperatures, though still below carbonization temperature, that under high electrostatic stress, as in the insulation of high-voltage apparatus, appreciable energy is represented by the leakage current through the insulation, and in this case rapid i^2r heating and final destruction of the material may result. That is, such materials, while excellent insulators at ordinary temperature, are unreliable at higher temperature.

It is quite probable that there is no essential difference between the true pyroelectric conductors, and the insulators, but the latter are merely pyroelectric conductors in which the initial resistivity and the voltage at the maximum point b are so high, that the change from the range (2) of the pyroelectrolyte, Fig. 4, to the range (3) can not be produced by increase of voltage. That is, the distinction between pyroelectric conductor and insulator would be the quantitative one, that in the former the maximum

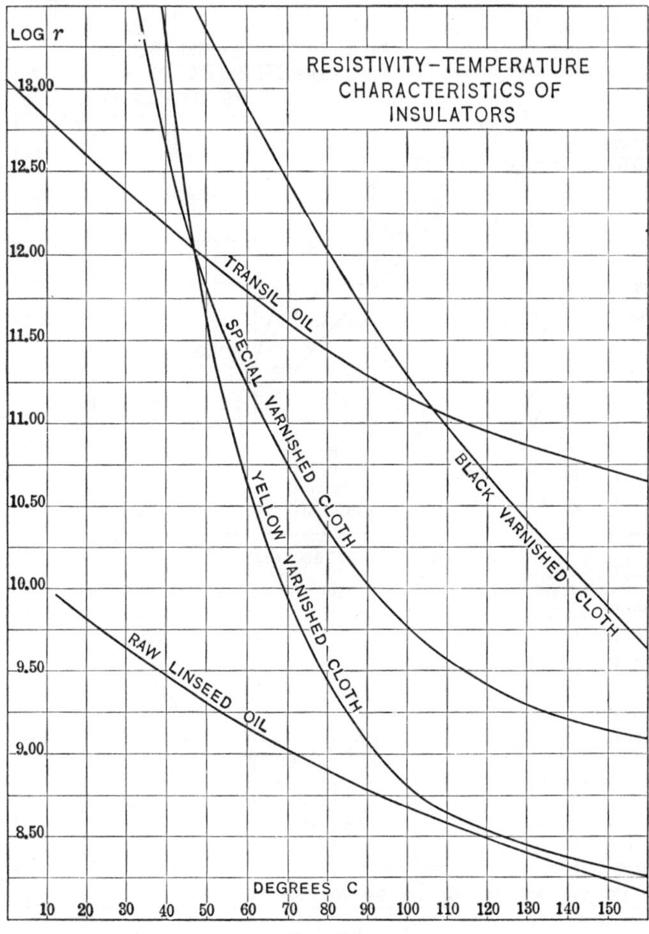

Fig. 14.

voltage point of the volt-ampere characteristic is within experimental reach, while with the latter it is beyond reach.

Whether this applies to all insulators, or whether among organic compounds as oils, there are true insulators, which are not pyroelectric conductors, is uncertain.

Positive temperature coefficient of resistivity is very often met in insulating materials such as oils, fibrous materials, etc. In this case, however, the rise of resistance at increase of temperature usually remains permanent after the temperature is again lowered,

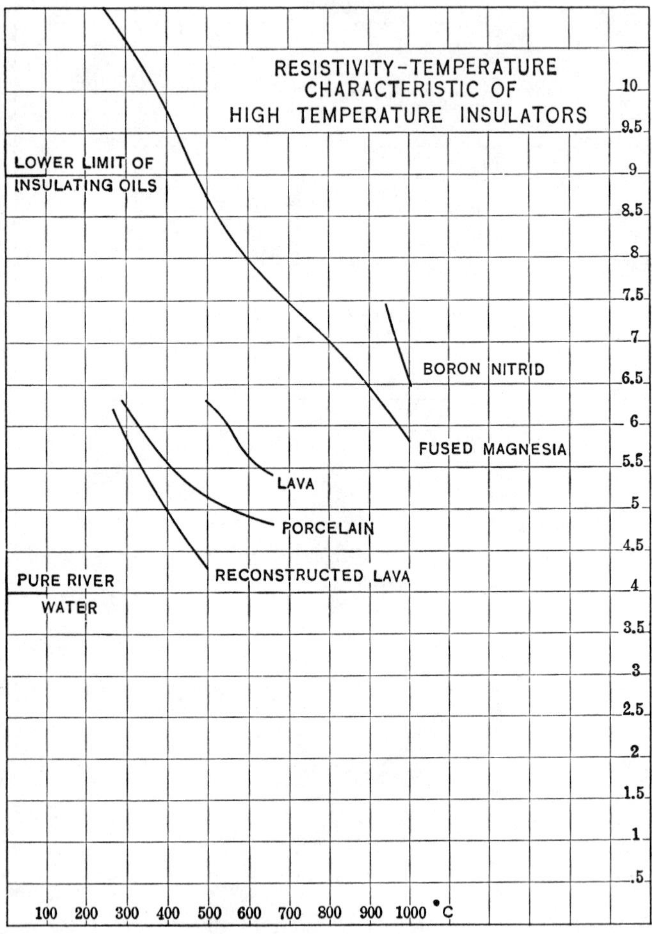

FIG. 15.

and the apparent positive temperature coefficient was due to the expulsion of moisture absorbed by the material. With insulators of very high resistivity, extremely small traces of moisture may decrease the resistivity many thousandfold, and the conductivity of insulating materials very often is almost entirely moisture conduction, that is, not due to the material proper, but due to the moisture absorbed by it. In such a case, prolonged drying may increase the resistivity enormously, and when dry, the material then shows the negative temperature coefficient of resistance, incident to pyroelectric conduction.

CHAPTER II

ELECTRIC CONDUCTION. GAS AND VAPOR CONDUCTORS

Gas, Vapor and Vacuum Conduction

18. As further, and last class may be considered vapor, gas and vacuum conduction. Typical of this is, that the volt-ampere characteristic is dropping, that is, the voltage decreases with increase of current, and that luminescence accompanies the conduction, that is, conversion of electric energy into light.

Thus, gas and vapor conductors are unstable on constant-potential supply, but stable on constant current. On constant potential they require a series resistance or reactance, to produce stability.

Such conduction may be divided into three distinct types: spark conduction, arc conduction, and true electronic conduction.

In spark conduction, the gas or vapor which fills the space between the electrodes is the conductor. The light given by the gaseous conductor thus shows the spectrum of the gas or vapor which fills the space, but the material of the electrodes is immaterial, that is, affects neither the light nor the electric behavior of the gaseous conductor, except indirectly, in so far as the section of the conductor at the terminals depends upon the terminal surface.

In arc conduction, the conductor is a vapor stream issuing from the negative terminal or cathode, and moving toward the anode at high velocity. The light of the arc thus shows the spectrum of the negative terminal material, but not that of the gas in the surrounding space, nor that of the positive terminal, except indirectly, by heat luminescence of material entering the arc conductor from the anode or from surrounding space.

In true electronic conduction, electrons existing in the space, or produced at the terminals (hot cathode), are the conductors. Such conduction thus exists also in a perfect vacuum, and may be accompanied by practically no luminescence.

Disruptive Conduction

19. Spark conduction at atmospheric pressure is the disruptive spark, streamers, and corona. In a partial vacuum, it is the Geissler discharge or glow discharge. Spark conduction is discontinuous, that is, up to a certain voltage, the "disruptive voltage," no conduction exists, except perhaps the extremely small true electronic conduction. At this voltage conduction begins and continues as long as the voltage persists, or, if the source of power is capable of maintaining considerable current, the spark conduction changes to arc conduction, by the heat developed at the negative terminal supplying the conducting arc vapor stream. The current usually is small and the voltage high. Especially at atmospheric pressure, the drop of the volt-ampere characteristic is extremely steep, so that it is practically impossible to secure stability by series resistance, but the conduction changes to arc conduction, if sufficient current is available, as from power generators, or the conduction ceases by the voltage drop of the supply source, and then starts again by the recovery of voltage, as with an electrostatic machine. Thus spark conduction also is called *disruptive conduction* and *discontinuous conduction.*

Apparently continuous—though still intermittent—spark conduction is produced at atmospheric pressure by capacity in series to the gaseous conductor, on an alternating-voltage supply, as corona, and as Geissler tube conduction at a partial vacuum, by an alternating-supply voltage with considerable reactance or resistance in series, or from a direct-current source of very high voltage and very limited current, as an electrostatic machine.

In the Geissler tube or vacuum tube, on alternating-voltage supply, the effective voltage consumed by the tube, at constant temperature and constant gas pressure, is approximately constant and independent of the effective current, that is, the volt-ampere characteristic a straight horizontal line. The Geissler tube thus requires constant current or a steadying resistance or reactance for its operation. The voltage consumed by the Geissler tube consists of a potential drop at the terminals, the "terminal drop," and a voltage consumed in the luminous stream, the "stream voltage." Both greatly depend on the gas pressure, and vary, with changing gas pressure, in opposite directions: the terminal drop decreases and the stream voltage increases with increasing gas pressure, and the total voltage consumed by the tube thus gives a minimum at some definite gas pressure. This pressure of minimum voltage depends on the length of the tube,

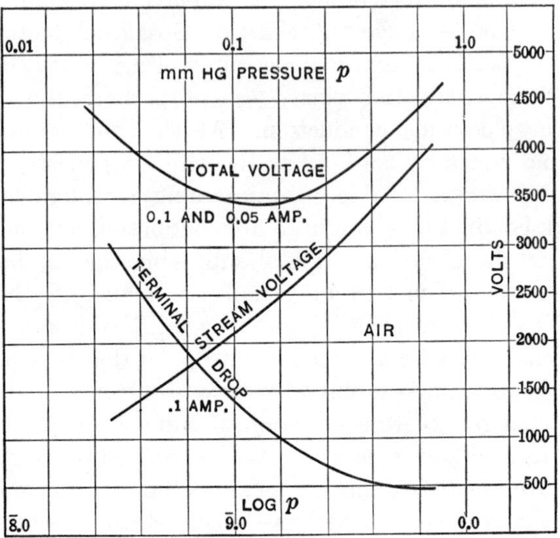

Fig. 16.

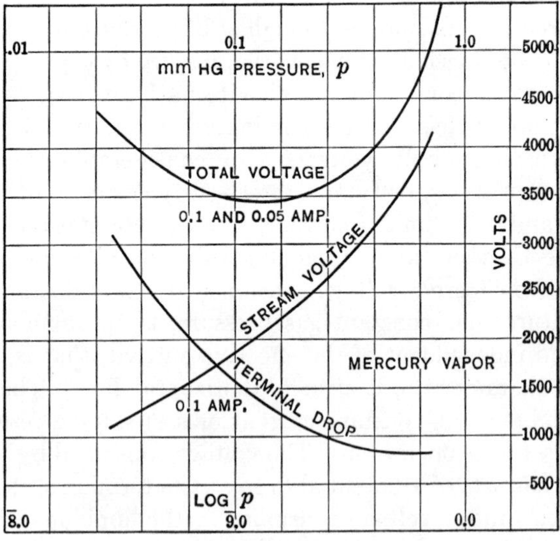

Fig. 17.

and the longer the tube, the lower is the gas pressure which gives minimum total voltage.

Fig. 16 shows the voltage-pressure characteristic, at constant current of 0.1 amp. and 0.05 amp., of a Geissler tube of 1.3 cm. internal diameter and 200 cm. length, using air as conductor, and Fig. 17 the characteristic of the same tube with mercury vapor as conductor. Figs. 16 and 17 also show the two component voltages, the terminal drop and the stream voltage, separately. As abscissæ are used the log of the gas pressure, in millimeter mercury column. As seen, the terminal drop decreases with increasing gas pressure, and becomes negligible compared with the stream voltage, at atmospheric pressure.

The voltage gradient, per centimeter length of stream, varies from 5 to 20 volts, at gas or vapor pressure from 0.06 to 0.9 mm. At atmospheric pressure (760 mm.) the disruptive voltage gradient, which produces corona, is 21,000 volts effective per centimeter. The specific resistance of the luminous stream is from 65 to 500 ohms per $cm.^3$ in the Geissler tube conduction of Figs. 16 and 17—though this term has little meaning in gas conduction. The specific resistance of the corona in air, as it appears on transmission lines at very high voltages, is still very much higher.

Arc Conduction

20. In the electric arc, the current is carried across the space between the electrodes or arc terminals by a stream of electrode vapor, which issues from a spot on the negative terminal, the so-called cathode spot, as a high-velocity blast (probably of a velocity of several thousand feet per second). If the negative terminal is fluid, the cathode spot causes a depression, by the reaction of the vapor blast, and is in a more or less rapid motion, depending on the fluidity.

As the arc conductor is a vapor stream of electrode material, this vapor stream must first be produced, that is, energy must be expended before arc conduction can take place. The arc, therefore, does not start spontaneously between the arc terminals, if sufficient voltage is supplied to maintain the arc (as is the case with spark conduction) but the arc has first to be started, that is, the conducting vapor bridge be produced. This can be done by bringing the electrodes into contact and separating them, or by a high-voltage spark or Geissler discharge, or by the vapor stream of another arc, or by producing electronic conduction, as by an incandescent filament. Inversely, if the current in the arc stopped even for a moment, conduction ceases, that is, the arc extinguishes and has to be restarted. Thus, arc conduction may also be called *continuous conduction.*

21. The arc stream is conducting only in the direction of its motion, but not in the reverse direction. Any body, which is reached by the arc stream, is conductively connected with it, if positive toward it, but is not in conductive connection, if negative or isolated, since, if this body is negative to the arc stream, an arc stream would have to issue from this body, to connect it conductively, and this would require energy to be expended on the body, before current flows to it. Thus, only if the arc stream is very hot, and the negative voltage of the body impinged by it very high, and the body small enough to be heated to high temperature, an arc spot may form on it by heat energy. If, therefore, a body touched by the arc stream is connected to an alternating voltage, so that it is alternately positive and negative toward the arc stream, then conduction occurs during the half-wave, when this body is positive, but no conduction during the negative half-wave (except when the negative voltage is so high as to give disruptive conduction), and the arc thus rectifies the alternating voltage, that is, permits current to pass in one direction only. The arc thus is a *unidirectional conductor*, and as such extensively used for *rectification* of alternating voltages. Usually vacuum arcs are employed for this purpose, mainly the mercury arc, due to its very great rectifying range of voltage.

Since the arc is a unidirectional conductor, it usually can not exist with alternating currents of moderate voltage, as at the end of every half-wave the arc extinguishes. To maintain an alternating arc between two terminals, a voltage is required sufficiently high to restart the arc at every half-wave by jumping an electrostatic spark between the terminals through the hot residual vapor of the preceding half-wave. The temperature of this vapor is that of the boiling point of the electrode material. The voltage required by the electrostatic spark, that is, by disruptive conduction, decreases with increase of temperature, for a 13-mm. gap about as shown by curve I in Fig. 18. The voltage required to maintain an arc, that is, the direct-current voltage, increases with increasing arc temperature, and therefore increasing radiation, etc., about as shown by curve II in Fig. 18. As seen, the curves I and II intersect at some very high temperature, and materials as carbon, which have a boiling point above this temperature, require a lower voltage for restarting than for maintaining the arc, that is, the voltage required to maintain the arc restarts it at every half-wave of alternating current, and such materials thus give a steady alternating arc. Even materials of a somewhat lower boiling point, in which the starting voltage is not much above the running voltage of the arc, maintain a steady alternating arc, as in starting the voltage consumed by the steadying resistance or reactance is available. Electrode materials of low

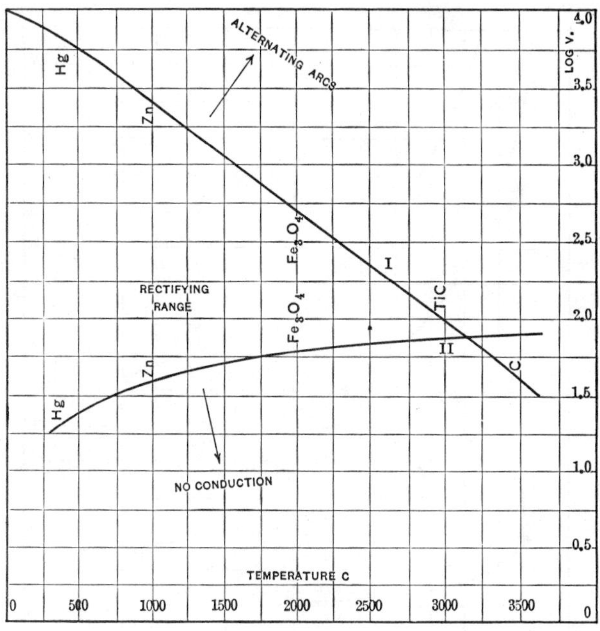

Fig. 18.

boiling point, however, can not maintain steady alternating arcs at moderate voltage.

The range in Fig. 18, above the curve I, thus is that in which alternating arcs can exist; in the range between I and II, an alternating voltage can not maintain the arc, but unidirectional current is produced from an alternating voltage, if the arc conductor is maintained by excitation of its negative terminals, as by an auxiliary arc. This, therefore, is the rectifying range of arc conduction. Below curve II any conduction ceases, as the voltage is insufficient to maintain the conducting vapor stream.

Fig. 18 is only approximate. As ordinates are used the logarithm of the voltage, to give better proportions. The boiling points of some materials are approximately indicated on the curves.

It is essential for the electrical engineer to thoroughly understand the nature of the arc, not only because of its use as illuminant, in arc lighting, but more still because accidental arcs are the foremost cause of instability and troubles from dangerous transients in electric circuits.

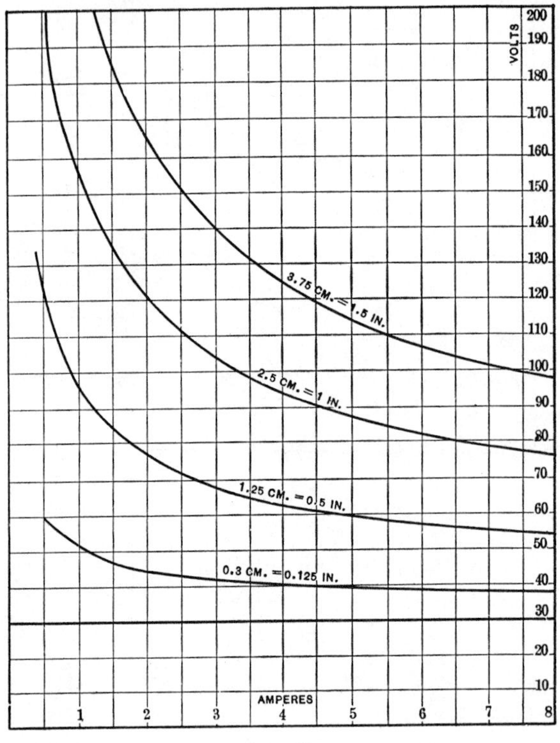

Fig. 19.

22. The voltage consumed by an arc stream, e_1, at constant current, i, is approximately proportional to the arc length, l, or rather to the arc length plus a small quantity, δ, which probably represents the cooling effect of the electrodes.

Plotting the arc voltage, e, as function of the current, i, at constant arc length, gives dropping volt-ampere characteristics, and the voltage increases with decreasing current the more, the longer the arc. Such characteristics are shown in Fig. 19 for the magnetite arcs of 0.3; 1.25; 2.5 and 3.75 cm. length.

These curves can be represented with good approximation by the equation

$$e = a + \frac{c(l + \delta)}{\sqrt{i}} \tag{4}$$

This equation, which originally was derived empirically, can also be derived by theoretical reasoning:

Assuming the amount of arc vapor, that is, the section of the conducting vapor stream, as proportional to the current, and the heat produced at the positive terminal as proportional to the vapor stream and thus the current, the power consumed at the terminals is proportional to the current. As the power equals the current times the terminal drop of voltage, it follows that this terminal drop, a, is constant and independent of current or arc length—similar as the terminal drop at the electrodes in electrolytic conduction is independent of the current.

The power consumed in the arc stream, $p_1 = e_1 i$, is given off from the surface of the stream, by radiation, conduction and convection of heat. The temperature of the arc stream is constant, as that of the boiling point of the electrode material. The power, therefore, is proportional to the surface of the arc stream, that is, proportional to the square root of its section, and therefore the square root of the current, and proportional to the arc length, l, plus a small quantity, δ, which corrects for the cooling effect of the electrodes. This gives

$$p_1 = e_1\, i = c\sqrt{i}\,(l + \delta)$$

or,

$$e_1 = \frac{c(l + \delta)}{\sqrt{i}} \tag{5}$$

as the voltage consumed in the arc stream.

Since a represents the coefficient of power consumed in producing the vapor stream and heating the positive terminal, and c the coefficient of power dissipated from the vapor stream, a and c are different for different materials, and in general higher for materials of higher boiling point and thus higher arc temperature. c, however, depends greatly on the gas pressure in the space in which the arc occurs, and decreases with decreasing gas pressure. It is, approximately, when l is given in centimeter at atmospheric pressure,

a = 13 volts for mercury,
= 16 volts for zinc and cadmium (approximately),
= 30 volts for magnetite,
= 36 volts for carbon;
c = 31 for magnetite,
= 35 for carbon;
δ = 0.125 cm. for magnetite,
= 0.8 cm. for carbon.

The least agreement with the equation (4) is shown by the carbon arc. It agrees fairly well for arc lengths above 0.75 cm., but for shorter arc lengths, the observed voltage is lower than given by equation (4), and approaches for $l = 0$ the value $e = 28$ volts.

It seems as if the terminal drop, $a = 36$ volts with carbon, consists of an actual terminal drop, $a_0 = 28$ volts, and a terminal drop of $a_1 = 8$ volts, which resides in the space within a short distance from the terminals.

Stability Curves of the Arc

23. As the volt-ampere characteristics of the arc show a decrease of voltage with increase of current, over the entire range of current, the arc is unstable on constant voltage supplied to its terminals, at every current.

Inserting in series to a magnetite arc of 1.8 cm. length, shown as curve I in Fig. 20, a constant resistance of $r = 10$ ohms, the voltage consumed by this resistance is proportional to the current, and thus given by the straight line II in Fig. 20. Adding this voltage II to the arc-voltage curve I, gives the total voltage consumed by the arc and its series resistance, shown as curve III. In curve III, the voltage decreases with increase of current, up to $i_0 = 2.9$ amp. and the arc thus is unstable for currents below 2.9 amp. For currents larger than 2.9 amp. the voltage increases with increase of current, and the arc thus is stable. The point $i_0 = 2.9$ amp. thus separates the unstable lower part of curve III, from the stable upper part.

With a larger series resistance, $r' = 20$ ohms, the stability range is increased down to 1.7 amp., as seen from curve III, but higher voltages are required for the operation of the arc.

With a smaller series resistance, $r'' = 5$ ohms, the stability range is reduced to currents above 4.8 amp., but lower voltages are sufficient for the operation of the arc.

At the stability limit, i_0, in curve III of Fig. 20, the resultant characteristic is horizontal, that is, the slope of the resistance curve II: $r = \frac{e'}{i}$, is equal but opposite to that of the arc charac-

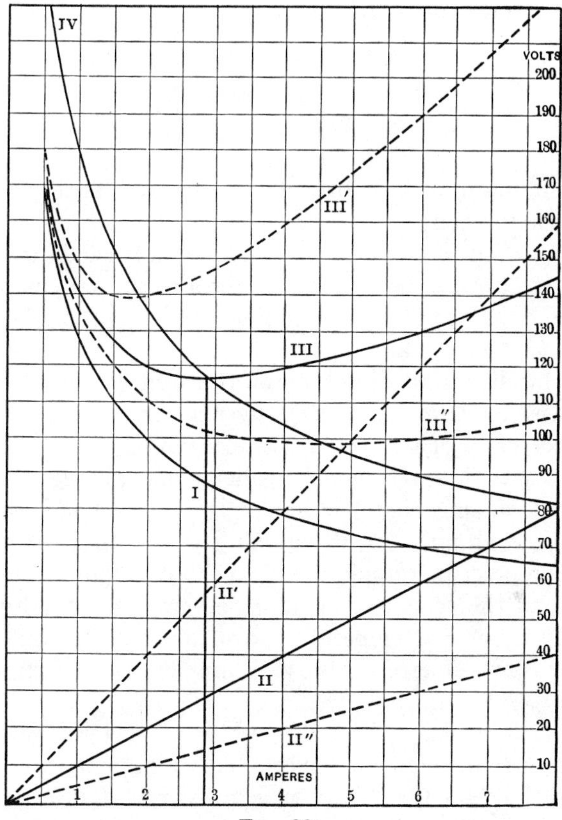

FIG. 20.

teristic I: $\frac{de}{di}$. The resistance, r, required to give the stability limit at current, i, thus is found by the condition

$$r = -\frac{de}{di} \tag{6}$$

Substituting equation (4) into (6) gives

$$r = \frac{c(l + \delta)}{2\,i\sqrt{i}} \tag{7}$$

as the minimum resistance to produce stability, hence,

$$ri = \frac{c(l + \delta)}{2\sqrt{i}} = 0.5\, e_1 \tag{8}$$

where e_1 = arc stream voltage, and

$$\begin{aligned} E &= e + ri \\ &= a + 1.5\,\frac{c(l + \delta)}{\sqrt{i}} \end{aligned} \tag{9}$$

is the minimum voltage required by arc and series resistance, to just reach stability.

(9) is plotted as curve IV in Fig. 20, and is called the *stability curve* of the arc. It is of the same form as the arc characteristic I, and derived therefrom by adding 50 per cent. of the voltage, e_1, consumed by the arc stream.

The stability limit of an arc, on constant potential, thus lies at an excess of the supply voltage over the arc voltage $e = a + e_1$, by 50 per cent. of the voltage, e_1, consumed in the arc stream. In general, to get reasonable steadiness and absence of drifting of current, a somewhat higher supply voltage and larger series resistance, than given by the stability curve IV, is desirable.

24. The preceding applies only to those arcs in which the gas pressure an the space surrounding the arc, and thereby the arc vapor pressure and temperature, are constant and independent of the current, as is the case with arcs in air, at "atmospheric pressure."

With arcs in which the vapor pressure and temperature vary with the current, as in vacuum arcs like the mercury arc, different considerations apply. Thus, in a mercury arc in a glass tube, if the current is sufficiently large to fill the entire tube, but not so large that condensation of the mercury vapor can not freely occur in a condensing chamber, the power dissipated by radiation, etc., may be assumed as proportional to the length of the tube, and to the current

$$p = e_1 i = cli$$

thus,

$$e_1 = cl \tag{10}$$

that is, the stream voltage of the tube, or voltage consumed by the arc stream (exclusive terminal drop) is independent of the current. Adding hereto the terminal drop, a, gives as the total voltage consumed by the mercury tube

$$e = a + cl \tag{11}$$

for a mercury arc in a vacuum, it is approximately

$$c = \frac{1.4}{d} \tag{12}$$

where $d=$ diameter of the tube, since the diameter of the tube is proportional to the surface and therefore to the radiation coefficient.

Thus,

$$e = 13 + \frac{1.4l}{d} \tag{13}$$

At high currents, the vapor pressure rises abnormally, due to incomplete condensation, and the voltage therefore rises, and

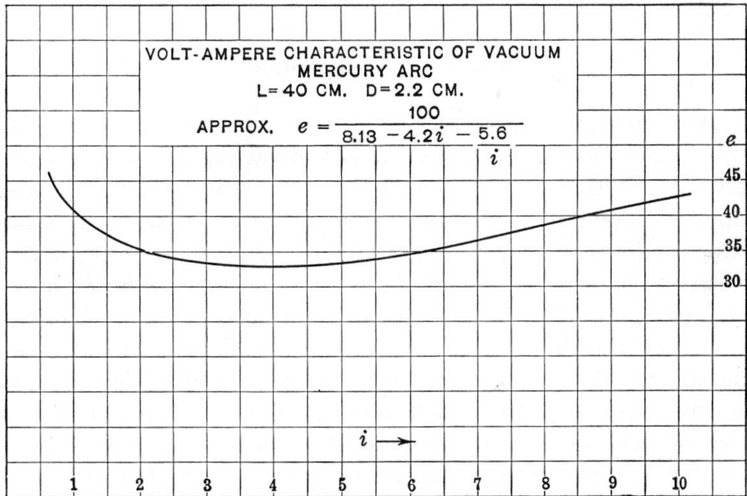

FIG. 21.

at low currents the voltage rises again, due to the arc not filling the entire tube. Such a volt-ampere characteristic is given in Fig. 21.

25. Herefrom then follows, that the voltage gradient in the mercury arc, for a tube diameter of 2 cm., is about ¾ volts per centimeter or about one-twentieth of what it is in the Geissler tube, and the specific resistance of the stream, at 4 amp., is about 0.2 ohms per cm.[3], or of the magnitude of one one-thousandth of what it is in the Geissler tube.

At higher currents, the mercury arc in a vacuum gives a rising volt-ampere characteristic. Nevertheless it is not stable on constant-potential supply, as the rising characteristic applies only to stationary conditions; the instantaneous characteristic is dropping. That is, if the current is suddenly increased, the voltage drops, regardless of the current value, and then gradually, with the increasing temperature and vapor pressure, increases again, to the permanent value, a lower value or a higher value, whichever may be given by the permanent volt-ampere characteristic.

In an arc at atmospheric pressure, as the magnetite arc, the voltage gradient depends on the current, by equation (1), and at 4 amp. is about 15 to 18 volts per centimeter. The specific resistance of the arc stream is of the magnitude of 1 ohm per cm.[3], and less with larger current arcs, thus of the same magnitude as in vacuum arcs.

Electronic Conduction

26. Conduction occurs at moderate voltages between terminals in a partial vacuum as well as in a perfect vacuum, if the terminals are incandescent. If only one terminal is incandescent, the conduction is unidirectional, that is, can occur only in that direction, which makes the incandescent terminal the cathode, or negative. Such a vacuum tube then rectifies an alternating voltage and may be used as rectifier. If a perfect vacuum exists in the conducting space between the electrodes of such a hot cathode tube, the conduction is considered as true electronic conduction. The voltage consumed by the tube is depending on the high temperature of the cathode, and is of the magnitude of arc voltages, hence very much lower than in the Geissler tube, and the current of the magnitude of arc currents, hence much higher than in the Geissler tube.

27. The complete volt-ampere characteristic of gas and vapor conduction thus would give a curve of the shape in Fig. 22. It consists of three branches separated by ranges of instability or discontinuity. The branch *a*, at very low current, electronic conduction; the branch *b*, discontinuous or Geissler tube conduction; and the branch *c*, arc conduction. The change from *a* to *b* occurs suddenly and abruptly, accompanied by a big rise of current, as soon as the disruptive voltage is reached. The change *b* to *c* occurs suddenly and abruptly, by the formation of a cathode spot, anywhere in a wide range of current, and is accompanied by a sudden drop of voltage. To show the entire range, as abscissæ are used $\sqrt[4]{i}$ and as ordinates $\sqrt{e}$.

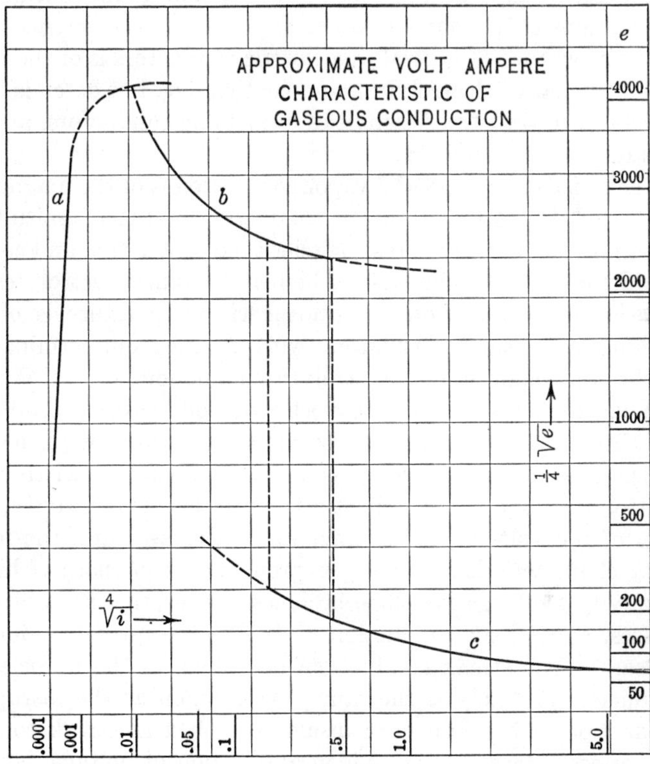

Fig. 22.

Review

28. The various classes of conduction: metallic conduction, electrolytic conduction, pyroelectric conduction, insulation, gas vapor and electronic conduction, are only characteristic types, but numerous intermediaries exist, and transitions from one type to another by change of electrical conditions, of temperature, etc.

As regards to the magnitude of the specific resistance or resistivity, the different types of conductors are characterized about as follows:

The resistivity of metallic conductors is measured in microhm-centimeters.

The resistivity of electrolytic conductors is measured in ohm-centimeters.

The resistivity of insulators is measured in megohm-centimeters and millions of megohm-centimeters.

The resistivity of typical pyroelectric conductors is of the magnitude of that of electrolytes, ohm-centimeters, but extends from this down toward the resistivities of metallic conductors, and up toward that of insulators.

The resistivity of gas and vapor conduction is of the magnitude of electrolytic conduction: arc conduction of the magnitude of lower resistance electrolytes, Geissler tube conduction and corona conduction of the magnitude of higher-resistance electrolytes.

Electronic conduction at atmospheric temperature is of the magnitude of that of insulators; with incandescent terminals, it reaches the magnitude of electrolytic conduction.

While the resistivities of pyroelectric conductors extend over the entire range, from those of metals to those of insulators, typical are those pyroelectric conductors having a resistivity of electrolytic conductors. In those with lower resistivity, the drop of the volt-ampere characteristic decreases and the instability characteristic becomes less pronounced; in those of higher resistivity, the negative slope becomes steeper, the instability increases, and streak conduction or finally disruptive conduction appears. The streak conduction, described on the pyroelectric conductor, probably is the same phenomenon as the disruptive conduction or breakdown of insulators. Just as streak conduction appears most under sudden application of voltage, but less under gradual voltage rise and thus gradual heating, so insulators of high disruptive strength, when of low resistivity by absorbed moisture, etc., may stand indefinitely voltages applied intermittently—so as to allow time for temperature equalization—while quickly breaking down under very much lower sustained voltage.

CHAPTER III

MAGNETISM

Reluctivity

29. Considering magnetism as the phenomena of a "magnetic circuit," the foremost differences between the characteristics of the magnetic circuit and the electric circuit are:

(*a*) The maintenance of an electric circuit requires the expenditure of energy, while the maintenance of a magnetic circuit does not require the expenditure of energy, though the starting of a magnetic circuit requires energy. A magnetic circuit, therefore, can remain "remanent" or "permanent."

(*b*) All materials are fairly good carriers of magnetic flux, and the range of magnetic permeabilities is, therefore, narrow, from 1 to a few thousands, while the range of electric conductivities covers a range of 1 to 10^{18}. The magnetic circuit thus is analogous to an uninsulated electric circuit immersed in a fairly good conductor, as salt water: the current or flux can not be carried to any distance, or constrained in a "conductor," but divides, "leaks" or "strays."

(*c*) In the electric circuit, current and e.m.f. are proportional, in most cases; that is, the resistance is constant, and the circuit therefore can be calculated theoretically. In the magnetic circuit, in the materials of high permeability, which are the most important carriers of the magnetic flux, the relation between flux, m.m.f. and energy is merely empirical, the "reluctance" or magnetic resistance is not constant, but varies with the flux density, the previous history, etc. In the absence of rational laws, most of the magnetic calculations thus have to be made by taking numerical values from curves or tables.

The only rational law of magnetic relation, which has not been disproven, is Fröhlich's (1882):

"*The permeability is proportional to the magnetizability*"

$$\mu = a(S - B) \tag{1}$$

where B is the magnetic flux density, S the saturation density, and $S - B$ therefore the magnetizability, that is, the still available increase of flux density, over that existing.

From (1) follows, by substituting,

$$\mu = \frac{B}{H} \tag{2}$$

and rearranging,

$$B = \frac{H}{\alpha + \sigma H} \tag{3}$$

where

$\sigma = \frac{1}{S}$ = *saturation coefficient*, that is, the reciprocal of the saturation value, S, of flux density, B, and

$$\alpha = \frac{1}{aS} = \frac{\sigma}{a},$$

for $B = O$, equation (1) gives

$$\mu_0 = aS = \frac{1}{\alpha}; \quad \alpha = \frac{1}{\mu_0} \tag{4}$$

that is, α is the reciprocal of the magnetic permeability at zero flux density.

A very convenient form of this law has been found by Kennelly (1893) by introducing the reciprocal of the permeability, as reluctivity ρ,

$$\rho = \frac{1}{\mu} = \frac{H}{B},$$

in the form, which can be derived from (3) by transposition.

$$\rho = \alpha + \sigma H \tag{5}$$

As α dominates the reluctivity at lower magnetizing forces, and thereby the initial rate of rise of the magnetization curve, which is characteristic of the "magnetic hardness" of the material, it is called the *coefficient of magnetic hardness*.

30. When investigating flux densities, B, at very high field intensities, H, it was found that B does not reach a finite saturation value, but increases indefinitely; that, however,

$$B_0 = B - H \tag{6}$$

reaches a finite saturation value S, which with iron usually is not far from 20 kilolines per cm.², and that therefore Fröhlich's and Kennelly's laws apply not to B, but to B_0. The latter, then, is usually called the *metallic magnetic density* or *ferromagnetic density.*

B_0 may be considered as the magnetic flux carried by the molecules of the iron or other magnetic material, in addition to the

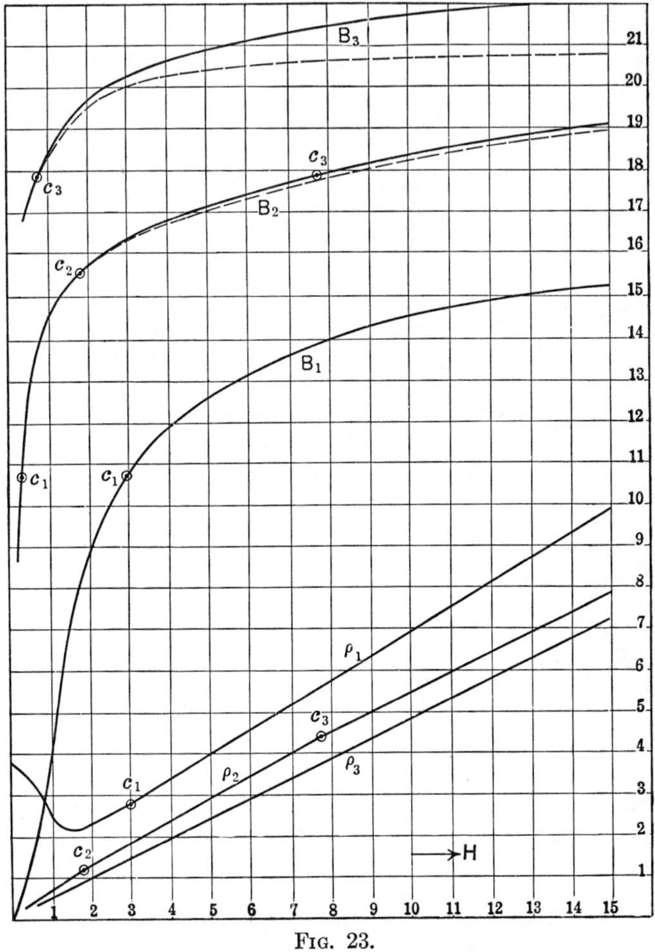

FIG. 23.

space flux, H, or flux carried by space independent of the material in space.

The best evidence seems to corroborate, that with the exception of very low field intensities (where the customary magnetization curve usually has an inward bend, which will be discussed later) in perfectly pure magnetic materials, iron, nickel, cobalt, etc., the linear law of reluctivity (5) and (3) is rigidly obeyed by the metallic induction B_0.

In the more or less impure commercial materials, however, the $\rho - H$ relation, while a straight line, often has one, and occasionally two points, where its slope, and thus the values of α and σ change.

Fig. 23 shows an average magnetization curve, of good standard iron, with field intensity, H, as abscissæ, and magnetic induction, B, as ordinates. The total induction is shown in drawn lines, the metallic induction in dotted lines. The ordinates are given in kilolines per cm.^{2}, the abscissæ in units for B_1, in tens for B_2, and in hundreds for B_3.

The reluctivity curves, for the three scales of abscissæ, are plotted as ρ_1, ρ_2, ρ_3, in tenths of milli-units, in milli-units and in tens of milli-units.

Below $H = 3$, ρ is not a straight line, but curved, due to the inward bend of the magnetization curve, B, in this range. The straight-line law is reached at the point c_1, at $H = 3$, and the reluctivity is then expressed by the linear law

$$\rho_1 = 0.102 + 0.059\, H \tag{7}$$

for

$$3 < H < 18,$$

giving an apparent saturation value,

$$S_1 = 16{,}950.$$

At $H = 18$, a bend occurs in the reluctivity line, marked by point c_2, and above this point the reluctivity follows the equation

$$\rho_2 = 0.18 + 0.0548\, H \tag{8}$$

for

$$18 < H < 80,$$

giving an apparent saturation value

$$S_2 = 18{,}250.$$

At $H = 80$, another bend occurs in the reluctivity line, marked by point c_3, and above this point, up to saturation, the reluctivity follows the equation

$$\rho_3 = 0.70 + 0.0477\, H \tag{9}$$

for

$$H > 80$$

giving the true saturation value,

$$S = 20{,}960.$$

Point c_2 is frequently absent.

Fig. 24 gives once more the magnetization curve (metallic induction) as B, and gives as dotted curves B_1, B_2 and B_3 the magnetization curves calculated from the three linear reluctivity equations (7), (8), (9). As seen, neither of the equations represents

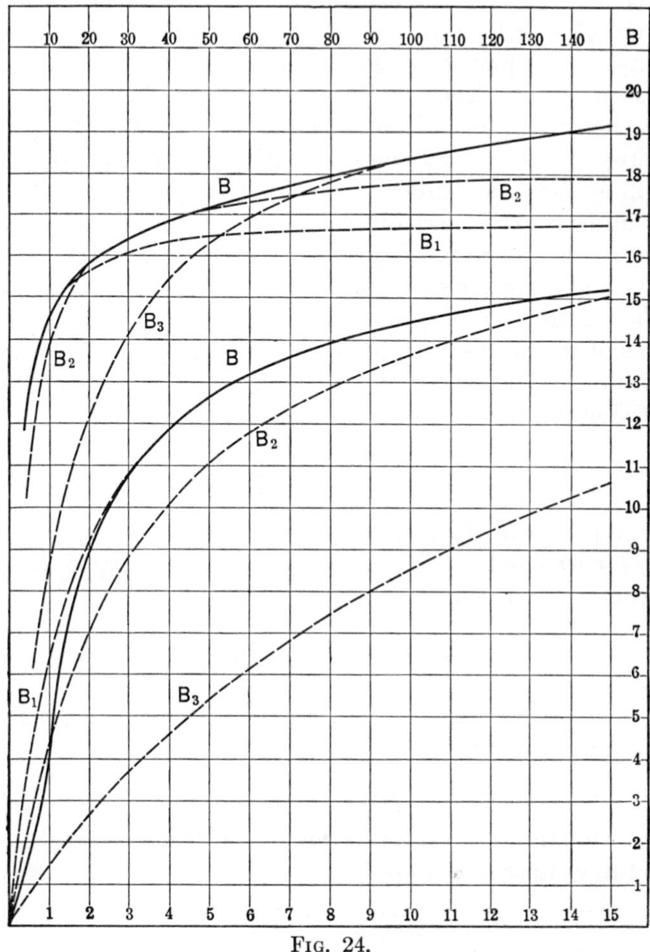

FIG. 24.

B even approximately over the entire range, but each represents it very accurately within its range. The first, equation (7), probably covers practically the entire industrially important range.

37. As these critical points c_2 and c_3 do not seem to exist in perfectly pure materials, and as the change of direction of the reluctivity line is in general the greater, the more impure the material, the cause seems to be lack of homogeneity of the material; that is, the presence, either on the surface as scale, or in the body, as inglomerate, of materials of different magnetic characteristics: magnetite, cementite, silicide. Such materials have a much greater hardness, that is, higher value of α, and thereby would give the observed effect. At low field intensities, H, the harder material carries practically no flux, and all the flux is carried by the soft material. The flux density therefore rises rapidly, giving low α, but tends toward an apparent low saturation value, as the flux-carrying material fills only part of the space. At higher field intensities, the harder material begins to carry flux, and while in the softer material the flux increases less, the increase of flux in the harder material gives a greater increase of total flux density and a greater saturation value, but also a greater hardness, as the resultant of both materials.

Thus, if the magnetic material is a conglomerate of fraction p of soft material of reluctivity ρ_1 (ferrite) and $q = 1 - p$ of hard material of reluctivity, ρ_2 (cementite, silicide, magnetite),

$$\left.\begin{aligned} \rho_1 &= \alpha_1 + \sigma_1 H \\ \rho_2 &= \alpha_2 + \sigma_2 H \end{aligned}\right\} \qquad (10)$$

at low values of H, the part p of the section carries flux by ρ_1, the part q carries flux by ρ_2, but as ρ_2 is very high compared with ρ_1, the latter flux is negligible, and it is

$$\rho^1 = \frac{\rho_1}{p} = \frac{\alpha_1}{p} + \frac{\sigma_1}{p} H \qquad (11)$$

At high values of H, the flux goes through both materials, more or less in series, and it thus is

$$\rho'' = p\rho_1 + q\rho_2 = (p\alpha_1 + q\alpha_2) + (p\sigma_1 + q\sigma_2)H \qquad (12)$$

if we assume the same saturation value, σ, for both materials, and neglect α_1 compared with α_2, it is

$$\rho'' = q\,\alpha_2 + \sigma H \qquad (13)$$

Substituting, as instance, (7) and (9) into (11) and (13) respectively, gives

$$\frac{\alpha_1}{p} = 0.102,$$

$$\frac{\sigma}{p} = 0.059,$$

$$q\alpha_2 = 0.70,$$
$$\sigma = 0.0477,$$

hence,

$$p = 0.80: \quad \rho_1 = 0.082 + 0.0477\,H,$$
$$q = 0.20: \quad \rho_2 = \quad 3.5 + 0.0477\,H.$$

However, the saturation coefficients, σ, of the two materials probably are usually not equal.

The deviation of the reluctivity equation from a straight line, by the change of slope at the critical points, c_2 and c_3, thus probably is only apparent, and is the outward appearance of a change of the flux carrier in an unhomogeneous material, that is, the result of a second and magnetically harder material beginning to carry flux.

Such bends in the reluctivity line have been artificially produced by Mr. John D. Ball in combining by superposition two different materials, which separately gave straight-line, ρ, curves, while combined they gave a curve showing the characteristic bend.

Very impure materials, like cast iron, may give throughout a curved reluctivity line.

32. For very low values of field intensity, $H < 3$, however, the straight-line law of reluctivity apparently fails, and the magnetization curve in Fig. 23 has an inward bend, which gives rise of ρ with decreasing H.

This curve is taken by ballistic galvanometer, by the step-by-step method, that is, H is increased in successive steps, and the increase of B observed by the throw of the galvanometer needle. It thus is a "rising magnetization curve."

The first part of this curve is in Fig. 25 reproduced, as B_1, in twice the abscissæ and half the ordinates, so as to give it an average slope of 45°, as with this slope curve shapes such as the inward bend of B_1 below $H = 2$, are best shown ("Engineering Mathematics," p. 286).

Suppose now, at some point, $B_0 = 13.15$, we stop the increase of H, and decrease again, down to 0. We do not return on the same magnetization curve, B_1, but on another curve, B'_1, the "decreasing magnetic characteristic," and at $H = 0$, we are not back to $B = 0$, but a residual or remanent flux is left, in Fig. 25: $R = 7.4$.

Where the magnetic circuit contains an air-gap, as the field circuits of electrical machinery, the decreasing magnetic characteristic, B'_1, is very much nearer to the increasing one, B_1, than in the closed magnetic circuit, Fig. 25, and practically coincides for higher values of H.

There appears no theoretical reason why the rising characteristic, B_1, should be selected as the representative magnetization curve, and not the decreasing characteristic, B'_1, except the incident, that B_1 passes through zero. In many engineering applications, for instance, the calculation of the regulation of a generator, that is, the decrease of voltage under increase of load, it is obviously the decreasing characteristic, B'_1, which is determining.

Suppose we continue B'_1 into negative values of H, to the point A_1, at $H = -1.5$, $B = -4$, and then again reverse, we get a rising magnetization curve, B'', which passes $H = 0$ at a negative remanent magnetism. Suppose we stop at point A_2, at $H = -1.12$, $B = -1.0$: the rising magnetization curve B''' then passes $H = 0$ at a positive remanent magnetism. There must thus be a point, A_0, between A_1 and A_2, such that the rising magnetization curve, B', starting from A_0, passes through the zero point $H = 0$, $B = 0$, and thereby runs into the curve, B_1.

The rising magnetization curve, or standard magnetic characteristic determined by the step-by-step method, B_1, thus is nothing but the rising branch of an unsymmetrical hysteresis cycle, traversed between such limits $+B_0$ and $-A_0$, that the rising branch of the hysteresis cycle passes through the zero point.

33. The characteristic shape of a hysteresis cycle is that it is a loop, pointed at either end and thereby having an inflexion point about the middle of either branch. In the unsymmetrical loop $+B_1$, $-A_0$ of Fig. 25, the zero point is fairly close to one extreme, A_0, and the inflexion point, characteristic of the hysteresis loop, thus lies between 0 and B_0, that is, on that part of the rising branch, which is used as the "magnetic characteristic," B_1, and thereby produces the inward bend in the magnetization curve at low fields, which has always been so puzzling.

If, however, we would stop the increase of H at B''_0, we would get the decreasing magnetization curve, B''_1, and still other curves for other starting points of the decreasing characteristic.

Thus, the relation between magnetic flux density, B, and magmetic field intensity, H, is not definite, but any point between the various rising and decreasing characteristics B'', B_1, B''', B''_1, B'_1, and for some distance outside thereof, is a possible B-H relation. B_1 has the characteristic that it passes through the zero point. But it is not the only characteristic which does this;

if we traverse the hysteresis cycle between the unsymmetrical limits $+A_0$ and $-B_0$, as shown in Fig. 26, its decreasing branch B_3 passes through the zero point, that is, has the same feature as B_1. It is interesting to note, that B_3 does not show an inward bend, and the reluctivity curve of B_3, given as ρ_3 in Fig. 28, apparently is a straight line.

Magnetic characteristics are frequently determined by the method of reversals, by reversing the field intensity, H, and observing the voltage induced thereby by ballistic galvanometer,

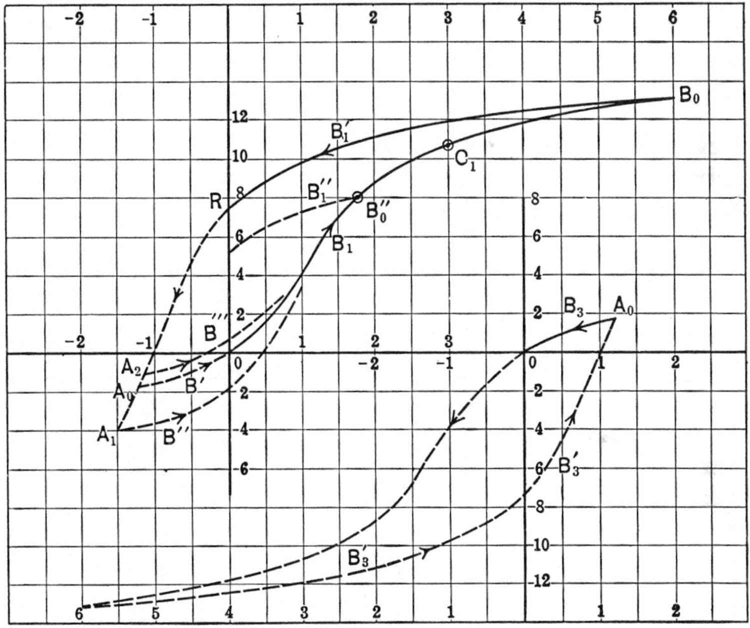

FIGS. 25 AND 26.

or using an alternating current for field excitation, and observing the induced alternating voltage, preferably by oscillograph to eliminate wave-shape error.

This "alternating magnetic characteristic" is the one which is of consequence in the design of alternating-current apparatus. It differs from the "rising magnetic characteristic," B_1 by giving lower values of B, for the same H, materially so at low values of H. It shows the inward bend at low fields still more pronounced than B_1 does. It is shown as curve B_2 in Fig. 27, and its reluctivity line given as ρ_2 in Fig. 28. At higher values of H: from $H = 3$ upward, B_1 and B_2 both coincide with the curve, B_0, representing the straight-line reluctivity law.

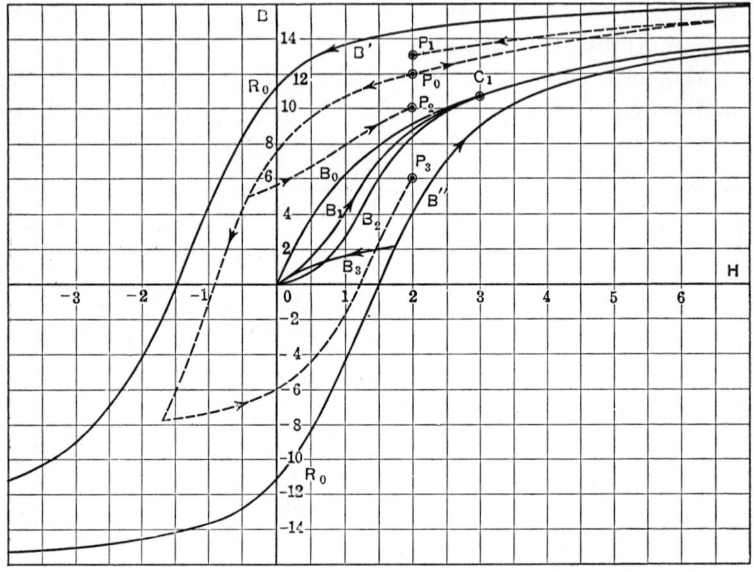

FIG. 27.

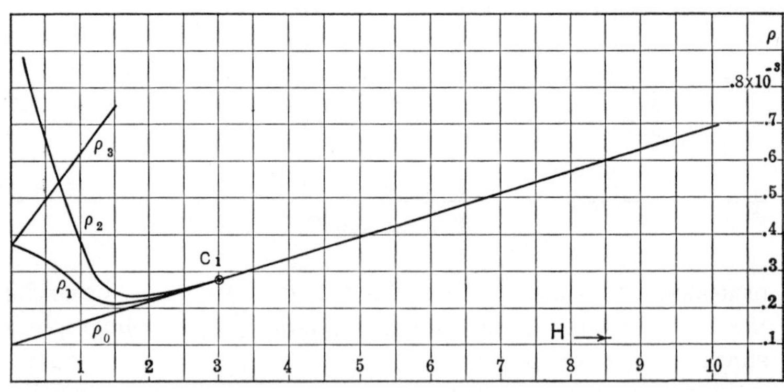

FIG. 28.

The alternating characteristic, B_2, is not a branch of any hysteresis cycle. It is reproducible and independent of the previous history of the magnetic circuit, except perhaps at extremely low values of H, and in view of its engineering importance as representing the conditions in the alternating magnetic field, it would appear the most representative magnetic characteristic, and is commonly used as such.

It has, however, the disadvantage that it represents an unstable condition.

Thus in Fig. 27, an alternating field $H = 1$ gives an alternating flux density, $B_2 = 2.6$. If, however, this field strength $H = 1$ is left on the magnetic circuit, the flux does not remain at $B_2 = 2.6$, but gradually creeps up to higher values, especially in the presence of mechanical vibrations or slight pulsations of the magnetizing current. To a lesser extent, the same occurs with the values of curve, B_1, to a greater extent with B_3. At very low densities, this creepage due to instability of the *B-H* relation may amount to hundreds of per cent. and continue to an appreciable extent for minutes, and with magnetically hard materials for many years. Thus steel structures in the terrestrial magnetic field show immediately after erection only a small part of the magnetization, which they finally assume, after many years.

Thus the alternating characteristic, B_2, however important in electrical engineering, can, due to its instability, not be considered as representing the true physical relation between B and H any more than the branches of hysteresis cycles B_1 and B_3.

34. Correctly, the relation between B and H thus can not be expressed by a curve, but by an area.

Suppose a hysteresis cycle is performed between infinite values of field intensity: $H = \pm \infty$, that is, practically, between very high values such as are given for instance by the isthmus method of magnetic testing (where values of H of over 40,000 have been reached. Very much lower values probably give practically the same curve). This gives a magnetic cycle shown in Fig. 5 as B', B''. Any point, H, B, within the area of this loop between B' and B'' of Fig. 27 then represents a possible condition of the magnetic circuit, and can be reached by starting from any other point, H_0, B_0, such as the zero point, by gradual change of H.

Thus, for instance, from point P_0, the points P_1, P_2, P_3, etc., are reached on the curves shown in the dotted lines in Fig. 27.

As seen from Fig. 27, a given value of field intensity, such as $H = 1$, may give any value of flux density between $B = -4.6$ and $B = +13.6$, and a given value of flux density, such as $B = 10$, may result from any value of field intensity, between $H = -0.25$ to $H = +3.4$

The different values of B, corresponding to the same value of H in the magnetic area, Fig. 27, are not equally stable, but the values near the limits B' and B'' are very unstable, and become more stable toward the interior of the area. Thus, the relation of point P_1, Fig. 27: $H = 2$, $B = 13$, would rapidly change, by the flux density decreasing, to P_0, slower to P_2 and then still slower, while from point P_3 the flux density would gradually creep up.

If thus follows, that somewhere between the extremes B' and B'', which are most unstable, there must be a value of B, which is stable, that is, represents the stationary and permanent relation between B and H, and toward this stable value, B_0, all other values would gradually approach. This, then, would give the true magnetic characteristic: the stable physical relation between B and H.

At higher field intensities, beyond the first critical point, c_1, this stable condition is rapidly reached, and therefore is given by all the methods of determining magnetic characteristics. Hence, the curves B_1, B_2, B_0 coincide there, and the linear law of reluctivity applies. Below c_1, however, the range of possible, B, values is so large, and the final approach to the stable value so slow, as to make it difficult of determination.

35. For $H = 0$, the magnetic range is from $-R_0 = -11.2$ to $+R_0 = 11.2$; the permanent value is zero. The method of reaching the permanent value, whatever may be the remanent magnetism, is well known; it is by "demagnetizing" that is, placing the material into a powerful alternating field, a demagnetizing coil, and gradually reducing this field to zero. That is, describing a large number of cycles with gradually decreasing amplitude.

The same can be applied to any other point of the magnetization curve. Thus for $H = 1$, to reach permanent condition, an alternating m.m.f. is superimposed upon $H = 1$, and gradually decreased to zero, and during these successive cycles of decreasing amplitude, with $H = 1$, as mean value, the flux density gradually approaches its permanent or stable value. (The only requirement is, that the initial alternating field must be higher than any unidirectional field to which the magnetic circuit had been exposed.)

This seems to be the value given by curve B_0, that is, by the straight-line law of reluctivity. In other words, it is probable that:

Fröhlich's equation, or Kennelly's linear law of reluctivity represent the permanent or stable relation between B and H, that is, the true magnetic characteristic of the material, over the entire range down to $H = 0$, and the inward bend of the magnetic characteristic for low field intensities, and corresponding increase of reluctivity ρ, is the persistence of a condition of magnetic instability, just as remanent and permanent magnetism are.

In approaching stable conditions by the superposition of an alternating field, this field can be applied at right angles to the unidirectional field, as by passing an alternating current lengthwise, that is, in the direction of the lines of magnetic force, through the material of the magnetic circuit. This superimposes a circular alternating flux upon the continuous-length flux, and permits observations while the circular alternating flux exists, since the latter does not induce in the exploring circuit of the former. Some 20 years ago Ewing has already shown, that under these conditions the hysteresis loop collapses, the inward bend of the magnetic characteristic practically vanishes, and the magnetic characteristic assumes a shape like curve B_0.

To conclude, then, it is probable that:

In pure homogeneous magnetic materials, the stable relation between field intensity, H, and flux density, B, is expressed, over the entire range from zero to infinity, by the linear equation of reluctivity

$$\rho = a + \sigma H,$$

where ρ applies to the metallic magnetic induction, $B - H$.

In unhomogeneous materials, the slope of the reluctivity line changes at one or more critical points, at which the flux path changes, by a material of greater magnetic hardness beginning to carry flux.

At low field intensities, the range of unstable values of B is very great, and the approach to stability so slow, that considerable deviation of B from its stable value can persist, sometimes for years, in the form of remanent or permanent magnetism, the inward bend of the magnetic characteristic, etc.

CHAPTER IV

MAGNETISM

Hysteresis

36. Unlike the electric current, which requires power for its maintenance, the maintenance of a magnetic flux does not require energy expenditure (the energy consumed by the magnetizing current in the ohmic resistance of the magnetizing winding being an electrical and not a magnetic effect), but energy is required to produce a magnetic flux, is then stored as potential energy in the magnetic flux, and is returned at the decrease or disappearance of the magnetic flux. However, the amount of energy returned at the decrease of magnetic flux is less than the energy consumed at the same increase of magnetic flux, and energy is therefore dissipated by the magnetic change, by conversion into heat, by what may be called *molecular magnetic friction*, at least in those materials, which have permeabilities materially higher than unity.

Thus, if a magnetic flux is periodically changed, between $+B$ and $-B$, or between B_1 and B_2, as by an alternating or pulsating current, a dissipation of energy by molecular friction occurs during each magnetic cycle. Experiment shows that the energy consumed per cycle and $cm.^3$ of magnetic material depends only on the limits of the cycle, B_1 and B_2, but not on the speed or wave shape of the change.

If the energy which is consumed by molecular friction is supplied by an electric current as magnetizing force, it has the effect that the relations between the magnetizing current, i, or magnetic field intensity, H, and the magnetic flux density, B, is not reversible, but for rising, H, the density, B, is lower than for decreasing H; that is, the magnetism lags behind the magnetizing force, and the phenomenon thus is called *hysteresis*, and gives rise to the *hysteresis loop*.

However, hysteresis and molecular magnetic friction are not

the same thing, but the hysteresis loop is the measure of the molecular magnetic friction only in that case, when energy is supplied to or abstracted from the magnetic circuit only by the magnetizing current, but not otherwise. Thus, if mechanical work is done by the magnetic cycle—as when attracting and dropping an armature—the hysteresis loops enlarge, representing not only the energy dissipated by molecular magnetic friction, but also that converted into mechanical work. Inversely, if mechanical energy is supplied to the magnetic circuit as by vibrating it mechanically, the hysteresis loop collapses or overturns, and its area becomes equal to the molecular magnetic friction minus the mechanical energy absorbed. The reaction machine, as synchronous motor and as generator, is based on this feature. See "Reaction Machine," in *Theory and Calculation of Electrical Apparatus* [McGraw-Hill edition, volume 6].

In general, when speaking of hysteresis, molecular magnetic friction is meant, and the hysteresis cycle assumed under the condition of no other energy conversion, and this assumption will be made in the following, except where expressly stated otherwise.

The hysteresis cycle is independent of the frequency within commercial frequencies and far beyond this range. Even at frequencies of hundred thousand cycles, experimental evidence seems to show that the hysteresis cycle is not materially changed, except in so far as eddy currents exert a demagnetizing action and thereby require a change of the impressed m.m.f., to get the same resultant m.m.f., and cause a change of the magnetic flux distribution by their screening effect.

A change of the hysteresis cycle occurs only at very slow cycles —cycles of a duration from several minutes to years—and even then to an appreciable extent only at very low magnetic densities. Thus at low values of B—below 1000—hysteresis cycles taken by ballistic galvanometer are liable to become irregular and erratic, by "magnetic creepage." For most practical purposes, however, this may be neglected.

37. As the industrially most important varying magnetic fields are the alternating magnetic fields, the hysteresis loss in alternating magnetic fields, that is, in symmetrical cycles, is of most interest.

In general, if a magnetic flux changes from the condition H_1, B_1: point P_1 of Fig. 29, to the condition H_2, B_2: point P_2, and we assume this magnetic circuit surrounded by an electric circuit of

n turns, the change of magnetic flux induces in the electric circuit the voltage, in absolute units,

$$e = n \frac{d\Phi}{dt} \tag{1}$$

it is, however,

$$\Phi = sB \tag{2}$$

where s = section of magnetic circuit. Hence

$$e = ns \frac{dB}{dt} \tag{3}$$

If i = current in the electric circuit, the m.m.f. is

$$F = ni \tag{4}$$

and the magnetizing force

$$f = \frac{ni}{l} \tag{5}$$

where l = length of the magnetic circuit.

And the field intensity

$$H = 4\pi f \tag{6}$$

hence, substituting (5) into (6) and transposing,

$$i = \frac{lH}{4\pi n} \tag{7}$$

is the magnetizing current in the electric circuit, which produces the flux density, B.

The power consumed by the voltage induced in the electric circuit thus is

$$p = ei = \frac{slH}{4\pi} \frac{dB}{dt} \tag{8}$$

or, per cm.³ of the magnetic circuit, that is, for $s = 1$ and $l = 1l$,

$$p = \frac{H}{4\pi} \frac{dB}{dt} \tag{9}$$

and the energy consumed by the change from H_1, B_1 to H_2, B_2, which is transferred from the electric into the magnetic circuit, or inversely,

$$w_{1,2} = \frac{1}{4\pi} \int_1^2 H dB \text{ ergs} \tag{10}$$

$$= \frac{A_{1,2}}{4\pi},$$

where $A_{1,2}$ is the area shown shaded in Fig. 29.

The energy consumed during a cycle, from H_0, B_0 to $-H_0, -B_0$ and back to H_0, B_0, thus is

$$w = \frac{1}{4\pi}\int_0^0 HdB \text{ ergs} \tag{11}$$

$$= \frac{A}{4\pi} \text{ ergs} \tag{12}$$

where

$\int_0^0 HdB = A$ is the area of the hysteresis loop, shown shaded in Fig. 30.

As the magnetic condition at the end of the cycle is the same as

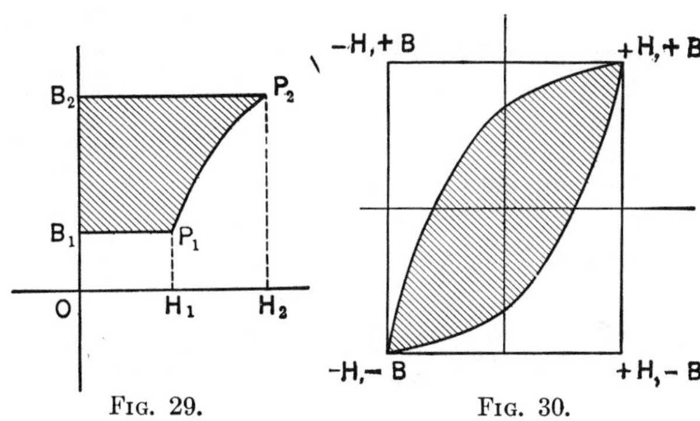

FIG. 29. FIG. 30.

at the beginning, all this energy, w, is dissipated as heat, that is, is the hysteresis energy which measures the molecular magnetic friction.

38. If in Fig. 30 the shaded area represents the hysteresis loop between $+H, +B$, and $-H, -B$, giving with a sinusoidal alternating flux the voltage and current waves, Fig. 31, the maximum area, which the hysteresis loop could theoretically assume, is given by the rectangle between $+H, +B; -H, +B; -H, -B; +H, -B$. This would mean, that the magnetic flux does not appreciably decrease with decreasing field intensity, until the field has reversed to full value. It would give the theoretical wave shape shown as Fig. 32. As seen, this is the extreme exaggeration of wave shape, Fig. 31.

The total energy of this rectangle, or maximum available magnetic energy, is

$$w_0 = \frac{4\,HB}{4\,\pi} = \frac{HB}{\pi} \tag{12}$$

or, if μ = permeability, thus $H = \frac{B}{\mu}$, it is

$$w_0 = \frac{B^2}{\pi\mu} \tag{13}$$

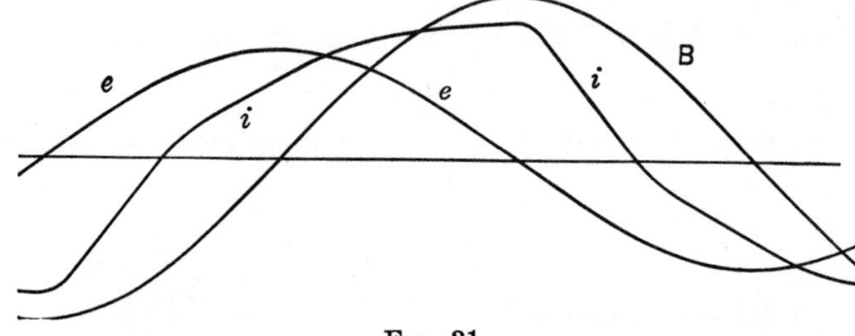

FIG. 31.

the maximum possible hysteresis loss.

The inefficiency of the magnetic cycle, or percentage loss of energy in the magnetic cycle, thus is

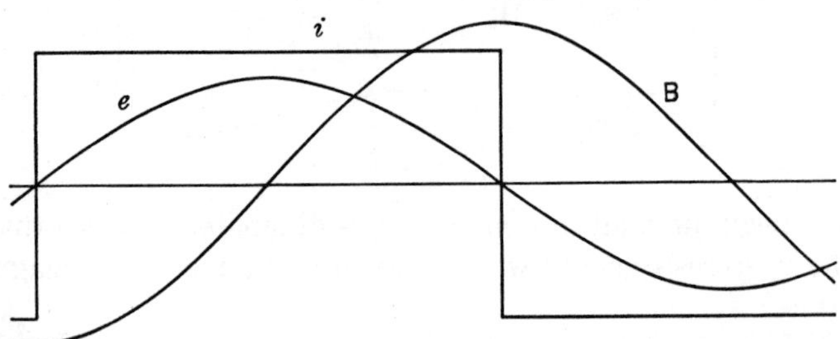

FIG. 32.

$$\left.\begin{aligned} \zeta &= \frac{w}{w_0} = \frac{\pi\mu w}{B^2} \\ &= \frac{\mu}{4\,B^2}\int_0^0 H dB \\ &= \frac{A\mu}{4\,B^2} \end{aligned}\right\} \tag{14}$$

39. Experiment shows that for medium flux density, that is, thoses values of B which are of the most importance industrially, from $B = 1000$ to $B = 12{,}000$, the hysteresis loss can with sufficient accuracy for most practical purposes be approximated by the empirical equation,

$$w = \eta B^{1.6} \tag{15}$$

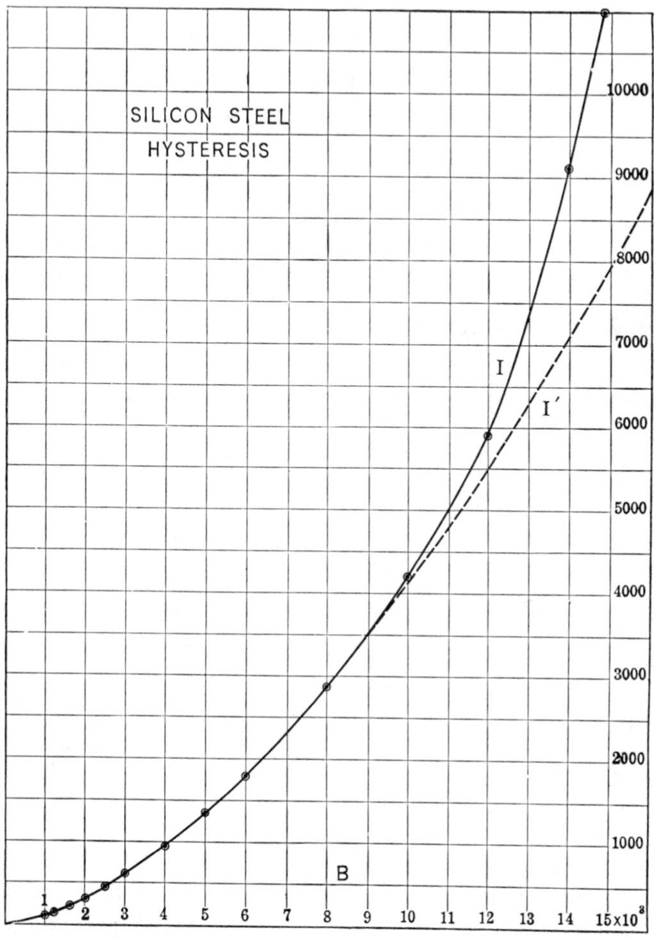

FIG. 33.

where η, the "coefficient of hysteresis," is of the magnitude of 1×10^{-3} to 2×10^{-3} for annealed soft sheet steel, if B is given in lines of force per cm.², and w is ergs per cm.³ and cycle.

Very often w is given in joules, or watt-seconds per cycle and per kilogram or pound of iron, and B in lines per square inch, or w is given in watts per kilogram or per pound at 60 cycles.

In Fig. 33 is shown, with B as abscissæ, the hysteresis loss, w, of a sample of silicon steel. The observed values are marked by circles. In dotted lines is given the curve calculated by the equation

$$w = 0.824 \times 10^{-3} B^{1.6} \tag{16}$$

As seen, the agreement the curve of 1.6^{th} power with the test values is good up to $B = 10{,}000$, but above this density, the observed values rise above the curve.

40. In Fig. 34 is plotted, with field intensity, H, as abscissæ, the magnetization curve of ordinary annealed sheet steel, in

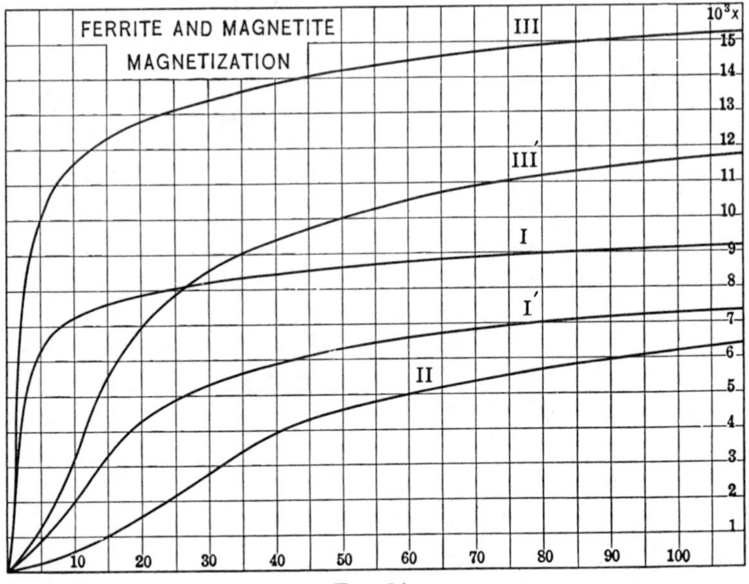

FIG. 34.

half-scale, as curve I, and the magnetization curve of magnetite, Fe_3O_4—which is about the same as the black scale of iron—in double-scale, as curve II. As III then is plotted, in full-scale, a curve taking 0.8 of I and 0.2 of II. This would correspond to the average magnetic density in a material containing 80 per cent. of iron and 20 per cent. (by volume) of scale. Curves I′ and III′ show the initial part of I and III, with ten times the scale of abscissæ and the same scale of ordinates.

Fig. 35 then shows, with the average magnetic flux density, B, taken from curve III of Fig. 34, as abscissæ, the part of the magnetic flux density which is carried by the magnetite, as curve I. As seen, the magnetite carries practically no flux up to $B = 10$, but beyond $B = 12$, the flux carried by the magnetite rapidly increases.

As curve II of Fig. 35 is shown the hysteresis loss in this inhomogeneous material consisting of 80 per cent. ferrite (iron) and 20 per cent. magnetite (scale) calculated from curves I and II of Fig.

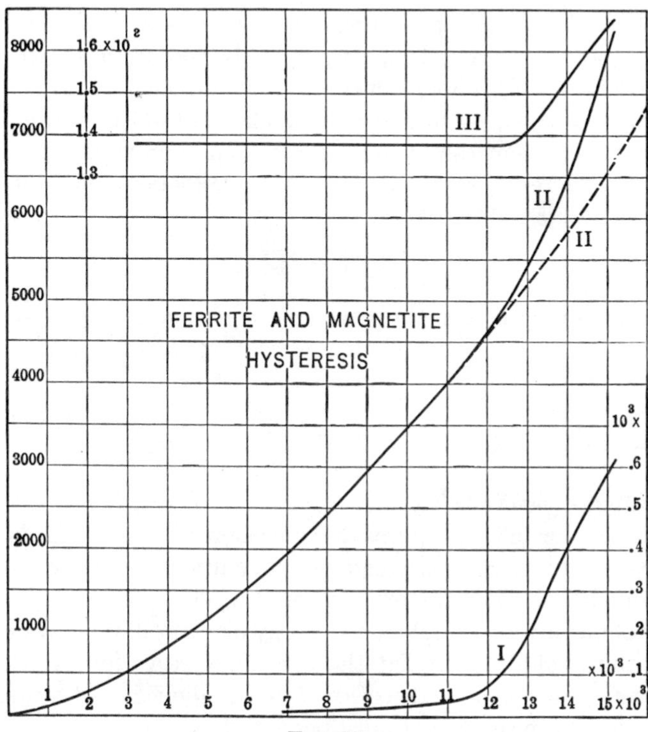

FIG. 35.

34 under the assumption that either material rigidly follows the 1.6^{th} power law up to the highest densities, by the equation,

Iron:

$$w_1 = 1.2\ B_I^{1.6} \times 10^{-3}.$$

Scale:

$$w_2 = 23.5\ B_{II}^{1.6} \times 10^{-3},$$

As curve II′ is shown in dotted lines the 1.6^{th} power equation,

$$w = 1.38\quad B^{1.6} \times 10^{-3}.$$

As seen, while either constituent follows the 1.6^{th} power law, the combination deviates therefrom at high densities, and gives an increase of hysteresis loss, of the same general characteristic as shown with the silicon steel in Fig. 33, and with most similar materials.

As curve III in Fig. 35 is then shown the increase of the hysteresis coefficient η, at high densities, over the value 1.38×10^{-3}, which it has at medium densities.

Thus, the deviation of the hysteresis loss at high densities, from the 1.6^{th} power law, may possibly be only apparent, and the result of lack of homogeneity of the material.

41. At low magnetic densities, the law of the 1.6^{th} power must cease to represent the hysteresis loss even approximately.

The hysteresis loss, as fraction of the available magnetic energy, is, by equation (14),

$$\zeta = \frac{\pi \mu w}{B^2} \tag{14}$$

Substituting herein the parabolic equation of the hysteresis loss,

$$w = \eta B^n \tag{17}$$

where $n = 1.6$, it is

$$\begin{aligned} \zeta &= \mu \pi \eta \quad B^{n-2} \\ &= \pi \mu \eta \quad B^{.4} \end{aligned} \tag{18}$$

With decreasing density B, B^{n-2} steadily increases, if $n < 2$, and as the permeability μ approaches a constant value, ζ, steadily increases in this case, thus would become unity at some low density, B, and below this, greater than unity. This, however, is not possible, as it would imply more energy dissipated, than available, and thus would contradict the law of conservation of energy. Thus, for low magnetic densities, if the parabolic law of hysteresis (17) applies, the exponent must be: $n \overline{\gtrless} 2$.

In the case of Fig. 33, for $\eta = 0.824 \times 10^{-3}$, assuming the permeability for extremely low density as

$$\mu = 1500,$$

ζ becomes unity, by equation (18), at

$$B = 30.$$

If $n > 2$, B^{n-2} steadily decreases with decreasing B, and the percentage hysteresis loss becomes less, that is, the cycle approaches reversibility for decreasing density; in other words, the hysteresis loss vanishes. This is possible, but not probable, and the probability is that for very low magnetic densities, the hysteresis losses approach proportionality with the square of the magnetic density, that is, the percentage loss approaches constancy.

From equation (17) follows

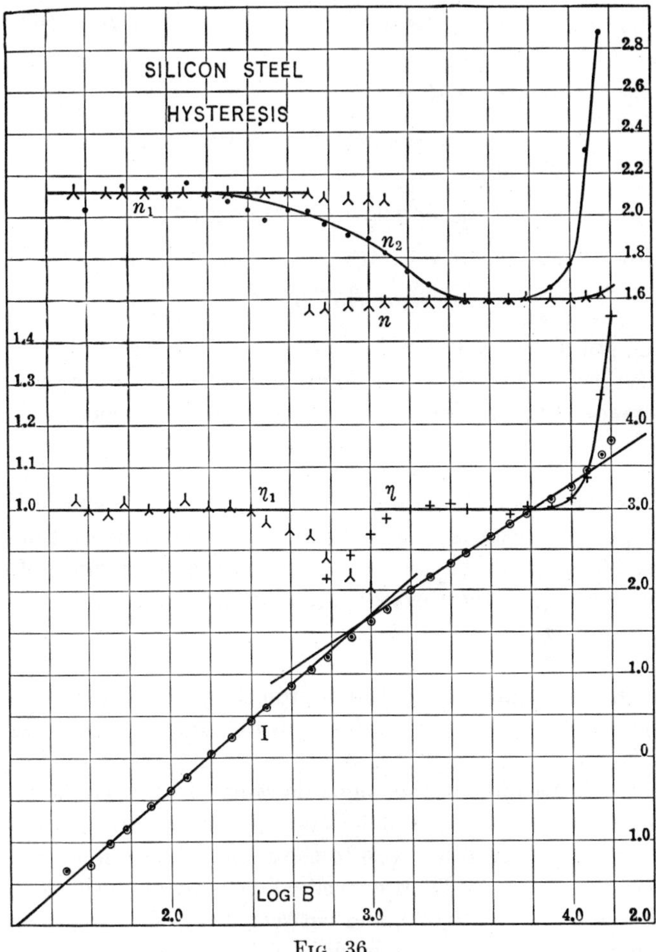

FIG. 36.

$$\log w = \log \eta + n \log B \tag{19}$$

That is:

"If the hysteresis loss follows a parabolic law, the curve plotted with log w against log B is a straight line, and the slope of this straight line is the exponent, n."

Thus, to investigate the hysteresis law, log w is plotted against log B. This is done for the silicon steel, Fig. 33, over the range from $B = 30$ to $B = 16{,}000$, in Fig. 36, as curve I.

Curve I contains two straight parts, for medium densities, from $\log B = 3; B = 1000$, to $\log B = 4; B = 10{,}000$, with slope 1.6006, and for low densities, up to $\log B = 2.6; B = 400$, with slope 2.11. Thus it is

For $1000 \leqq B \leqq 10{,}000$:

$$w = 0.824\, B^{1.6} \times 10^{-3}$$

For $B \leqq 400$:

$$w = 0.00257\, B^{2.11} \times 10^{-3}$$

However, in this lower range, $n = 2$ gives a curve:

$$w = 0.0457\, B^2 \times 10^{-3}$$

which still fairly well satisfies the observed values.

As the logarithmic curve for a sample of ordinary annealed sheet steel, Fig. 37, gives for the lower range the exponent,

$$n = 1.923,$$

and as the difficulties of exact measurements of hysteresis losses increase with decreasing density, it is quite possible that in both, Figs. 36 and 37 the true exponent in the lower range of magnetic densities is the theoretically most probable one,

$$n = 2,$$

that is, that at about $B = 500$, in iron the point is reached, below which the hysteresis loss varies with the square of the magnetic density.

42. As over most of the magnetic range the hysteresis loss can be expressed by the parabolic law (17), it appears desirable to adapt this empirical law also to the range where the logarithmic curve, Figs. 36 and 37, is curved, and the parabolic law does not apply, above $B = 10{,}000$, and between $B = 500$ and $B = 1000$, or thereabouts. This can be done either by assuming the coefficient η as variable, or by assuming the exponent n as variable.

(*a*) Assuming η as constant,

$$\eta = 0.824 \times 10^{-3} \text{ for the medium range, where } n \cong 1.6$$
$$\eta_1 = 0.0457 \times 10^{-3} \text{ for the low range, where } n_1 \cong 2$$

The coefficients n and n_1 calculated from the observed values of w, then, are shown in Fig. 36 by the three-cornered stars in the upper part of the figure.

(*b*) Assuming n as constant,

$$n = 1.6 \text{ for the medium range, where } \eta \cong 0.0824 \times 10^{-3}$$
$$n_1 = 2 \text{ for the low range, where } \eta_1 \cong 0.0457 \times 10^{-3}$$

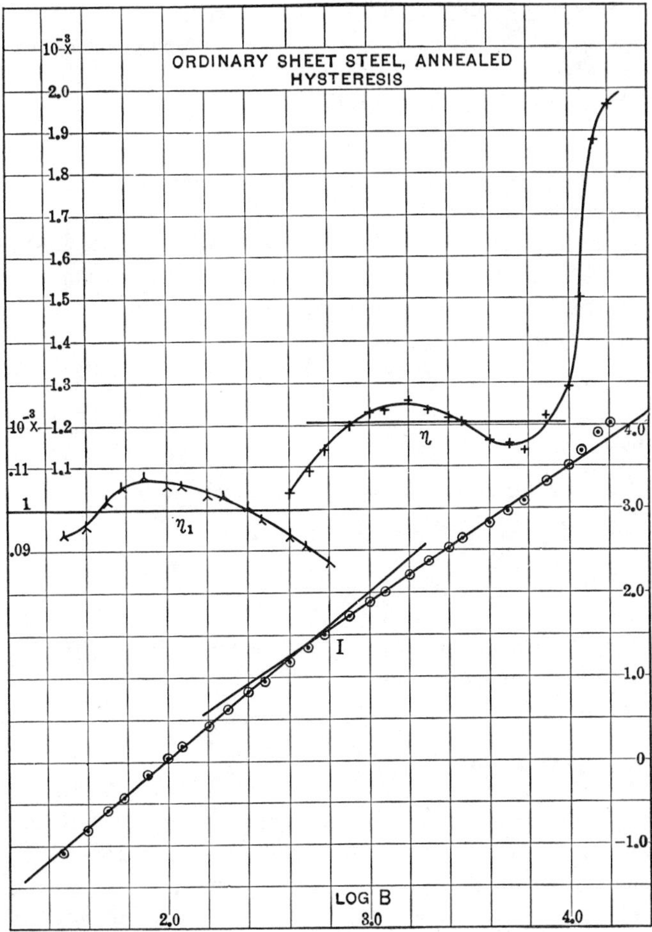

FIG. 37.

The variation of η and η_1, from the values in the constant range, then, are best shown in per cent., that is, the loss w calculated from the parabolic equation and a correction factor applied for values of B outside of the range.

Fig. 37 shows the values of η and η_1, as calculated from the parabolic equations with $n = 1.6$ and $n_1 = 2$, and Fig. 36 shows the percentual variation of η and η_1.

The latter method, (*b*), is preferable, as it uses only one exponent, 1.6, in the industrial range, and uses merely a correction factor. Furthermore, in the method (*a*), the variation of the exponent is very small, rising only to 1.64, or by 2.5 per cent., while in method (*b*) the correction factor is 1.46, or 46 per cent., thus a much greater accuracy possible.

43. If the parabolic law applies,

$$w = \eta B^n \tag{17}$$

the slope of the logarithmic curve is the exponent n.

If, however, the parabolic law does not rigidly apply, the slope of the logarithmic curve is not the exponent, and in the range, where the logarithmic curve is not straight, the exponent thus can not even be approximately derived from the slope.

From (17) follows

$$\log w = \log \eta + n \log B, \tag{19}$$

differentiating (19), gives, in the general case, where the parabolic law does not strictly apply,

$$d \log w = d \log \eta + nd \log B + \log Bdn,$$

hence, the slope of the logarithmic curve is

$$\frac{d \log w}{d \log B} = n + \left(\log B \frac{dn}{d \log B} + \frac{d \log \eta}{d \log B}\right) \tag{20}$$

If n = constant, and η = constant, the second term on the right-hand side disappears, and it is

$$\frac{d \log w}{d \log B} = n \tag{21}$$

that is, the slope of the logarithmic curve is the exponent.

If, however, η and n are not constant, the second term on the right-hand side of equation (20) does not in general disappear, and the slope thus does not give the exponent.

Assuming in this latter case the slope as the exponent, it must be

$$\log B \frac{dn}{d \log B} + \frac{d \log \eta}{d \log B} = 0.$$

Or,

$$\frac{d \log \eta}{dn} = - \log B \tag{22}$$

In this case, n and much more still η show a very great variation, and the variation of η is so enormous as to make this representation valueless.

As illustration is shown, in Fig. 36, the slope of the curve as n_2. As seen, n_2 varies very much more than n or n_1.

To show the three different representations, in the following table the values of n and η are shown, for a different sample of iron.

TABLE.

$B\ 10^3$	(a) η = const. = 1.254	(b) n = const. = 1.6	(c) $n_2 = \frac{d \log w}{d \log B}$	η_2
below 10.00	$n = 1.6$	$\eta = 1.254 \times 100^{-3}$	$n_2 = 1.6$	$\eta_2 = 1.254 \times 10^{-6}$
10.00	$= 1.601$	$= 1.268$	$= 1.79$	230.00
11.23	$= 1.604$	$= 1.302$	$= 2.23$	3.68
12.63	$= 1.617$	$= 1.468$	$= 2.66$	0.0488
13.30	$= 1.624$	$= 1.570$	$= 2.83$	0.0133
14.00	$= 1.630$	$= 1.668$	$= 2.98$	0.0032
14.65	$= 1.634$	$= 1.738$	$= 3.15$	0.00069

As seen, to represent an increase of hysteresis loss by $\frac{1.738}{1.254} =$ 1.39, or 39 per cent., under (c), n_2 is nearly doubled, and η_2 reduced to $\frac{1}{1,800,000}$ of its initial value.

44. The equation of the hysteresis loss at medium densities,

$$W = \eta B^n; \quad n = 1.6$$

is entirely empirical, and no rational reason has yet been found why this approximation should apply. Calculating the coefficient n from test values of B and W, shows usually values close to 1.6, but not infrequently values of n are found, as low as 1.55, and even values below 1.5, and values up to 1.7 and even above 1.9 In general, however, the more accurate tests give values of n which do not differ very much from 1.6, so that the losses can still be represented by the curve with the exponent $n = 1.6$, without serious error. This is desirable, as it permits comparing different materials by comparing the coefficients η. This would not be the case, if different values of n were used, as even a small change of n makes a very large change of η: a change of n by 1 per cent., at $B = 10{,}000$, changes η by about 16 per cent.

Thus in Fig. 37 is represented as I the logarithmic curve of a sample of ordinary annealed sheet steel, which at medium density gives the exponent $n = 1.556$, at low densities the exponent $n_1 = 1.923$. Assuming, however, $n = 1.6$ and $n_1 = 2.0$, gives the average values $\eta = 1.21 \times 10^{-3}$ and $\eta_1 = 0.10 \times 10^{-3}$, and the

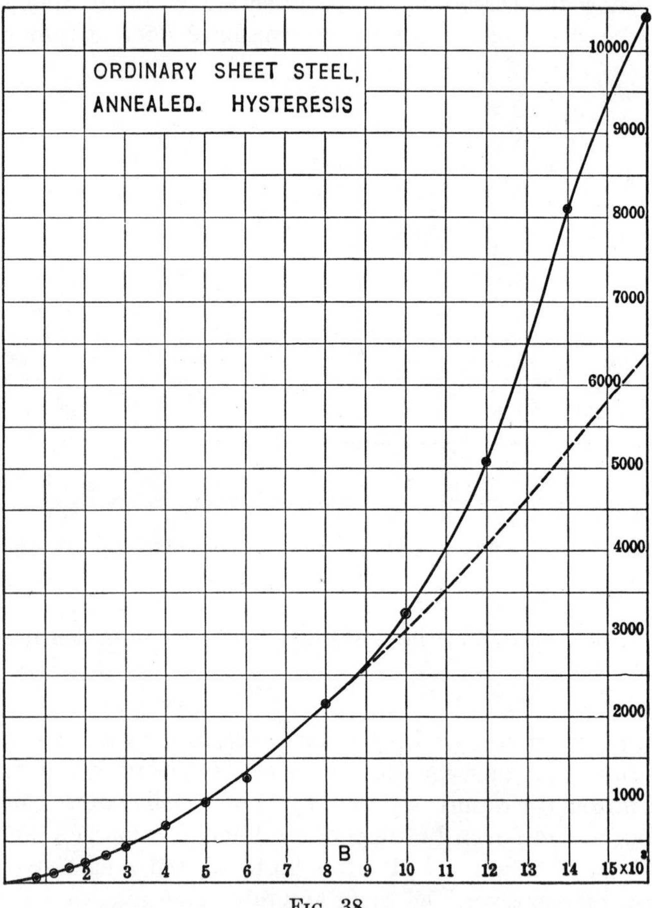

FIG. 38.

individual calculated values of η and η_1 are then shown on Fig. 37 by crosses and three-pointed stars, respectively.

Fig. 38 then shows the curve of observed loss, in drawn line, and the 1.6^{th} power curve calculated in dotted line, and Fig. 39 the lower range of the calculated curve, with the observations marked by circles. Fig. 40 shows, for the low range, the curve of $\eta_1 B^2$, in two different scales, with the observed values marked by cycles. As seen, although in this case the deviation of n from 1.6 respectively 2 is considerable, the curves drawn with n = 1.6 and n_1 = 2 still represent the observed values fairly well in

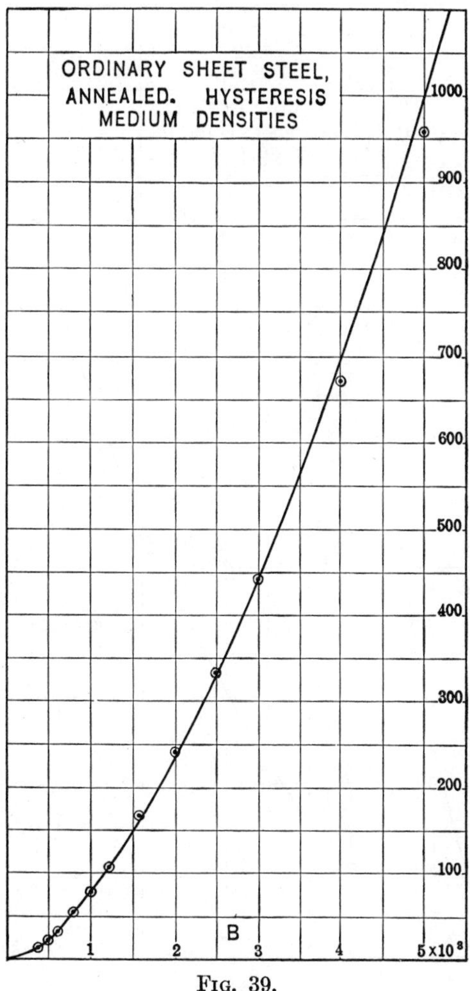

FIG. 39.

the range of B from 500 to 10,000, and below 500, respectively, so that the 1.6th power equation for the medium, and the quadratic equation for the low values of B can be assumed as sufficiently accurate for most purposes, except in the range of high densities in those materials, where the increase of hysteresis loss occurs there.

While the measurement of the hysteresis loss appears a very simple matter, and can be carried out fairly accurately over a

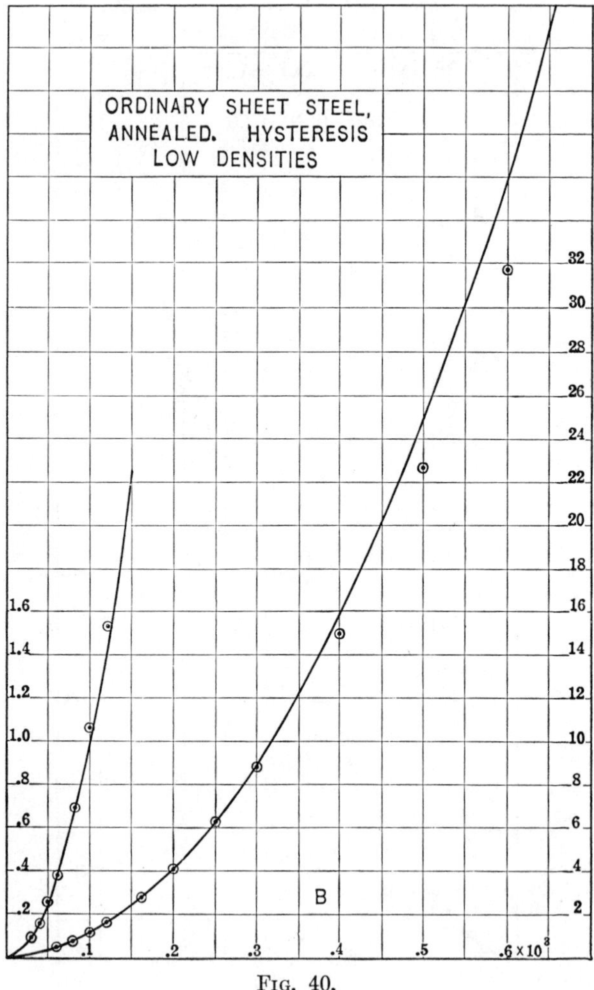

Fig. 40.

narrow range of densities, it is one of the most difficult matters to measure the hysteresis loss over a wide range of densities with such accuracy as to definitely determine the exact value of the exponent n, due to varying constant errors, which are beyond control. While true errors of observations can be eliminated by multiplying data, with a constant error this is not the case, and if the constant error changes with the magnetic density, it results in an apparent change of n. Such constant errors, which increase or decrease, or even reverse with changing B, are in the Ballistic galvanometer method the magnetic creepage at lower B, and at higher B the sharp-pointed shape of the hysteresis loop, which makes the area between rising and decreasing characteristic difficult to determine. In the wattmeter method by alternating current, varying constant errors are the losses in the instruments, the eddy-current losses which change with the changing flux distribution by magnetic screening in the iron, with the temperature, etc., by wave-shape distortion, the unequality of the inner and outer length of the magnetic circuit, etc.

45. Symmetrical magnetic cycles, that is, cycles performed between equal but opposite magnetic flux densities, $+B$ and $-B$, are industrially the most important, as they occur in practically all alternating-current apparatus. Unsymmetrical cycles, that is, cycles between two different values of magnetic flux density, B_1 and B_2, which may be of different, or may be of the same sign, are of lesser industrial importance, and therefore have been little investigated until recently.

However, unsymmetrical cycles are met in many cases in alternating- and direct-current apparatus, and therefore are of importance also.

In most inductor alternators the magnetic flux in the armature does not reverse, but pulsates between a high and a low value in the same direction, and the hysteresis loss thus is that of an unsymmetrical non-reversing cycle.

Unsymmetrical cycles occur in transformers and reactors by the superposition of a direct current upon the alternating current, as discussed in Chapter VIII, "Shaping of Waves," or by the equivalent thereof, such as the suppression of one-half wave of the alternating current. Thus, in the transformers and reactors of many types of rectifiers, as the mercury-arc rectifier, the magnetic cycle is unsymmetrical.

Unsymmetrical cycles occur in certain connections of transformers (three-phase star-connection) feeding three-wire synchronous converters, if the direct-current neutral of the converter is connected to the transformer neutral.

They may occur and cause serious heating, if several transformers with grounded neutrals feed the same three-wire distribution circuit, by stray railway return current entering the three-wire a ternating distribution circuit over one neutral and leaving it over another one.

Two smaller unsymmetrical cycles often are superimposed on an alternating cycle, and then increase the hysteresis loss. Such occurs in transformers or reactors by wave shapes of impressed voltage having more than two zero values per cycle, such as that shown in Fig. 51 of Chapter VIII, "Shaping of Waves."

They also occur sometimes in the armatures of direct-current motors at high armature reaction and low field excitation, due to the flux distortion, and under certain conditions in the armatures of regulating pole converters.

A large number of small unsymmetrical cycles are sometimes superimposed upon the alternating cycle by high-frequency pulsation of the alternating flux due to the rotor and stator teeth, and then may produce high losses. Such, for instance, is the case in induction machines, if the stator and rotor teeth are not proportioned so as to maintain uniform reluctance, or in alternators or direct-current machines, in which the pole faces are slotted to receive damping windings, or compensating windings, etc., if the proportion of armature and pole-piece slots is not carefully designed.

46. The hysteresis loss in an unsymmetrical cycle, between limits B_1 and B_2, that is, with the amplitude of magnetic variation $B = \frac{B_1 - B_2}{2}$, follows the same approximate law of the 1.6^{th} power,

$$w_0 = \eta_0 B^{1.6}$$

as long as the average value of the magnetic flux variation,

$$B_0 = \frac{B_1 + B_2}{2},$$

is constant.

With changing B_0, however, the coefficient η_0 changes, and increases with increasing average flux density, B_0.

John D. Ball has shown, that the hysteresis coefficient of the unsymmetrical cycle increases with increasing average density, B_0, and approximately proportional to a power of B_0. That is,

$$\eta_0 = \eta + \beta\eta \, B_0^{1.9}.$$

Thus, in an unsymmetrical cycle between limits B_1 and B_2 of magnetic flux density, it is

$$w = \left\{\eta + \beta \left(\frac{B_1 + B_2}{2}\right)^{1.9}\right\} \left(\frac{B_1 - B_2}{2}\right)^{1.6} \tag{23}$$

where η is the coefficient of hysteresis of the alternating-current cycle, and for $B_2 = -B_1$, equation (23) changes to that of the symmetrical cycle.

Or, if we substitute,

$$B_0 = \frac{B_1 + B_2}{2} \tag{24}$$

= average value of flux density, that is, average of maximum and minimum.

$$B = \frac{B_1 - B_2}{2} \tag{25}$$

= amplitude of unsymmetrical cycle,

it is

$$w = (\eta + \beta B_0^{1.9})B^{1.6} \tag{26}$$

or,

$$w = \eta_0 B^{1.6} \tag{27}$$

where

$$\eta_0 = \eta + \beta B_0^{1.9} \tag{28}$$

or, more general,

$$w = \eta_0 B^n \tag{29}$$

$$\eta_0 = \eta + \beta B_0^m \tag{30}$$

For a good sample of ordinary annealed sheet steel, it was found,

$$\eta = 1.06 \times 10^{-3} \tag{31}$$

$$\beta = 0.344 \times 10^{-10}$$

For a sample of annealed medium silicon steel,

$$\left.\begin{aligned}\eta &= 1.05 \times 10^{-3}\\ \beta &= 0.32 \times 10^{-10}\end{aligned}\right. \tag{32}$$

Fig. 41 shows, with B_0 as abscissæ, the values of η_0, by equations (30) and (32).

As seen, in a moderately unsymmetrical cycle, such as between $B_1 = +12{,}000$ and $B_2 = -4000$, the increase of the hysteresis loss over that in a symmetrical cycle of the same amplitude, is moderate, but the increase of hysteresis loss becomes very large

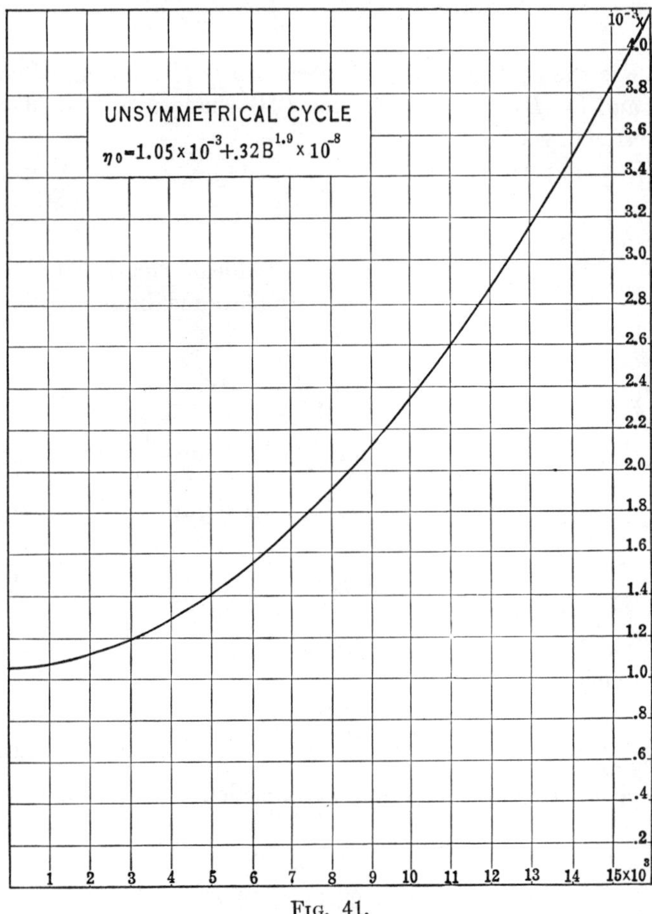

Fig. 41.

in highly unsymmetrical cycles, such as between $B_1 = 16{,}000$ and $B_2 = 12{,}000$.

CHAPTER V

MAGNETISM

Magnetic Constants

47. With the exception of a few ferromagnetic substances, the magnetic permeability of all materials, conductors and dielectrics, gases, liquids and solids, is practically unity for all industrial purposes. Even liquid oxygen, which has the highest permeability, differs only by a fraction of a per cent. from non-magnetic materials.

Thus the permeability of neodymium, which is one of the most paramagnetic metals, is $\mu = 1.003$; the permeability of bismuth, which is very strongly diamagnetic, is $\mu = 1 - 0.00017 = 0.99983$.

The magnetic elements are iron, cobalt, nickel, manganese and chromium. It is interesting to note that they are in atomic weight adjoining each other, in the latter part of the first half of the first large series of the periodic system:

	Ti	V	Cr	Mn	Fe	Co	Ni	Cu	Zn
Atomic weight..................	48	51	52	55	56	58	59	61	65

The most characteristic, because relatively most constant, is the metallic magnetic saturation, S, or its reciprocal, the saturation coefficient, σ, in the reluctivity equation. The saturation density seems to be little if any affected by the physical condition of the material. By the chemical composition, such as by the presence of impurities, it is affected only in so far as it is reduced approximately in proportion to the volume occupied by the non-magnetic materials, except in those cases where new compounds result.

It seems, that the saturation value is an absolute limit of the element, and in any mixture, alloy or compound, the saturation value reduced to the volume of the magnetic metal contained therein, can not exceed that of the magnetic metal, but may be lower, if the magnetic metal partly or wholly enters a compound of lower intrinsic saturation value. Thus, if $S = 21 \times 10^3$ is the saturation value of iron, an alloy or compound containing

72 per cent. by volume of iron can have a maximum saturation value of $S = 0.72 \times 21 \times 10^3 = 15.1 \times 10^3$ only, or a still lower saturation value.

The only known exception herefrom seems to be an iron-cobalt alloy, which is alleged to have a saturation value about 10 per cent. higher than that of iron, though cobalt is lower than iron.

The coefficient of magnetic hardness, α, however, and the coefficient of hysteresis, η, vary with the chemical, and more still with the physical characteristic of the magnetic material, over an enormous range.

Thus, a special high-silicon steel, and the chilled glass hard tool steel in the following tables, have about the same percentage of non-magnetic constituents, 4 per cent., and about the same saturation value, $S = 19.2 \times 10^3$, but the coefficient of hardness of chilled tool steel, $\alpha = 8 \times 10^{-3}$, is 200 times that of the special silicon steel, $\alpha = 0.04 \times 10^{-3}$, and the coefficient of hysteresis of the chilled tool steel, $\eta = 75 \times 10^{-3}$, is 125 times that of the silicon steel, $\eta = 0.6 \times 10^{-3}$. Hardness and hysteresis loss seem to depend in general on the physical characteristics of the material, and on the chemical constitution only as far as it affects the physical characteristics.

Chemical compounds of magnetic metals are in general not ferromagnetic, except a few compounds as magnetite, which are ferromagnetic.

With increasing temperature, the magnetic hardness α, decreases, that is, the material becomes magnetically softer, and the saturation density, S, also slowly decreases, until a certain critical temperature is reached (about 760°C. with iron), at which the material suddenly ceases to be magnetizable or ferromagnetic, but usually remains slightly paramagnetic.

As the result of the increasing magnetic softness and decreasing saturation density, with increasing temperature the density, B, at low field intensities, H, increases, at high field intensities decreases. Such B-temperature curves at constant H, however, have little significance, as they combine the effect of two changes, the increase of softness, which predominates at low H, and the decrease of saturation, which predominates at high H.

Heat treatment, such as annealing, cooling, etc., very greatly changes the magnetic constants, especially α and η—more or less in correspondence with the change of the physical constants brought about by the heat treatment.

Very extended exposure to moderate temperature—100 to 200°C.—increases hardness and hysteresis loss with some materials, by what is called ageing, while other materials are almost free of ageing.

48. The most important, and therefore most completely investigated magnetic metal is iron.

Its saturation value is probably between $S = 21.0 \times 10^3$ and $S = 21.5 \times 10^3$, the saturation coefficient thus $\sigma = 0.047$.

As all industrially used iron contains some impurities, carbon, silicon, manganese, phosphorus, sulphur, etc., usually saturation values between 20×10^3 and 21×10^3 are found on sheet steel or cast steel, etc., lower values, 19 to 19.5×10^3, in silicon steels containing several per cent. of Si, and still much lower values, 12 to 15×10^3, in very impure materials, such as cast iron.

Two types of iron alloys seem to exist:

1. Those in which the alloying material does not directly affect the magnetic qualities, but only indirectly, by reducing the volume of the iron and thereby the saturation value, and by changing the physical characteristics and thereby the hardness and hysteresis loss.

Such apparently are the alloys with carbon, silicon, titanium, chromium, molybdenum and tungsten, etc., as cast iron, silicon steel, magnet steel, etc.

2. Those in which the alloying material changes the magnetic, characteristics.

Such apparently are the alloys with nickel, manganese, mercury, copper, cobalt, etc.

In this class also belong the chemical compounds of the magnetic materials.

Thus, a manganese content of 10 to 15 per cent. makes the iron practically non-magnetic, lowers the permeability to $\mu = 1.4$. However, even here it is not certain whether this is not an extreme case of magnetic hardness, and at extremely high magnetic fields the normal saturation value of the iron would be approached.

Some nickel steels (25 per cent. Ni) may be either magnetic, or non-magnetic. However, pure iron, when heated to high incandescence, becomes non-magnetic at a certain definite temperature, and when cooling down, becomes magnetizable again at another definite, though lower temperature, and between these two temperatures, iron may be magnetic or unmagnetic, depending whether it has reached this temperature from lower, or from higher temperatures. Apparently, for these nickel steels, the critical temperature range, within which they can be magnetic or unmagnetic, is within the range of atmospheric temperature, and thus, after heating, they become non-magnetic, after cooling to sufficiently low temperature, they become magnetizable again. Thus, a steel containing 17 per cent. nickel, 4.5 per cent. chromium, 3 per cent. manganese, has permeability 1.004, that is, is almost completely unmagnetic.

Heterogeneous mixtures, such as powdered iron incorporated in resin, or iron filings in air, seem to give saturation densities not far different from those corresponding to their volume percentage of iron, but give an enormous increase of hardness, α, and hysteresis, η, as is to be expected.

Most chemical compounds of iron are non-magnetic. Ferromagnetic is only magnetite, which is the intermediate oxide and may be considered as ferrous ferrite. There also is an alleged magnetic sulphide of iron, though I have never seen it, magnetkies, Fe_7S_8 or Fe_8S_9.

As magnetite, Fe_3O_4, contains 72 per cent. of Fe, by weight, and has the specific weight 5.1, its volume per cent. of iron would be 48 per cent., and the saturation density $S = 10 \times 10^3$.

Observations on the magnetic constants of magnetite give a saturation density of 4.7×10^3 to 5.91×10^3, so that magnetite would fall in the second class of iron compounds, those in which the saturation density is affected, and lowered, by the composition.

Not only *magnetite*, which may be considered as ferrous ferrite, but numerous other ferrites, that is, salts of the acid $Fe_2O_4H_2$, are to some extent ferromagnetic, such as copper and cobalt ferrite, calcium ferrite, etc.

49. *Cobalt*, next adjoining to iron in the periodic system of elements, is the magnetic metal which has been least investigated. Its saturation value probably is between $S = 12 \times 10^3$ and $S = 14 \times 10^3$, and its magnetic characteristic looks very similar to that of cast iron. Partly this is due to the similar saturation value, partly probably due to the feature that most of the available data were taken on cast cobalt.

It is interesting to note that Cobalt retains its magnetizability up to much higher temperatures than iron or any other material, so that above 800 degrees C., Cobalt is the only magnetic material.

More information is available on *nickel*, the metal next adjoining to cobalt in the periodic system of elements. Its saturation density is the lowest of the magnetic metals, probably between $S = 6 \times 10^3$ and $S = 7 \times 10^3$.

Some data on nickel and nickel alloys are given in the following table. In general, nickel seems to show characteristics very similar to those of iron, except that all the magnetic densities are reduced in proportion to the lower saturation density; but the effect of the physical characteristics on the magnetic constants appears to be the same. Interesting is, that nickel seems to be least sensitive to impurities in their effect on the reluctivity curve.

Nickel ceases to be magnetizable already below red heat.

The next metal beyond nickel, in the periodic system of elements, is copper, and this is non-magnetic, as far as known.

On the other side of iron, in the periodic system, is *manganese*.

This is very interesting in so far as it has never been observed in a strongly magnetic state, but many of the alloys of manganese are more or less strongly magnetic, and estimating from the saturation values of manganese alloys, the saturation value of manganese as pure metal should be about $S = 30 \times 10^3$. This would make it the most magnetic metal.

In favor of manganese as magnetic metal also is the unusual behavior of its alloys with iron: the alloys of nickel, and of cobalt with iron also show unusual characteristics, and this seems to be a characteristic of alloys between magnetic metals.

The best known magnetic manganese alloys are the Heusler alloys, of manganese with copper and aluminum, and the characteristics of three such alloys are given in the following table. The most magnetic shows about the same saturation value as magnetite, but higher saturation values, equal to those of nickel, have been observed.

A curious feature of some Heusler alloys is, that when slowly cooled from high temperatures, they are very little magnetic, and have low saturation values. The quicker they are cooled, the higher their permeability and their saturation value, and the best values have been reached by dropping the molten alloy into water, so suddenly chilling it.

In general, the Heusler alloys are especially sensitive to heat treatment, and some of them show the ageing in a most pronounced degree, so that maintaining the alloy for a considerable time at moderate temperature, increases hardness and hysteresis loss more than tenfold.

Magnetic alloys of manganese also are known with antimony, arsenic, phosphorus, bismuth, boron, with zinc and with tin, etc. Usually, the best results are given by alloys containing 20 to 30 per cent. of manganese. Little is known of these magnetic alloys, except that they may be in a magnetic state, or in an unmagnetic stage. They are most conveniently produced by dissolving manganese metal in the superheated alloying metal, or in this metal with the addition of some powerful reducing metal, as sodium or aluminum, but the alloy is only sometimes magnetic, sometimes practically unmagnetic, and the conditions of the formation of the magnetic state are unknown.

Apparently, there also exists an intermediary oxide of manganese, or a compound oxide of manganese with that of the other metal, which is strongly magnetic. The black slag, appearing in the fusion of manganese with other metals such as antimony, zinc, tin, without flux, often is strongly magnetic, more so than the alloy itself.

A mixture of about 25 per cent. powdered manganese metal, and 75 per cent. powdered antimony metal, heated together to a moderate temperature—in a test-tube—gives a strongly magnetic black powder, which can be used like iron filings, to show the lines of forces of the magnetic field, but has not further been investigated.

A considerable number of such magnetic manganese alloys have been investigated by Heusler and others, and their constants are given in the following table.

It is supposed that these magnetic manganese alloys are chemical compounds, similar as magnetite or magnetkies. Thus the copper-aluminum-manganese alloy of Heusler is a compound of 1 atom of aluminum with 3 atoms of copper or manganese: Al-(Mn or Cu)$_3$, usually $AlMnCu_2$. Other magnetic manganese compounds then are:

With antimony.	$MnSb$ and Mn_2Sb
With bismuth.	$MnBi$
With arsenic.	$MnAs$
With boron.	MnB
With phosphorus.	MnP
With tin.	Mn_4Sn and Mn_2Sn

Next adjacent to manganese in the periodic system of elements is *chromium.* Neither the metal, nor any of its alloys (except those with magnetic metals) have ever been observed in the magnetic state. There is, however, an intermediary oxide of chromium, alleged to be Cr_5O_9 (a basic chromic chromate?) which is strongly magnetic. It forms, in black scales, in a narrow range of temperature, by passing CrO_2Cl_2 with hydrogen through a heated tube.

A second strongly magnetic chromium oxide is Cr_4O_9 (a basic chromic bichromate?). It is easily produced by rapidly heating CrO_3, but the product is not always the same. Their magnetic characteristics have never been investigated, and they are the only indication which would point to chromium having potentially magnetic qualities.

The metal next to chromium in the periodic system of elements, vanadium, is non-magnetic, as far as known.

50. On attached tables are given the magnetic constants of the better known magnetic materials, metals, alloys, mixtures and compounds:

The first tables give the saturation density, S, and the demagnetization temperature, that is, temperature at which the material ceases to be ferromagnetic, and its specific gravity.

It is interesting to note that with some magnetic materials the demagnetization temperature is very close to, or within the range of, atmospheric temperature.

The second table gives more complete data of those materials, of which such data are available. It gives:

S = saturation density, or value of $B - H$ for infinitely high H;

α = coefficient of magnetic hardness;

σ = coefficient of magnetic saturation.

Where the reluctivity line shows a bend at some critical point, α and σ are given for the lower range—which is the one industrially most useful—together with the range of field intensity, for which this value applies, and are given also for the highest range observed, together with the value of field intensity H, above which the latter values of α and σ apply.

η = coefficient of hysteresis, in the 1.6^{th} power law.

β = coefficient of unsymmetrical cycle, for the two cases where this is known.

Demagnetization temperature, that is, temperature at which ferromagnetism ceases.

ρ = electrical resistivity of the material—which refers to the eddy-current losses in magnetic cycles.

Sp. gr. = specific gravity of the material.

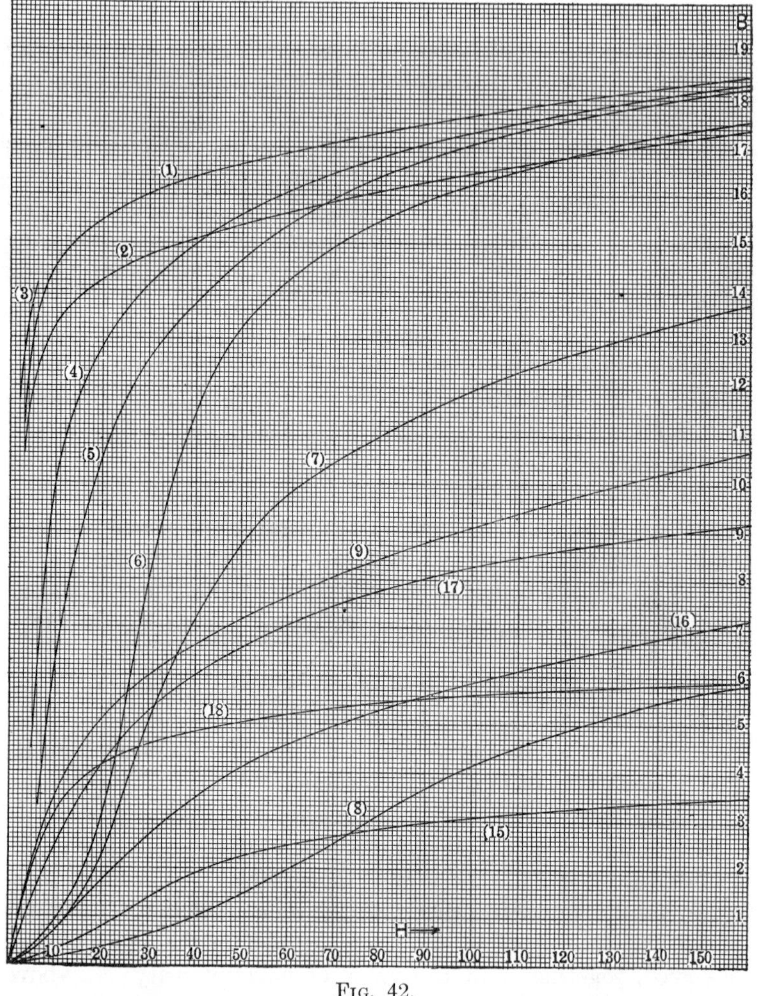

FIG. 42.

Fig. 42 gives the magnetic characteristics, up to $H = 160$ (beyond this, the linear law of reluctivity usually applies), for a number of magnetic materials of higher values of saturation densities.

Fig. 43 gives, with twice the scale of ordinates, but the same abscissæ, the magnetic characteristic of some materials of low saturation density.

Fig. 44 gives, with ten times the scale of abscissæ, and the same scale of ordinates, the initial part of the magnetic characteristic,

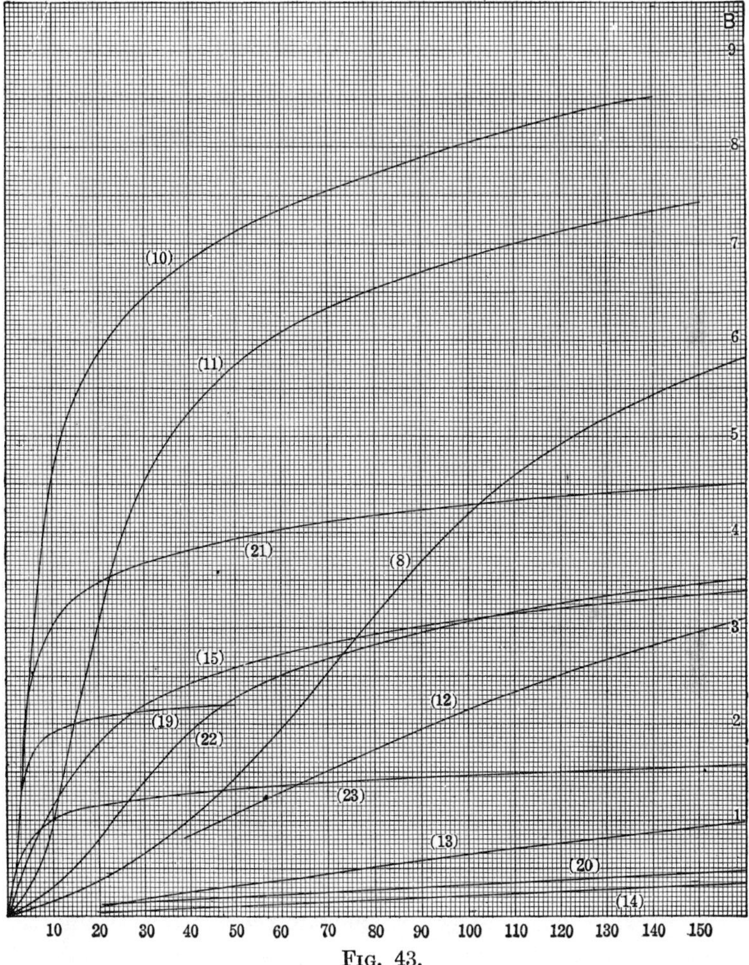

FIG. 43.

up to $H = 16$, for the magnetically soft materials of Fig. 42, that is, materials with low value of α, which rise so rapidly to high values of density that the initial part of their characteristic is not well shown in the scale of Fig. 42.

The magnetic characteristics in Figs. 42, 43 and 44 are denoted by numbers, and these numbers refer to the materials given in the table of "Magnetic Constants" under the same numbers.

With regards to the magnetic data, it must be realized, however, that the numerical values, especially of the less-investigated materials, are to some extent uncertain, due to the great difficulty of exact magnetic measurements.

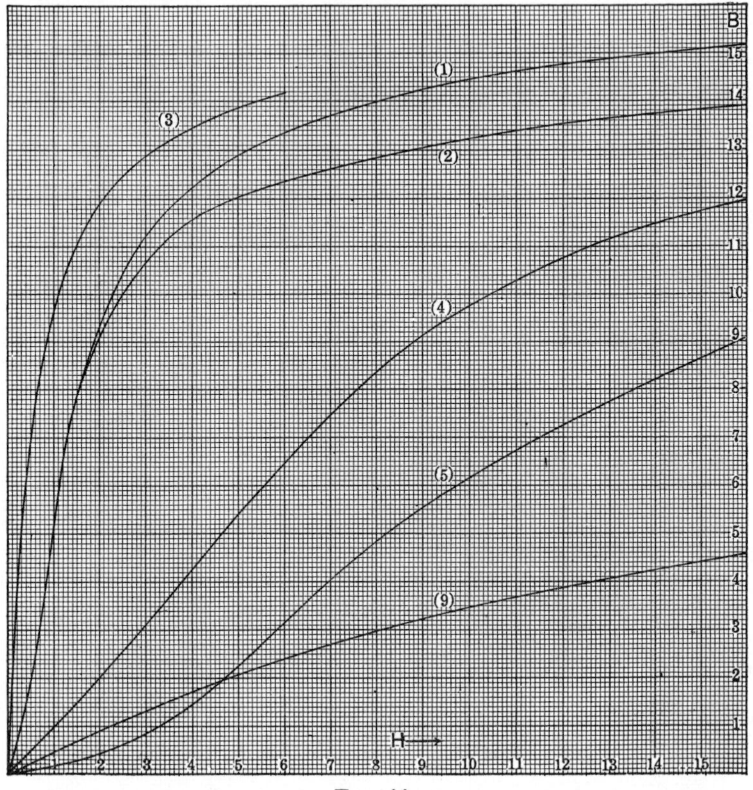

FIG. 44.

The saturation density, S, which is the most constant and most definite and permanent magnetic quantity, can be measured either directly, by measuring B in such very high fields, $-H=$ 10,000 and over—that $B-H$ does not further increase, or indirectly, by observing the B, H curve up to moderately high fields, therefrom derive the reluctivity curve: ρ, H, and from the straight-line law of the latter curve determine σ and therewith B.

TABLE I.—SATURATION DENSITY

$S = (B - H)_{H=\infty}$

Material	Authority	Saturation density, S	Demagnetization temperature, °C.	Specific gravity
		$\times 10^{+3}$		
Iron:				
Most probable value	1916	21.0–21.5	760	7.70
Best standard sheet steel, annealed	1915–16	20.70		
Average standard sheet steel, annealed	1915–16	20.20		
Pure iron	Wedekind	21.10	765	
Swedish wrought iron	Ewing	21.25		
Iron, 99.88 per cent	Hatfield	21.15		
Iron	Gumlich	21.60?		
Electrolytic iron	Gumlich	21.70?		
Commercial steel	Williams	22.00?		
Vacuum-melted electrolytic iron	Williams	22.60?		
Pure iron	DuBois	23.20?	756	7.86
Average sheet iron	1892	20.10		
Average medium silicon steel, annealed, 2.5 per cent	1915–16	19.25		
Average soft steel castings	1915–16	20.20		
Special tungsten steel	....	20.30		
Magnet steel	....	18.50		
Average cast iron	1915–16	15.00		
Fe_2Co, cobalt iron	Williams	22.50?	520	
Fe_2Co, cobalt iron, vacuum-melted and forged	Williams	25.80?		
Fe_2Co, cobalt-iron, probable value, about	....	23.50		
Scale of silicon steel	....	9–10		
Iron amalgam, 11 per cent	1892	0.90		
Magnetite, Fe_3O_4	1892	4.70	....	5.10
Magnetite, Fe_3O_4	DuBois	5.46–5.91	536–589	
Magnetkies, Fe_7S_8 or Fe_8S_9	DuBois	0.88	....	4.60
$CuFe_2O_4$, copper ferrite	....	....	280	
$CoFe_2O_4$, cobalt ferrite	....	....	280–290	
Cobalt:				
Probable value	1916	12–14		
Cast cobalt	....	11.10		
Cobalt	....	12.10		
Cobalt	Wedekind	13.30		
Cobalt, 1.66 per cent. Fe	Ewing	16.45?		
Cobalt	DuBois	17.20?	1075	8.70
Cobalt, pure	Stifler	17.85?		
Cobalt, vacuum-melted	Williams	18.85?		
Nickel:				
Probable value	1916	6–7		
Nickel, 99 per cent. pure	....	6.15		
Nickel wire, soft	1892	5.88		
Nickel, cast	....	6.52		
Nickel	DuBois	7.27?	340–376	8.93
Nickel	....	8.17?		
Monel metal	....	2.22		
Binel metal	....	0.26		
Manganese, Heusler alloys:				
$AlMnCu_2$, soft, high permeability	....	4.67		
$AlMnCu_2$, hard, high permeability	....	3.92		
$AlMnCu_2$, soft, low permeability	....	1.56		
$AlMnCu_2$, highest values	....	7.00	310	
Manganese-antimony, MnSb	....	7.00	310–330	
Manganese-antimony, Mn_2Sb	....	3.85		
Manganese-boron, MnB	....	3.10		
Manganese-phosphorus, MnP	....	0.70	18–26	
Manganese-bismuth, MnB	....	....	360–380	
Manganese-arsenic, MnAs	....	....	40–50	
Manganese-tin, Mn_4Sn				
Chromium:				
Cr_4O_9, chromic bichromate				
Cr_5O_9, chromic chromate				

TABLE II.—MAGNETIC CONSTANTS OF MATERIALS

Reference number	Material	Saturation density, S	High range: Hardness, α	High range: Saturation, σ	High range: Range, H	Low range: Hardness, α	Low range: Saturation, σ	Low range: Range, H	Hysteresis: Coefficient, η	Hysteresis: Asymmetry coefficient, β	Demagnetization temperature, °C.	Electrical resistivity, ρ	Specific gravity, Sp. gr.
		$\times 10^3$	$\times 10^{-3}$	$\times 10^{-3}$		$\times 10^{-3}$	$\times 10^{-3}$		$\times 10^{-3}$	$\times 10^{-10}$		$\times 10^{-6}$	
	Iron: S = 21.0 to 21.5												
(1)	Standard sheet steel, annealed, average 1915–16	20.2	0.76	0.0494	>100	0.10	0.0588	2.5–50	1.2	0.344	......	12	7.7
	Standard sheet steel, annealed, best	20.7	0.74	0.0483	>100	0.06	0.0583	2.5–50					
	Soft iron wire (Ewing)	......	......	......	......	0.2	0.0635	......	2.0				
	Pure iron (Wedekind)	21.1	1.5	0.0474	150–845	......	......	......	......	......	765	......	7.86
	Average sheet iron, 1892	20.1	......	......	......	0.3	0.05	......	3.5				
(2)	Medium silicon sheet steel, annealed, 2.5 per cent. Si, average 1915–16	19.25	0.93	0.0519	>100	0.10	0.0639	2.5–50	0.8	0.32	......	40	7.5
(3)	Special alloy steel, best values	......	......	......	......	0.04	0.064	7–6	0.3	......	......	60	
(4)	Soft steel castings, average 1915–16	20.2	0.85	0.0495	>100	0.465	0.0547	10–50					
	Soft steel castings, 1892	......	......	......	......	0.44–0.9	0.051–0.059	......	9–12				
(5)	Steel shafting	20.6	1.0	0.0486	>100	0.876	0.0509	20–50					
(6)	Tungsten steel, special	20.3	1.2	0.0493	>60								
(7)	Magnet steel	18.5	2.92	0.0541	>60	......	......	......	48.0				
(8)	Chilled tool steel, 1892	......	......	......	......	......	......	......	75.0				
(9)	Cast iron, average 1915–16	15.0	4.2	0.0667	>80	1.93	0.102	20–50	8–20	......	......	80–85	6.7
	Cast iron, average 1892	......	......	......	......	2.0–2.9	0.094–0.098	......	12–16				
(10)	Scale of silicon steel, vacuum-annealed, 5.5–5.7 per cent. Si; 78.5 per cent. Fe	9.7	2.15	0.103	>90	1.01	0.121	40–90	7.0	......	......	30–80	6.0
(11)	Scale of silicon steel, furnace-annealed, 5.3–5.5 per cent. Si; 76 per cent. Fe	8.9	3.30	0.112	>90	2.89	0.117	40–90	21.6	......	......	100–160	5.64
(12)	Powdered iron in resin 10 ÷ 1 by weight	10.1	39.0	0.0993						......	......		
(13)	Powdered iron in resin, 3 ÷ 1 by weight, 1 ÷ 1 by volume	8.4	173.0	0.119	......	......	......	......	24.0	......	......	80×10^6	3.2–3.6
	Iron filings in air, 30 per cent. volume, 1892	3.5	70.0	0.290	......	......	......	......	56.0				
(14)	Iron amalgam, 11 per cent. (Hg_5Fe_2), 1892	0.9	500.0	1.12	......	......	......	......	231.0				
(15)	Magnetite, Fe_2O_3, 1892	4.7	8.9	0.213	......	......	......	......	23.5				
	Magnetite, (DuBois)	5.46	......	......	......	......	......	......	......	......	536–589	......	5.1–5.2
	Magnetite, (DuBois)	5.91	......	......	......	......	......	......	......	......		......	

TABLE II.—MAGNETIC CONSTANTS OF MATERIALS—*Continued*

Reference number	Material	Saturation density, S	High range			Low range			Hysteresis		Demagnetization temperature, °C.	Electrical resistivity, ρ	Specific gravity, Sp. gr.
			Hardness, α	Saturation, σ	Range, H	Hardness, α	Saturation, σ	Range, H	Coefficient, η	Asymmetry coefficient, β			
		$\times 10^3$	$\times 10^{-3}$	$\times 10^{-8}$		$\times 10^{-8}$	$\times 10^{-3}$		$\times 10^{-3}$	$\times 10^{-10}$		10^{-6}	
	Cobalt: Probably S = 12 to 14												
(16)	Cast cobalt	11.1	8.7	0.09	>40	......	......	......	10–20	......	1075	......	8.7
	Cast cobalt	13.6	3.0	0.073	>30								
(17)	Soft cobalt	11.1	5.0	0.09									
	Cobalt (Wedekind)	13.3	35.0	0.075	370–785								
	Nickel: Probably S = 6 to 7												
(18)	Nickel, 99 per cent	6.15	0.61	0.163	>5	......	......	......	20–25	...... }	340–376	......	8.93
	Soft nickel wire, 1892	5.88	1.0	0.17	......	......	......	......	12.5	...... }			
	Hard nickel wire, 1892	......	......	......	......	......	......	......	38.5				
	Cast nickel	6.52	3.5	0.153									
(19)	Monel metal	2.22	1.51	0.45	>70	0.855	0.455	5–70	44.0	......	......	44–53	
(20)	Binel metal, Fe, 70; Ni, 23.8; Cr, 3.68; Mn, 2.47; P, 0.06; Si, 0.04; S, 0.019	0.26	0.564	3.84	>140	0.35	0.522	60–140	......	......	......	88	
	Manganese: (In alloys, S = 28 to 30)												
(21)	Heusler alloy, $Al(Mn\ or\ Cu)_3$, soft, high permeability	4.67	2.39	0.214	>40	0.9	0.242	5–40	3–30.0	......	615 (310)		
(22)	Heusler alloy, $Al(Mn\ or\ Cu)_3$, hard, high permeability	3.92	6.4	0.256	......	......	......	......					
(23)	Heusler alloy, $Al(Mn\ or\ Cu)_3$, soft low permeability	1.56	8.0	0.643	>40	2.4	0.765	5–40					
	Heusler alloy, $Al(Mn\ or\ Cu)_3$, best values	7.0											
	Manganese-antimony, MnSb	7.0	73.0	0.143	180–530	......	......	......	......	......	310–330		
	Manganese-antimony, Mn_2Sb	3.85	208.0	0.26	190–790								
	Manganese-boron, MnB	3.1	182.0	0.32	85–775								
	Manganese-phosphorus, MnP	0.7	420.0	1.44	335–775	......	......	......	......	......	18–26		

Such extremely high fields, as to reach complete magnetic saturation, are produced only between the conical pole faces of a very powerful large electromagnet. The area of the field then is very small, and it is difficult to get perfect uniformity of the field. The tendency is to underestimate the field, and this gives too high values of S. Thus, in the following table those values of S, which appear questionable for this case, have been marked by the interrogation sign.

The indirect method, from the straight-line reluctivity curve, gives more accurate values of S, as S is derived from a complete curve branch, and this method thus is preferable. However, the value derived in this manner is based on the assumption that there is no further critical point in the reluctivity curve beyond the observed range. This is correct with iron, as the best tests by the direct method check. With cobalt, there may be a critical point in the reluctivity curve beyond the observed range, as there are several observations by the direct method, which give very much higher, though erratic, values of saturation, S.

The value of the magnetic hardness, a, also is difficult to determine for very soft materials, especially where the method of observation requires correction for joints, etc., and the extremely high values of permeability—over 15,000—therefore appear questionable.

CHAPTER VI

MAGNETISM

MECHANICAL FORCES

1. General

51. Mechanical forces appear wherever magnetic fields act on electric currents. The work done by all electric motors is the result of these forces. In electric generators, they oppose the driving power and thereby consume the power which finds its equivalent in the electric power output. The motions produced by the electromagnet are due to these forces. Between the primary and the secondary coils of the transformer, between conductor and return conductor of an electric circuit, etc., such mechanical forces appear.

The electromagnet, and all electrodynamic machinery, are based on the use of these mechanical forces between electric conductors and magnetic fields. So also is that type of transformer which transforms constant alternating voltage into constant alternating current. In most other cases, however, these mechanical forces are not used, and therefore are often neglected in the design of the apparatus, under the assumption that the construction used to withstand the ordinary mechanical strains to which the apparatus may be exposed, is sufficiently strong to withstand the magnetic mechanical forces. In the large apparatus, operating in the modern, huge, electric generating systems, these mechanical forces due to magnetic fields may, however, especially under abnormal, though not infrequently occurring, conditions of operation (as short-circuits), assume such formidable values, so far beyond the normal mechanical strains, as to require consideration. Thus generators and large transformers on big generating systems have been torn to pieces by the magnetic mechanical forces of short-circuits, cables have been torn from their supports, disconnecting switches blown open, etc.

In the following, a general study of these forces will be given. This also gives a more rational and thereby more accurate design of the electromagnet, and permits the determination of what may be called the efficiency of an electromagnet.

Investigations and calculations dealing with one form of energy only, as electromagnetic energy, or mechanical energy, usually are relatively simple and can be carried out with very high accuracy. Difficulties, however, arise when the calculation involves the relation between several different forms of energy, as electric energy and mechanical energy. While the elementary relations between different forms of energy are relatively simple, the calculation involving a transformation from one form of energy to another, usually becomes so complex, that it either can not be carried out at all, or even only approximate calculation becomes rather laborious and at the same time gives only a low degree of accuracy. In most calculations involving the transformation between different forms of energy, it is therefore preferable not to consider the relations between the different forms of energy at all, but to use the *law of conservation of energy* to relate the different forms of energy, which are involved.

Thus, when mechanical motions are produced by the action of a magnetic field on an electric circuit, energy is consumed in the electric circuit, by an induced e.m.f. At the same time, the stored magnetic energy of the system may change. By the law of conservation of energy, we have:

Electric energy consumed by the induced e.m.f. = *mechanical energy produced,* + *increase of the stored magnetic energy.* (1)
The consumed electric energy, and the stored magnetic energy, are easily calculated, as their calculation involves one form of energy only, and this calculation then gives the mechanical work done, $= Fl$, where F = mechanical force, and l = distance over which this force moves.

Where mechanical work is not required, but merely the mechanical forces, which exist, as where the system is supported against motions by the mechanical forces—as primary and secondary coils of a transformer, or cable and return cable of a circuit—the same method of calculation can be employed, by assuming some distance l of the motion (or dl); calculating the mechanical energy $w_0 = Fl$ by (1), and therefrom the mechanical force as $F = \frac{w_0}{l}$, or $F = \frac{dw_0}{dl}$.

Since the induced e.m.f., which consumes (or produces) the electric energy, and also the stored magnetic energy, depend on the current and the inductance of the electric circuit, and in alternating-current circuits the impressed voltage also depends on the inductance of the circuit, the inductance can frequently be expressed by supply voltage and current; and by substituting this in equation (1), the mechanical work of the magnetic forces can thus be expressed, in alternating-current apparatus, by supply voltage and current.

In this manner, it becomes possible, for instance, to express the mechanical work and thereby the pull of an alternating electromagnet, by simple expressions of voltage and current, or to give the mechanical strains occurring in a transformer under short-circuits, by an expression containing only the terminal voltage, the short-circuit current, and the distance between primary and secondary coils, without entering into the details of the construction of the apparatus.

This general method, based on the law of conservation of energy, will be illustrated by some examples, and the general equations then given.

2. The Constant-current Electromagnet

52. Such magnets are most direct-current electromagnets, and also the series operating magnets of constant-current arc lamps on alternating-current circuits.

Let i_0 = current, which is constant during the motion of the armature of the electromagnet, from its initial position 1, to its final position 2, l = the length of this motion, or the stroke of the electromagnet, in centimeters, and n = number of turns of the magnet winding.

The magnetic flux Φ, and the inductance

$$L = \frac{n\Phi}{i_0} 10^{-8} \tag{2}$$

of the magnet, vary during the motion of its armature, from a minimum value,

$$\Phi_1 = \frac{i_0 L_1}{n} 10^8 \tag{3}$$

in the initial position, to a maximum value,

$$\Phi_2 = \frac{i_0 L_2}{n} 10^8 \tag{4}$$

in the end position of the armature.

Hereby an e.m.f. is induced in the magnet winding,

$$e' = n \frac{d\Phi}{dt} 10^{-8} = i_0 \frac{dL}{dt} \tag{5}$$

This consumes the power

$$p = i_0 e' = i_0^2 \frac{dL}{dt} \tag{6}$$

and thereby the energy

$$w = \int_2^1 pdt = ni_0^2(L_2 - L_1) \tag{7}$$

Assuming that the inductance, in any fixed position of the armature, does not vary with the current, that is, that magnetic saturation is absent,[1] the stored magnetic energy is:

In the initial position, 1,

$$w_1 = \frac{i_0^2 L_1}{2} \tag{8}$$

in the end position, 2,

$$w_2 = \frac{i_0^2 L_2}{2} \tag{9}$$

The increase of the stored magnetic energy, during the motion of the armature, thus is

$$w' = w_2 - w_1 = \frac{i_0^2}{2}(L_2 - L_1) \tag{10}$$

The mechanical work done by the electromagnet thus is, by the law of conservation of energy,

$$\begin{aligned} w_0 &= w - w' \\ &= \frac{i_0^2}{2}(L_2 - L_1) \text{ joules.} \end{aligned} \tag{11}$$

If l = length of stroke, in centimeters, F = average force, or pull of the magnet, in gram weight, the mechanical work is

$$Fl \text{ gram-cm.}$$

Since

$$g = 981 \text{ cm.-sec.} \tag{12}$$

= acceleration of gravity, the mechanical work is, in absolute units,

$$Flg$$

[1] If magnetic saturation is reached, the stored magnetic energy is taken from the magnetization curve, as the area between this curve and the vertical axis, as discussed before.

and since 1 joule = 10^7 absolute units, the mechanical work is

$$w_0 = Flg\ 10^{+7} \text{ joules.} \tag{13}$$

From (11) and (12) then follows,

$$Fl = \frac{i_0^2}{2g}(L_2 - L_1)10^{+7} \text{ gram-cm.} \tag{14}$$

as the *mechanical work of the electromagnet,* and

$$F = \frac{i_0^2}{2g}\frac{L_2 - L_1}{l}10^7 \text{ grams} \tag{15}$$

as the average force, or pull of the electromagnet, during its stroke l.

Or, if we consider only a motion element dl,

$$F = \frac{i_0^2}{2g}\frac{dL}{dl}10^{+7} \text{ grams} \tag{16}$$

as the force, or pull of the electromagnet in any position l.

Reducing from gram-centimeters to foot-pounds, that is, giving the stroke l in feet, the pull F in pounds, we divide by

$$454 \times 30.5 = 13{,}850$$

which gives, after substituting for g from (12)

$$(14)\colon\ Fl = 3.68\, i_0^2(L_2 - L_1) \text{ ft.-lb} \tag{17}$$

$$(15)\colon\ F = 3.68\, i_0^2\frac{L_2 - L_1}{l} \text{ lb.} \tag{18}$$

$$(16)\colon\ F = 3.68\, i_0^2\frac{dL}{dl} \text{ lb.} \tag{19}$$

These equations apply to the direct-current electromagnet as well as to the alternating-current electromagnet.

In the alternating-current electromagnet, if i_0 is the effective value of the current, F is the effective or average value of the pull, and the pull or force of the electromagnet pulsates with double frequency between 0 and $2F$.

53. In the alternating-current electromagnet usually the voltage consumed by the resistance of the winding, i_0r, can be neglected compared with the voltage consumed by the reactance of the winding, i_0x, and the latter, therefore, is practically equal to the terminal voltage, e, of the electromagnet. We have then, by the general equation of self-induction,

$$e = 2\pi\, fLi_0 \tag{20}$$

where f = frequency, in cycles per second.

From which follows,

$$i_0 L = \frac{e}{2\pi f} \tag{21}$$

and substituting (21) in equations (14) to (19), gives as the equation of the *mechanical work, and the pull of the alternating-current electromagnet.*

In the metric system:

$$Fl = \frac{i_0(e_2 - e_1)10^7}{4\pi f g} \text{ gram-cm.} \tag{22}$$

$$F = \frac{i_0(e_2 - e_1)\,10^7}{4\pi f g l} = \frac{i_0}{4\pi f g}\frac{de}{dl}10^7 \text{ grams} \tag{23}$$

In foot-pounds:

$$Fl = \frac{0.586\, i_0(e_2 - e_1)}{f} \text{ ft.-lb.} \tag{24}$$

$$F = \frac{0.586\, i_0(e_2 - e_1)}{fl} = \frac{0.586 i_0}{f}\frac{de}{dl} \text{ lb.} \tag{25}$$

Example.—In a 60-cycle alternating-current lamp magnet, the stroke is 3 cm., the voltage, consumed at the constant alternating current of 3 amp. is 8 volts in the initial position, 17 volts in the end position. What is the average pull of the magnet?

$$l = 3 \text{ cm.}$$
$$e_1 = 8$$
$$e_2 = 17$$
$$f = 60$$
$$i_0 = 3$$

hence, by (23),

$$F = 122 \text{ grams } (= 0.27 \text{ lb.})$$

The work done by an electromagnet, and thus its pull, depend, by equation (22), on the current i_0 and the difference in voltage between the initial and the end position of the armature, $e_2 - e_1$; that is, depend upon the difference in the volt-amperes consumed by the electromagnet at the beginning and at the end of the stroke. With a given maximum volt-amperes, $i_0 e_2$, available for the electromagnet, the maximum work would thus be done, that is, the greatest pull produced, if the volt-amperes at the beginning of the stroke were zero, that is, $e_1 = 0$, and the theoretical maximum output of the magnet thus would be

$$F_m l = \frac{i_0 e_2 10^7}{4\pi f g} \tag{26}$$

and the ratio of the actual output, to the theoretically maximum output, or the efficiency of the electromagnet, thus is, by (22) and (26),

$$\eta = \frac{F}{F_m} = \frac{e_2 - e_1}{e_2} \tag{27}$$

or, using the more general equation (14), which also applies to the direct-current electromagnet,

$$\eta = \frac{L_2 - L_1}{L_2} \tag{28}$$

The efficiency of the electromagnet, therefore, is the difference between maximum and minimum voltage, divided by the maximum voltage; or the difference between maximum and minimum volt-ampere consumption, divided by the maximum volt-ampere consumption; or the difference between maximum and minimum inductance, divided by the maximum inductance.

As seen, this expression of efficiency is of the same form as that of the thermodynamic engine,

$$\frac{T_2 - T_1}{T_2}$$

From (26) it also follows, that the maximum work which can be derived from a given expenditure of volt-amperes, i_0e_2, is limited. For $i_0e_2 = 1$, that is, for 1 volt-amp. the maximum work, which could be derived from an alternating electromagnet, is, from (26),

$$F_m l = \frac{10^7}{4\,\pi f g} = \frac{810}{f} \text{ gram-cm.} \tag{29}$$

That is, a 60-cycle electromagnet can never give more than 13.5 gram-cm., and a 25-cycle electromagnet never more than 32.4 gram-cm. pull per volt-ampere supplied to its terminals.

Or inversely, for an average pull of 1 gram over a distance of 1 cm., a minimum of $\frac{1}{13.5}$ volt-amp. is required at 60 cycles, and a minimum of $\frac{1}{32.4}$ volt-amp. at 25 cycles.

Or, reduced to pounds and inches:

For an average pull of 1 lb. over a distance of 1 in., at least 86 volt-amp. are required at 60 cycles, and at least 36 volt-amp. at 25 cycles.

This gives a criterion by which to judge the success of the design of electromagnets.

3. The Constant-potential Alternating Electromagnet

54. If a constant alternating potential, e_0, is impressed upon an electromagnet, and the voltage consumed by the resistance, ir, can be neglected, the voltage consumed by the reactance, x, is constant and is the terminal voltage, e_0, thus the magnetic flux, Φ, also is constant during the motion of the armature of the electromagnet. The current, i, however, varies, and decreases from a maximum, i_1, in the initial position, to a minimum, i_2, in the end position of the armature, while the inductance increases from L_1 to L_2.

The voltage induced in the electric circuit by the motion of the armature,

$$e' = n\frac{d\Phi}{dt} 10^8 \tag{30}$$

then is zero, and therefore also the electrical energy expended,

$$w = 0.$$

That is, the electric circuit does no work, but the mechanical work of moving the armature is done by the stored magnetic energy.

The increase of the stored magnetic energy is

$$w' = \frac{i_2{}^2L_2 - i_1{}^2L_1}{2} \tag{31}$$

and since the mechanical energy, in joules, is by (13),

$$w_0 = Flg\ 10^7$$

the equation of the law of conservation of energy,

$$w = w' + w_0 \tag{32}$$

then becomes

$$0 = \frac{i_2{}^2L_2 - i_1{}^2L_1}{2} + Flg\ 10^{-7},$$

or

$$Fl = \frac{i_1{}^2L_1 - i_2{}^2L_2}{2g} 10^7 \text{ gram-cm.} \tag{33}$$

Since, from the equation of self-induction, in the initial position,

$$e_0 = 2\ \pi f L_1 i_1 \tag{34}$$

in the end position

$$e_0 = 2\ \pi f L_2 i_2 \tag{35}$$

substituting (34) and (35) in (33), gives the equation of the constant-potential alternating electromagnet.

$$Fl = \frac{e_0(i_1 - i_2)}{4\,\pi f g} 10^7 \text{ gram-cm.} \tag{36}$$

and

$$F = \frac{e_0(i_1 - i_2)}{4\,\pi f g l} 10^7 = \frac{e_0}{4\,\pi f g} \frac{di}{dl} 10^7 \text{ grams} \tag{37}$$

or, in foot-pounds,

$$Fl = \frac{0.586\, e_0(i_1 - i_2)}{f} \text{ ft.-lb.} \tag{38}$$

$$F = \frac{0.586\, e_0(i_1 - i_2)}{fl} = \frac{0.586\, e_0}{f} \frac{di}{dl} \text{ lb.} \tag{39}$$

Substituting $Q = ei =$ volt-amperes, in equations (36) to (39) of the constant-potential alternating electromagnet, and equations (22) to (25) of the constant-current alternating magnet, gives the same expression of mechanical work and pull:

In metric system:

$$Fl = \frac{\Delta Q}{4\,\pi f g} 10^7 \text{ gram-cm.} \tag{40}$$

$$F = \frac{\Delta Q}{4\,\pi f g l} 10^7 = \frac{1}{4\,\pi f g} \frac{dQ}{dl} 10^7 \text{ grams} \tag{41}$$

In foot-pounds:

$$Fl = \frac{0.586\, \Delta Q}{f} \text{ ft.-lb.} \tag{42}$$

$$F = \frac{0.586\, \Delta Q}{fl} = \frac{0.586}{f} \frac{dQ}{dl} \text{ lb.} \tag{43}$$

where ΔQ = difference in volt-amperes consumed by the magnet in the initial position, and in the end position of the armature.

Both types of alternating-current magnet, then, give the same expression of efficiency,

$$\eta = \frac{\Delta Q}{Q_m} \tag{44}$$

where Q_m is the maximum volt-amperes consumed, corresponding to the end position in the constant-current magnet, to the initial position in the constant-potential magnet.

4. Short-circuit Stresses in Alternating-current Transformers

55. At short-circuit, no magnetic flux passes through the secondary coils of the transformer, if we neglect the small voltage consumed by the ohmic resistance of the secondary coils. If the supply system is sufficiently large to maintain constant voltage at the primary terminals of the transformer even at short-circuit, full magnetic flux passes through the primary coils.[1] In this case the total magnetic flux passes between primary coils and secondary coils, as self-inductive or leakage flux. If then x = self-inductive or leakage reactance, e_0 = impressed e.m.f., $i_0 = \frac{e_0}{x}$ is the short-circuit current of the transformer. Or, if as usual the reactance is given in per cent., that is, the ix (where i = full-load current of the transformer) given in per cent. of e, the short-circuit current is equal to the full-load current divided by the percentage reactance. Thus a transformer with 4 per cent. reactance would give a short-circuit current, at maintained supply voltage, of 25 times full-load current.

To calculate the force, F, exerted by this magnetic leakage flux on the transformer coils (which is repulsion, since primary and secondary currents flow in opposite direction) we may assume, at constant short-circuit current, i_0, the secondary coils moved against this force, F, and until their magnetic centers coincide with those of the primary coils; that is, by the distance, l, as shown diagrammatically in Fig. 45, the section of a shell-type transformer. When brought to coincidence, no magnetic flux passes between primary and secondary coils, and during this motion, of length, l, the primary coils thus have cut the total magnetic flux, Φ, of the transformer.

Hereby in the primary coils a voltage has been induced,

$$e' = n \frac{d\Phi}{dt} 10^{-8}$$

where n = effective number of primary turns.

The work done or rather absorbed by this voltage, e', at current, i_0, is

$$w = \int e' i_0 dt = n i_0 \Phi \, 10^{-8} \text{ joules.} \tag{45}$$

[1] If the terminal voltage drops at short-circuit on the transformer secondaries, the magnetic flux through the transformer primaries drops in the same proportion, and the mechanical forces in the transformer drop with the square of the primary terminal voltage, and with a great drop of the terminal voltage, as occurs for instance with large transformers at the end of a transmission line or long feeders, the mechanical forces may drop to a small fraction of the value, which they have on a system of practically unlimited power.

If L = leakage inductance of the transformer, at short-circuit, where the entire flux, Φ, is leakage flux, we have

$$\Phi = \frac{Li_0}{n} 10^8 \tag{46}$$

hence, substituted in (45)

$$w = i_0^2 L \tag{47}$$

The stored magnetic energy at short-circuit is

$$w_1 = \frac{i_0^2 L}{2} \tag{48}$$

and since at the end of the assumed motion through distance, l, the leakage flux has vanished by coincidence between primary and secondary coils, its stored magnetic energy also has vanished, and the change of stored magnetic energy therefore is

$$w' = w_1 = \frac{i_0^2 L}{2} \tag{49}$$

Hence, the mechanical work of the magnetic forces of the short-circuit current is

$$w_0 = w - w' = \frac{i_0^2 L}{2} \tag{50}$$

It is, however, if F is the force, in grams, l, the distance between the magnetic centers of primary and secondary coils,

$$w_1 = Flg\ 10^{-7} \text{ joules.}$$

Hence,

$$Fl = \frac{i_0^2 L}{2\,g} 10^7 \text{ gram-cm.} \tag{51}$$

and

$$F = \frac{i_0^2 L}{2\,gl} 10^7 \text{ grams} \tag{52}$$

the mechanical force existing between primary and secondary coils of a transformer at the short-circuit current, i_0.

Since at short-circuit, the total supply voltage, e_0, is consumed by the leakage inductance of the transformer, we have

$$e_0 = 2\,\pi f L i_0 \tag{53}$$

hence, substituting (53) in (52), gives

$$F = \frac{e_0 i_0\, 10^7}{4\,\pi f g l} \text{ grams}$$

$$= \frac{810\, e_0 i_0}{fl} \text{ grams} \tag{54}$$

Example.—Let, in a 25-cycle 1667-kw. transformer, the supply voltage, $e_0 = 5200$, the reactance $= 4$ per cent. The transformer contains two primary coils between three secondary coils, and the distance between the magnetic centers of the adjacent coils or half coils is 12 cm., as shown diagrammatically in Fig. 45. What force is exerted on each coil face during short-circuit, in a system which is so large as to maintain constant terminal voltage?

At 5200 volts and 1667 kw., the full-load current is 320 amp. At 4 per cent. reactance the short-circuit current therefore, $i_0 = \frac{320}{0.04} = 8000$ amp. Equation (54) then gives, for $f = 25$, $l = 12$,

$$F = 112 \times 10^6 \text{ grams}$$
$$= 112 \text{ tons.}$$

This force is exerted between the four faces of the two primary coils, and the corresponding faces of the secondary coils, and on every coil face thus is exerted the force

$$\frac{F}{4} = 28 \text{ tons}$$

This is the average force, and the force varies with double frequency, between 0 and 56 tons, and is thus a large force.

56. Substituting $i_0 = \frac{e_0}{x}$ in (54), gives as the short-circuit force of an alternating-current transformer, at maintained terminal voltage, e_0, the value

$$F = \frac{e_0^2\, 10^7}{4\, \pi f g l x} = \frac{810\, e_0^2}{f l x} \text{ grams} \tag{55}$$

That is, the short-circuit stresses are inversely proportional to the leakage reactance of the transformer, and to the distance, l, between the coils.

In large transformers on systems of very large power, safety therefore requires the use of as high reactance as possible.

High reactance is produced by massing the coils of each circuit.

Let in a transformer

$$n = \text{number of coil groups}$$

(where one coil is divided into two half coils, one at each end of the coil stack, as one secondary coil in Fig. 45, where $n = 2$) the mechanical force per coil face then is, by (55),

$$F_0 = \frac{F}{2n} = \frac{e_0^2\, 10^7}{8\, \pi f g n l x} = \frac{810\, e_0^2}{2\, f n l x} \text{ grams} \tag{56}$$

Let x = leakage reactance of transformer;
l_0 = distance between coil surfaces;
l_1 = thickness of primary coil;
l_2 = thickness of secondary coil.

Between two adjacent coils, P and S in Fig. 45, the leakage flux density is uniform for the width l_0 between the coil surfaces,

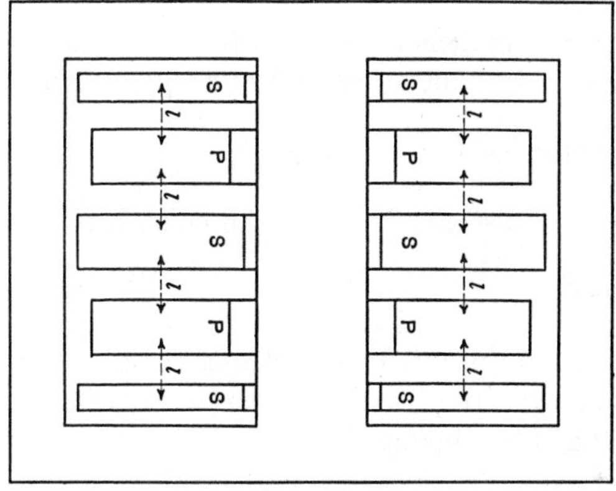

FIG. 45.

and then decreases toward the interior of the coils, over the distance $\frac{l_1}{2}$ respectively $\frac{l_2}{2}$, to zero at the coil centers. All the coil turns are interlinked with the leakage flux in the width, l_0, but toward the interior of the coils, the number of turns interlinked with the leakage flux decreases, to zero at the coil center, and as the leakage flux density also decreases, proportional to the distance from the coil center, to zero in the coil center, the interlinkages between leakage flux and coil turns decrease over the space $\frac{l_1}{2}$ respectively $\frac{l_2}{2}$, proportional to the square of the distance from the coil center, thus giving a total interlinkage distance,

$$\int_0^{\frac{l_1}{2}} u^2 du = \frac{l_1}{6},$$

where u is the distance from the coil center.

Thus the total interlinkages of the leakage flux with the coil turns are the same as that of a uniform leakage flux density over the width $l_0 + \frac{l_1}{6} + \frac{l_2}{6}\cdot$ This gives the effective distance between coil centers, for the reactance calculation,

$$l = l_0 + \frac{l_1 + l_2}{6} \tag{57}$$

Assuming now we regroup the transformer coils, so as to get m primary and m secondary coils, leaving, however, the same iron structure.

The leakage flux density between the coils is hereby changed in proportion to the changed number of ampere-turns per coil, that is, by the factor $\frac{n}{m}\cdot$

The effective distance between the coils, l, is changed by the same factor $\frac{n}{m}\cdot$

The number of interlinkages between leakage flux and electric circuits, and thus the leakage reactance, x, of the transformer, thus is changed by the factor

$$\left(\frac{n}{m}\right)^2.$$

That is, by regrouping the transformer winding within the same magnetic circuit and without changing the number of turns of the electric circuit, the leakage reactance, x, changes inverse proportional to the square of the number of coil groups.

As by equation (56) the mechanical force is inverse proportional to x, l and n, and x changes proportional to $\left(\frac{n}{m}\right)^2$, l proportional to $\frac{n}{m}$, the mechanical force per coil thus changes proportional to

$$\left(\frac{n}{m}\right)^2 \times \frac{n}{m} \times \frac{m}{n} = \left(\frac{n}{m}\right)^2$$

That is, regrouping the transformer winding in the same winding space changes the mechanical force inverse proportional to the square of the coil groups, thus inverse proportional to the change of leakage reactance.

However, the distance l_0 between the coils is determined by insulation and ventilation. Thus its decrease, when increasing the number of coil groups, would usually not be permissible, but more winding space would have to be provided by changing the magnetic circuit, and inversely, with a reduction of the number of coil groups, the winding space, and with it the magnetic circuit, would be reduced.

Assuming, then, that at the change from n to m coil groups, the distance between the coils, l_0, is left the same.

The effective leakage space then changes from

$$l = l_0 + \frac{l_1 + l_2}{6},$$

to

$$l' = l_0 + \frac{n}{m}\frac{l_1 + l_2}{6} = l\,\frac{l_0 + \frac{n}{m}\frac{l_1 + l_2}{6}}{l_0 + \frac{l_1 + l_2}{6}},$$

and the leakage reactance thus changes from

$$x$$

to

$$x' = \frac{n}{m}\frac{l'}{l}x;$$

hence the mechanical force per coil, from

$$F_0 = \frac{F}{2\,n} = \frac{e_0^2\,10^7}{8\,\pi f n g l x},$$

to

$$\begin{aligned} F'_0 = \frac{F'}{2m} &= \frac{e_0^2\,10^7}{8\,\pi f n g l' x'} \\ &= F_0\,\frac{nlx}{ml'x'} \\ &= F_0\left(\frac{l}{l'}\right)^2 \\ &= F_0\left(\frac{l_0 + \frac{n}{m}\frac{l_1 + l_2}{6}}{l_0 + \frac{l_1 + l_2}{6}}\right)^2. \end{aligned} \qquad (58)$$

Thus, if $\frac{l_1 + l_2}{6}$ is large compared with l_0,

$$F'_0 = \left(\frac{n}{m}\right)^2 F_0,$$

that is, the mechanical forces vary with the square of the number of coil groups.

If $\frac{l_1 + l_2}{6}$ is small compared with l_0,

$$F_0{}^1 = F_0$$

that is, the mechanical forces are not changed by the change of the number of coil groups.

In actual design, decreasing the number of coil groups usually materially decreases the mechanical forces, but materially less than proportional to the square of the number of coil groups.

5. Repulsion between Conductor and Return Conductor

57. If i_0 is the current flowing in a circuit consisting of a conductor and the return conductor parallel thereto, and l the distance between the conductors, the two conductors repel each other by the mechanical force exerted by the magnetic field of the circuit, on the current in the conductor.

As this case corresponds to that considered in section 2, equation (16) applies, that is,

$$F = \frac{i_0{}^2}{2\,g}\,\frac{dL}{dl}\,10^7 \text{ grams},$$

The inductance of two parallel conductors, at distance l from each other, and conductor diameter l_d is, per centimeter length of conductor,

$$L = \left(4 \log \frac{2\,l}{l_d} + \mu\right) 10^{-9} \text{ henrys} \tag{59}$$

Hence, differentiated,

$$\frac{dL}{dl} = \frac{4 \times 10^{-9}}{l}$$

and, substituted in (16),

$$F = \frac{i_0{}^2}{50\,gl} \text{ grams} \tag{60}$$

or substituting (12),

$$F = \frac{20.4\, i_0^2\, 10^{-6}}{l} \text{ grams} \tag{61}$$

If $l = 150$ cm. (5 ft.)

$$i_0 = 200 \text{ amp.}$$

this gives

$F = 0.0054$ grams per centimeter length of circuit, hence it is inappreciable.

If, however, the conductors are close together, and the current very large, as the momentary short-circuit current of a large alternator, the forces may become appreciable.

For example, a 2200-volt 4000-kw. quarter-phase alternator feeds through single conductor cables having a distance of 15 cm. (6 in.) from each other. A short-circuit occurs in the cables, and the momentary short-circuit current is 12 times full-load current. What is the repulsion between the cables?

Full-load current is, per phase, 910 amp. Hence, short-circuit current, $i_0 = 12 \times 910 = 10{,}900$ amp. $l = 15$. Hence,

$$F = 160 \text{ grams per centimeter.}$$

Or multiplied by $\frac{30.5}{454}$

$$F = 10.8 \text{ lb. per feet of cable.}$$

That is, pulsating between 0 and 21.6 lb. per foot of cable. Hence sufficient to lift the cable from its supports and throw it aside.

In the same manner, similar problems, as the opening of disconnecting switches under short-circuit, etc., can be investigated.

6. General Equations of Mechanical Forces in Magnetic Fields

58. In general, in an electromagnetic system in which mechanical motions occur, the inductance, L, is a function of the position, l, during the motion. If the system contains magnetic material, in general the inductance, L, also is a function of the current, i, especially if saturation is reached in the magnetic material.

Let, then, L = inductance, as function of the current, i, and position, l;

L_1 = inductance, as function of the current, i, in the initial position 1 of the system;

L_2 = inductance, as function of the current, i, in the end position 2 of the system.

If then Φ = magnetic flux, n = number of turns interlinked with the flux, the induced e.m.f. is

$$e' = n \frac{d\Phi}{dt} 10^{-8} \tag{62}$$

We have, however,

$$n\Phi = iL\, 10^8;$$

hence,

$$e' = \frac{d(iL)}{dt} \tag{63}$$

the power of this induced e.m.f. is

$$p = ie' = i\frac{d(iL)}{dt},$$

and the energy

$$w = \int^2 pdt = \int_1^2 id(iL)$$

$$= \int_1^2 i^2 dL + \int_1^2 iLdi \tag{64}$$

The stored magnetic energy in the initial position 1 is

$$w_1 = \int_0^1 id(iL_1) \tag{65}$$

In the end position 2,

$$w_2 = \int_0^2 id(iL_2) \tag{66}$$

and the mechanical work thus is, by the law of conservation of energy

$$w_0 = w - w_2 + w_1$$

$$= \int_1^2 id(iL) + \int_0^1 id(iL_1) - \int_0^2 id(iL_2) \tag{67}$$

and since the mechanical work is

$$w_0 = Flg\, 10^{-7} \tag{68}$$

We have:

$$Fl = \frac{10^7}{g}\left\{\int_1^2 id(iL) + \int_0^1 id(iL_1) - \int_0^2 id(iL_2)\right\} \text{ gram-cm.} \tag{69}$$

If L is not a function of the current, i, but only of the position, that is, if saturation is absent, L_1 and L_2 are constant, and equation (69) becomes,

$$Fl = \frac{10^7}{g}\left\{\int_1^2 id(iL) + \frac{i_1^2L_1 - i_2^2L_2}{2}\right\} \text{ gram-cm.} \qquad (70)$$

(*a*) If i = constant, equation (70) becomes,

$$Fl = \frac{10^7}{g}\,\frac{i^2(L_2 - L_1)}{2}$$

(Constant-current electromagnet.)

(*b*) If L = constant, equation (70) becomes,

$$Fl = 0.$$

That is, mechanical forces are exerted only where the inductance of the circuit changes with the mechanical motion which would be produced by these forces.

(*c*) If iL = constant, equation (70) becomes,

$$Fl = \frac{10^7}{g}\,\frac{iL(i_1 - i_2)}{2}$$

(Constant-potential electromagnet.)

In the general case, the evaluation of equation (69) can usually be made graphically, from the two curves, which give the variation of L_1 with i in the initial position, of L_2 with i in the final position, and the curve giving the variation of L and i with the motion from the initial to the final position.

In alternating magnetic systems, these three curves can be determined experimentally by measuring the volts as function of the amperes, in the fixed initial and end position, and by measuring volts and amperes, as function of the intermediary positions, that is, by strictly electrical measurement.

As seen, however, the problem is not entirely determined by the two end positions, but the function by which i and L are related to each other in the intermediate positions, must also be given. That is, in the general case, the mechanical work and thus the average mechanical force, are not determined by the end positions of the electromagnetic system. This again shows an analogy to thermodynamic relations.

If then in case of a cyclic change, the variation from position

1 to 2 is different from that from position 2 back to 1, such a cyclic change produces or consumes energy.

$$w = \int_1^2 id(iL) + \int_2^1 id(iL) = \oint_1^1 id(iL)$$

Such a case is the hysteresis cycle. The reaction machine is based on such a cycle.

SECTION II

CHAPTER VII

SHAPING OF WAVES: GENERAL

59. In alternating-current engineering, the sine wave, as shown in Fig. 46, is usually aimed at as the standard. This is not due to any inherent merit of the sine wave.

For all those purposes, where the energy developed by the current in a resistance is the object, as for incandescent lighting, heating, etc., any wave form is equally satisfactory, as the energy of the wave depends only on its effective value, but not on its shape.

With regards to insulation stress, as in high-voltage systems, a flat-top wave of voltage and current, such as shown in Fig. 47, would be preferable, as it has a higher effective value, with the same maximum value and therefore with the same strain on the insulation, and therefore transmits more energy than the sine wave, Fig. 46.

Inversely, a peaked wave of voltage, such as Fig. 48, and such as the common saw-tooth wave of the unitooth alternator, is superior in transformers and similar devices, as it transforms the energy with less hysteresis loss. The peaked voltage wave, Fig. 48, gives a flat-topped wave of magnetism, Fig. 47, and thereby transforms the voltage with a lesser maximum magnetic flux, than a sine wave of the same effective value, that is, the same power. As the hysteresis loss depends on the maximum value of the magnetic flux, the reduction of the maximum value of the magnetic flux, due to a peaked voltage wave, results in a lower hysteresis loss, and thus higher efficiency of transformation. This reduction of loss may amount to as much as 15 to 25 per cent of the total hysteresis loss, in extreme cases.

Inversely, a peaked voltage wave like Fig. 48 would be objectionable in high-voltage transmission apparatus, by giving an unnecessary high insulation strain, and a flat-top wave of voltage like Fig. 47, when impressed upon a transformer, would give a peaked wave of magnetism and thereby an increased hysteresis loss.

The advantage of the sine wave is, that it remains unchanged in shape under most conditions, while this is not the case with any other wave shape, and any other wave shape thus introduces the danger, that under certain conditions, or in certain parts of the circuit, it may change to a shape which is undesirable or even

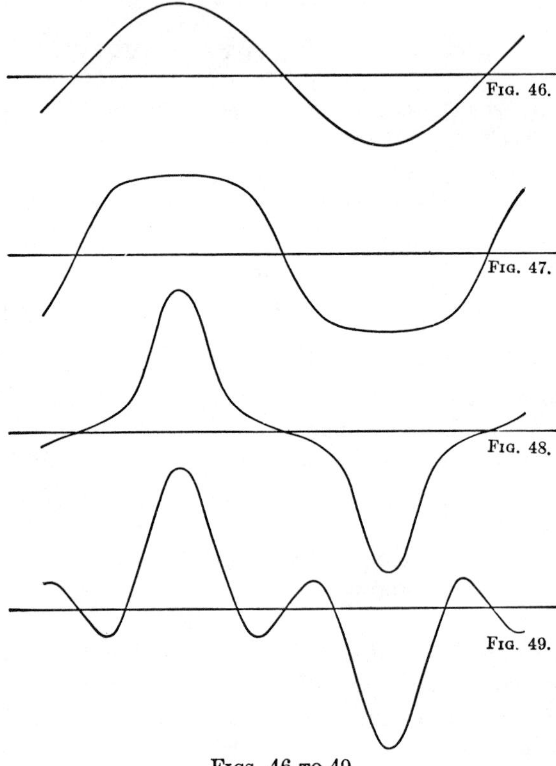

FIGS. 46 TO 49.

dangerous. Voltage, e, and current, i, are related to each other by proportionality, by differentiation and by integration, with resistance, r, inductance, L, and capacity, C, as factors,

$$e = ri,$$

$$e = L\frac{di}{dt},$$

$$e = C\int idt,$$

and as the differentials and integrals of sines are sines, as long as r, L and C are constant—which is mostly the case—sine waves of voltage produce sine waves of current and inversely, that is, the sine wave shape of the electrical quantities remains constant.

A flat-topped current wave like Fig. 47, however, would by differentiation give a self-inductive voltage wave, which is peaked, like Fig. 48. A voltage wave like Fig. 48, which is more efficient in transformation, may by further distortion, as by intensification of the triple harmonic by line capacity, assume the shape,

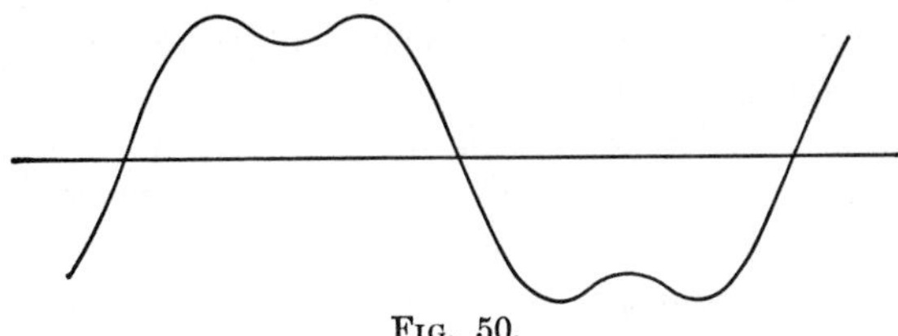

Fig. 50.

Fig. 49, and the latter then would give, when impressed upon a transformer, a double-peaked wave of magnetism, Fig. 50, and such wave of magnetism gives a magnetic cycle with two small

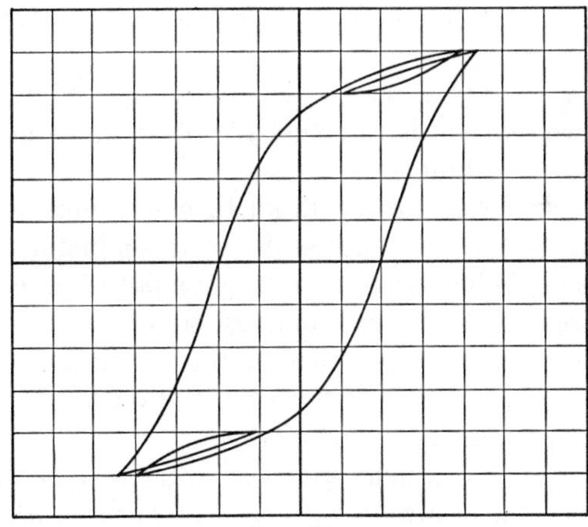

Fig. 51.

secondary loops at high density, as shown in Fig. 51, and an additional energy loss by hysteresis in these two secondary loops, which is considerable due to the high mean magnetic density, at which the secondary loop is traversed, so that in spite of the reduced maximum flux density, the hysteresis loss may be increased.

Therefore, in alternating-current engineering, the aim generally is to produce and use a wave which is a sine wave or nearly so.

60. In an alternating-current generator, synchronous or induction machine, commutating machine, etc., the wave of voltage induced in a single armature conductor or "face conductor" equals the wave of field flux distribution around the periphery of the magnet field, modified, however, by the reluctance pulsations of the magnetic circuit, where such exist. As the latter produce higher harmonics, they are in general objectionable and to be avoided as far as possible.

By properly selecting the length of the pole arc and the length of the air-gap between field and armature, a sinusoidal field flux distribution and thereby a sine wave of voltage induced in the armature face conductor could be produced. In this direction, however, the designer is very greatly limited by economic consideration: length of pole arc, gap length, etc., are determined within narrow limits by the requirement of the economic use of the material, questions of commutation, of pole-face losses, of field excitation, etc., so that as a rule the field flux distribution and with it the voltage induced in a face conductor differs materially from sine shape.

The voltage induced in a face conductor may contain even harmonics as well as odd harmonics, and often, as in most inductor alternators, a constant term.

The constant term cancels in all turn windings, as it is equal and opposite in the conductor and return conductor of each turn. Direct-current induction (continuous, or pulsating current) thus is possible only in half-turn windings, that is, windings in which each face conductor has a collector ring at either end, so-called unipolar machines (see *Theory and Calculation of Electrical Apparatus*, McGraw-Hill edition, volume 6).

In every winding, which repeats at every pole or 180 electrical degrees, as is almost always the case, the even harmonics cancel, even if they existed in the face conductor. In any machine in which the flux distribution in successive poles is the same, and merely opposite in direction, that is, in which the poles are symmetrical, no even harmonics are induced, as the field flux distribution contains no even harmonics. Even harmonics would, however, exist in the voltage wave of a machine designed as shown diagrammatically in Fig. 52, as follows:

The south poles S have about one-third the width of the north poles N, and the armature winding is a unitooth 50 per cent. pitch winding, shown as A in Fig. 52.

Assuming sinusoidal field flux distribution in the air-gaps under the poles N and S of Fig. 52, curve I in Fig. 53 shows the field flux distribution and thus the voltage induced in a single-face conductor. Curve II shows the voltage wave in a 50 per cent. pitch turn and therewith that of the winding A. As seen, this contains a pronounced second harmonic in addition to the fundamental. If, then, a second 50 per cent. pitch winding is located on the armature, shown as B in Fig. 52, by connecting B and A in series with each other in such direction that the fundamentals cancel (that is, in opposition for the fundamental wave), we get voltage wave III of Fig. 53, which contains only the even harmonics, that is, is of double frequency. Connecting A and B in series so that the fundamentals add and the second harmonics cancel, gives the wave IV. If the machine is a three-phase Y-connected alternator, with curve IV as the voltage per phase, or Y voltage, the delta or terminal voltage, derived by combination of two Y voltages under 60°, then is given by the curve V of Fig. 53. Fig. 54 shows the corresponding curves for the flux distribution of uniform density under the pole and tapering off at the pole corners, curve I, such as would approximately correspond to actual con-

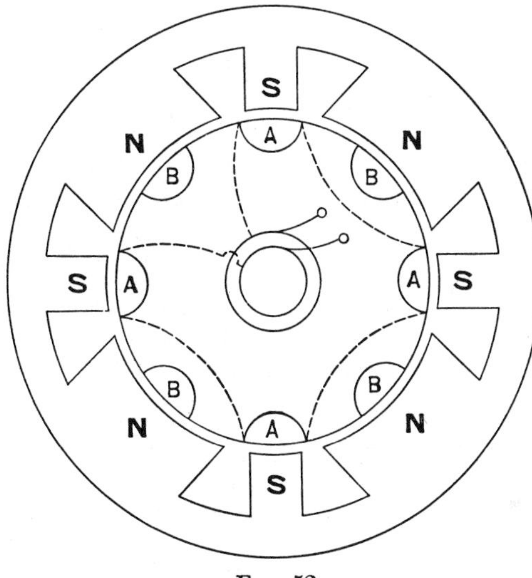

Fig. 52.

ditions. As seen, curve III as well as V are approximately sine waves, but the one of twice the frequency of the other. Thus, such a machine, by reversing connections between the two windings A and B, could be made to give two frequencies, one double the other, or as synchronous motor could run at two speeds, one one-half the other.

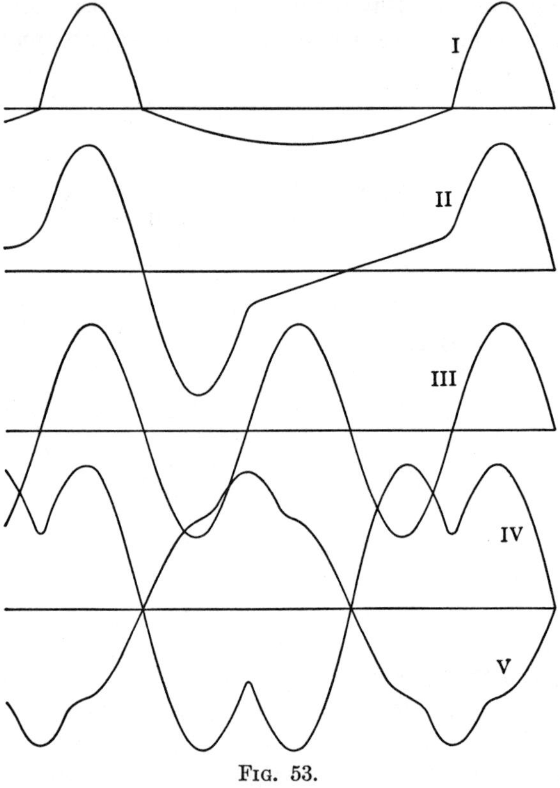

Fig. 53.

61. Distribution of the winding over an arc of the periphery of the armature eliminates or reduces the higher harmonics, so that the terminal voltage wave of an alternator with distributed winding is less distorted, or more nearly sine-shaped, than that of a single turn of the same winding (or that of a unitooth alternator). The voltage waves of successive turns are slightly out of phase with each other, and the more rapid variations due to higher harmonics thus are smoothed out. In two armature turns different in position on the armature circumference by δ electrical degrees ("electrical degrees" means counting the pitch of two poles as 360°), the fundamental waves are δ degrees out of phase, the third harmonics 3δ degrees, the fifth harmonics 5δ degrees, and so on, and their resultants thus get less and less, and becomes zero for that harmonic n, where $n\delta = 180°$.

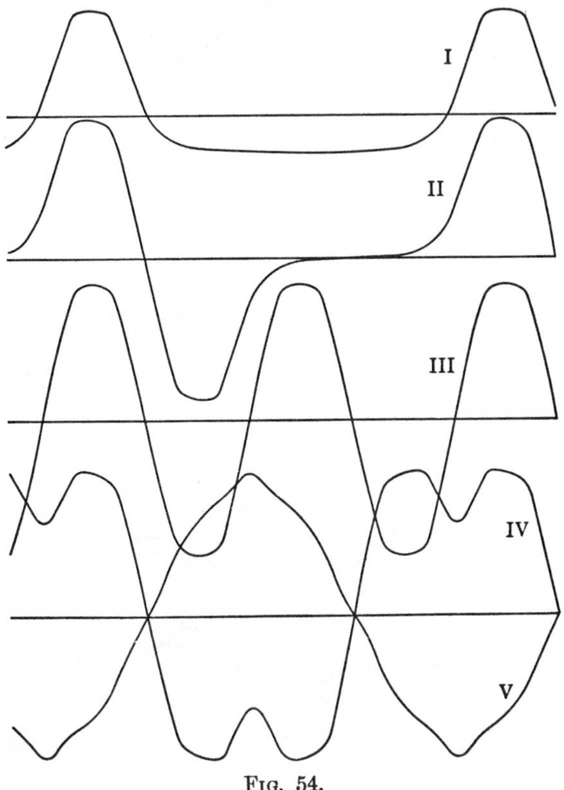

FIG. 54.

If

$$e = e_1 \sin \Phi + e_3 \sin 3 (\Phi - \alpha_3) + e_5 \sin 5 (\Phi - \alpha_5) + e_7 \sin 7 (\Phi - \alpha_7) + \ldots \quad (1)$$

is the voltage wave of a single turn, and the armature winding of m turns covers an arc of ω electrical degrees on the armature periphery (per phase), the coefficients of the harmonics of the resultant voltage wave are

$$E_n = me_n \text{ avg. } \cos \begin{Bmatrix} +\frac{n\omega}{2} \\ \\ -\frac{n\omega}{2} \end{Bmatrix} \tag{2}$$

or, since

$$\text{avg. } \cos \begin{Bmatrix} +\frac{n\omega}{2} \\ \\ -\frac{n\omega}{2} \end{Bmatrix} = \frac{2}{n\omega} \sin \frac{n\omega}{2}$$

$$E_n = \frac{2m}{n\omega} e_n \sin \frac{n\omega}{2} \tag{3}$$

and

$$E = \frac{2m}{\omega} \left\{ e_1 \sin \frac{\omega}{2} \sin \phi + \frac{e_3}{3} \sin \frac{3\omega}{2} \sin 3(\phi - \alpha_3) + \frac{e_5}{5} \sin \frac{5\omega}{2} \sin 5(\phi - \alpha_5) + \ldots \right\} \tag{4}$$

Thus, in a three-phase winding like that of the three-phase synchronous converter, in which each phase covers an arc of 120° $= \frac{2\pi}{3}$, it is $\frac{\omega}{2} = \frac{\pi}{3}$, hence,

$$E = \frac{3m\sqrt{3}}{2\pi} \left\{ e_1 \sin \phi - \frac{e_5}{5} \sin 5(\phi - \alpha_5) + \frac{e_7}{7} \sin 7(\phi - \alpha_7) - + \ldots \right\} \tag{5}$$

that is, the third harmonic and all its multiples, the ninth, fifteenth, etc., cancel, all other harmonics are greatly reduced, the more, the higher their order.

In a three-phase Y-connected winding, in which each phase covers 60° $= \frac{\pi}{3}$ of the periphery, as commonly used in induction and synchronous machines, it is $\frac{\omega}{2} = \frac{\pi}{6}$, hence,

$$E = \frac{3m}{\pi} \left\{ e_1 \sin \phi + \frac{2}{3} e_3 \sin 3(\phi - \alpha_3) + \frac{1}{5} e_5 \sin 5(\phi - \alpha_5) - \frac{1}{7} e_7 \sin 7(\phi - \alpha_7) - \frac{2}{9} e_9 \sin 9(\phi - \alpha_9) - \frac{1}{11} e_{11} \sin 11(\phi - \alpha_{11}) + \frac{1}{13} e_{13} \sin 13(\phi - \alpha_{13}) + - \ldots \right\} \tag{6}$$

Here the third harmonics do not cancel, but are especially large. Thus in a Y-connected three-phase machine of the usual 60° winding, the Y voltage may contain pronounced third harmonics, which, however, cancel in the delta voltage.

Thus with the distributed armature winding, which is now almost exclusively used, the wave-shape distortion due to the non-sinusoidal distribution of the field flux is greatly reduced, that is, the higher harmonics in the voltage wave decreased, the more so, the higher their order, and very high harmonics, such as the seventeenth, thirty-fifth, etc., therefore do not exist in such machines to any appreciable extent, except where produced by other causes. Such are a pulsation of the magnetic reluctance of the field due to the armature slots, or a pulsation of the armature reactance, as discussed in Chapter I of Dover edition Volume II, Part I, or a space resonance of the armature conductors with some of the harmonics. The latter may occur if the field flux distribution contains a harmonic of such order, that the voltages induced by it are in phase in the successive armature conductors, and therefore add, that is, when the spacing of the armature conductors coincides with a harmonic of the field flux, and the armature turn pitch and winding pitch are such that this harmonic does not cancel.

Inversely, if two turns are displaced from each other on the armature periphery by $\frac{1}{n}$ of the pole pitch, or $\frac{\pi}{n}$, and are connected in series, then in the resultant voltage of these two turns, the nth harmonics are out of phase by n times $\frac{\pi}{n}$, or by $\pi = 180°$, that is, are in opposition and so cancel.

Thus in a unitooth Y-connected three-phase alternator, while each phase usually contains a strong third harmonic, the terminal voltage can contain no third harmonic or its multiples: the two phases, which are in series between each pair of terminals, are one-third pole pitch, or 60 electrical degrees displaced on the armature periphery, and their third harmonic voltages therefore $3 \times 60 = 180°$ displaced, or opposite, that is, cancel, and no third harmonic can appear in the terminal voltage wave, or delta voltage, but a pronounced third harmonic may exist—and give trouble—in the voltage between each terminal and the neutral, or the Y voltage.

62. By the use of a fractional-pitch armature winding, higher harmonics can be eliminated. Assume the two sides of the armature turn, conductor and return conductor, are not separated from each other by the full pitch of the field pole, or 180 electrical degrees, but by less (or more); that is, each armature turn or coil covers not the full pitch of the pole, but the part p less (or more), that is, covers $(1 \pm p)$ 180°. The coil then is said to be $(1 \pm p)$ fractional pitch, or has the pitch deficiency p. The voltages induced in the two sides of the coil then are not equal and in phase, but are out of phase by 180 p for the fundamental, and by 180 np for the nth harmonic. Thus, if $np = 1$, for this n^{th} harmonic the voltages in the two sides of the coil are equal and opposite, thus cancel, and this harmonic is eliminated.

Therefore, two-thirds pitch winding eliminates the third harmonic, four-fifths pitch winding the fifth harmonic, etc.

Peripherally displacing half the field poles against the other half by the fraction q of the pole pitch, or by 180 q electrical degrees, causes the voltages induced by the two sets of field poles to be out of phase by 180 nq for the nth harmonic, and thereby eliminates that harmonic, for which $nq = 1$.

By these various means, if so desired, a number of harmonics can be eliminated. Thus in a Y-connected three-phase alternator with the winding of each phase covering 60 electrical degrees, with four-fifths pitch winding and half the field poles offset against the other by one-seventh of the pole pitch, the third, fifth, and seventh harmonic and their multiples are eliminated, that is, the lowest harmonic existing in the terminal voltage of such a machine is the eleventh, and the machine contains only the eleventh, thirteenth, seventeeth, ninteenth, twenty-third, twenty-ninth, thirty-first, thirty-seventh, etc. harmonics. As by the distributed winding these harmonics are greatly decreased, it follows that the terminal voltage wave would be closely a sine, irrespective of the field flux distribution, assuming that no slot harmonics exist.

63. In modern machines, the voltage wave usually is very closely a sine, as the pronounced lower harmonics, caused by the field flux distribution, which gave the saw-tooth, flat-top, peak or multiple-peak effects in the former unitooth machines, are greatly reduced by the distributed winding and the use of fractional pitch. Individual high harmonics, or pairs of high harmonics, are occasionally met, such as the seventeenth and ninteenth, or the thirty-fifth and thirty-seventh, etc. They are due to the pulsation of the magnetic field flux caused by the pulsation of the field reluctance by the passage of the armature slots, and occasionally, under load, by magnetic saturation of the armature self-inductive flux, that is, flux produced by the current in an armature slot and surrounding this slot, in cases where very many ampere conductors are massed in one slot, and the slot opening bridged or nearly so.

The low harmonics, third, fifth, seventh, are relatively harmless, except where very excessive and causing appreciable increase of the maximum voltage, or the maximum magnetic flux and thus hysteresis loss. The very high harmonics as a rule are relatively harmless in all circuits containing no capacity, since they are necessarily fairly small and still further suppressed by the inductance of the circuit. They may become serious and even dangerous, however, if capacity is present in the circuit, as the current taken by capacity is proportional to the frequency, and even small voltage harmonics, if of very high order, that is, high frequency, produce very large currents, and these in turn may cause dangerous voltages in inductive devices connected in series into the circuit, such as current transformers, or cause resonance effects in transformers, etc. With the increasing extent of very high-voltage transmission, introducing capacity into the systems, it thus becomes increasingly important to keep the very high harmonics practically out of the voltage wave.

Incidentally it follows herefrom, that the specifications of wave shape, that it should be within 5 per cent. of a sine wave, which is still occasionally met, has become irrational: a third harmonic of 5 per cent. is practically negligible, while a thirty-fifth harmonic of 5 per cent., in the voltage wave, would hardly be permissible. This makes it necessary in wave-shape specifications, to discriminate against high harmonics. One way would be, to specify not the wave shape of the voltage, but that of the current taken by a small condenser connected across the voltage. In the condenser current, the voltage harmonics are multiplied by their order. That is, the third harmonic is increased three times, the fifth harmonic five times, the thirty-fifth harmonic 35 times, etc. However, this probably overemphasizes the high harmonics, gives them too much weight, and a better way appears to be, to specify the current wave taken by a small condenser having a specified amount of non-inductive resistance in series.

Thus for instance, if $x = 1000$ ohms = capacity reactance of the condenser, at fundamental frequency, $r = 100$ ohms = resistance in series to the condenser, the impedance of this circuit, for the nth harmonic, would be

$$Z_n = r - j\frac{x}{n} = 100 - \frac{1000}{n}j \tag{7}$$

or, absolute, the impedance,

$$z_n = 1000\sqrt{\frac{1}{n^2} + 0.01} \tag{8}$$

and, the admittance,

$$y_n = \frac{0.001\,n}{\sqrt{1 + 0.01\,n^2}} \tag{9}$$

and therefore, the multiplying factor,

$$f = \frac{y_n}{y_1} = \frac{1.005\,n}{\sqrt{1 + 0.01\,n^2}} \tag{10}$$

this gives, for

n	f	n	f
1	1.0	13	8.0
3	2.9	15	8.4
5	4.5	25	9.3
7	5.8	35	9.6
9	6.7	45	9.8
11	7.4	∞	10.0

Thus, with this proportion of resistance and capacity, the maximum intensification is tenfold, for very high harmonics. By using a different value of the resistance, it can be made anything desired.

A convenient way of judging on the joint effect of all harmonics of a voltage wave is by comparing the current taken by such a condenser and resistance, with that taken by the same condenser and resistance, at a sine wave of impressed voltage, of the same effective value.

Thus, if the voltage wave

$$\begin{aligned} e &= 600 + 18_3 + 12_5 + 9_7 + 4_9 + 2_{11} + 3_{13} + 30_{23} + 24_{25} \\ &= 600\,\{\,1 + 0.03_3 + 0.02_5 + 0.015_7 + 0.0067_9 + 0.0033_{11} \\ &\qquad + 0.005_{13} + 0.05_{23} + 0.04_{25}\,\} \end{aligned}$$

(where the indices indicate the order of the harmonics) of effective value

$$e = \sqrt{600^2 + 18^2 + 12^2 + 9^2 + 4^2 + 2^2 + 3^2 + 30^2 + 24^2}$$
$$= 601.7$$

is impressed upon the condenser resistance of the admittance, y_n, the current wave is

$$i = 0.603 \{ 1 + 0.087_3 + 0.09_5 + 0.087_7 + 0.0445_9 + 0.0247_{11} + 0.04_{13} + 0.46_{23} + 0.37_{25} \}$$
$$= 0.603 \times 1.173$$
$$= 0.707$$

while with a sine wave of voltage, of $e_0 = 601.7$, the current would be

$$i_0 = 0.599,$$

giving a ratio

$$\frac{i}{i_0} = 1.18,$$

or 18 per cent. increase of current due to wave-shape distortion by higher harmonics.

64. While usually the sine wave is satisfactory for the purpose for which alternating currents are used, there are numerous cases where waves of different shape are desirable, or even necessary for accomplishing the desired purpose. In other cases, by the internal reactions of apparatus, such as magnetic saturation, a wave-shape distortion may occur and requires consideration to avoid harmful results.

Thus in the regulating pole converter (so-called "split-pole converter") variations of the direct-current voltage are produced at constant alternating-current voltage input, by superposing a third harmonic produced by the field flux distribution, as discussed under "Regulating Pole Converter" in *Theory and Calculation of Electrical Apparatus* [McGraw-Hill edition, volume 6]. In this case, the third harmonic must be restricted to the local or converter circuit by proper transformer connections: either three-phase connection of the converter, or Y or double-delta connections of the transformers with a six-phase converter.

The appearance of a wave-shape distortion by the third harmonic and its multiples, in the neutral voltage of Y-connected transformers, and its intensifications by capacity in the secondary circuit, and elimination by delta connection, has been discussed in Chapter I of Dover edition, Volume II, Part I.

In the flickering of incandescent lamps, and the steadiness of arc lamps at low frequencies, a difference exists between the flat-top wave of current with steep zero, and the peaked wave with flat zero, the latter showing appreciable flickering already at a somewhat higher frequency, as is to be expected.

In general, where special wave shapes are desirable, they are usually produced locally, and not by the generator design, as with the increasing consolidation of all electric power supply in large generating stations, it becomes less permissible to produce a desired wave shape within the generator, as this is called upon to supply power for all purposes, and therefore the sine wave as the standard is preferable.

One of the most frequent causes of very pronounced wave-shape distortion, and therefore a very convenient means of producing certain characteristic deviations from sine shape, is magnetic saturation, and as instance of a typical wave-shape distortion, its causes and effects, this will be more fully discussed in the following.

CHAPTER VIII

SHAPING OF WAVES BY MAGNETIC SATURATION

65. The wave shapes of current or voltage produced by a closed magnetic circuit at moderate magnetic densities, such as are commonly used in transformers and other induction apparatus, are

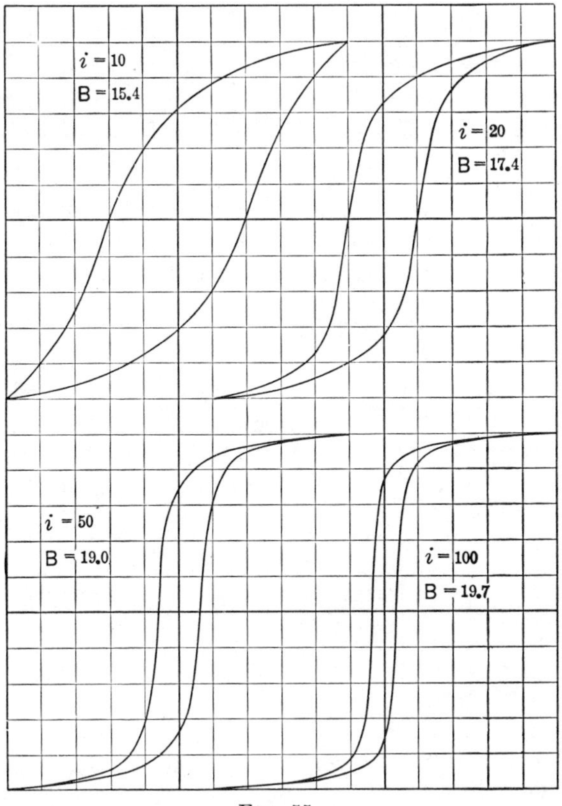

FIG. 55.

discussed in Volume II, *Electric Waves and Impulses*, Dover edition.

The characteristic of the wave-shape distortion by magnetic

saturation in a closed magnetic circuit is the production of a high peak and flat zero, of the current with a sine wave of impressed voltage, of the voltage with a sine wave of current traversing the circuit.

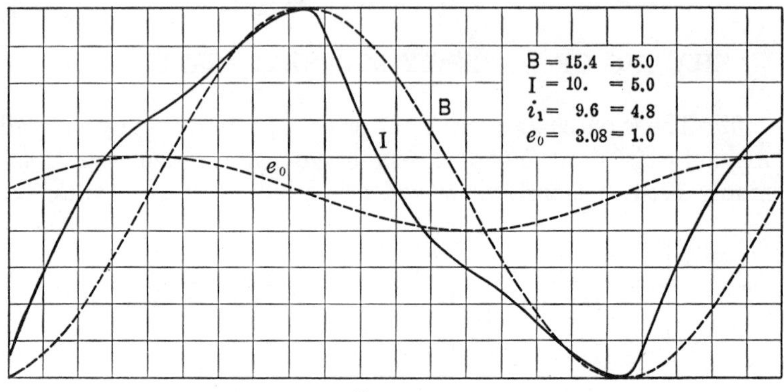

FIG. 56.

In Fig. 55 are shown four magnetic cycles, corresponding respectively to beginning saturation: $B = 15.4$ kilolines per cm.², $H = 10$; moderate saturation: $B = 17.4$, $H = 20$; high saturation:

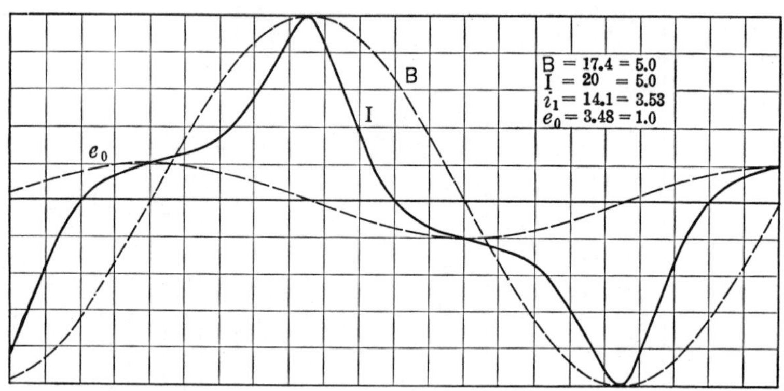

FIG. 57.

$B = 19.0$, $H = 50$; and very high saturation: $B = 19.7$, $H = 100$. Figs. 56, 57, 58 and 59 show the four corresponding current waves I, at a sine wave of impressed voltage e_0, and therefore sine wave of magnetic flux, B (neglecting ir drop in the winding, or rather, e_0 is the voltage induced by the alternating magnetic flux density B). In these four figures, the maximum values of e_0, B and I are chosen of the same scale, for wave-shape comparison, though in reality, in Fig. 59, very high saturation, the maximum of current, I, is ten times as high as in Fig. 56, beginning saturation. As seen, in Fig. 56 the current is the usual saw-tooth wave of transformer-exciting current, but slightly peaked, while in Fig. 59 a high peak exists. The numerical values are given in Table I.

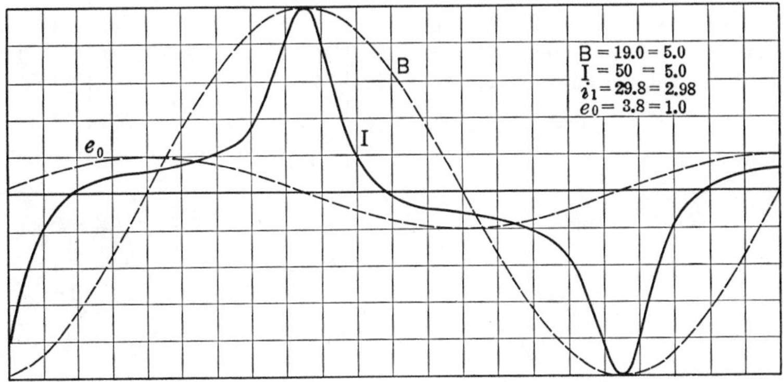

FIG. 58.

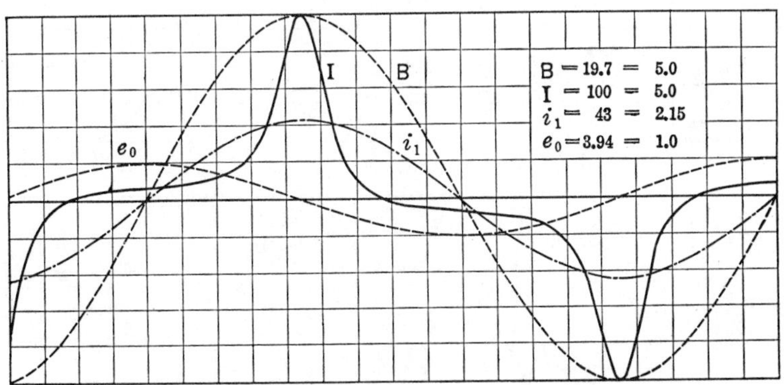

FIG. 59.

That is, at beginning saturation, the maximum value of the saw-tooth wave of current differs little from what it would be with a sine wave of the same effective value, being only 4 per cent. higher. At moderate saturation, however, the current peak is already 42 per cent. higher than in a sine wave of the same effective value, and becomes 132 per cent. higher than in a sine wave, at the very high saturation of Fig. 59.

Inversely, while the maximum values of current at the higher

TABLE I

	Beginning saturation, $B=15.4$	Moderate saturation, $B=17.4$	High saturation, $B=19.0$	Very high saturation, $B=19.7$
Sine wave of voltage, e_0, maximum.......	3.08	3.48	3.80	3.94
Maximum value of current, I...........	10.00	20.00	50.00	100.00
Effective value of current, $\times\sqrt{2} : i_1$.....	9.6	14.1	29.8	43.0
Form factor of current wave $\frac{I}{i_1}$	1.04	1.42	1.68	2.32
Ratio of effective currents..............	1.00	1.47	3.11	4.48

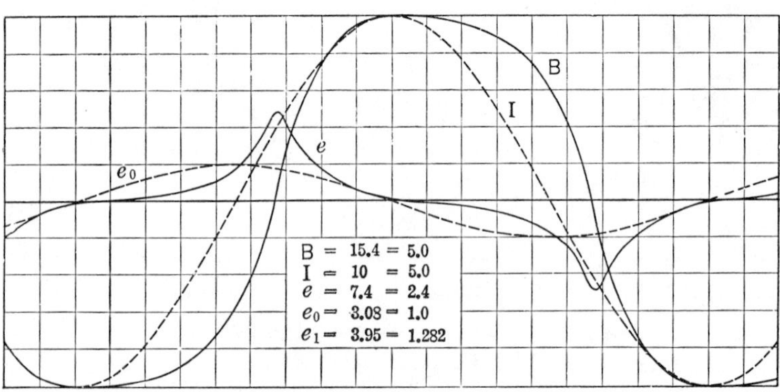

FIG. 60.

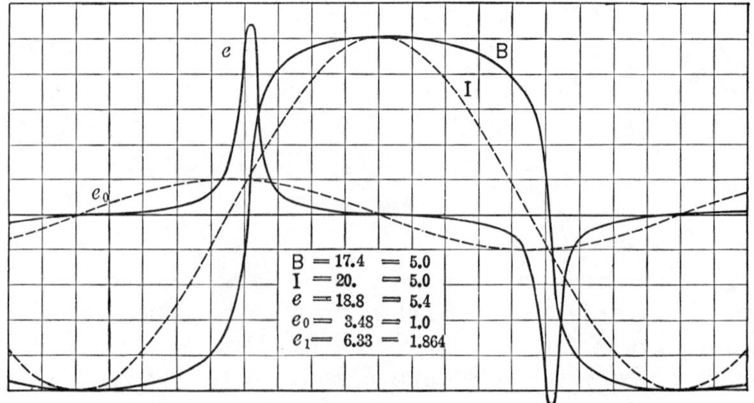

FIG. 61.

saturations are two, five and ten times the maximum current value at beginning saturation, the effective values are only 1.47, 3.1 and 4.47 times higher. Thus, with increasing magnetic saturation, the effective value of current rises much less than the maximum value, and when calculating the exciting current of a saturated magnetic circuit, as an overexcited transformer, from the magnetic characteristic derived by direct current, under the as-

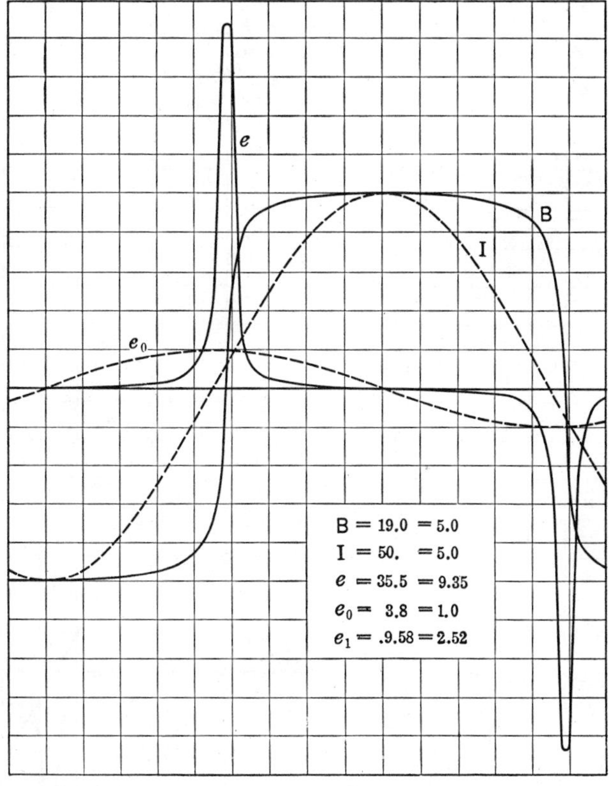

FIG. 62.

sumption of a sine wave, the calculated exciting current may be more than twice as large as the actual exciting current.

66. Figs. 60 to 63 show, for a sine wave of current, I, traversing a closed magnetic circuit, and the same four magnetic cycles given in Fig. 55, the waves of magnetic flux density, B, of induced voltage, e, the sine wave of voltage, e_0, which would be induced if the magnetic density, B, were a sine wave of the same maximum value, and Fig. 63 also shows the equivalent sine wave, e_1, of the (distorted) induced voltage wave, e.

As seen, already at beginning saturation, Fig. 60, the voltage peak is more than twice as high as it would be with a sine wave,

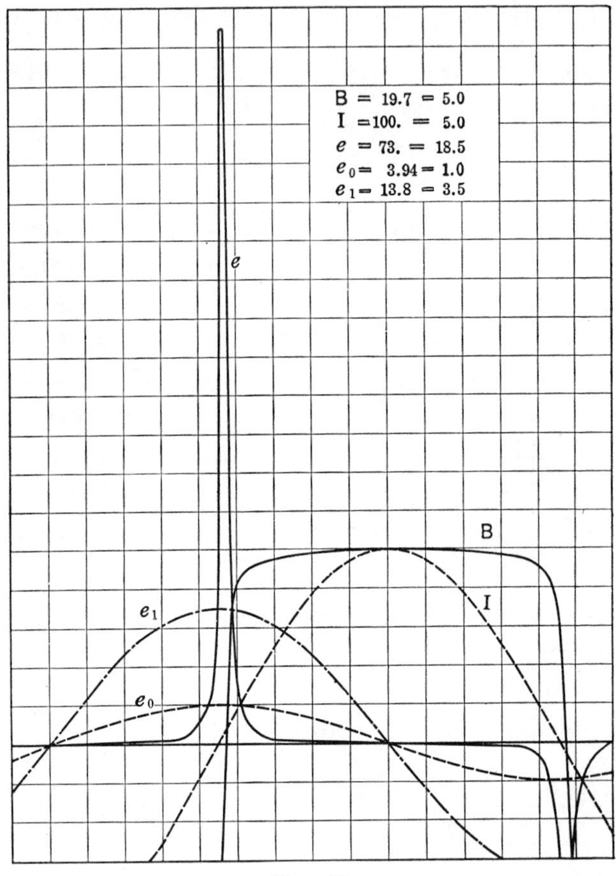

Fig. 63.

and rises at higher saturations to enormous values: 18.5 times the sine wave value in Fig. 63.

The magnetic flux wave, B, becomes more and more flat-topped with increasing saturation, and finally practically rectangular, in Fig. 63.

The curves 60 to 63 are drawn with the same maximum values

of current, I, flux density, B, and sine wave voltage, e_0, for better comparison of their wave shapes.

The numerical values are:

TABLE II

	Beginning saturation, $B=15.4$	Moderate saturation, $B=17.4$	High saturation, $B=19.0$	Very high saturation, $B=19.7$
Sine wave of current, I, maximum	10.0	20.0	50.0	100.0
Flat-top wave of magnetic density, B, maximum	15.4	17.4	19.0	19.7
Peaked voltage wave e, maximum	7.4	18.8	35.5	73.0
Ratio	1.00	2.56	4.80	9.88
Sine wave of voltage, e_0, maximum, for same maximum flux	3.08	3.48	3.80	3.94
Ratio	1.00	1.13	1.23	1.28
Form factor of voltage wave, $\frac{e}{e_0}$	2.40	5.40	9.35	18.50
Equivalent sine wave of voltage, e_1, maximum	3.95	6.33	9.58	13.80
Ratio	1.00	1.60	2.42	3.50
$\frac{e_1}{e_0}$ (maxima)	1.282	1.864	2.520	3.500
$\frac{e}{e_1}$ (maxima)	1.87	2.97	3.70	5.28

As seen, the wave-shape distortion due to magnetic saturation is very much greater with a sine wave of current traversing the closed magnetic circuit, than it is with a sine wave of voltage impressed upon it.

With increasing magnetic saturation, with a sine wave of current, the effective value of induced voltage increases much more rapidly than the magnetic flux increases, and the maximum value of voltage increases still much more rapidly than the effective value: an increase of flux density, B, by 28 per cent., from beginning to very high saturation, gives an increase of the effective value of induced voltage (as measured by voltmeter) by 250 per cent., or 3.5 times, and an increase of the peak value of voltage (which makes itself felt by disruption of insulation, by danger to life, etc.) by 888 per cent., or nearly ten times.

At very high saturation, the voltage wave practically becomes one single extremely high and very narrow voltage peak, which occurs at the reversal of current.

At the very high saturation, Fig. 63, the effective value, e_1, of the voltage is 3.5 times as high as it would be with a sine wave of magnetic flux; the maximum value, e, is more than five times as high as it would be with a sine wave of the same effective value, e_1, that is, more than five times as high, as would be expected from the voltmeter reading, and it is 18.5 times as high as it would be with a sine wave of magnetic flux.

Thus, an oversaturated closed magnetic circuit reactance, which consumes $e_0 = 50$ volts with a sine wave of voltage, e_0, and thus of magnetic density, B, would, at the same maximum magnetic density, that is, the same saturation, with a sine wave of current—as would be the case if the reactance is connected in series in a constant-current circuit—give an effective value of terminal voltage of $e_1 = 3.5 \times 50 = 175$ volts, and a maximum peak voltage of $e = 18.8 \times 50 \times \sqrt{2} = 1330$ volts.

Thus, while supposed to be a low-voltage reactance, $e_0 = 50$ volts, and even the voltmeter shows a voltage of only $e_1 = 175$, which, while much higher, is still within the limit that does not endanger life, the actual peak voltage $e = 1330$ is beyond the danger limit.

Thus, magnetic saturation may in supposedly low-voltage circuits produce dangerously high-voltage peaks.

A transformer, at open secondary circuit, is a closed magnetic circuit reactance, and in a transformer connected in series into a circuit—such as a current transformer, etc.—at open secondary circuit unexpectedly high voltages may appear by magnetic saturation.

67. From the preceding, it follows that the relation of alternating current to alternating voltage, that is, the reactance of a closed magnetic circuit, within the range of magnetic saturation, is not constant, but varies not only with the magnetic density, B, but for the same magnetic density B, the reactance may have very different values, depending on the conditions of the circuit: whether constant potential, that is, a sine wave of voltage impressed upon the reactance; or constant current, that is, a sine wave of current traversing the circuit; or any intermediate condition, such as brought about by the insertion of various amounts of resistance, or of reactance or capacity, in series to the closed magnetic circuit reactance.

The numerical values in Table III illustrate this.

I gives the magnetic field intensity, and thus the direct current, which produces the magnetic density, B—that is, the B–H curve of the magnetic material. An alternating current of maximum value, I, thus gives an alternating magnetic flux of maximum flux density B. If I and B, were both sine waves, that is, if

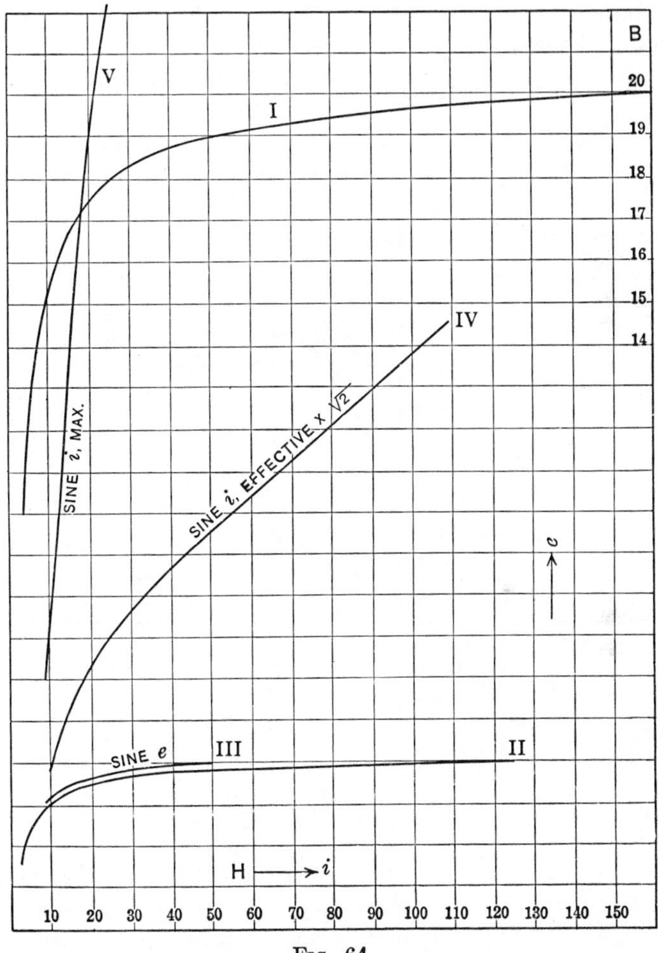

FIG. 64.

during the cycle current and magnetic flux were proportional to each other, as in an unsaturated open magnetic circuit, e_0, as given in the third column, would be the maximum value of the induced voltage, and $x_0 = \frac{e_0}{I}$ the reactance. This reactance varies with the density, and greatly decreases with increasing magnetic saturation, as well known.

However, if e_0 and thus B are sine waves, I can not be a sine wave, but is distorted as shown in Figs. 56 to 59, and the effective value of the current, that is, the current as it would be read by an alternating ammeter, multiplied by $\sqrt{2}$ (that is, the maximum value of the equivalent sine waves of exciting current) is given as i_1. The reactance is then found as $x_p = \frac{e_0}{i_1}$. This is the reactance

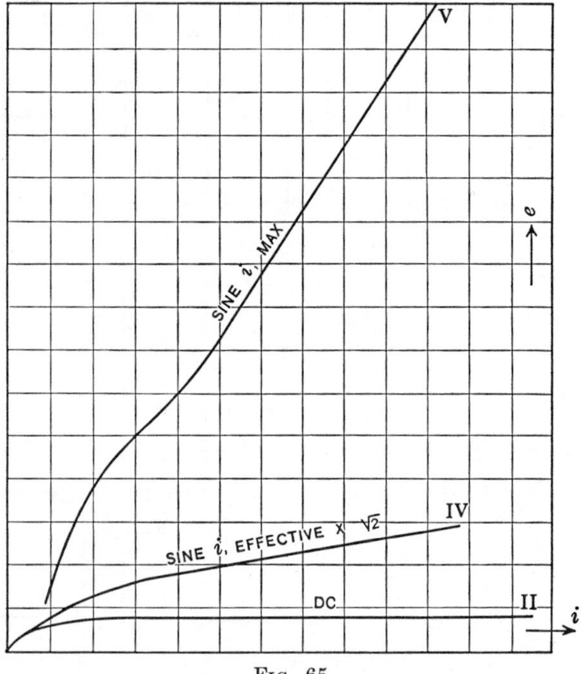

FIG. 65.

of the closed magnetic circuit on constant potential, that is, on a sine wave of impressed voltage, and, as seen, is larger than x_0.

If, however, the current, I, which traverses the reactance, is a sine wave, then the flux density, B, and the induced voltage are not sines, but are distorted as in Figs. 60 to 63, and the effective value of the induced voltage (that is, the voltage as read by alternating voltmeter), multiplied by $\sqrt{2}$ (that is; the maximum of the equivalent sine wave of voltage) is given as e_1 in Table III, and the true maximum value of the induced voltage wave is e.

The reactance, as derived by voltmeter and ammeter readings under these conditions. that is, on a constant-current circuit, or with a sine wave of current traversing the magnetic circuit, is $x_c = \frac{e_1}{I}$, thus larger than the constant-potential reactance, x_p.

Much larger still is the reactance derived from the actual maximum values of voltage and current: $x_m = \frac{e}{I}$:

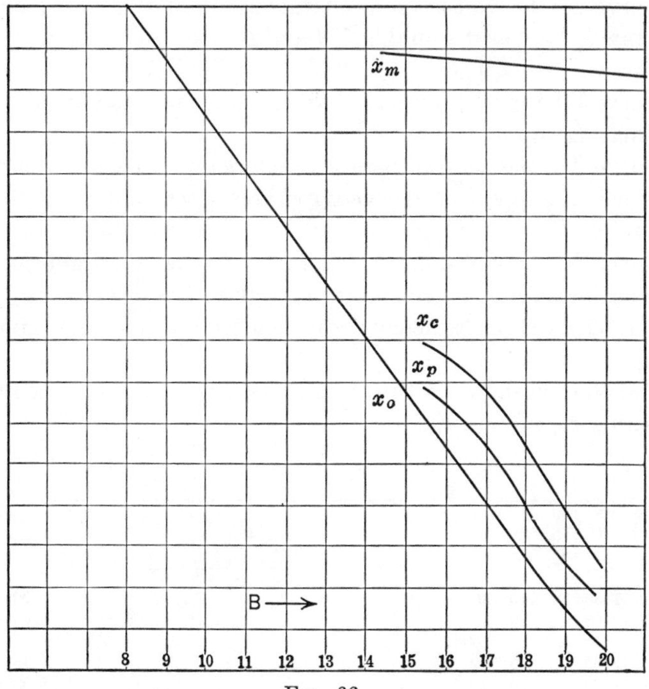

FIG. 66.

It is interesting to note that x_m, the peak reactance, is approximately constant, that is, does not decrease with increasing magnetic saturation. (The higher value at beginning saturation, for $I = 20$, may possibly be due to an inaccuracy in the hysteresis cycle of Fig. 55, a too great steepness near the zero value, rather than being actual.)

It is interesting to realize, that when measuring the reactance of a closed magnetic circuit reactor by voltmeter and ammeter readings, it is not permissible to vary the voltage by series resistance, as this would give values indefinite between x_p and x_c, depending on the relative amount of resistance. To get x_p, the generated supply voltage of a constant-potential source must be varied; to get x_c, the current in a constant-current circuit must be varied. As seen, the differences may amount to several hundred per cent.

As graphical illustration, Fig. 64 shows:

As curve I the magnetic characteristic, as derived with direct current.

Curve II the volt-ampere characteristic of the closed circuit reactance, I, e_0, as it would be if I and B, that is, e_0, both were sine waves.

Curve III the volt-ampere characteristic on constant-potential alternating supply, i_1, e_0.

Curve IV the volt-ampere characteristic on constant-current alternating supply, as derived by voltmeter and ammeter, I, e_1, and as

Curve V the volt-ampere characteristic on constant-current alternating supply, as given by the peak values of I and e.

Fig. 65 gives the same curves in reduced scale, so as to show V completely.

Fig. 66 then shows, with B as abscissæ, the values of the reactances x_0, x_p, x_c, and x_m.

TABLE III

I	B	e_0	$x_0 = \frac{e_0}{I}$	i_1	$x_p = \frac{e_0}{i_1}$	e_1	$x_c = \frac{e_1}{I}$	e	$x_m = \frac{e}{I}$	p	p_0
2.0	7.30	.....	0.7300	......	....	......	....	......	......	1.00	
3.0	10.00	.....	0.6670	......	....	......	....	......	......	1.09	
4.0	11.50	.....	0.5750	......	....	......	....	......	......	1.27	
5.0	12.50	.....	0.5000	......	....	......	....	......	......	1.46	
7.5	14.30	.....	0.3810	......	....	......	....	......	......	1.92	
10.0	15.40	3.08	0.3080	9.0	0.342	3.95	0.395	7.4	0.74	2.37	2.40
15.0	16.70	.....	0.2230	......	....	......	....	......	......	3.27	
20.0	17.40	3.48	0.1740	14.1	0.247	6.33	0.316	18.8	0.94	4.20	5.40
30.0	18.30	.....	0.1220	......	....	......	....	......	......	6.00	
40.0	18.70	.....	0.0930	......	....	......	....	......	......	7.85	
50.0	19.00	3.80	0.0760	29.8	0.127	9.58	0.912	35.5	0.71	9.60	9.35
75.0	19.35	.....	0.0520	......	....	......	....	......	......	14.10	
100.0	19.70	3.94	0.0394	43.0	0.092	13.80	0.138	73.0	0.73	18.50	18.5
125.0	19.85	.....	0.0320	......	....	......	....	......	......	22.80	
150.0	19.95	.....	0.0270	......	....	......	....	......	......	27.00	

68. Another way of looking at the phenomenon is this: while with increasing current traversing a closed magnetic circuit, the magnetic flux density is limited by saturation, the induced voltage peak is not limited by saturation, as it occurs at the current reversal, but it is proportional to the rate of change of the magnetic flux density at the current reversal, and thus approximately proportional to the current.

Thus, approximately, within the range of magnetic saturation, with increasing current traversing the closed magnetic circuit (like that of a series transformer):

The magnetic flux density, and therefore the mean value of induced voltage remains constant;

The peak value of induced voltage increases proportional to the current, and therefore;

The effective value of induced voltage increases proportional to the square root of the current.

Thus, if the exciting current of a series transformer is 5 per cent. of full-load current, and the secondary circuit is opened, while the primary current remains the same, the effective voltage consumed by the transformer increases approximately $\sqrt{20} = 4.47$ times, and the maximum voltage peak 20 times above the full-load voltage of the transformer.

As the shape of the magnetic flux density and voltage waves are determined by the current and flux relation of the hysteresis cycles, and the latter are entirely empirical and can not be expressed mathematically, therefore it is not possible to derive an exact mathematical equation for these distorted and peaked voltage waves from their origin. Nevertheless, especially at higher saturation, where the voltage peaks are more pronounced, the equation of the voltage wave can be derived and represented by a Fourier series with a fair degree of accuracy. By thus deriving the Fourier series which represents the peaked voltage waves, the harmonics which make up the wave, and their approximate values can be determined and therefrom their probable effect on the system, as resonance phenomena, etc., estimated.

The characteristic of the voltage-wave distortion due to magnetic saturation in a closed magnetic circuit traversed by a sine wave of current is, that the entire voltage wave practically contracts into a single high peak, at, or rather shortly after, the moment of current reversal, as shown in Figs. 63, 62, etc.

With the same maximum value of magnetic density, B, and thus of flux, Φ, the area of the induced voltage wave, and thus the mean value of the voltage, is the same, whatever may be the wave of magnetism and thus of voltage, since $\Phi = \int e\,dt$, and the area of the peaked voltage wave of the saturated magnetic circuit, e, thus is the same as that of a sine wave of voltage, e_0. Neglecting then the small values of voltage, e, outside of the voltage peak, if this voltage peak of e is p times the maximum value of the sine wave, e_0, its width is $\frac{1}{p}$ of that of the sine wave, and if the sine wave of voltage, e_0, is represented by the equation

$$e_0 \cos \phi \tag{11}$$

the peak of the distorted voltage wave is represented, in first approximation, by assuming it as of sinusoidal shape, by

$$pe_0 \cos p\phi \tag{12}$$

That is, the distorted voltage wave, e, can be considered as represented by $pe_0 \cos p\phi$ within the angle

$$-\frac{\pi}{2\,p} < \phi < \frac{\pi}{2\,p} \tag{13}$$

and by zero outside of this range.

The value of p follows, approximately, from the consideration that the peak reactance, x_m, is independent of the saturation, or constant, since it depends on the rate of change of magnetism with current near the zero value, where there is no saturation, and the ratio $\frac{dB}{dI}$ thus (approximately) constant.

Or, in other words, if below saturation, in the range where the magnetic permeability is a maximum, the current, i, produces the magnetic flux, Φ, and thereby induces the voltage, e', the reactance is

$$x' = \frac{e'}{i} \tag{14}$$

This is the maximum reactance, below saturation, of the magnetic circuit, and can be calculated from the dimensions and the magnetic characteristic, in the usual manner, by assuming sine waves of i and B.

The peak reactance, x_m, of the saturated magnetic circuit is approximately equal to x', and thus can be calculated with reasonable approximation, from the dimensions of the magnetic circuit and the magnetic characteristic.

If now, in the range of magnetic saturation, a sine wave of current, of maximum value I, traverses the closed magnetic circuit, the peak value of the (distorted) induced voltage is

$$e = x_m I \tag{15}$$

where

$$x_m = x' = \frac{e'}{i} \tag{16}$$

is the maximum reactance of the magnetic circuit below saturation, derived by the assumption of sine waves, e' and i.

If B is the maximum value of the magnetic density produced by the sine wave of current of maximum value, I, and, e_0, the maximum value of the sine wave of voltage induced by a sinusoidal variation of the magnetic density, B, the "form factor" of the peaked voltage wave of the saturated magnetic circuit is

$$p = \frac{e}{e_0} = \frac{x_m I}{e_0} \tag{17}$$

thus determined, approximately.

As illustrations are given, in the second last column of Table III, the form factors, p, calculated in this manner, and in the last column are given the actual form factors, p_0, derived from the curves 60 to 63. As seen, the agreement is well within the uncertainty of observation of the shape of the hysteresis cycles, except perhaps at $I = 20$, and there probably the calculated value is more nearly correct.

69. The peaked voltage wave induced by the saturated closed magnetic circuit can, by assuming it as symmetrical and counting the time from the center of the peak, be represented by the Fourier series.

$$\left.\begin{aligned} e &= a_1 \cos \phi + a_3 \cos 3\,\phi + a_5 \cos 5\,\phi + a_7 \cos 7\,\phi + \ldots \\ &= \Sigma\, a_n \cos n\phi \end{aligned}\right\} \tag{18}$$

where

$$a_n = \frac{4}{\pi}\int_0^{\frac{\pi}{2}} e \cos n\phi\, d\phi \tag{19}$$

$$= 2\,\text{avg}(e \cos n\phi)_0^{\frac{\pi}{2}} \tag{20}$$

The slight asymmetry of the peak would introduce some sine terms, which might be evaluated, but are of such small values as to be negligible.

(*a*) For the lower harmonics, where n is small compared to p,

$\cos n\phi$ is practically constant and $= 1$ during the short voltage peak $e = pe_0 \cos p\phi$, and it is, therefore,

$$\begin{aligned} a_n &= 2 \operatorname{avg}(e)_0^{\frac{\pi}{2}} \\ &= 2 \operatorname{avg}(pe_0 \cos p\phi)_0^{\frac{\pi}{2}} \\ &= \frac{2}{p} \operatorname{avg}(pe_0 \cos p\phi)_0^{\frac{\pi}{2}p} \\ &= 2\, e_0 \operatorname{avg} \cos = \frac{4}{\pi} e_0. \end{aligned}$$

(*b*) For the harmonic, where $n = p$, it is

$$\begin{aligned} a_p &= 2 \operatorname{avg}(pe_0 \cos^2 p\phi)_0^{\frac{\pi}{2}} \\ &= \frac{2}{p} \operatorname{avg}(pe_0 \cos^2 p\phi)_0^{\frac{\pi}{2}} \\ &= 2\, e_0 \operatorname{avg} \cos^2 = e_0. \end{aligned}$$

(*c*) For still higher harmonics than $n = p$, $\cos n\phi$ assumes negative values within the range of the voltage peak, and a_n thereby rapidly decreases, finally becomes zero and then negative, at $n = 3\, p$, positive again at $n = 5\, p$, etc., but is practically negligible.

Thus, the coefficients of the Fourier series decrease gradually, with increasing order, n,

from $\frac{4}{\pi} e_0$ as maximum, to e_0 for $n = p$, and then with increasing rapidity fall off to negligible values.

Their exact values can easily be derived by substituting (12) into (19),

$$a_n = \frac{4}{\pi} \int_0^{\frac{\pi}{2p}} pe_0 \cos p\phi \cos n\phi \, d\phi \tag{21}$$

here the integration is extended to $\frac{\pi}{2p}$ only, as beyond this, the voltage, e, is not given by equation (12) any more, but is zero.

(21) integrates by

$$\begin{aligned} a_n &= \frac{2\, pe_0}{\pi} \Big/ \frac{\sin(p+n)\phi}{p+n} + \frac{\sin(p-n)\phi}{p-n} \Big/_0^{\frac{\pi}{2p}} \\ &= \frac{2\, pe_0}{\pi} \left\{ \frac{\sin \frac{\pi}{2}\left(1 + \frac{n}{p}\right)}{p+n} + \frac{\sin \frac{\pi}{2}\left(1 - \frac{n}{p}\right)}{p-n} \right\}, \end{aligned}$$

but since $\sin \frac{\pi}{2}\left(1 + \frac{n}{p}\right) = \sin \frac{\pi}{2}\left(1 - \frac{n}{p}\right)$, it is

$$a_n = \frac{4\, e_0 \sin \frac{\pi}{2}\left(1 - \frac{n}{p}\right)}{\pi\left(1 - \frac{n^2}{p^2}\right)} \tag{22}$$

and

$$e = \frac{4\, e_0}{\pi} \sum \frac{\sin \frac{\pi}{2}\left(1 - \frac{n}{p}\right)}{1 - \frac{n^2}{p^2}} \cos n\phi \tag{23}$$

as the equations of the voltage wave distorted by magnetic saturation.

70. These coefficients, a_n, are very easily calculated, and as instances are given in Table IV, the coefficients of the distorted voltage wave of Fig. 62, which has the form factor $p = 9.35$.

TABLE IV

$p = 9.35$ $\qquad a_n = \frac{4\, e_0}{\pi} \frac{\sin \frac{\pi}{2}\left(1 - \frac{n}{p}\right)}{1 - \frac{n^2}{p^2}}$

n = 1	3	5	7	9	11	13	15	17	19
$\frac{a_n}{e_0}$ = 1.270	1.242	1.188	1.114	1.018	0.906	0.786	0.658	0.528	0.406
n = 21	23	25	27	29	31	33			
$\frac{a_n}{e_0}$ = 0.292	0.189	0.101	0.031	−0.023	−0.060	−0.082			

As seen, after $n = 9$, the values of a_n rapidly decrease, and become negative, though of negligible value, after $n = 27$.

In Fig. 67 the successive values of $\frac{a_n}{e_0}$ are shown as curve.

In reality, the peaked voltage wave of magnetic saturation, as shown in Figs. 61 to 63, is not half a sine wave, but is rounded off at the ends, toward the zero values. Physically, the meaning of the successive harmonics is, that they raise the peak and cut off the values outside of the peak. It is the high harmonics, which sharpen the edge of the peak, and the rounded edge of the peak in the actual wave thus means that the highest harmonics, which give very small or negative values of a_n, are lower than given by equations (23), or rather are absent.

Thus, by omitting the highest harmonics, the wave is rounded off and brought nearer to its actual shape. Thus, instead of following the curve, a_n, as calculated and given in Fig. 67, we cut it off before the zero value of a_n, about at $n = 23$, and follow the curve line, a'_n, which is drawn so that $\Sigma \frac{a'_n}{e_0} = 9.35$, that is, that the voltage peak has the actual value.

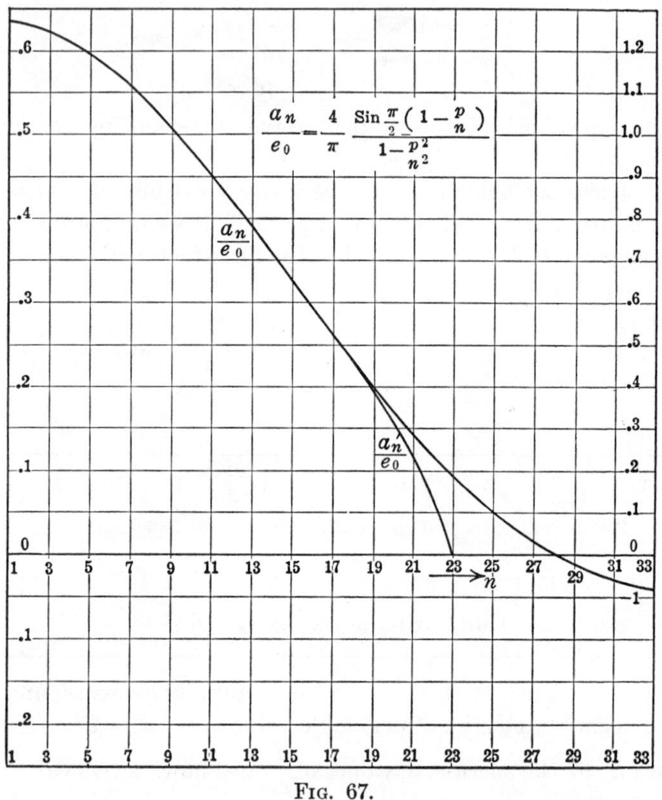

FIG. 67.

The equation of the peaked voltage in Fig. 62 then becomes

$$e = e_0 \{1.270 \cos \phi + 1.242 \cos 3\phi + 1.188 \cos 5\phi + 1.114 \cos 7\phi + 1.018 \cos 9\phi + 0.906 \cos 11\phi + 0.786 \cos 13\phi + 0.658 \cos 15\phi + 0.529 \cos 17\phi + 0.400 \cos 19\phi + 0.240 \cos 21\phi\}.$$

Or, in symbolic writing,

$$e = e_0\{1.270_1 + 1.242_3 + 1.188_5 + 1.114_7 + 1.018_9 + 0.906_{11} + 0.786_{13} + 0.658_{15} + 0.529_{17} + 0.400_{19} + 0.240_{21}\}$$

$$= 1.270\, e_0 \{1_1 + 0.978_3 + 0.953_5 + 0.877_7 + 0.800_9 + 0.713_{11} + 0.617_{13} + 0.517_{15} + 0.416_{17} + 0.315_{19} + 0.189_{21}\}.$$

It is of interest to note how extended a series of powerful harmonics is produced. It is easily seen that in the presence of capacity, these large and very high harmonics may be of considerable danger. In any reactance, which is intended for use in series to a high-voltage circuit, the use of a closed magnetic circuit thus constitutes a possible menace from excessive voltage peaks if saturation occurs.

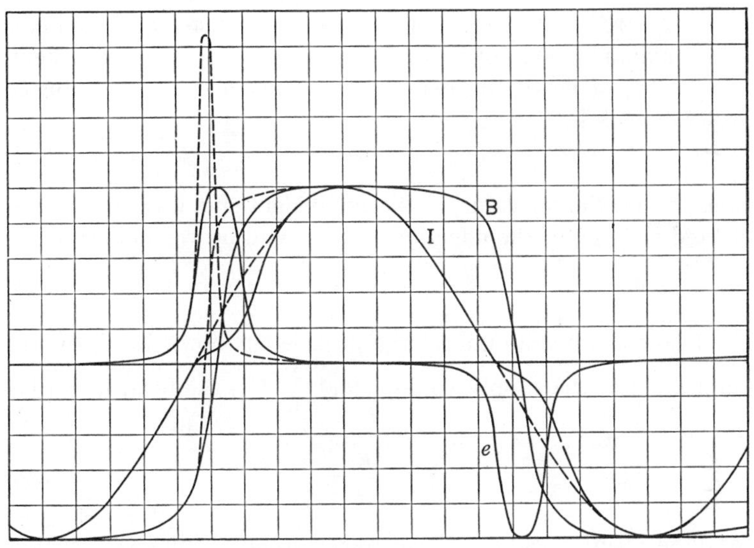

FIG. 68.

71. Such high-voltage peaks by magnetic saturation in a closed magnetic circuit traversed by a sine wave of current can occur only if the available supply voltage is sufficiently high. If the total supply voltage of the circuit is less than the voltage peak produced by magnetic saturation, obviously this voltage peak must be reduced to a value below the voltage available in the supply circuit, and in this case simply the current wave can not remain a sine, but is flattened at the zero values, and with it the wave of magnetic density.

Thus, if in Fig. 62 the maximum supply voltage is $E = 19.0$, the maximum peak voltage can not rise to $e = 35.5$, but stops at

$e \leqq E$, and when this value is reached, the rate of change of flux density, B, and thus of current, I, decreases, as shown in Fig. 68, in drawn lines. In dotted lines are added the curves corresponding to unlimited supply voltage. The voltage peak is thereby reduced, correspondingly broadened, and retarded, and the current is flattened at and after its zero value, the more, the lower the maximum supply voltage.

The reactance is reduced hereby also, from $x_c = 0.192$, in Fig. 62, to $x_c = 0.140$.

In other words, if p is the form factor of the distorted voltage wave, which would, with unlimited supply voltage, be induced by the saturated magnetic circuit of maximum density, B, and e_0 is the maximum value of the sine wave of voltage, which a sinusoidal flux of maximum density, B, would induce, the distorted voltage peak is

$$e = pe_0 \tag{24}$$

and the maximum value of the equivalent sine wave of the distorted voltage, or the effective voltage read by voltmeter, is

$$e_1 = \sqrt{p}\, e_0 \tag{25}$$

If now the maximum voltage peak is cut down to E, by the limitation of the supply voltage, and $\frac{e}{E} = q$, the form factor becomes

$$p' = \frac{E}{e_0} = \frac{p}{q}, \tag{26}$$

and the effective value of the distorted voltage, times $\sqrt{2}$, that is, the maximum of the equivalent sine wave, is

$$e'_1 = \sqrt{p'}\, e_0 = \frac{e_1}{\sqrt{q}} = \sqrt{e_0 E}, \tag{27}$$

thus varies with the supply voltage, E.

The reactance then is

$$x'_c = \frac{e'_1}{I} = \frac{x_c}{\sqrt{q}} \tag{28}$$

Thus, for $e = 35.5$, $E = 19.0$, it is

$$q = 1.87,$$

and as $e_0 = 3.80$; $p = 9.35$, it is

$$p' = \frac{p}{q} = 5.0,$$

$$e'_1 = \frac{e_1}{\sqrt{q}} = \frac{9.58}{1.37} = 7.0,$$

$$x'_c = \frac{x_c}{\sqrt{q}} = \frac{0.192}{1.37} = 1.40.$$

These values, however, are only fair approximations, as they are based on the assumption of sinusoidal shape of the peaks.

72. In the preceding, the assumption has been made, that the magnetic flux passes entirely within the closed magnetic circuit, that is, that there is no magnetic leakage flux, or flux which closes through non-magnetic space outside of the iron conduit.

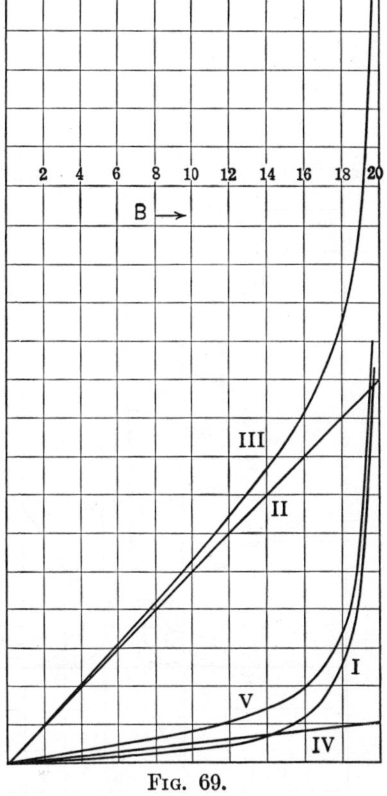

Fig. 69.

If there is a magnetic leakage flux—and there must always be some—it somewhat reduces the voltage peak, the more, the greater is the proportion of the leakage flux to the main flux. The leakage flux, in open magnetic circuit, is practically proportional to the current, and that part of the voltage, which is induced by the leakage flux, therefore, is a sine wave, with a sine wave of current, hence does not contribute to the voltage peak.

Such high magnetic saturation peaks occur only in a closed magnetic circuit. If the magnetic circuit is not closed, but contains an air-gap, even a very small one, the voltage peak, with a sine wave of current, is very greatly reduced, since in the air-gap magnetic flux and magnetizing current are proportional.

Thus, below saturation and even at beginning saturation, an air-gap in the magnetic circuit, of one-hundredth of its length, makes the voltage wave practically a sine wave, with a sine wave of current, as discussed in Volume I, Part I, Chapter I, Dover edition.

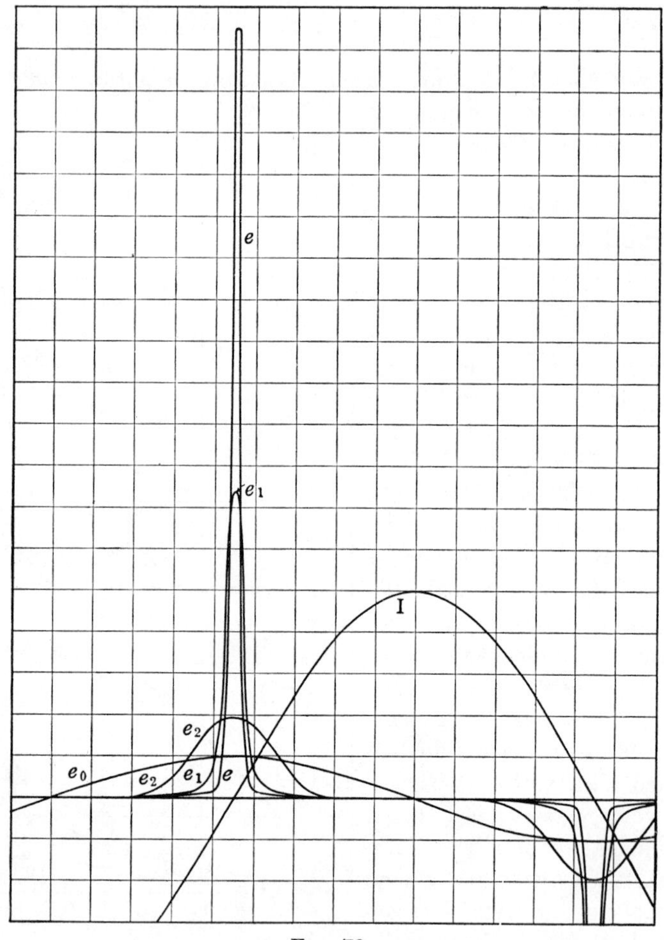

FIG. 70.

The enormous reduction of the voltage peak by an air-gap of 1 per cent. of the length of the magnetic circuit is shown in Figs. 69 and 70.

In Fig. 69, with the magnetic flux density, B, as abscissæ, the m.m.f. of the iron part of the magnetic circuit is shown as curve I. This would be the magnetizing current if the magnetic circuit were closed. Curve II show the m.m.f. consumed in an air-gap of 1 per cent. of the length of the magnetic circuit of curve I, and curve III, therefore, shows the total m.m.f. of the magnetizing current of the magnetic circuit with 1 per cent. air-gap.

Choosing 'as instance the very high saturation $B = 19.7$, the same as illustrated in Fig. 63, and neglecting the hysteresis—which is permissible, as the hysteresis does not much contribute to the wave-shape distortion—the corresponding voltage waves are plotted in Fig. 70, in the same scale as Figs. 56 to 63: for a sine wave of current, curves Fig. 69 give the corresponding values of magnetic flux, and from the magnetic flux wave is derived, as $\frac{dB}{d\phi}$, the voltage wave. The waves of magnetism are not plotted. e_0 is the sine wave of voltage, which would be induced by a sinusoidal variation of magnetic flux; e is the peaked voltage wave induced in a closed magnetic circuit of the same maximum values of magnetism, of form factor $p = 18.5$ (the same as Fig. 63), and e_2 is the voltage wave induced in a magnetic circuit having an air-gap of 1 per cent. of its length. As seen, the excessive peak of e has vanished, and e_2 has a moderate peak only, of form factor $p = 1.9$.

Even a much smaller air-gap has a pronounced effect in reducing the voltage peak. Thus curves IV and V show the m.m.fs. of the air-gap and of the total magnetic circuit, respectively, when containing an air-gap of one-thousandth of the length of magnetic circuit. e_1 in Fig. 70 then shows the voltage wave corresponding to V in Fig. 69: of form factor $p = 7.4$.

Thus, while excessive voltage peaks are produced in a highly saturated closed magnetic circuit, even an extremely small air-gap, such as given by some butt-joints, materially reduces the peak: from form factor $p = 18.5$ to 7.4 at one-thousandth gap length, and with an air-gap of 1 per cent. length, only a moderate peakedness remains at the highest saturation, while at lower saturation the voltage wave is practically a sine.

73. Even a small air-gap in the magnetic circuit of a reactor greatly reduces the wave-shape distortion, that is, makes the voltage wave more sinusoidal, and cuts off the saturation peak. The latter, however, is the case only with a complete air-gap. A partial air-gap or bridged gap, while it makes the wave shape more sinusoidal elsewhere, does not reduce but greatly increases the voltage peak, and produces excessive peaks even below saturation, with a sine wave of current, and such bridged gaps are, therefore, objectionable with series reactors in high-voltage circuits. In shunt reactors, or reactors having a constant sine wave of impressed voltage, the bridged gap merely produces a short flat zero of the current wave, thus is harmless, and for these purposes the bridged gap reactance—shown diagrammatically in Fig. 71 —is extensively used, due to its constructive advantages: greater

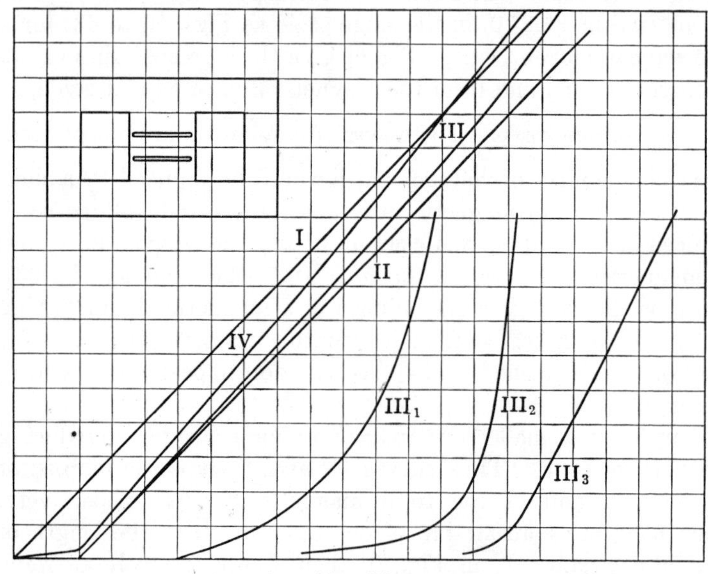

Fig. 71.

rigidity or structure and, therefore, absence of noise, and reduced magnetic stray fields and eddy-current losses resulting therefrom.

Assuming that one-tenth of the gap is bridged, and that the length of the gap is one one-hundredth that of the entire magnetic circuit, as shown diagrammatically in Fig. 71. With such a bridged gap, with all but the lowest m.m.fs. the narrow iron bridges of the gap are saturated, thus carry the flux density $S + H$, where S = metallic saturation density, = 20 kilolines per cm.² in these figures, and H the magnetizing force in the gap.

For one-tenth of the gap, the flux density thus is $H + S$, for the other nine-tenths, it is H, and the average flux density in the gap thus is

$$B = H + 0.1\,S = H + 2,$$

or, if p = bridged fraction of gap,

$$B = H + pS.$$

Curve II in Fig. 71 shows, with the average flux density, B, as abscissæ, the m.m.f. required by the gap,

$$H = B - 0.1\,S$$
$$= B - 2,$$

while curve I shows the m.m.f. which an unbridged gap would require.

Adding to the ordinates of II the values of the m.m.f. required for the iron part of the magnetic circuit, or the other 99 per cent., gives as curve III the total m.m.f. of the reactor.

The lower part of curve III is once more shown, with five times the abscissæ B, and 1000, 100 and 10 times, respectively, the ordinates H, as III_1. III_2, III_3.

74. From $B = 2$ upward, curve III is practically a straight line, and plotting herefrom for a sine wave of current, I and thus m.m.f., H, the wave of magnetism, B, and of voltage, e, these curves become within this range similar to a sine wave as shown as B and e in Fig. 72. Below $B = 2$, however, the slope of the B–H curve and with this their wave shapes change enormously. The B wave becomes practically vertical, that is, B abruptly reverses, and corresponding thereto, the voltage abruptly rises to an excessive peak value, that is, a high and very narrow voltage peak appears on top of the otherwise approximately sine-shaped voltage wave, e.

Choosing the same value as in Fig. 60, $B = 15.4$ or beginning saturation, as the maximum value of flux density: at this, in an entirely closed magnetic circuit the voltage peak is still moderate. On the B–H curve III of Fig. 71, the flux density, $B = 15.4$, requires the m.m.f., $H = 14.4$ If then B and H would vary sinusoidally, giving a sine wave of voltage, e_0, the average value of this voltage wave, e_0, would be proportional to the average rate of magnetic change, or to $\frac{B}{H} = \frac{15.4}{14.4} = 1.07$, and the maximum value of the sine wave of voltage would be $\frac{\pi}{2}$ as high, or,

$$e_0 = \frac{\pi}{2}\frac{B}{H} = \frac{1.07\,\pi}{2} = 1.68.$$

The maximum value of the actual voltage curve, e, occurs at the moment where B passes through zero, and is, from curve III_1,

$$e = \left[\frac{B}{H}\right]_0 = \frac{290}{5} = 580.$$

This, then, is the peak voltage of the actual wave, while, if it were a sine wave, with the same maximum magnetic flux, the maximum voltage would be $e_0 = 1.68$.

The voltage peak produced by the bridged gap and the form factor thus is

$$p = \frac{e}{e_0} = \frac{580}{1.68} = 345,$$

that is, 345 times higher than it would be with a sine wave.

Obviously, such peak can hardly ever occur, as it is usually beyond the limit of the available supply voltage. It thus means, that during the very short moment of time, when during the current reversal the flux density in the iron bridge of the gap changes from saturation to saturation in the reverse direction, a voltage peak rises up to the limits of voltage given by the supply system. This peak is so narrow that even the oscillograph usually does not completely show it.

However, such practically unlimited peaks occur only in a perfectly closed magnetic circuit, containing a bridged gap. If, in addition to the bridged gap of 1 per cent., an unbridged gap of 0.1 per cent.—such as one or several butt-joints—is present, giving the B–H curve IV of Fig. 71, the voltage peak is greatly reduced. It is

$$e_0 = \frac{\pi}{2}\frac{B}{H} = \frac{\pi}{2}\frac{15.4}{15.95} = 1.51,$$

$$e = \left[\frac{B}{H}\right]_0 = \frac{1000}{100} = 10,$$

hence, the relative voltage peak, or form factor,

$$p = \frac{e}{e_0} = 6.6.$$

That is, by this additional gap of one one-thousandth of the length of magnetic circuit, the peak voltage is reduced from 345 times that of the sine wave, to only 6.6 times, or to less than 2 per cent. of its previous value.

As seen from the reasoning in paragraph and Fig. 67, the

peaked wave of Fig. 72 contains very pronounced harmonics up to about the 701th, which at 60 cycles of fundamental frequency, gives frequencies up to 42,000, or well within the range of the danger frequencies of high-voltage power transformers, that is,

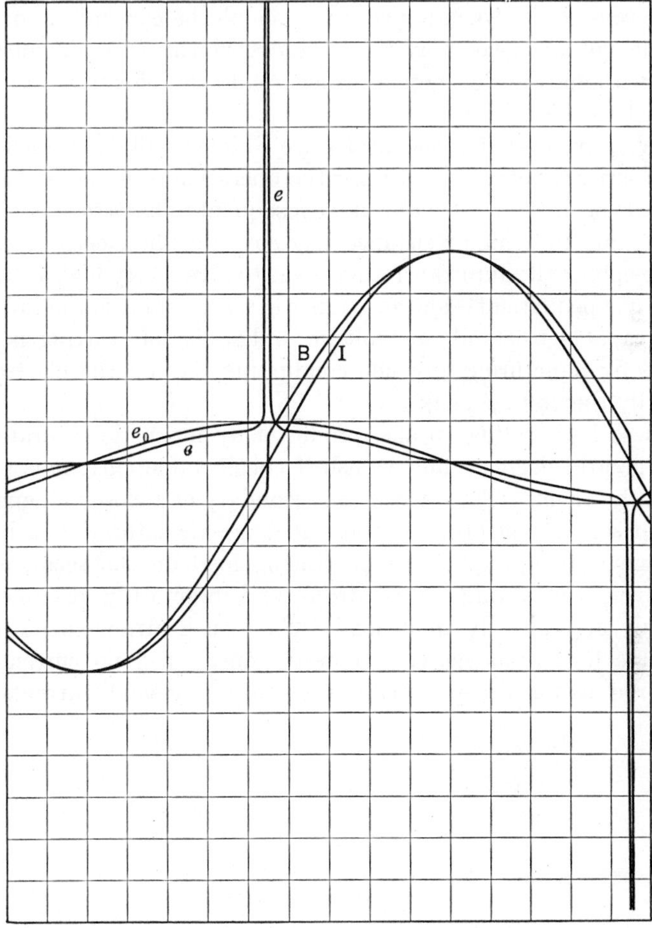

FIG. 72.

frequencies with which the high-voltage coils of transformers, as circuits of distributed capacity, can resonate.

75. Magnetic saturation, and closed or partly closed magnetic circuits thus are a likely source of wave-shape distortion, resulting in high voltage peaks, and where they are liable to occur, as in current transformers, series transformers at open secondary circuit, autotransformers or reactors, etc., they may be guarded against by using a small air-gap in the magnetic circuit, or by providing the extra insulation required to stand the voltage, and the secondary circuit, even if of an effective voltage which is not dangerous to life when a sine wave, should be carefully handled as the voltage peak may reach values which are dangerous to life, without the voltmeter—which reads the effective value—indicating this.

Inversely, such voltage peaks are intentionally provided in some series autotransformers for the operation of individual arcs of the type, in which slagging and consequent failures to start may occur, due to a high-resistance slag covering the electrode tips. By designing the autotransformer so as to give a very high voltage peak at open circuit—and providing in the apparatus the insulation capable to stand this voltage—reliability of starting is secured by puncturing any non-conducting slag on the electrode tips, by the voltage peak.

These high voltage peaks, produced by magnetic saturation, etc., greatly decrease and vanish if considerable current is produced by them. Thus, when the secondary of a closed magnetic circuit series transformer is open, at magnetic saturation, a high voltage peak appears; with increasing load on the secondary, however, the voltage peak drops and practically disappears already at relatively small load. Thus such arrangements are suitable for producing voltage peaks only when no current is required, as for disruptive effects, or only very small currents.

CHAPTER IX

WAVE SCREENS. EVEN HARMONICS

76. The elimination of voltage and current distortion, and production of sine waves from any kind of supply wave, that is, the reverse procedure from that discussed in the preceding chapter, is accomplished by what has been called "wave screens."

Series reactance alone acts to a considerable extent as wave screen, by consuming voltage proportional to the frequency and the current, and thereby reducing the harmonics of voltage in the rest of the circuit the more, the higher their order.

Let the voltage impressed upon the circuit be denoted symbolically by

$$e = e_1 + e_3 + e_5 + e_7 + \ldots$$

$$= \Sigma\, e_n \tag{29}$$

where n denotes the order of the harmonic of absolute numerical value e_n.

If, then, the reactance x (at fundamental frequency) is inserted into the circuit of resistance, r, the impedance is

$$z_1 = \sqrt{r^2 + x^2} \text{ for the fundamental frequency, and}$$

$$z_n = \sqrt{r^2 + n^2x^2} \text{ for the } n\text{th harmonic,} \tag{30}$$

and the current thus is

$$i = \frac{e}{z} = \sum \frac{e}{\sqrt{r^2 + n^2x^2}}, \tag{31}$$

or, denoting

$$\frac{r}{x} = c, \tag{32}$$

it is

$$i = \frac{1}{x}\sum \frac{e_n}{\sqrt{n^2 + c^2}} = \frac{e_1}{x\sqrt{1 + c^2}} + \frac{e_3}{x\sqrt{9 + c^2}} + \frac{e_5}{x\sqrt{25 + c^2}} + \ldots \tag{33}$$

if r is small compared with x, c^2 is negligible compared with 1, 9, 25, etc., and it is

$$i = \frac{1}{x}\left\{ e_1 + \frac{e_3}{3} + \frac{e_5}{5} + \frac{e_7}{7} + \ldots \right\},$$

that is, the current, i, and thus the voltage across the resistance, r, shows the harmonics of the supply voltage, e, reduced in proportion to their order, n.

Even if r is large compared with x, and thus $c^2 > 1$, finally c^2 becomes negligible with n^2, and the harmonics decrease with their order.

77. The screening effect of the series reactance is increased by shunting a capacity, C, beyond the inductance, L, that is, across the resistance, r, as shown in Fig. 73. By consuming current

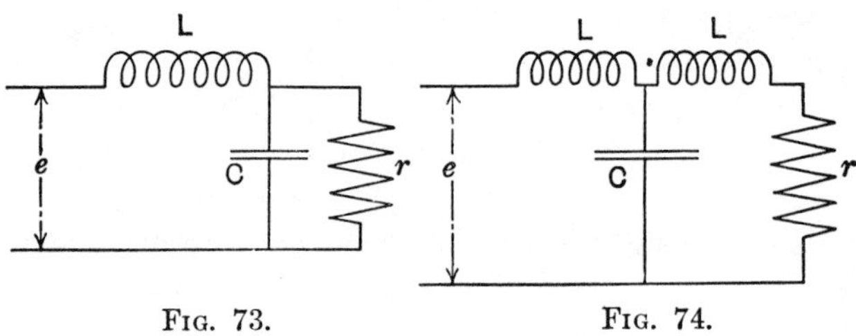

FIG. 73. FIG. 74.

proportional to frequency and voltage, the condenser shunts the more of the current passing through the reactance, the higher the frequency, and thereby still further reduces the higher harmonics of current in the resistance, r, and thus of voltage across this resistance. Its effect is limited, however, by the decreasing voltage distortion at r and thus at the condenser, C.

Thus the screening effect is still further increased by inserting a second inductance, L, beyond the condenser, C, in series to the resistance, r, as shown in Fig. 74. By making the second inductance equal to the first one, and making the condenser, C, of the same reactance, for the fundamental wave, as each of the two inductances, we get what probably is the most effective wave screen.

Under the condition, that the two inductive reactances and the capacity reactance are equal, the equation of the current in the resistance, r, is, for the nth harmonic,

$$I = \frac{je_0}{xn(n^2 - 2) - jr(n^2 - 1)}. \tag{34}$$

or, absolute,

$$i = \frac{e_0}{x} \times \frac{1}{\sqrt{n^2(n^2 - 2)^2 + c^2(n^2 - 1)^2}} \tag{35}$$

where

$$c = \frac{r}{x} \tag{36}$$

If c is small, that is, r small compared with x, the current becomes

$$i = \frac{e_0}{xn\,(n^2 - 2)} \tag{37}$$

or, for higher values of n,

$$i = \frac{e_0}{xn^3}, \tag{38}$$

that is, it decreases with increasing order of harmonic, and proportional to the *cube* of the order n, thus shows an extremely rapid decrease.

If c is not negligible, the denominator in (35) is larger, and i, therefore, still smaller.

As illustration may be shown the current, i_0, and thus the voltage, e_0, across a resistance, r, under the very greatly distorted and peaked voltage of Fig. 62:

(*a*) for a series reactance, x, equal to r, that is, $c = 1$;

(*b*) for the complete wave screen of two inductances and one capacity.

It is

impressed voltage,

$$e = 1.27 \quad e_0 \{ 1_1 + 0.978_3 + 0.935_5 + 0.877_7 + 0.800_9 + 0.713_{11} + 0.617_{13} + 0.517_{15} + 0.416_{17} + 0.315_{19} + 0.189_{21}\}.$$

(*a*) Reduction factor of the nth harmonic,

$$\frac{1}{\sqrt{n^2 + c^2}} = \frac{1}{\sqrt{n^2 + 1}},$$

hence,

$$e_1 = \frac{1.27}{\sqrt{2}} e_0 \{ 1_1 + 0.442_3 + 0.258_5 + 0.175_7 + 0.125_9 + 0.091_{11} + 0.067_{13} + 0.049_{15} + 0.034_{17} + 0.023_{19} + 0.013_{21}\}.$$

(b) Reduction factor of the nth harmonic,

$$\frac{1}{n(n^2 - 2)},$$

hence,

$$e_2 = 1.27\ e_0 \{1_1 + 0.047_3 + 0.008_5 + 0.003_7 + 0.001_9 + 0.001_{11}\}.$$

That is, the third harmonic is reduced to less than 5 per cent., the fifth to less than 1 per cent., and the higher ones are practically entirely absent.

While in the supply voltage wave, e, the voltage peak (by adding the numerical values of all the harmonics: $1 + 0.978 + 0.935 + \ldots$) is 7.36 times that of the fundamental wave, it is reduced by series reactance to less than 2.28 times the maximum of the fundamental wave, that is, very greatly reduced, and by the complete wave screen to less than 1.06 times the maximum of the fundamental. That is, in the last case the voltage is practically a perfect sine wave.

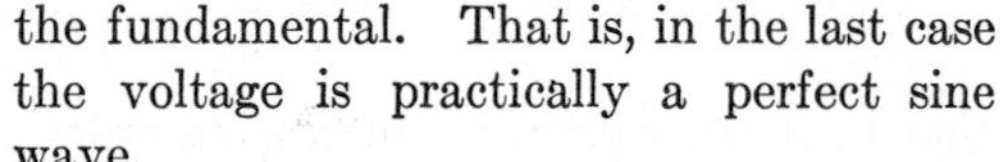

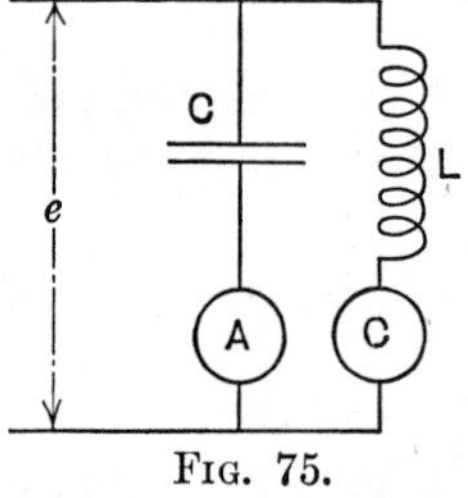

Fig. 75.

78. By "wave screens" the separation of pulsating currents into their alternating and their continuous component, or the separation of complex alternating currents—and thus voltages—into their constituent harmonics can be accomplished, and inversely, the combination of alternating and continuous currents or voltages into resultant complex alternating or pulsating currents.

The simplest arrangement of such a wave screen for separating, or combining alternating and continuous currents into pulsating ones, is the combination, in shunt with each other, of a capacity, C, and an inductance, L, as shown in Fig. 75. If, then, a pulsating voltage, e, is impressed upon the system, the pulsating current, i, produced by it divides, as the continuous component can not pass through the condenser, C, and the alternating component is barred by the inductance, L, the more completely, the higher this inductance. Thus the current, i_1, in the apparatus, A, is a true alternating current, while the current, i_0, in the apparatus, C, is a slightly pulsating direct current.

Inversely, by placing a source of alternating voltage, such as an alternator or the secondary of a transformer, at A, and a source of continuous voltage, such as a storage battery or direct-current generator, at C, in the external circuit a pulsating voltage, e, and pulsating current, i, result.

If the capacity, C, is so large as to practically short-circuit the alternating voltage, and the inductance, L, so high as to practically open-circuit the alternating voltage, the separation—of combination—is practically complete, and independent of the frequency of the alternating wave.

Wave screens based on resonance for a definite frequency by series connection of capacity and inductance, can be used to separate the current of this frequency from a complex current or voltage wave, such as those given in Figs. 56 to 63, and thus can be used for separation of complex waves into their components, by "harmonic analysis."

Thus in Fig. 76, if the successive capacities and inductances are chosen such that

$$2\,\pi f L_1 = \frac{1}{2\,\pi f C_1},$$

$$6\,\pi f L_3 = \frac{1}{6\,\pi f C_3},$$

$$10\,\pi f L_5 = \frac{1}{10\,\pi f C_5},$$

$$2n\,\pi f L_n = \frac{1}{2\,\pi f n C_n} \qquad (39)$$

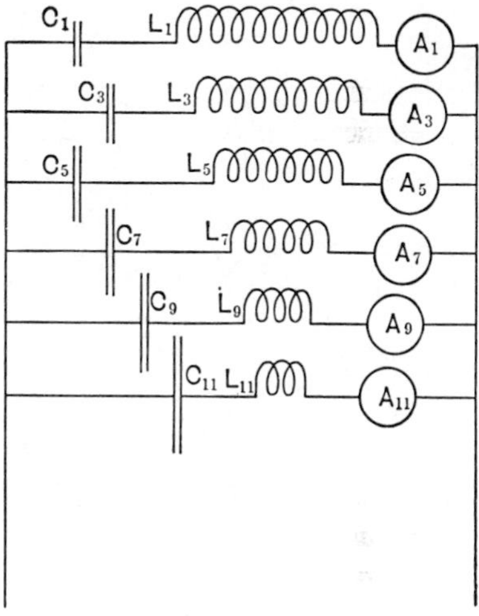

FIG. 76.

where f = frequency of the fundamental wave.

Then, through any of the branch circuits C_n, L_n, only the nth harmonic, i_n, can pass to an appreciable extent.

Such resonant wave screen, however, has the serious disadvantage to require very high constancy of f, since the resonance condition between C_n and L_n depends on the square of f,

$$\frac{1}{C_n} = 4\,\pi^2 f^2 L_n.$$

79. Even harmonics are produced in a closed magnetic circuit by the superposition of a continuous current upon the alternating wave. With an alternating sine wave impressed upon an iron magnetic circuit, saturation, or in general the lack of proportionality between magnetic flux and m.m.f., produces a wave-shape distortion, that is, higher harmonics, of voltage with a sine wave of current, of current with a sine wave of impressed voltage. The constant term of a wave, however, is the first even harmonic, and thus, if the impressed wave comprises a fundamental sine and a

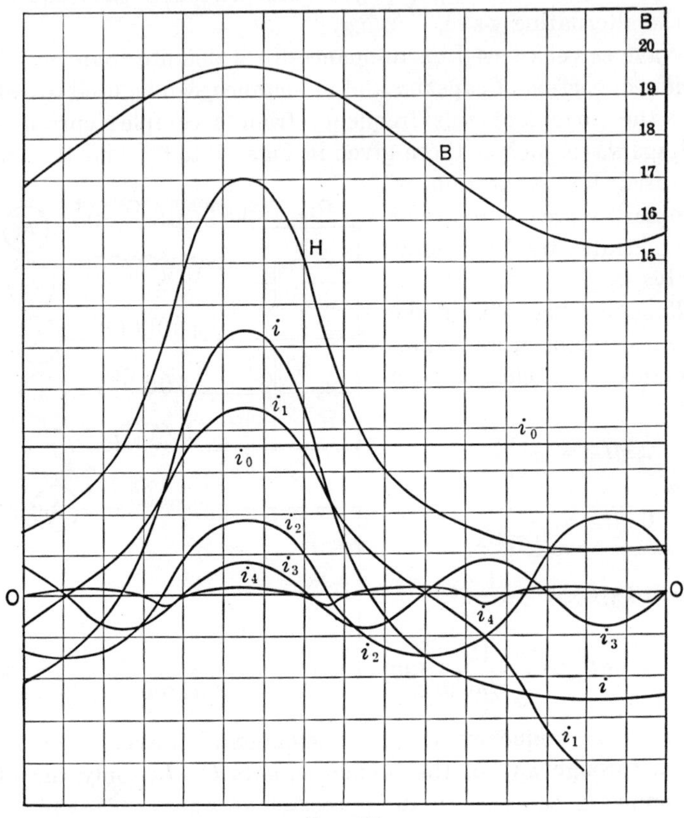

Fig. 77.

constant term, the former gives rise to the odd harmonics, the latter to the even harmonics.

Let, then, on the alternating sine wave of impressed voltage a continuous current by superimposed. The magnetic flux then oscillates sinusoidally, not between equal and opposite values, but between two unequal values, which may be of the same, or of opposite signs. That is, it performs an unsymmetrical magnetic cycle. Neglecting again hysteresis, that is, assuming the rising and the decreasing magnetization curve as coincident—which is permissible as approximation, since the hysteresis contributes little to distortion—and choosing the same magnetization curve as in the preceding, curve I in Fig. 64, we may as an instance consider a sinusoidal magnetic pulsation between the limits +15.4 and +19.7, corresponding to a variation of the m.m.f. between $H = +10$ and $H = +100$.

Fig. 77 then gives, as curve B, the sinusoidally pulsating magnetic flux density. Taking from curve I, Fig. 64, the values of H corresponding to the values in curve B, Fig. 77, gives curve H. This, resolved (*Engineering Mathematics*, paragraph 92) gives the constant term $i_0 = 36$, and the alternating current, i. The latter is unsymmetrical, having one short half-wave of a peak value 64, and one long half-wave of maximum value 26. It thus resolves into the odd harmonics, i_1, alternating between ± 45, and the even harmonics, mainly the second harmonic, alternating between maximum values +18 and −15. i_1 is peaked with flat zero, thus showing a third harmonic, which is separated as i_3, and i_2 is unsymmetrical, showing further even harmonics, which are separated as i_4, but are rather small.

Thus the pulsating exciting current of the sinusoidally varying unidirectional magnetic flux

$$B = 17.55 + 2.15 \cos \phi$$

is given by

$$H = 36 + 37 \cos \phi + 16.5 \cos 2\phi + 8 \cos 3\phi + 2 \cos 4\phi + \ldots$$

Instead of superimposing a direct current upon an alternating wave, as by connecting in series an alternator and a direct-current generator or storage battery, two separate coils can be used on the magnetic circuit, one energized by an alternating impressed voltage, the other by a direct current. A high inductive reactance would then be connected in the latter circuit, to eliminate the current pulsation which would be caused by the alternating voltage induced in this coil.

Connecting two such magnetic circuits with their direct-current magnetizing coils in series, but in opposition (without the use of a series reactance) eliminates the induced fundamental wave, but leaves the second harmonic in the direct-current circuit, which thus can be separated. Numerous arrangements can then be devised by two magnet cores energized by separate alternating-current exciting coils and saturated by one common direct-current exciting coil, surrounding both cores, or their common return, etc.

80. The preceding may illustrate some of the numerous wave-shape distortions which are met in electrical engineering, their characteristics, origin, effects, use and danger. Numerous other wave distortions, such as those produced by arcs, by unidirectional conductors, by dielectric effects such as corona, by Y connection of transformers for reactors, by electrolytic polarization, by pulsating resistance or reactance, etc., are discussed in other chapters or may be studied in a similar manner.

CHAPTER X

INSTABILITY OF CIRCUITS: THE ARC

A. General

81. During the earlier days of electrical engineering practically all theoretical investigations were limited to circuits in stable or stationary condition, and where phenomena of instability occurred, and made themselves felt as disturbances or troubles in electric circuits, they either remained ununderstood or the theoretical study was limited to the specific phenomenon, as in the case of lightning, dropping out of step of induction motors, hunting of synchronous machines, etc., or, as in the design of arc lamps and arc-lighting machinery, the opinion prevailed that theoretical calculations are impossible and only design by trying, based on practical experience, feasible.

The first class of unstable phenomena, which was systematically investigated, were the transients, and even today it is questionable whether a systematic theoretical classification and investigation of the conditions of instability in electric circuits is yet feasible. Only a preliminary classification and discussion of such phenomena shall be attempted in the following.

Three main types of instability in electric systems may be distinguished:

I. The transients of readjustment to changed circuit conditions.

II. Unstable electrical equilibrium, that is, the condition in which the effect of a cause increases the cause.

III. Permanent instability resulting from a combination of circuit constants which can not coexist.

I. Transients

82. Transients are the phenomena by which, at the change of circuit conditions, current, voltage, etc., readjust themselves from the values corresponding to the previous condition to the values corresponding to the new condition of the circuit. For instance, if a switch is closed, and thereby a load put on the circuit, the current can not instantly increase to the value corresponding to the increased load, but some time elapses, during which the increase of the stored magnetic energy corresponding to the increased current, is brought about. Or, if a motor switch is closed, a period of acceleration intervenes before the flow of current becomes stationary, etc.

The characteristic of transients therefore is, as implied in the term, that they are of limited, usually very short duration, intervening between two periods of stationary conditions.

Considerable theoretical work has been done, more or less systematically, on transients, and a great mass of information is thus available in the literature. These transients are more extensively treated in *Transient Electrical Phenomena*, Volume III, Dover edition, and on pp. 127–274, Volume II, Dover edition, and therefore will be omitted in the following. However, to some extent, the transients of our theoretical literature still are those of the "phantom circuit," that is, a circuit in which the constants r, L, C, g, are assumed as constant. The effect of the variation of constants, as found more or less in actual circuits: the change of L with the current in circuits containing iron; the change of C and g with the voltage (corona, etc.); the change of r and g with the frequency, etc., has been studied to a limited extent only, and in specific cases.

In the application of the theory of transients to actual electric circuits, considerable judgment thus is often necessary to allow and correct for these "secondary" phenomena which are not included in the theoretical equations.

Especially deficient is our knowledge of the conditions under which the attenuation constant of the transient becomes zero or negative, and the transient thereby becomes permanent, or becomes a cumulative surge, and the phenomenon thereby one of unstable equilibrium.

II. Unstable Electrical Equilibrium

83. If the effect brought about by a cause is such as to oppose or reduce the cause, the effect must limit itself and stability be finally reached. If, however, the effect brought about by a cause increases the cause, the effect continues with increasing intensity, that is, instability results.

This applies not to electrical phenomena alone, but equally to all other phenomena.

Instability of an electric circuit may assume three different forms:

1. Instability leading up to stable conditions.

For instance, in a pyroelectric conductor of the volt-ampere characteristic given in Fig. 78, at the impressed voltage, e_0, three different values of current are possible: i_1, i_2 and i_3. i_1 and i_3 are stable, i_2 unstable. That is, at current, i_2, passing through the conductor under the constant impressed voltage, e_0, a momentary increase of current would give an excess voltage beyond that required by the conductor, thereby increase the current still

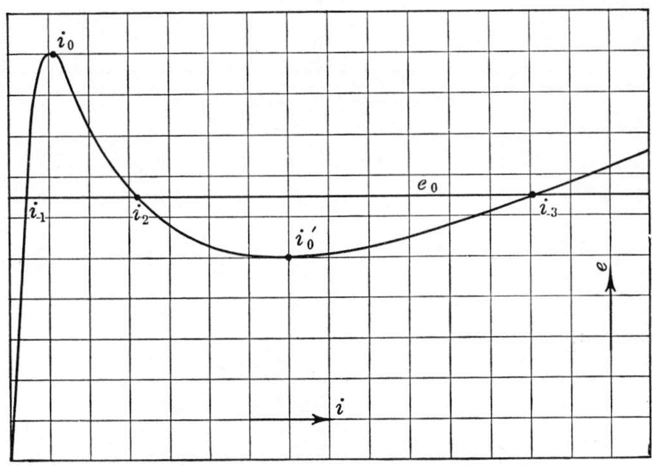

FIG. 78.

further, and with increasing rapidity the current would rise, until it becomes stable at the value, i_3. Or, a momentary decrease of current, by requiring a higher voltage than available, would further decrease the current, and with increasing rapidity the current would decrease to the stable value, i_1.

2. Instability putting the circuit out of service.

An instance is the arc on constant-potential supply. With the volt-ampere characteristic of the arc shown as A, in Fig. 79, a current of 4 amp. would require 80 volts across the arc terminals. At a constant impressed voltage of 80, the current could not remain at 4 amp., but the current would either decrease with increasing rapidity, until the arc goes out, or the current would increase with increasing rapidity, up to short-circuit, that is, until the supply source limits the current.

3. Instability leading again to instability, and thus periodically repeating the phenomena.

For instance, if an arc of the volt-ampere characteristic, A, in Fig. 79 is operated in a constant-current circuit of sufficiently high direct voltage to restart the arc when it goes out, and the arc

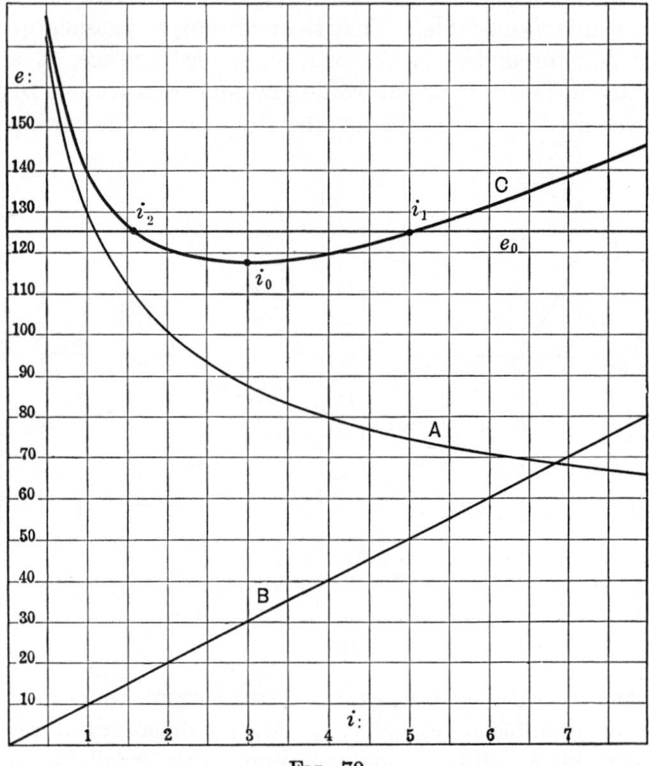

Fig. 79.

is shunted by a condenser, the condenser makes the arc unstable and puts it out; the available supply voltage, however, starts it again, and so periodically the arc starts and extinguishes, as an "oscillating arc."

84. There are certain circuit elements which tend to produce instability, such as arcs, pyroelectric conductors, condensers, induction and synchronous motors, etc., and their recognition therefore is of great importance to the engineer, in guarding

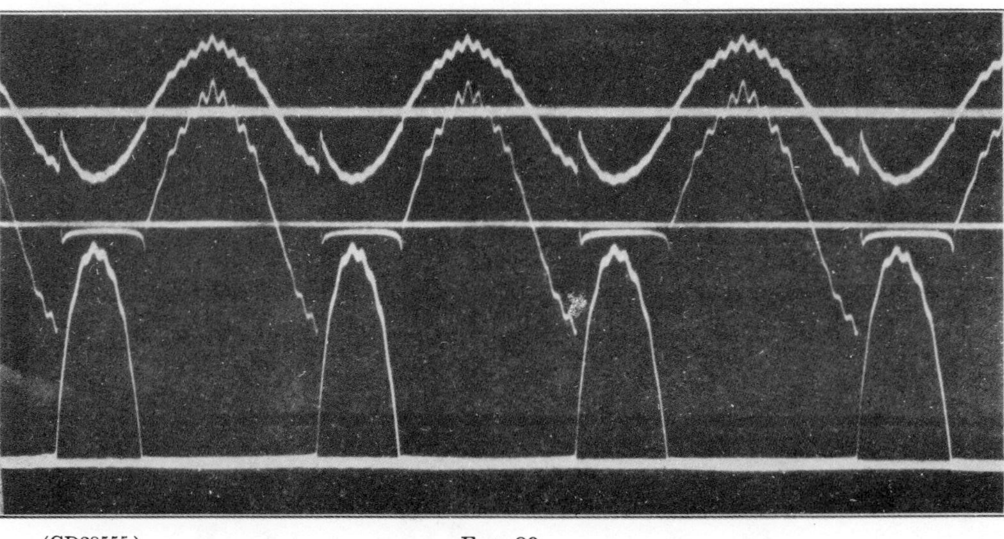

(CD28555.)

FIG. 80.

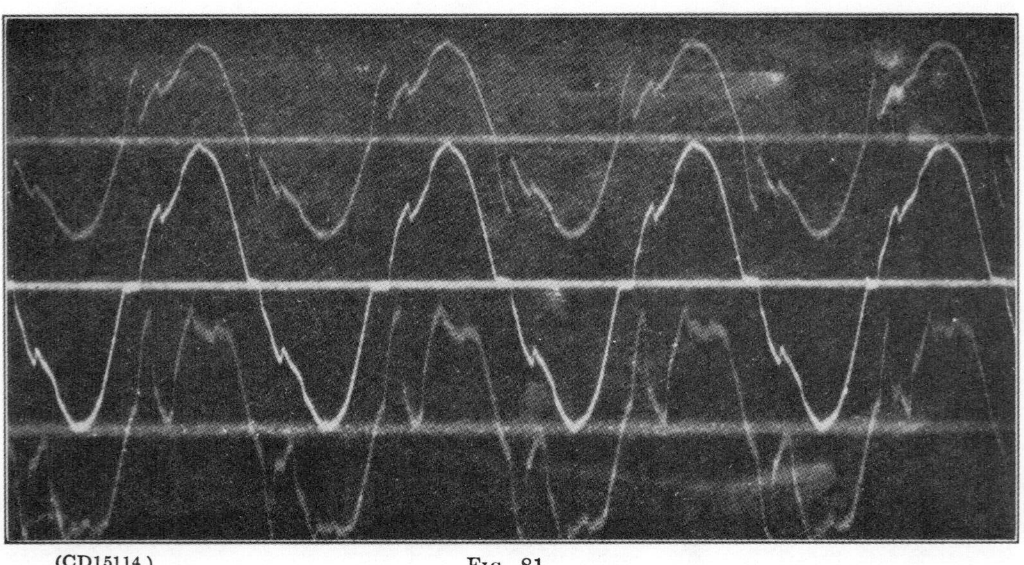

(CD15114.)

FIG. 81.

against instability. Whether instability results, and what form it assumes, depends, however, not only on the "exciting element," as we may call the cause of the instability, but on all the elements of the circuit. Thus an arc is unstable, form (2), on constant-voltage supply at its terminals; it is stable on constant-current supply. But when shunted by a condenser, it becomes unstable on constant current, and the instability may be form (2) or form (3), depending on the available voltage. With a resistance, r, of volt-ampere characteristic ir shown as B, in Fig. 79, the arc is stable on constant-voltage supply for currents above $i_0 = 3$ amp., unstable below 3 amp., and therefore, with a constant-supply voltage, e_0, two current values, i_1 and i_2, exist, of which the former one is stable, the latter one unstable. That is, current, i_2, can not persist, but the current either runs up to i_1 and the arc then gets stable (form 1), or the current decreases and the arc goes out, instability form (2).

Thus it is not feasible to separately discuss the different forms of instability, but usually all three may occur, under different circuit conditions.

The electric arc is the most frequent and most serious cause of instability of electric circuits, and therefore should first be suspected, especially if the instability assumes the form of high-frequency disturbances or abrupt changes of current or voltage, such as is shown for instance in the oscillograms, Figs. 80 and 81.

Somewhat similar effects of instability are produced by pyro-electric conductors.

Induction motors and synchronous motors may show instability of speed: dropping out of step, etc.

III. Permanent instability

85. If the constants of an electric circuit, as resistance, inductance, capacity, disruptive strength, voltage, speed, etc., have values, which can not coexist, the circuit is unstable, and remains so as long as these constants remain unchanged.

Case (3) of II, unstable equilibrium, to some extent may be considered as belonging in this class.

The most interesting class in this group of unstable electric systems are the oscillations resulting sometimes from a change of circuit conditions (switching, change of load, etc.), which continue indefinitely with constant intensity, or which steadily increase in intensity, and may thus be called permanent and cumulative surges, hunting, etc. They may be considered as transients in which the attenuation constant is zero or negative.

In the transient resulting from a change of circuit conditions, the energy which represents the difference of stored energy of the circuit before and after the change of circuit condition, is dissipated by the energy loss in the circuit. As energy losses always occur, the intensity of a true transient thus must always be a maximum at the beginning, and steadily decrease to zero or permanent condition. An oscillation of constant intensity, or of increasing intensity, thus is possible only by an energy supply to the oscillating system brought about by the oscillation. If this energy supply is equal to the energy dissipation, constancy of the phenomenon results. If the energy supply is greater than the energy dissipation, the oscillation is cumulative, and steadily increases until self-destruction of the system results, or the increasing energy loss becomes equal to the energy supply, and a stationary condition of oscillation results. The mechanism of this energy supply to an oscillating system from a source of energy differing in frequency from that of the oscillation is still practically unknown, and very little investigating work has been done to clear up the phenomenon. It is not even generally realized that the phenomenon of a permanent or cumulative line surge involves an energy supply or energy transformation of a frequency equal to that of the oscillation.

Possibly the oldest and best-known instance of such cumulative oscillation is the hunting of synchronous machines.

Cumulative oscillations between electromagnetic and electrostatic energy have been observed by their destructive effects in high-voltage electric circuits on transformers and other apparatus, and have been, in a number of instances where their frequency was sufficiently low, recorded by the oscillograph. They obviously are the most dangerous phenomena in high-voltage electric circuits. Relatively little exact knowledge exists of their origin. Usually—if not always—an arc somewhere in the system is instrumental in the energy supply which maintains the oscillation. In some instances, as in wireless telegraphy, they have found industrial application. A systematic theoretical investigation of these cumulative electrical oscillations probably is one of the most important problems before the electrical engineer today.

The general nature of these permanent and cumulative oscillations and their origin by oscillating energy supply from the transient of a change of circuit condition, is best illustrated by the instance of the hunting of synchronous machines, and this will, therefore, be investigated somewhat more in detail.

B. The Arc as Unstable Conductor

86. The instability of the arc is the result of its dropping volt-ampere characteristic, as discussed in paragraphs 18 to 27 of the

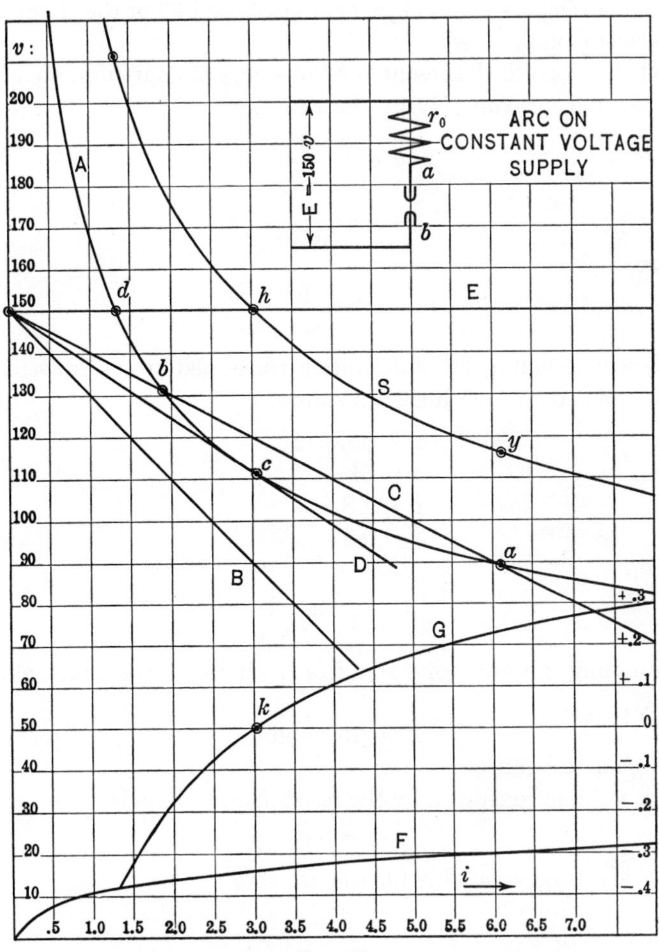

FIG. 82.

chapter on "Electric Conduction" [Dover, Volume I, Part III, Chapter II, pp. 356–369]. As shown there, the arc is always unstable on constant voltage impressed upon it. Series resistance or reactance produces stability for currents above a certain critical value of current, i_0. Such curves, giving the voltage consumed by the arc and its series resistance as function of the current, thus may be termed stability curves of the arc. Their minimum values, that is, the stability limits corresponding to the different resistances, give the stability characteristic of the arc. The equations of the arc, and of its stability curves and stability characteristic, are given in paragraphs 22 and 23 of the chapter on "Electric Conductors."

Let, in Fig. 82, A present the volt-ampere characteristic of an arc, given approximately by the equation

$$\left.\begin{aligned} e &= a + \frac{c(l + \delta)}{\sqrt{i}} \\ &= a + \frac{b}{\sqrt{i}} \end{aligned}\right\} \tag{1}$$

where

$$e' = \frac{b}{\sqrt{i}} \tag{2}$$

is the stream voltage, that is, voltage consumed by the arc stream.

Fig. 82 is drawn with the constants,

$$\begin{aligned} a &= 35, \\ c &= 51, \\ l &= 1.8, \\ \delta &= 0.8, \end{aligned}$$

hence,

$$e = 35 + \frac{133}{\sqrt{i}}.$$

Assuming this arc is operated from a circuit of constant-voltage supply,

$$E = 150 \text{ volts},$$

through a resistance, r_0

The voltage consumed by the resistance, r_0, then is

$$e_2 = r_0 i, \tag{3}$$

and the voltage available for the arc thus

$$e_1 = E - r_0 i \tag{4}$$

Lines B, C and D of Fig. 82 give e_1, for the values of resistance,

$$\begin{aligned} r_0 &= 20 \text{ ohms } (B) \\ &= 10 \text{ ohms } (C) \\ &= 13 \text{ ohms } (D). \end{aligned}$$

As seen, line B does not intersect the volt-ampere characteristic, A, of the arc, that is, with 20 ohms resistance in series, this l = 2.5 cm. arc can not be operated from E = 150 volt supply.

Line C intersects A at a and b, i = 6.1 and 1.9 amp. respectively.

At a, i = 6.1 amp., the arc is stable;

At b, i = 1.9 amp., the arc is unstable;

for the reasons discussed before: an increase of current decreases the voltage consumed by the circuit, $e + e_2$, and thus still further increases the current, and inversely. Thus the arc either goes out, or the current runs up to i = 6.1 amp., where the arc gets stable.

Line D is drawn tangent to A, and the contact point, c, thus gives the minimum current, i = 3.05 amp., of operation of the arc on E = 150 volts, that is, the value of current or of series resistance, at which the arc ceases to be stable: a point of the stability characteristic, S, of the arc.

This stability characteristic is determined by the condition

$$\frac{de_0}{di} = 0, \tag{5}$$

where

$$e_0 = e + r_0 i \tag{6}$$

$$= a + \frac{b}{\sqrt{i}} + r_0 i,$$

this gives

$$r_0 = \frac{b}{2\, i\sqrt{i}} = \frac{e_1}{2\, i} \tag{7}$$

and

$$\left.\begin{aligned} e_0 &= a + \frac{1.5\, b}{\sqrt{i}} \\ &= a + 1.5\, e_1 \end{aligned}\right\} \tag{8}$$

as the equation of the stability characteristic of the arc on a constant-voltage circuit.

87. In general, the condition of stability of a circuit operated on constant-voltage supply, is

$$\frac{de}{di} > 0 \tag{9}$$

where e is the voltage consumed by the current, i, in the circuit.

The ratio of the change of voltage, de, as fraction of the total voltage, e, brought about by a change of current, di, as fraction of the total current, i, thus may be called the *stability coefficient* of the circuit,

$$\left.\begin{aligned}\delta &= \frac{\dfrac{de}{e}}{\dfrac{di}{i}}\\ &= \frac{\dfrac{de}{di}}{\dfrac{e}{i}}\end{aligned}\right\}. \qquad (10)$$

In a circuit of constant resistance, r, it is

$$\frac{e}{i} = r,$$

$$\frac{de}{di} = r,$$

hence,

$$\delta = 1,$$

that is, the stability coefficient of a circuit of constant resistance, r, is unity.

In general, if the effective resistance, r, is not constant, but varies with the current, i, it is

$$e = ri,$$

$$\frac{de}{dt} = r + i\frac{dr}{di};$$

hence, the stability coefficient

$$\delta = 1 + \frac{\dfrac{dr}{di}}{\dfrac{r}{i}} \qquad (11)$$

thus in a circuit, in which the resistance increases with the current, the stability coefficient is greater than 1. Such is that of a conductor with positive temperature coefficient of resistance, in which the temperature rise due to the increase of current increases the resistance. A conductor with negative temperature coefficient of resistance gives a stability coefficient less than 1, but as long as δ is still positive, that is, the decrease of resistance slower than the increase of current, the circuit is stable.

$$\delta > 0 \qquad (12)$$

is the condition of stability of a circuit on constant-voltage supply, and

$$\delta < 0 \tag{13}$$

is the condition of instability, and

$$\delta = 0 \tag{14}$$

thus gives the stability characteristic of the circuit.

In the arc,

$$e = a + \frac{b}{\sqrt{i}},$$

the stability coefficient is, by (10),

$$\delta = -\frac{b}{2\,e\sqrt{i}} = -\frac{e'}{2\,e} \tag{15}$$

that is, equals half the stream voltage, $\frac{e'}{2}$, divided by the arc voltage, e.

Or, substituting for e in (15), and rearranging,

$$\delta = -\frac{1}{2\left(1 + \frac{a}{b}\sqrt{i}\right)} \tag{16}$$

$$= -\frac{1}{2(1 + 0.2625\,\sqrt{i})}$$

in Fig. 82.

For $i = 0$, it is $\delta = -0.5$;

$i = \infty$, it is $\delta = 0$.

The stability coefficient of the arc having the volt-ampere characteristic, A, in Fig. 82 is shown as F in Fig. 82.

88. On constant-voltage supply, $E = 150$ volts, the arc having the characteristic, A, Fig. 82, can not be operated at less than 3.05 amperes. At $i = 3.05$ is its stability limit, that is, the stability coefficient of arc plus series resistance, r_0, required to give 150 volts, changes from negative for lower currents, to positive for higher currents.

The stability coefficient of such arcs, operated on constant-voltage supply through various amounts of series resistance, r_0, then would be given by

$$\delta_0 = \frac{\frac{de_0}{di}}{\frac{e_0}{i}},$$

where

$$e_0 = a + \frac{b}{\sqrt{i}} + r_0 i \tag{17}$$

and the resistance r_0 chosen so as to give

$$e_1 = 150 \text{ volts,}$$

from (17) follows,

$$\delta_0 = \frac{-\frac{b}{2\sqrt{i}} + r_0 i}{e_0},$$

and, substituting from (17),

$$ir_0 = e_0 - a - \frac{b}{\sqrt{i}}$$

gives

$$\delta_0 = 1 - \frac{a + \frac{1.5\,b}{\sqrt{i}}}{e_0} \tag{18}$$

or,

$$\delta_0 = 1 - \frac{e_0'}{e_0} \tag{19}$$

where e_0 is the supply voltage, e_0' the voltage given by the stability characteristic, S.

δ_0, the stability characteristic of the arc, A, on $E = 150$ volt constant-potential supply, is given as curve, G, in Fig. 82. As seen, it passes from negative—instability—to positive—stability—at the point, k, corresponding to c and h on the other curves.

89. On a constant-current supply, an arc is inherently stable. Instability, however, may result by shunting the arc by a resistance, r_1. Thus in Fig. 83, let $I = 5$ amp. be the constant supply current. The volt-ampere characteristic of the arc is given by A, and shows that on this 5-amp. circuit, the arc consumes 94 volts, point d.

Let now the arc be shunted by resistance, r_1. If e = voltage consumed by the arc, the current shunted by the resistance, r_1, is

$$i_1 = \frac{e}{r_1} \tag{20}$$

and the current available for the arc thus is

$$i = I - i_1 \tag{21}$$

$$= I - \frac{e}{r_1}$$

or

$$e = r_1(I - i). \tag{22}$$

Curves B, C and D of Fig. 83 show the values of equation (22) for

$$\begin{aligned} r_1 &= 32 \text{ ohms: line } B \\ &= 48 \text{ ohms: line } C \\ &= 40.8 \text{ ohms: line } D. \end{aligned}$$

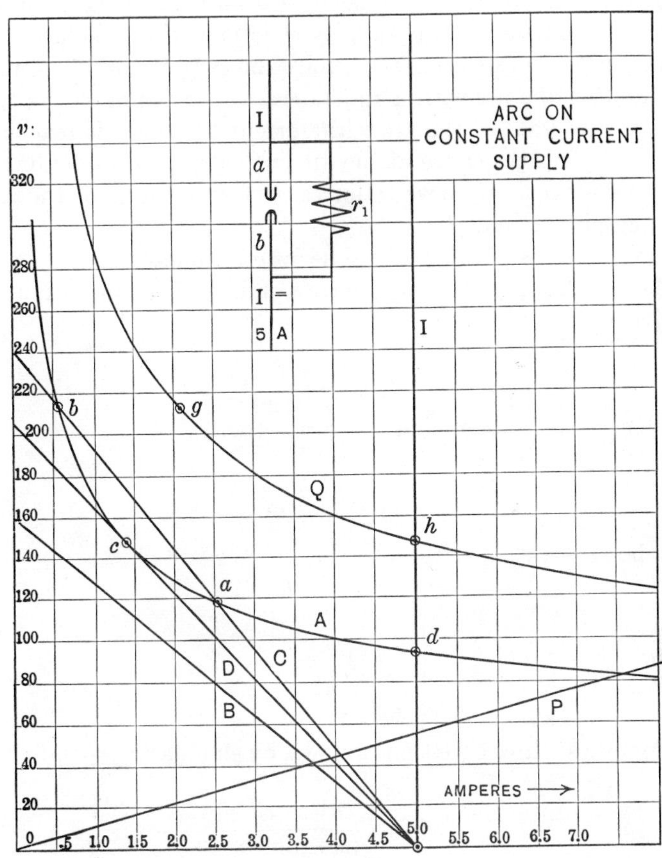

FIG. 83.

Line B does not intersect the arc characteristic, A, that is, with a resistance as low as $r_1 = 32$, no arc can be maintained on the 5-amp. constant-current circuit.

Line C intersects A at two points:

(a) $i = 2.55$ amp., $e = 118$ volts, stable condition;

(b) $i = 0.55$ amp., $e = 214$ volts, unstable condition.

Line D is drawn tangent to A, touches at c: $i = 1.4$ amp.,

$e = 148$ volts, the limit of stability. At $I = 5$ amp., the point h, at $e = 148$ volts, thus gives the voltage consumed by an arc when by shunting it with a resistance the stability limit is reached.

Drawing then from the different points of the abscissæ, i, tangents on A, and transferring their contact points, c, b, to the abscissæ, from which the tangent is drawn, gives the points h, g, of the constant-current stability characteristic of the arc, that is, the curve of arc voltages in a constant-current circuit, I, when by shunting the arc with a resistance, r_1, consuming current, i_1, the stability limit of the arc with current $i = I - i_1$ is reached.

P then gives the curve of the arc currents, i, corresponding to the arc voltage, e, of curve Q, for the different values of the constant-circuit current, I.

The equations of Q and P are derived as follows:

The stability limit, point c, corresponding to circuit current, I, as given by

$$\frac{de}{di} = -r_1,$$

where e = arc voltage, and i = arc current.
Or,

$$r_1 = \frac{b}{2\,i\sqrt{i}}. \tag{23}$$

It is, however,

and

$$\begin{cases} e = a + \dfrac{b}{\sqrt{i}} \\ \dfrac{e}{I - i} = r_1. \end{cases}$$

From these three equations follows, by eliminating r_1 and i or e, Q,

$$I = \frac{b^2(3\,e - a)}{(e - a)^3} \tag{24}$$

P,

$$I = \frac{i(3\,b + 2\,a\sqrt{i})}{b}. \tag{25}$$

These curves are of lesser interest than the constant-voltage stability curve of the arc, S in Fig. 82.

It is interesting to note, that the resistance, r_1 (23), which makes an arc unstable as shunting resistance in a constant-current circuit, has the same value as the resistance, r_0, (7), which as series resistance makes it unstable in a constant-voltage supply circuit.

90. Due to the dropping volt-ampere characteristic, two arcs can not be operated in parallel, unless at least one of them has a sufficiently high resistance in series.

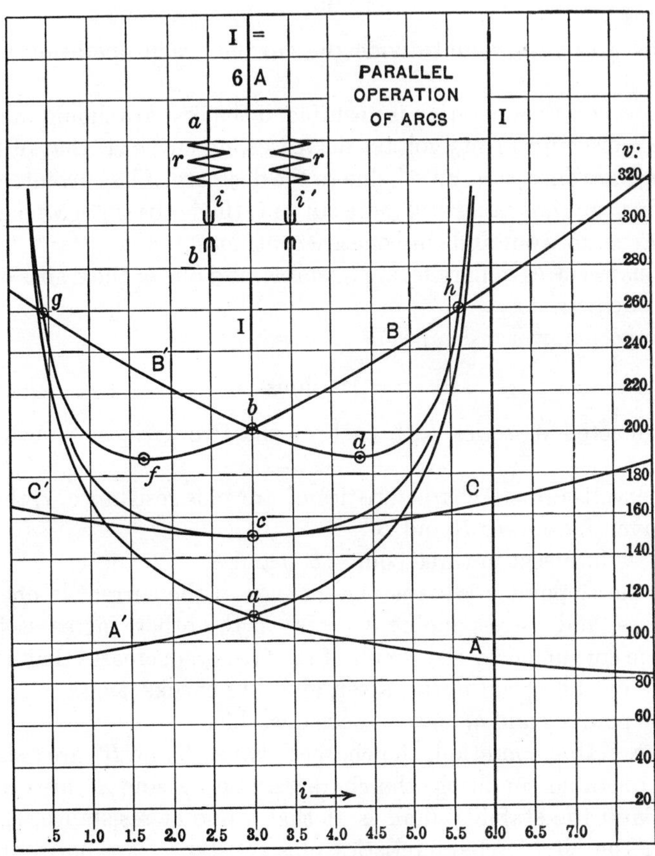

FIG. 84.

Let, as shown in Fig. 84, two arcs be connected in parallel into the circuit of a constant current

$$I = 6 \text{ amp.}$$

Assume at first both arcs of the same length and same electrode material, that is, the same volt-ampere characteristic.

Let i = current in the first arc, thus $i' = I - i$ = current in the second arc.

The volt-ampere characteristic of the first arc, then, is given by A in Fig. 84, that of the second arc by A'.

As the two parallel arcs must have the same voltage, the operating point is the point, a, of the intersection of A and A' in Fig. 84.

The arcs thus would divide the current, each operating at 3 amp.

However, the operation is unstable: if the first arc should take a little more current, its voltage decreases, on curve A, that of the second arc increases, on A', due to the decrease of its current, and the first arc thus takes still more current, thus robs the second arc, the latter goes out and only one arc continues.

Thus two arcs in parallel are unstable, and one of them goes out, only one persists.

Suppose now a resistance of

$$r = 30 \text{ ohms}$$

is connected in series with each of the two arcs, as shown in Fig. 84.

The volt-ampere characteristics of arc plus resistance, r, then, are given by curves B and B'.

These intersect in three points: b, g and h.

Of these, point b is stable: an increase of the current in one of the arcs, and corresponding decrease in the other, increases the voltage consumed by the circuit of the former, decreases that consumed by the circuit of the latter, and thus checks itself.

The points g and h, however, are unstable.

At b, stable condition, the characteristics, B and B', are rising; at a, unstable condition, the characteristics, A and A', are dropping, and the stability limit is at that value of resistance, r, at which the circuit characteristics plus resistance, are horizontal, the point c, where the characteristics, C and C', touch each other.

c is the stability limit of C or C', thus a point of the stability characteristic of either arc, or given by the equation

$$e = a + \frac{1.5\, b}{\sqrt{i}}.$$

Fig. 85 shows the case of two parallel arcs, which are not equal and do not have equal resistances, r, in series, one being a long arc, having no resistance in series, the other a short arc with a resistance $r = 40$ ohms in series.

The volt-ampere characteristic of the long arc is given by A, that of the short arc by B, and that of the short arc plus resistance, r, by C.

A and C intersect at three points, a, b and c. Of these, only the point a is stable, as any change of current from this point limits

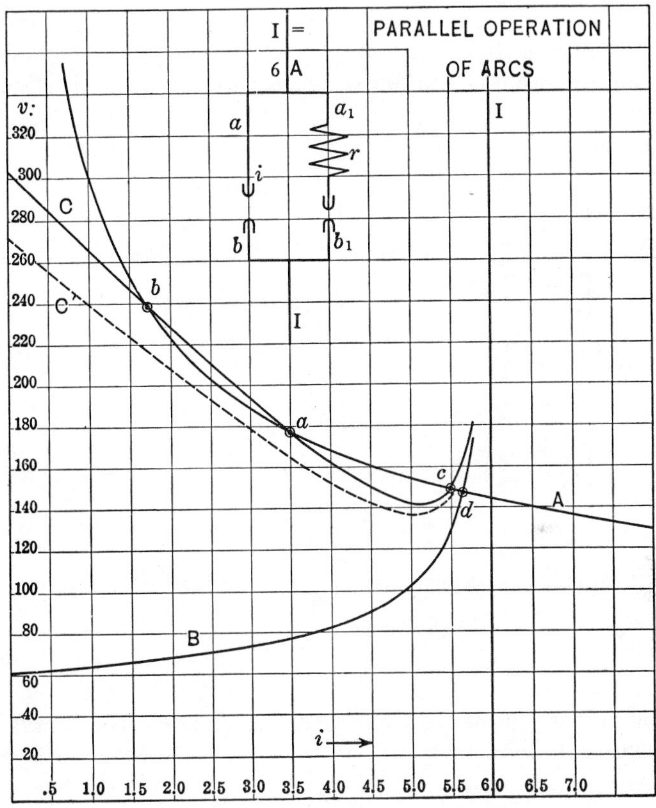

FIG. 85.

itself; b and c, however, are unstable. Thus, at the latter points, the arcs can not run, but the current changes until either one arc has gone out and one only persists, or both run at point a.

However, the angle under which the two curves, A and C, intersect at a is so small, that even at a the two arcs are not very stable.

Furthermore, a small change in either of the two curves, A or C, results in the two points of intersection a and b vanishing. Thus, if r is reduced from 40 ohms to 35 ohms, the curve C changes to C', shown dotted in Fig. 85, and as the latter does not intersect A except at the unstable point c, parallel operation is not possible.

That is, two such arcs can be operated in parallel only over a limited range of conditions, and even then the parallel operation is not very stable.

The preceding may illustrate the effect of resistance on the stability of operation of arcs.

Similarly, other conditions can be investigated, as the stability

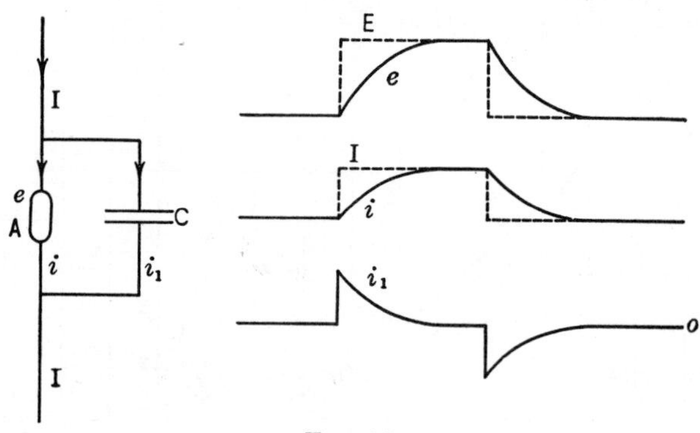

FIG. 86.

condition of arcs with resistance in series and in shunt, on constant, voltage supply, etc.

91. Let

$$e = E$$

be the voltage consumed by a circuit, A, Fig. 86, when traversed by a current

$$i = I.$$

If, then, in this circuit the current changes by δI, to

$$i = I + \delta I,$$

the voltage consumed by the circuit changes by δE, to

$$e = E \pm \delta E,$$

and the change of voltage is of the same sign as that of the current producing it, if A is a resistance or other circuit in which the voltage rises with the current, or is of opposite sign, if the circuit, A, has a dropping volt-ampere characteristic, as an arc.

Suppose now the circuit, A, is shunted by a condenser, C. As long as current, i, and voltage, e, in the circuit, A, are constant, no current passes through the condenser, C. If, however, the voltage of A changes, a current, i_1, passes through the condenser, given by the equation

$$i_1 = C\frac{de}{dt}. \tag{26}$$

If, then, the supply current, I, suddenly changes by δI, from I to $I + \delta I$, and the circuit, A, is a dead resistance, r, without the condenser, C, the voltage of A would just as suddenly change, from E to $E + \delta E$. By (26) this would, however, give an infinite current, i_1, in the condenser. However, the current in the condenser can not exceed δI, as with

$$i_1 = \delta I$$

at the moment of supply current change, the total excess current would in the first moment flow through the condenser, and the circuit, A, thus in this moment not change in current or voltage.

A finite current in the condenser, C, requires a finite rate of change of e in the circuit, A, starting from the previous value, E, at the starting moment, the time, $t = 0$.

Thus, if i = current, e = voltage of circuit, A, at time, t, after the increase of the supply current, I, by δI, it is
current in condenser,

$$i_1 = C\frac{de}{dt}.$$

current in circuit, A,

$$i = I + \delta I - i_1 \tag{27}$$

thus, voltage of circuit, A, of resistance, r,

$$\begin{aligned} e &= ri \\ &= rI + r\delta I - ri_1 \end{aligned} \tag{28}$$

substituting (26) into (28), gives

$$e = r(I + \delta I) - rC\frac{de}{dt},$$

or,

$$\frac{de}{r(I + \delta I) - e} = \frac{1}{rC}\,dt \tag{29}$$

integrated by

$$\left.\begin{aligned} e &= rI - r\delta I\,(1 - \epsilon^{-\frac{t}{rC}}) \\ &= E - \delta E\,(1 - \epsilon^{-\frac{t}{rC}}) \end{aligned}\right\} \tag{30}$$

since $e = E$ for $t = 0$ is the terminal condition which determines the integration constant.

With a sudden change of the supply current, I, by δI, as shown by the dotted lines, I, in Fig. 86, the voltage, e, and current, i, in the circuit, A, and the current, i_1, in the condenser, C, thus change by the exponential transients shown in Fig. 86 as e, i and i_1.

92. Suppose now, however, that the circuit, A, has a dropping volt-ampere characteristic, is an arc.

A sudden decrease of the supply current, I by δI, to $I - \delta I$, would by the arc characteristic, $e = a + \frac{b}{\sqrt{i}}$, cause an increase of the voltage of circuit, A, from E to $E + \delta E$. Such a sudden increase of E would send an infinite current through C, that is, all the supply current would momentarily go through the condenser, C, none through the arc, A, and the latter would thus go out, and that, no matter how small the condenser capacity, C. Thus, with the condenser in shunt to the circuit, A, the voltage, A, can not vary instantly, but at a decrease of the supply current, I, by δI, the voltage of A at the first moment must remain the same, E, and the current in A thus must remain also, and as the supply current has decreased by δI, the condenser, C, thus must feed the current, δI, back into the arc, A. This, however, requires a decreasing voltage rating of A, at decreasing supply current, and this is not the case with an arc.

Inversely, a sudden increase of I, by δI, decreases the voltage of A, thus causes the condenser, C, to discharge into A, still further decreases its voltage, and the condenser momentarily short-circuits through the arc, A; but as soon as it has discharged and the arc voltage again rises with the decreasing current, the condenser, C, robs the arc, A, and puts it out.

Thus, even a small condenser in shunt to an arc makes it unstable and puts it out.

If a resistance, r_0, is inserted in series to the arc in the circuit, A, stability results if the resistance is sufficient to give a rising volt-ampere characteristic, as discussed previously.

Resistance in series to the condenser, C, also produces stability, if sufficiently large: with a sudden change of voltage in the arc circuit, A, the condenser acts as a short-circuit in the first moment, passing the current without voltage drop, and the voltage thus has to be taken up by the shunt resistance, r_1, giving the same condition of stability as with an arc in a constant-current circuit, shunted by a resistance, paragraph 89.

If, in addition to the capacity, C, an inductance, L, and some resistance, r, are shunted across the circuit, A, of a rising volt-ampere characteristic, as shown in Fig. 87, the readjustment occurring at a sudden change of the supply current, I, is not exponential, as in Fig. 86, but oscillatory, as in Fig. 87. As in the circuit, A, assuming it consists of a resistance, r, current and voltage vary simultaneously or in phase, current and voltage in the condenser branch circuit also must be in phase with each other, that is, the frequency of the oscillation in Fig. 87 is that at which capacity, C, and inductance, L, balance, or is the resonance frequency.

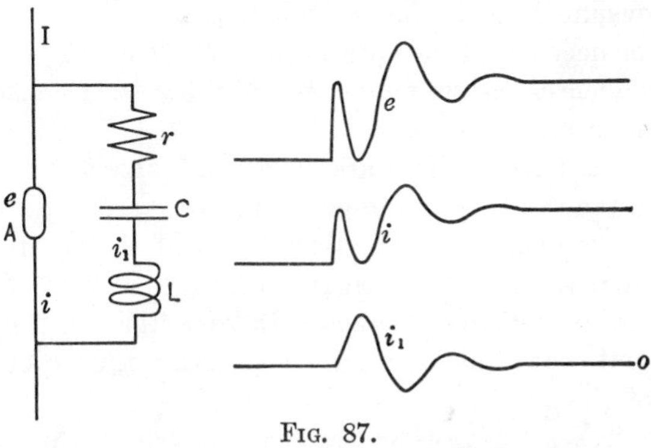

FIG. 87.

If circuit, A, in Fig. 87 is an arc circuit, and the resistance, r, in the shunt circuit small, instability again results, in the same manner as discussed before.

93. Another way of looking at the phenomena resulting from a condenser, C, shunting a circuit, A, is:

Suppose in Fig. 86 at constant-supply current, I, the current in the circuit, A, should begin to decrease, for some reason or another. Assuming as simplest case, a uniform decrease of current.

The current in the circuit, A, then can be represented by

$$i = I\left(1 - \frac{t}{t_0}\right) \tag{31}$$

where t_0 is the time which would be required for a uniform decrease down to nothing.

At constant-supply current, I, the condenser thus must absorb the decrease of current in A, that is, the condenser current is

$$i_1 = I \frac{t}{t_0}. \tag{32}$$

With decrease of current, i, if A is a circuit with rising characteristic, for instance, an ohmic resistance, the voltage of A decreases. The voltage at the condenser increases by the increasing charging current, i_1, thus the condenser voltage tends to rise over the circuit voltage of A, and thus checks the decrease of the voltage and thus of the current in A. Thus, the conditions are stable.

Suppose, however, A is an arc.

A decrease of the current in A then causes an increase of the voltage consumed by A, the arc voltage, e_0.

The same decrease of the current in A, by deflecting the current into the condenser, causes an increase of the voltage consumed by C, the condenser voltage, e_1.

If, now, at a decrease of the arc current, i, the arc voltage, e_0, rises faster than the condenser voltage, e_1, the increase of e_0 over e_1 deflects still more current from A into C, that is, the arc current decreases and the condenser current increases at increasing rate, until the arc current has decreased to zero, that is, the arc has been put out. In this case, the condenser thus produces instability of the arc.

If, however, e_0 increases slower than e_1, that is, the condenser voltage increases faster than the arc voltage, the condenser, C, shifts current over into the arc circuit, A, that is, the decrease of current in the arc circuit checks itself, and the condition becomes stable.

The voltage rise at the condenser is given by

$$\frac{de}{dt} = \frac{1}{C} i_1;$$

hence, by (32),

$$\frac{de}{dt} = \frac{tI}{t_0 C} \tag{33}$$

from the volt-ampere characteristic of the arc,

$$e = a + \frac{b}{\sqrt{i}} \tag{34}$$

follows,

the voltage rise at the arc terminals,

$$\frac{de}{dt} = -\frac{b}{2i\sqrt{i}}\frac{di}{dt} \tag{35}$$

and, by (31),

$$\frac{di}{dt} = -\frac{I}{t_0};$$

hence, substituted into (34),

$$\frac{de}{dt} = \frac{bI}{2\,t_0 i\sqrt{i}}. \tag{36}$$

The condition of stability is, that the voltage rise at the condenser, (33), is greater than that at the arc, (36), thus,

$$\frac{tI}{t_0C} > \frac{bI}{2\,t_0 i\sqrt{i}},$$

or,

$$\frac{2\,ti\sqrt{i}}{bC} > 1 \tag{37}$$

or, substituting for t from equation (31), gives

$$\frac{2\,t_0 i\sqrt{i}\,(I-i)}{bC} > 1 \tag{38}$$

as the condition of stability, and

$$\frac{2\,t_0 i\sqrt{i}\,(I-i)}{bC} = 1 \tag{39}$$

thus is the stability limit.

94. Integrating (33) and substituting the terminal condition: $t = 0$; $e = E$, gives

$$e_1 = E + \frac{t^2 I}{2\,t_0 C} \tag{40}$$

as the equation of the voltage at the condenser terminals.

Substitute (31) into (34) gives

$$e_0 = a + \frac{b}{\sqrt{I}\sqrt{1-\frac{t}{t_0}}} \tag{41}$$

as the equation of the arc voltage.

For,

$$a = 35,$$
$$b = 200,$$
$$I = 3,$$

hence,

$$E = 151,$$

and

$$t_0 = 10^{-4} \text{ sec.},$$

and, for the three values of capacity,

$$C = 10^{-6} \qquad (e_1)$$
$$0.75 \times 10^{-6} \qquad (e_2)$$
$$0.5 \times 10^{-6} \qquad (e_3)$$

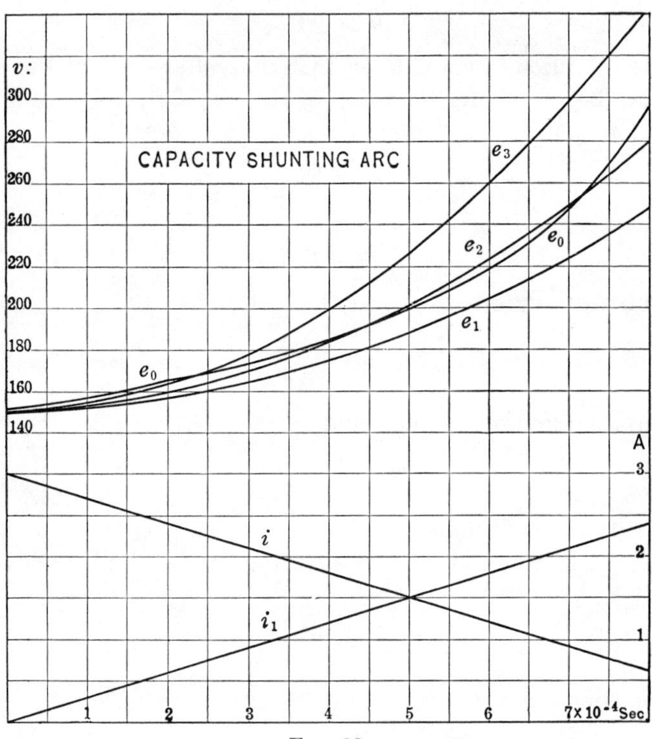

FIG. 88.

the curves of the arc voltage, e_0,
and of the condenser voltage, e_1, e_2, e_3,
are shown on Fig. 88,
together with the values of i and i_1.

As seen, e_1 is below e_0 over the entire range. That is, 1 mf. makes the arc unstable over the entire range. 0.5 mf., e_3, gives instability up to about $t = 0.25 \times 10^{-4}$ sec., then stability results. With 0.75 mf., e_2, there is a narrow range of stability, between

$4\frac{3}{4}$ and $7\frac{1}{4} \times 10^{-4}$ sec., before and after this instability exists.

From equation (37), the condition of stability, it follows that for small values of t, that is, small current fluctuations, the conditions are always unstable. That is, no matter how small a condenser is, it always has an effect in increasing the current fluctuations in the arc, the more so, the higher the capacity, until conditions become entirely unstable.

From equations (40) and (41) follows as the stability limit

$$e_0 = e_1,$$

$$a + \frac{b}{\sqrt{I}\sqrt{1 - \frac{t}{t_0}}} = E + \frac{t^2 I}{2\, t_0 C},$$

or, expanded into a series,

$$a + \frac{b}{\sqrt{I}} \left\{ 1 + \frac{t}{2\, t_0} + \ . \ . \ . \right\} = E + \frac{t^2 I}{2\, t_0 C},$$

cancelling $E = a + \frac{b}{\sqrt{I}}$ and rearranging, gives

$$t_1 = \frac{bC}{I\sqrt{I}} \tag{42}$$

thus, at the time,

$$t_1 = \frac{bC}{I\sqrt{I}},$$

the condition changes from unstable to stable.

As t_1 must be smaller than t_0, the total time of change, it follows:

$$t_0 > \frac{bC}{I\sqrt{I}} \tag{43}$$

or,

$$C < \frac{t_0 I \sqrt{I}}{b} \tag{44}$$

are expressions of the (approximate) stability limit of an arc with condenser shunt.

As seen from (44),

the larger t_0 is, that is, the slower the arc changes, the larger is the permissible shunted capacity, and inversely.

As an instance, let

$$b = 200,$$
$$I = 3,$$

and

$$(a) \quad t_0 = 10^{-3},$$

which is probably the approximate magnitude in the carbon arc. This gives

$$C < 26 \text{ mf.}$$

Let:

$$(b) \quad t_0 = 10^{-5},$$

which is probably the approximate magnitude in the mercury arc. This gives

$$C < 0.26 \text{ mf.}$$

95. Consider the case of a circuit, A, Fig. 87, supplied by a constant current, I, but shunted by a capacity, C, inductance, L, and resistance, r, in series.

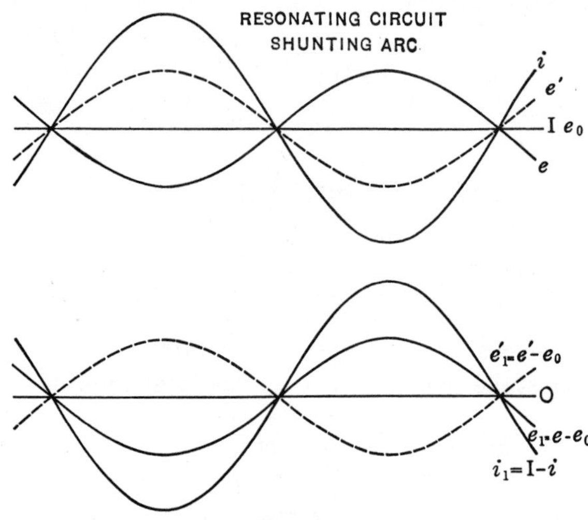

FIG. 89.

As long as the current in the circuit, A—whether resistance or arc—is steady, no current passes the condenser circuit, and the current and voltage in A thus are constant, $i = I$, $e = e_0$.

Suppose now a pulsation of the current, i, should be produced in circuit, A, as shown as i in Fig. 89. Then, with constant-supply current, I, an alternating current,

$$i_1 = I - i,$$

would traverse the condenser circuit, C, since the continuous component of current can not traverse the condenser, C.

Due to the pulsation of current, i in A, the voltage, e, of circuit, A, would pulsate also. These voltage pulsations are in the same direction as the current pulsation, if A is a resistance, in opposite direction, if A is an arc; in either case, however, they are in phase with the current pulsation, and the alternating voltage on the condenser,

$$e_1 = e_0 - e,$$

thus is in phase with the alternating current, i_1, that is, capacity, C, and inductance, L, neutralize.

Thus, the only pulsation of current and voltage, which could occur in a circuit, A, shunted by capacity and inductance, is that of the resonance frequency of capacity and inductance.

Suppose the circuit, A, is a dead resistance. The voltage pulsation produced by a current pulsation, i, in this circuit then would be in the same direction as i, that is, would be as shown in dotted line by e' in Fig. 89. In the condenser circuit, C, the alternating component of voltage thus would be

$$e'_1 = e' - e_0,$$

thus would be in opposition to the alternating current, i_1, as shown in Fig. 89 in dotted line. That is, it would require a supply of power to maintain such pulsation.

Thus, with a dead resistance as circuit, A, or in general with A as a circuit of rising volt-ampere characteristic, the maintenance of a resonance pulsation of current and voltage between A and C, at constant current, I, requires a supply of alternating-current power in the condenser circuit, and without such power supply the pulsation could not exist, hence, if started, would rapidly die out, as oscillation, as shown in Fig. 87.

96. Suppose, however, A is an arc. A current pulsation, i, then gives a voltage pulsation in opposite direction, as shown by e in Fig. 89, and the alternating current, $i_1 = I - i$, and the alternating voltage, $e_1 = e - e_0$, in the condenser circuit, thus would be in phase with each other, as shown by i_1 and e_1 in Fig. 89. That is, they would represent power generation, or rather transformation of power from the constant direct-current supply, I, into the alternating-current resonating condenser circuit, C.

Thus, such a local pulsation of the arc current, i, and corresponding alternating current, i_1, in the condenser circuit, if once started, would maintain itself without external power supply, and would even be able to supply the power represented by voltage, e_1, with current, i_1, into an external circuit, as the resistance, r, shown in Fig. 87, or through a transformer into a wireless sending circuit, etc.

Thus, due to the dropping arc characteristic, an arc shunted by capacity and inductance, on a constant-current supply, becomes a generator of alternating-current power, of the frequency set by the resonance of C and L.

If the resistance, r, or in general, the load on the oscillating circuit, C, is greater than $r_1 = \frac{e_1}{i_1}$, that is, if a higher voltage would be required to send the current, i_1, through the resistance, r, than the voltage, e_1, generated by the oscillating arc, A, the pulsations die out as oscillations.

If r is less than $\frac{e_1}{i_1}$, the pulsations increase in amplitude, that is, current, i_1, and voltage, e_1, increase, until either, by the internal reaction in the arc, the ratio, $\frac{e_1}{i_1}$, drops to equality with the effective resistance of the load, r, and stability of oscillation is reached, or, if $\frac{e_1}{i_1}$ never falls to equality with r—for instance, if $r = 0$, the oscillations increase up to the destruction of the circuit: the extinction of the arc.

If, in the latter case, the voltage back of the supply current, I, is sufficiently high to restart the arc, A, the phenomena repeats, and we have a series of successive arc oscillations, each rising until it puts the arc out, and then the arc restarts.

We thus have here the mechanism which produces a *cumulative oscillation*, that is, a transient, which does not die out, but increases in amplitude, until the increasing energy losses limit its further increase, or until it destroys the circuit, and in the latter case, it may become recurrent.

It is very important to realize in electrical engineering, that any electric circuit with dropping volt-ampere characteristic is capable of transforming power into a cumulative oscillation, and thereby is able under favorable conditions to produce cumulative oscillations, such as hunting, etc.

Where the arc oscillations limit themselves, and the alternating current and voltage in the condenser circuit thus reach a constant value, the arc often is called a "*singing arc*," due to the musical note given by the alternating wave. Where the arc oscillations

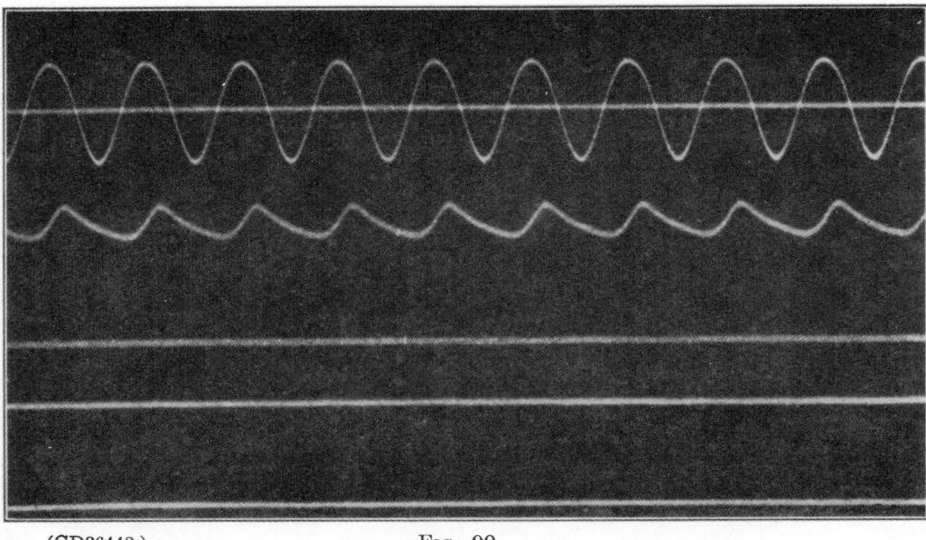

(CD26442.)

Fig. 90.

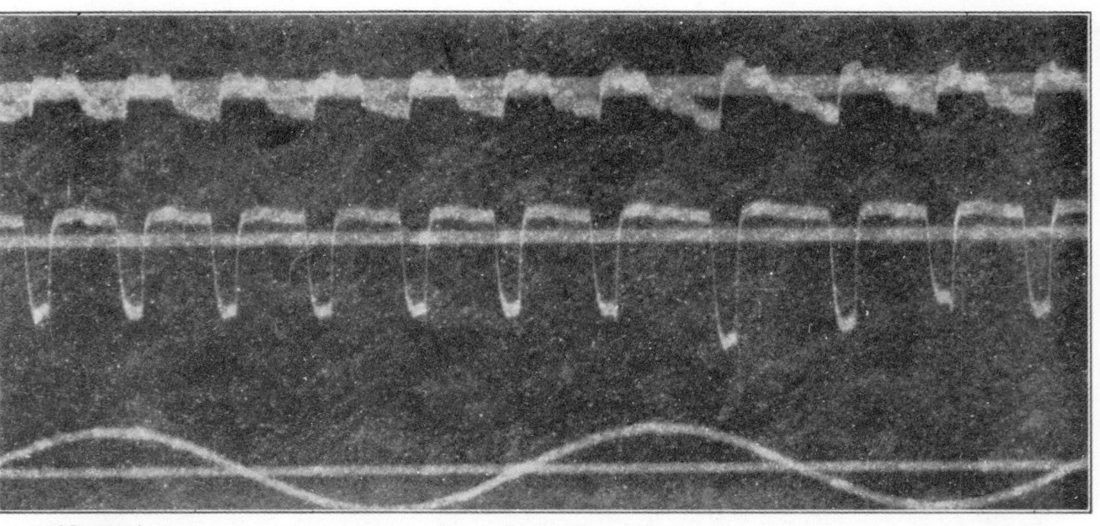

(CD15024.)

FIG. 91.

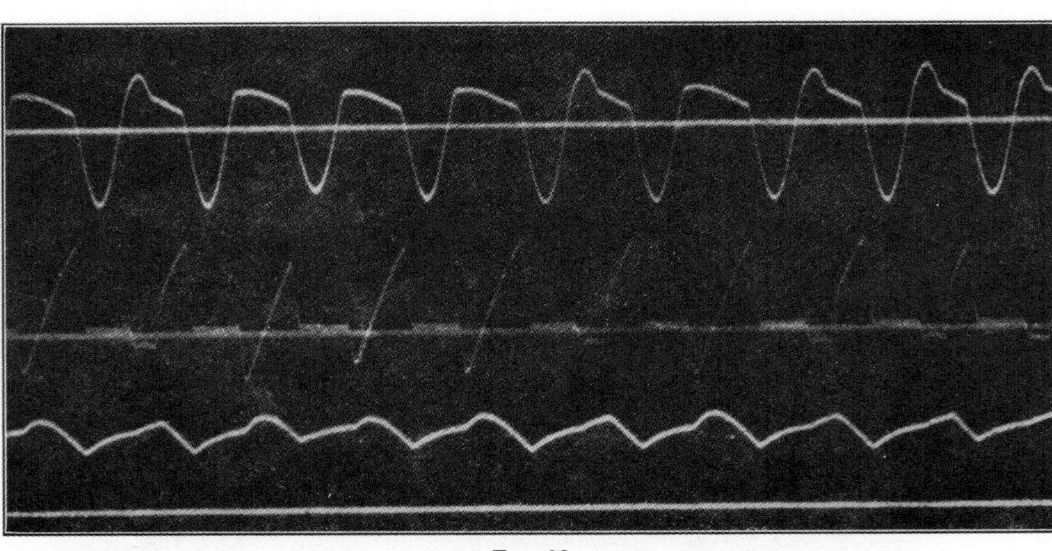

(CD26439.) FIG. 92.

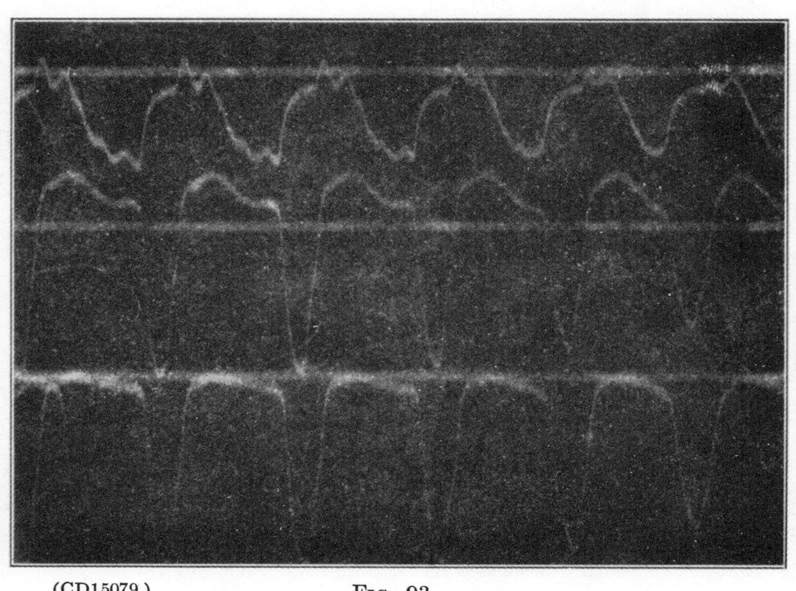

(CD15079.)

Fig. 93.

rise cumulatively to interruption, and the arc then restarts by the supply voltage and repeats the same phenomenon, it may be called a "*rasping arc*," by the harsh noise produced by the interrupted cumulative oscillation.

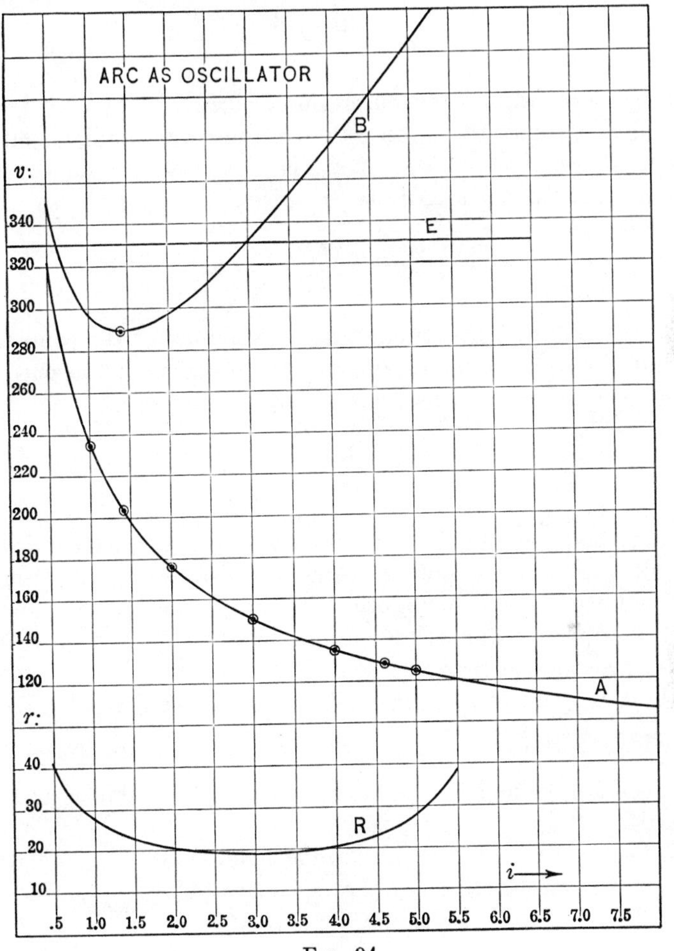

FIG. 94.

Figs. 90 and 91 give oscillograms of singing arcs; Figs. 92 and 93, of rasping arcs, 90 and 92 in circuits with massed constants, 91 and 93 in transmission lines.

97. As an illustration, let curve, A, in Fig. 94 represent the volt-ampere characteristic of an arc, and assume that this arc is operating steadily at I amp., consuming e_0 volts.

Suppose this arc is shunted by capacity, C, inductance, L, and resistance, r, as shown in Fig. 87.

For a small pulsation of the arc current around its average value $i = I$, the corresponding voltage pulsation is given by

$$\frac{de}{di} = -\frac{b}{2i\sqrt{i}}.$$

Or, in general, for any pulsation of current, i, by δi, between i' and i'', around the mean value I, the corresponding voltage pulsation δe, between e' and e'', is given by the volt-ampere characteristic of the arc, A, as

$$\frac{\delta e}{\delta i} = -\frac{e'' - e'}{i'' - i'}.$$

$e_1 = \delta e$ thus is the voltage, made available for the condenser circuit, by the arc pulsation, and in phase with the current, $i_1 = -\delta i$ in the condenser circuit, and

$$R = -\frac{\delta e}{\delta i} = \frac{e'' - e'}{i'' - i'},$$

thus is the permissible effective resistance in the condenser circuit, that is, the maximum value of resistance, through which the pulsating arc can maintain its alternating power supply: with a larger resistance, the oscillations die out; with a smaller resistance, they increase.

From the arc characteristic, A, thus can be derived a curve of effective resistances, R, as the values of $\frac{\delta e}{\delta i}$, for pulsations between $i + \delta i$ and $i - \delta i$, and such a curve is shown as R in Fig. 94.

We may say, that the arc, when shunted by an oscillating circuit, has an effective negative resistance,

$$\frac{\delta e}{\delta i},$$

and thereby generates alternating power, from the consumed direct-current power, and is able to supply alternating power through an effective resistance of the oscillating circuit, of $R = -\frac{\delta e}{\delta i}$.

The arc characteristic in Fig. 94 is drawn with the equation

$$e = 35 + \frac{200}{\sqrt{i}}$$

and for

$$i = I = 3 \text{ amp. as mean value,}$$

the values of the effective resistance, R, increase from

$$R = -\frac{de}{di} = 18.5 \text{ ohms}$$

for very small oscillations, to

$$R = 20.3 \text{ ohms}$$

for oscillations of 1 amp., between $i = 2$ and $i = 4$, to

$$R = 27.5 \text{ ohms}$$

for oscillations of 2 amp., between $i = 1$ and $i = 5$, etc.

Thus, if with this oscillating arc, Figs. 87 and 94, a load resistance $r < 18.5$ ohms is used, oscillation starts immediately, and cumulatively increases.

If the resistance, r, is greater than 18.5 ohms, for instance, is

$$r = 22.5 \text{ ohms,}$$

then no oscillation starts spontaneously, but the arc runs steady, and no appreciable current passes through the condenser circuit. But if once the current in the arc is brought below 1.5 amp., or above 4.5 amp., the oscillation begins and cumulatively increases, since for oscillations of an amplitude greater than between 1.5 and 4.5 amp., the effective resistance, R, is greater than 22.5 ohms.

In either case, however, as soon as an oscillation starts, it cumulatively increases, since the effective resistance, R, steadily increases with increase of the amplitude of oscillation. That is, stability of oscillation, or a "singing arc" can not be reached, but an oscillation, once started, proceeds to the extinction of the arc, and only a "rasping arc" could be produced.

98. However, the arc characteristic, A, of Fig. 94 is the stationary characteristic, that is, the volt-ampere relation at constant current, i, and voltage, e.

If current, i, and thus voltage, e, rapidly fluctuate, the arc characteristic, A, changes, and more or less flattens out. That is, for any value of the current, i, the volume of the arc stream and the temperature of the arc terminals, still partly correspond to previous values of current, thus are lower for rising, higher for decreasing current, and as the result, the arc voltage, e, which de-

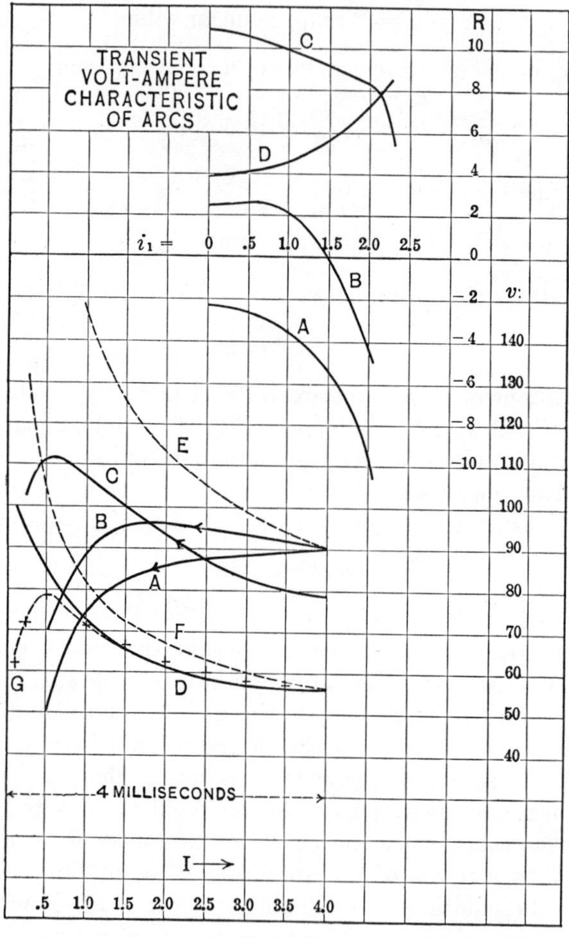

FIG. 95.

pends on the resistance of the arc stream and the potential drop of the terminals, is different, the variation of voltage, for the same variation of current, is less, and the effective negative arc resistance thereby is lowered, or may entirely vanish.

Fig. 95 shows a number of such transient arc characteristics, estimated from oscillographic tests of alternating arcs, and their corresponding effective resistances, R.

They are:

(A) Carbon.

(B) Hard carbon.

(C) Acheson graphite.

(D) Titanium carbid.

(E) Hard carbon, stationary characteristic.

(F) Titanium carbid, stationary characteristic.

As seen from the curves of R in the upper part of Fig. 95, the effective resistances, R, which represent the alternating power generated by the oscillating arc, are much lower with the transient arc characteristic, than would be with the permanent arc characteristic in Fig. 94.

Curve D, titanium carbide, gives under these conditions an unstable or "rasping" arc. That is, with a resistance in the condenser circuit of less than $R = 3.8$ ohms, the oscillation starts spontaneously and cumulatively increases to the extinction of the arc; with a resistance of more than 3.8 ohms, the oscillation does not spontaneously start, but if once started with an amplitude which brings the value of R from curve, D, above that of the resistance in the condenser circuit, cumulative oscillation occurs.

With the carbon arc, A, no oscillations can occur under any condition, the effective resistance, R, is negative, and the arc characteristic rising.

With the hard carbon arc, B, an oscillation starts with a resistance less than 2.4 ohms, cumulatively increases, but its amplitude finally limits itself, to 1.45 amp. if the resistance in the oscillating circuit is zero, to 1.05 amp. with 2 ohms resistance, etc., as seen from the curve, B, in the upper part of Fig. 95. Even with more than 2.4 ohms resistance, up to 2.6 ohms resistance, an oscillation can exist, if once started, as the curve of R, starting from $R = 2.4$ ohms at $i_1 = 0$, rises to $R = 2.6$ ohms at $i_1 = 0.75$, and then drops to zero at $i_1 = 1.45$ ohms, and beyond this becomes negative.

The curve, C, of Acheson graphite, starts with a resistance $R =$ 10.8 ohms, but the resistance, R, steadily drops with increasing oscillating current, i_1, down to zero at $i_1 = 2.4$ amp. Thus, with a resistance in the condenser circuit, of 10 ohms, the oscillations would have an amplitude of $i_1 = 0.9$ amp.; with 8 ohms resistance an amplitude of 2.1 amp., etc.

From these curves of R, Fig. 95, the regulation curves of the alternating-current generation could now be constructed.

It is interesting to note, that in many of these transient arc characteristics, Fig. 95, the voltage does not indefinitely rise with decreasing current, but reaches a maximum and then decreases again, in B and C, and the oscillation resistance, that is, the resistance through which an alternating current can be maintained by the oscillating arc, thus decreases with increasing amplitude of the oscillation. Thus, if the resistance in the oscillating condenser circuit is less than the permissible maximum, an oscillation starts, cumulatively increases, but finally limits itself in amplitude.

The decrease of the arc voltage with decreasing current, for low values of current in a rapidly fluctuating arc, is due to the time lag of the arc voltage behind the current.

99. The arc voltage, e, consists of the arc terminal drop, a, and the arc stream voltage, e_1:

$$e = a + e_1.$$

The stream voltage, e_1, is the voltage consumed in the effective resistance of the arc stream; but as the arc stream is produced by the current, the volume of the arc stream and its resistance thus depends on the current, i, in the arc, that is, the stream voltage is

$$e_1 = \frac{b}{\sqrt{i}}$$

and the resistance of the arc stream thus

$$r_1 = \frac{e_1}{i} = \frac{b}{i\sqrt{i}}.$$

Thus, if,

$$a = 35$$
$$b = 200,$$

for

$$i = 2 \text{ amp., it is}$$
$$r_1 = 70.7 \text{ ohms,}$$
$$e_1 = 141.4 \text{ volts,}$$
$$e = 176.4 \text{ volts.}$$

But, if the arc current rapidly varies, for instance decreases, then, when the current in the arc is i_1, the volume of the arc stream and thus its resistance is still that corresponding to the previous current, i'_1.

If thus, at the moment where the current in the arc has become

$$i_1 = 2 \text{ amp.},$$

the arc stream still has the volume and thus the resistance corresponding to the previous current,

$$i'_1 = 3 \text{ amp.},$$

this resistance is

$$r'_1 \doteq \frac{200}{3\sqrt{3}} = 38.5 \text{ ohms},$$

and the stream voltage, at the current

$$i_1 = 2 \text{ amp.},$$

but with the stream resistance, r'_1, corresponding to the previous current, $i'_1 = 3$ amp., thus is

$$\begin{aligned} e'_1 &= r'_1 i_1 \\ &= 77 \text{ volts}, \end{aligned}$$

instead of $e_1 = 141.4$ volts, as it would be under stationary conditions.

That is, the stream voltage and thus the total arc voltage at rapidly decreasing current is lower, at rapidly increasing current higher than at stationary current.

With a periodically pulsating current, it follows herefrom, that at the extreme values of current—maximum and minimum— the voltage has not yet reached the extreme values corresponding to these currents, that is, the amplitude of voltage pulsation is reduced. This means the transient volt-ampere characteristic of the arc is flattened out, compared with the permanent characteristic, and caused to bend downward at low currents, as shown by C and B in Fig. 95.

Assuming a sinusoidal pulsation of the current in the arc and assuming the arc stream resistance to lag behind the current by a suitable distance, we then get, from the stationary volt-ampere characteristic of the arc, the transient characteristics.

Thus in Fig. 96, from the stationary arc characteristic, S, the transient arc characteristic, T, is derived. In this figure is shown as S and T the effective resistance corresponding to the stationary characteristic, S, respectively the transient characteristic, T.

As seen, the stationary characteristic, S, gives an arc oscillation which is cumulative and self-destructive, that is, the effective resistance, R, rises indefinitely with increasing amplitude of pulsation. The transient characteristic, however, gives an effective resistance, R, which with increasing amplitude of pulsation

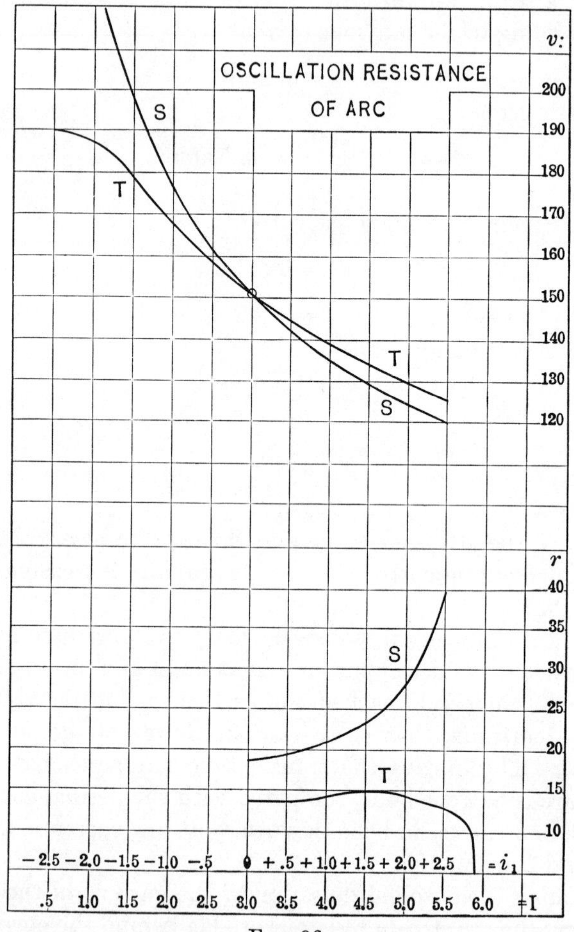

Fig. 96.

first increases, but then decreases again, down to zero, so that the cumulative oscillations produced by this arc are self-limiting, increase in amplitude only up to the value, where the effective resistance, R, has fallen to the value corresponding to the load on the oscillating circuit.

As further illustration, from the stationary volt-ampere characteristic of the titanium arc, shown as *F* in Fig. 95, values of the transient characteristic have been calculated and are shown in Fig. 95 by crosses. As seen, they fairly well coincide with the transient volt-ampere characteristic, *D*, of the titanium arc, at least for the larger currents.

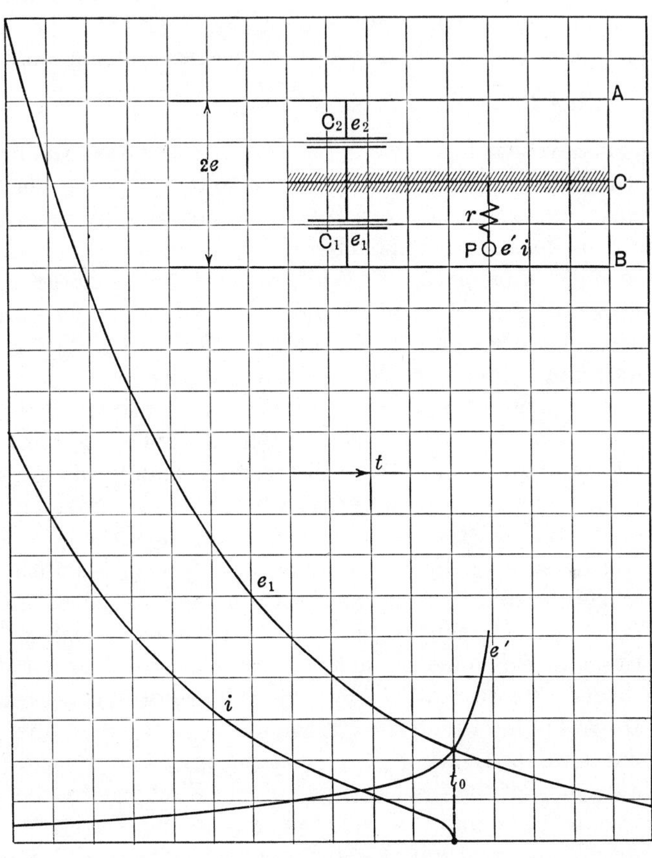

FIG. 97.

In the electric arc we thus have an electric circuit with dropping volt-ampere characteristic. Such a circuit is unstable under various conditions which may occur in industrial circuits, and thereby may be, and frequently is, the source of instability of electric circuits, and of cumulative oscillations appearing in such circuits.

100. For instance, let, in Fig. 97, A and B be two conductors of an ungrounded high-potential transmission line, and $2\ e$ the voltage impressed between these two conductors. Let C represent the ground.

The capacity of the conductors, A and B, against ground, then, may be represented diagrammatically by two condensers, C_1 and C_2, and the voltages from the lines to ground by e_1 and e_2. In general, the two line capacities are equal, $C_1 = C_2$, and the two voltages to ground thus equal also, $e_1 = e_2 = e$, with a single-phase; $= \frac{2\ e}{\sqrt{3}}$ with a three-phase line.

Assume now that a ground, P, is brought near one of the lines, A, to within the striking distance of the voltage, e. A discharge then occurs over the conductor, P. Such may occur by the puncture of a line insulator as not infrequently the case. Let r = resistance of discharge path, P. While without this discharge path, the voltage between A and C would be $e_1 = e$ (assuming single-phase circuit) with a grounded conductor, P, approaching line A within striking distance of voltage, e, a discharge occurs over P forming an arc, and the circuit of the impressed voltage, $2\ e$, now comprises the condenser, C_2, in series to the multiple circuit of condenser, C_1, and arc, P, and the condenser, C_1, rapidly discharges, voltage, e_1, decreases, and the voltage, e_2, increases. With a decrease of voltage, e_1, the discharge current, i, also decreases, and the voltage consumed by the discharge arc, e', increases until the two voltages, e_1 and e', cross, as shown in the curve diagram of Fig. 97. At this moment the current, i, in the arc vanishes, the arc ceases, and the shunt of the condenser, C_1, formed by the discharge over P thus ceases. The voltage, e_1, then rises, e_2 decreases and the two voltages tend toward equality, $e_1 = e_2 = e$. Before this point is reached, however, the voltage, e_1, has passed the disruptive strength of the discharge gap, P, the discharge by the arc over P again starts, and the cycle thus repeats indefinitely.

In Fig. 97 are diagrammatically sketched voltage, e_1, of condenser, C_1, the voltage, e', consumed by the discharge arc over P, and the current, i, of this arc, under the assumption that r is sufficiently high to make the discharge non-oscillatory. If r is small, each of these successive discharges is an oscillation.

Such an unstable circuit gives a continuous series of successive discharges, which are single impulses, as in Fig. 97, or more commonly are oscillations.

If the line conductors, A and B, in Fig. 97 have appreciable inductance, as is the case with transmission lines, in the charge of the condenser, C_1, after it has been discharged by the arc over P, the voltage, e_1, would rise beyond e, approaching $2\ e$, and the discharge would thus start over P, even if the disruptive strength of this gap is higher than e, provided that it is still below the voltage momentarily reached by the oscillatory charge of the line condenser, P_1.

This combination of two transmission line conductors and the ground conductor, P, approaching near line, A, to a distance giving a striking voltage above e, but below the momentary charging voltage, of C_1, then constitutes a circuit which has two permanent conditions, one of stability and one of instability. If the voltage is gradually applied, $e_1 = e_2 = e$, the condition is stable, as no discharge occurs over P. If, however, by some means, as a momentarily overvoltage, a discharge is once produced over the spark-gap, P, the unstable condition of the circuit persists in the form of successive and recurrent discharges.

101. Usually, the resistance, r, of the discharge path is, or after a number of recurrent discharges, becomes sufficiently low to make the discharge oscillatory, and a series of recurrent oscillations then result, a so-called "arcing ground." Oscillograms of such an arcing grounds on a 30-mile 30-kv. transmission line are shown in Figs. 98, 99 and 100.

If, however, the resistance of the discharge path is very low, a sustained or cumulative oscillation results, as discussed in the preceding, that is, the arcing ground becomes a stationary oscillation of constant-resonance frequency, increasing cumulatively in current and voltage amplitude until limited by increasing losses or by destruction of apparatus.

In transmission lines, usually the resistance is too high to produce a cumulative oscillation; in underground cables, usually the inductance is too low and thus no cumulative oscillation results, except perhaps sometimes in single-conductor cables, etc. In the high-potential windings of large high-voltage power transformers, however, as circuits of distributed capacity, inductance and resistance, the resistance commonly is below the value through which a cumulative oscillation can be produced and maintained, and in high-potential transformers, destruction by high voltages resulting from the cumulative oscillation of some arc in the system, and building up to high stationary waves, have frequently been observed.

The "arcing ground" as recurrent single impulses, the "arcing ground oscillation" as more or less rapidly damped recurrent oscillations in transmission lines—of frequencies from a few hundred to a few thousand cycles—and the "stationary oscillations" causing destruction in high-potential transformer windings, at frequencies of 10,000 to 100,000 cycles, thus are the same phenomena of the dropping arc characteristic, causing permanent instability of the electric circuit, and differ from each other merely by the relative amount of resistance in the discharge path.

CHAPTER XI

INSTABILITY OF CIRCUITS: INDUCTION AND SYNCHRONOUS MOTORS

C. Instability of Induction Motors

102. Instability of electric circuits may result from causes which are not electrical: thus, mechanical relations between the torque given by a motor and the torque required by its load, may lead to instability.

Let

D = torque given by a motor at speed, S, and

D' = torque required by the load at speed, S.

The motor, then, could theoretically operate, that is, run at constant speed, at that speed, S, where

$$D = D' \tag{1}$$

However, at this speed and load, the operation may be stable, that is, the motor continue to run indefinitely at constant speed, or the condition may be unstable, that is, the speed change with increasing rapidity, until stability is reached at some other speed, or the motor comes to a standstill, or it destroys itself.

In general, the motor torque, D. and the load torque, D', change with the speed, S.

If, then,

$$\frac{dD'}{dS} > \frac{dD}{dS} \tag{2}$$

the conditions are stable, that is, any change of speed, S, changes the motor torque less than the load torque, and inversely, and thus checks itself.

If, however,

$$\frac{dD'}{dS} < \frac{dD}{dS} \tag{3}$$

the operation is unstable, as a change of speed, S, changes the motor torque, D, more than the load torque, D', and thereby further increases the change of speed, etc.

$$\frac{dD'}{dS} = \frac{dD}{dS} \tag{4}$$

thus is the expression of the stability limit.

For instance, assuming a load requiring a constant torque at all speeds. The load torque thus is given by a horizontal line

$$D' = \text{const.} \tag{5}$$

in Fig. 101.

Let then the speed-torque curve of the motor be represented by the curve, D, in Fig. 101. D approximately represents the torque curve of a series motor. At the constant-load torque, D', the motor runs at the speed, $S = 0.6$, point a of Fig. 101, and the speed is stable, as any tendency to change of speed, checks itself. If

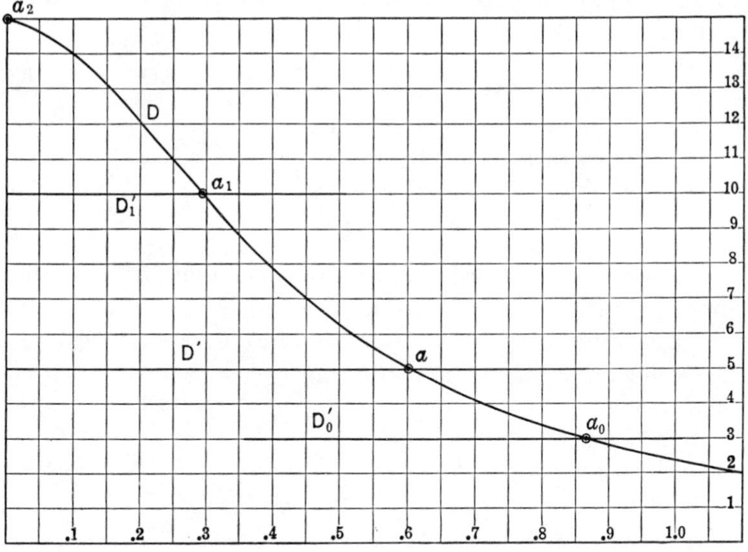

Fig. 101.

the load torque decreases to D'_0, the speed rises to $S = 0.865$, point a_0; if the load torque increases to D'_1, the speed drops to $S = 0.29$, point a_1, but the conditions are always stable, until finally with increasing load torque, D', and decreasing speed, standstill is reached at point a_2.

Let now the speed-torque curve of a motor be represented by D in Fig. 102: the curve of a squirrel-cage induction motor with moderately high resistance secondary. The horizontal line, D', corresponding to a load torque of $D' = 10$, intersects D at two points, a and b.

At a, $S = 0.905$, the speed is stable. At b, however, $S = 0.35$, the conditions are unstable, and the motor thus can not run at b, but either—if the speed should drop or the load rise ever so little—the motor begins to slow down, thereby, on curve, D, its torque falls below that of the load, D', thus it slows down still more, and so, with increasing rapidity the motor comes to a standstill. Or, if the motor speed should be a little higher, or the load momentarily a little lower, the motor speed rises, until stability is reached at point a.

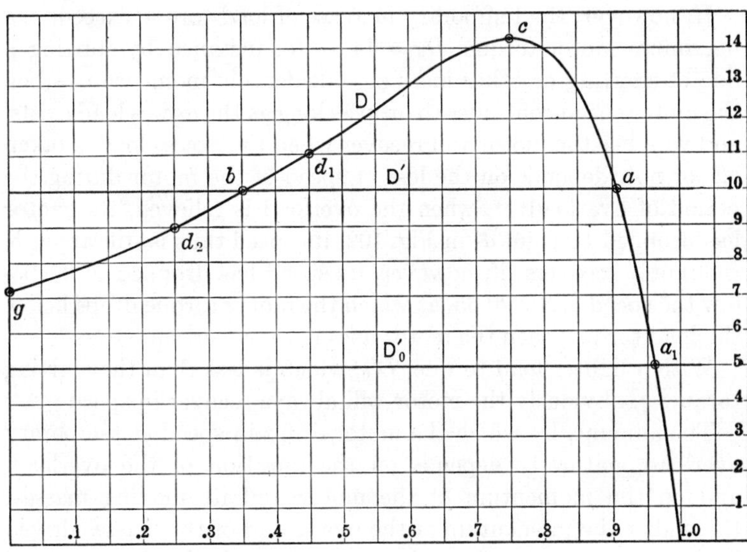

Fig. 102.

With increasing load torque, D', the speed gradually drops, from $S = 0.905$ at $D' = 10$, point a, down to point c, at $S = 0.75$, $D' = 14.3$; from there, however, the speed suddenly drops to standstill, that is, it is not possible to operate the motor at speeds less than $S = 0.75$, at constant load-torque, and the branch of the motor characteristic from the starting point, g, up to the maximum torque point, c, is unstable on a load requiring constant torque.

At load torque, $D' = 10$, the motor can not start the load, can not carry it below b, $S = 0.35$; at speeds from b to a, $S = 0.35$ to 0.905, the motor speeds up; at speeds above a, $S = 0.905$, the motor slows down, and drops into stable condition at a.

With a load torque, $D'_0 = 5$, the motor starts and runs up to speed a_1, $S = 0.96$.

$D' = 7.2$, point g, thus, is the maximum load torque which the motor can start.

103. Suppose now, while running in stable condition, at point a, with the load torque, $D' = 10$, the load torque is momentarily increased. If this increase leaves D' lower than the maximum motor torque, $D_0 = 14.3$, the motor speed slows down, but remains above c, and thus when the increase of load is taken off, the motor again speeds up to a.

If, however, the temporary increase of load torque exceeds the maximum motor torque, $D_0 = 14.3$—for instance by starting a line of shafting or other mass of considerable momentum—then the motor speed continues to drop as long as the excess load exists, and whether the motor will recover when the excess load is taken off, or not, depends on the loss of speed of the motor during the period of overload: if, when the overload is relieved, the motor has dropped to point d_1 in Fig. 102, its speed thus is still above b, the motor recovers; if, however, its speed has dropped to d_2, below the speed b, $S = 0.35$, at which the motor torque drops below the load torque, then the motor does not recover, but stops.

With a lighter load torque, D'_0, which is less than the starting torque, g, obviously the motor will always recover in speed

The amount, by which the motor drops in speed at temporary overload, naturally depends on the duration of the overload, and on the momentum of the motor and its moving masses: the higher the momentum of the motor and of the masses driven by it at the moment of overload, the slower is the drop of speed of the motor, and the higher thus the speed retained by it at the moment when the overload is relieved.

Thus a motor of low starting torque, that is, high speed regulation, may be thrown out of step by picking up a load of high momentum rapidly, while by adding a flywheel to the motor, it would be enabled to pick up this load. Or, it may be troublesome to pick up the first load of high momentum, while the second load of this character may give no trouble, as, due to the momentum of the load already picked up, the speed would drop less.

Thus a motor carrying no load, may be thrown out of step by a load which the same motor, already partly loaded (with a load of considerable momentum), would find no difficulty to pick up.

The ability of an induction motor, to carry for a short time without dropping out of step a temporary excessive overload, naturally also depends on the excess of the maximum motor torque (at c in Fig. 102) over the normal load torque of the motor. A motor, in which the maximum torque is very much higher—several hundred per cent.—than the rated torque, thus could momentarily carry overloads which a motor could not carry, in which the maximum torque exceeds the rated torque only by 50 per cent., as was the case with the early motors. However, very high maximum torque means low internal reactance and thus high exciting current, that is, low power-factor at partial loads, and of the two types of motors:

(*a*) High overload torque, but poor power-factor and efficiency at partial loads;

(*b*) Moderate overload torque, but good power-factor and efficiency at partial loads;

the type (*b*) gives far better average operating conditions, except in those rare cases of operation at constant full-load, and is therefore preferable, though a greater care is necessary to avoid momentary excessive overloads.

Gradually the type (*a*) had more and more come into use, as the customers selected the motor, and the power supply company neglected to pay much attention to power-factor, and it is only in the last few years, that a realization of the harmful effects of low power-factors on the economy of operation of the systems is again directing attention to the need of good power-factors at partial loads, and the industry thus is returning to type (*b*), especially in view of the increasing tendency toward maximum output rating of apparatus.

In distributing transformers, the corresponding situation had been realized by the central stations since the early days, and good partial load efficiencies and power-factors secured.

104. The induction motor speed-torque curve thus has on a constant-torque load a stable branch, from the maximum torque point, c, Fig. 102, to synchronism; and an unstable branch, from standstill to the maximum torque point.

However, it would be incorrect to ascribe the stability or instability to the induction motor-speed curve; but it is the character of the load, the requirement of constant torque, which makes a part of the speed curve unstable, and on other kinds of load no instability may exist, or a different form of instability.

Thus, considering a load requiring a torque proportional to

the speed, such as would be given, approximately, by an electric generator at constant field excitation and constant resistance as load.

The load-torque curves, then, would be straight lines going through the origin, as shown by D'_1, D'_2, D'_3, etc., for increasingly larger values of load, in Fig. 103. The motor-torque curve, D, is the same as in Fig. 102. As seen, all the lines, D', intersect D at points, a_1, a_2, a_3 . . . ., at which the speed is stable, since

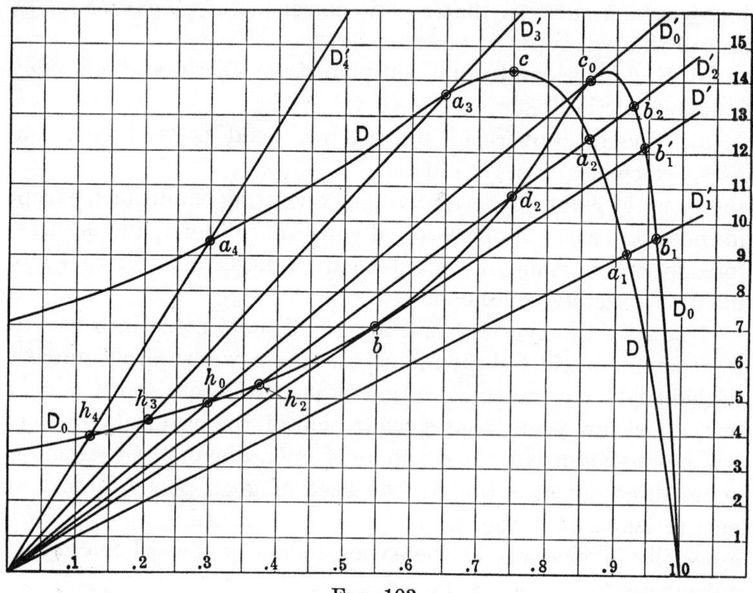

FIG. 103.

$$\frac{dD'}{dS} > \frac{dD}{dS}.$$

Thus, with this character of load, a torque required proportional to the speed, and the motor-torque curve, D, no instability exists, but conditions are stable from standstill to synchronism, just as in Fig. 101. That is, with increasing load, the speed decreases and increases again with decreasing load.

If, however, the motor curve is as shown by D_0 in Fig. 103, that is, low starting torque and a maximum torque point close to synchronism, as corresponds to an induction motor with low resistance secondary, then for a certain range of load, between

D' and D'_0, the load-torque line, D'_2, intersects the motor curve, D_0, in three points b_2, d_2, h_2.

At b_2, $S = 0.925$, and at h_2, $S = 0.375$, conditions are stable; at d_2, $S = 0.75$, instability exists.

Thus with this load, D'_2, the motor can run at two different speeds in stable conditions: a high speed, above c_0, and a low speed, below b; while there is a third, theoretical speed, d_2, which is unstable. In the range below h_2, the motor speeds up to h_2; in the range between h_2 and d_2, the motor slows down to h_2; in the range between d_2 and b_2, the motor speeds up to b_2, and in the range above b_2, the motor slows down to b_2.

There is thus a (fairly narrow) range of loads between D' and D'_0, in which an unstable branch of the induction motor-torque curve exists, at intermediate speeds; at low speed as well as at high speed conditions are stable.

For loads less than D', conditions are stable over the entire range of speed; for loads above D'_0, the motor can run only at low speeds, h_3, h_4, but not at high speeds; but there is no load at which the motor would not start and run up to some speed.

Obviously, at the lower speeds, the current consumed by the motor is so large, that the operation would be very inefficient.

It is interesting to note, that with this kind of load, the "maximum torque point," c, is no characteristic point of the motor-torque curve, but two points, c_0 and b, exist, between which the operation of the motor is unstable, and the speed either drops down below b, or rises above c_0.

105. With a load requiring a torque proportional to the square of the speed, such as a fan, or a ship propeller, conditions are almost always stable over the entire range of speed, from standstill to synchronism, and an unstable range of speed may occur only in motors of very low secondary resistance, in which the drop of torque below the maximum torque point, c, of the motor characteristic is very rapid, that is, the torque of the motor decreases more rapidly than with the square of the speed. This may occur with very large motors, such as used on ship propellers, if the secondary resistance is made too low.

More frequently instability with such fan or propeller load or other load of similar character may occur with single-phase motors, as in these the drop of the torque curve below maximum torque is much more rapid, and often a drop of torque with increasing speed occurs, especially with the very simple and cheap starting devices economically required on very small motors, such as fan motors.

Instability and dropping out of step of induction motors also may be the result of the voltage drop in the supply lines, and furthermore may result from the regulation of the generator voltage being too slow. Regarding hereto, however, see *Theory and Calculation of Electrical Apparatus*, in the chapter on "Stability of Induction Machines" [McGraw-Hill edition, volume 6].

D. Hunting of Synchronous Machines

106. In induction-motor circuits, instability almost always assumes the form of a steady change, with increasing rapidity, from the unstable condition to a stable condition or to standstill, etc.

Oscillatory instability in induction-motor circuits, as the result of the relation of load to speed and electric supply, is rare. It has been observed, especially in single-phase motors, in cases of considerable oversaturation of the magnetic circuit.

Oscillatory instability, however, is typical of the synchronous machine, and the hunting of synchronous machines has probably been the first serious problem of cumulative oscillations in electric circuits, and for a long time has limited the industrial use of synchronous machines, in its different forms:

(*a*) Difficulty and failure of alternating-current generators to operate in parallel.

(*b*) Hunting of synchronous converters.

(*c*) Hunting of synchronous motors.

While considerable theoretical work has been done, practically all theoretical study of the hunting of synchronous machines has been limited to the calculation of the frequency of the transient oscillation of the synchronous machine, at a change of load, frequency or voltage, at synchronizing, etc. However, this transient oscillation is harmless, and becomes dangerous only if the oscillation ceases to be transient, but becomes permanent and cumulative, and the most important problem in the study of hunting thus is the determination of the cause, which converts the transient oscillation into a cumulative one, that is, the determination of the source of the energy, and the mechanism of its transfer to the oscillating system. To design synchronous machines, so as to have no or very little tendency to hunting, obviously requires a knowledge of those characteristics of design which are instrumental in the energy transfer to the oscillating system, and thereby cause hunting, so as to avoid them and produce the greatest possible inherent stability.

If, in an induction motor running loaded, at constant speed, the load is suddenly decreased, the torque of the motor being in excess of the reduced load causes an acceleration, and the speed increases. As in an induction motor the torque is a function of the speed, the increase of speed decreases the torque, and thereby decreases the increase of speed until that speed is reached at which the motor torque has dropped to equality with the load, and thereby acceleration and further increase of speed ceases, and the motor continues operation at the constant higher speed, that is, the induction motor reacts on a decrease of load by an increase of speed, which is gradual and steady without any oscillation.

If, in a synchronous motor running loaded, the load is suddenly decreased, the beginning of the phenomenon is the same as in the induction motor, the excess of motor torque causes an acceleration, that is, an increase of speed. However, in the synchronous motor the torque is not a function of the speed, but in stationary condition the speed must always be the same, synchronism, and the torque is a function of the relative position of the rotor to the impressed frequency. The increase of speed, due to the excess torque resulting from the decreased load, causes the rotor to run ahead of its previous relative position, and thereby decreases the torque until, by the increased speed, the motor has run ahead from the relative position corresponding to the previous load, to the relative position corresponding to the decreased load. Then the acceleration, and with it the increase of speed, stops. But the speed is higher than in the beginning, that is, is above synchronism, and the rotor continues to run ahead, the torque continues to decrease, is now below that required by the load, and the latter thus exerts a retarding force, decreases the speed and brings it back to synchronism. But when synchronous speed is reached again, the rotor is ahead of its proper position, thus can not carry its load, and begins to slow down, until it is brought back into its proper position. At this position, however, the speed is now below synchronism, the rotor thus continues to drop back, and the motor torque increases beyond the load, thereby accelerates again to synchronous speed, etc., and in this manner conditions of synchronous speed, with the rotor position behind or ahead of the position corresponding to the load, alternate with conditions of proper relative position of the rotor, but below or above synchronous speed, that is, an oscillation results which usually dies down at a rate depending on the energy losses resulting from the oscillation.

107. As seen, the characteristic of the synchronous machine is, that readjustment to a change of load requires a change of relative position of the rotor with regard to the impressed frequency, without any change of speed, while a change of relative position can be accomplished only by a change of speed, and this results in an over-reaching in position and in speed, that is, in an oscillation.

Due to the energy losses caused by the oscillation, the successive swings decrease in amplitude, and the oscillation dies down. If, however, the cause which brings the rotor back from the position ahead or behind its normal position corresponding to the changed load (excess or deficiency of motor torque over the torque required by the load) is greater than the torque which opposes the deviation of the rotor from its normal position, each swing tends to exceed the preceding one in amplitude, and if the energy losses are insufficient, the oscillation thus increases in amplitude and becomes cumulative, that is, hunting.

In Fig. 104 is shown diagrammatically as p, the change of the relative position of the rotor, from p_1 corresponding to the previous load to p_2 the position further forward corresponding to the decreased load.

v then shows the oscillation of speed corresponding to the oscillation of position.

The dotted curve, w_1, then shows the energy losses resulting from the oscillation of speed (hysteresis and eddies in the pole faces, currents in damper windings), that is, the damping power, assumed as proportional to the square of the speed.

If there is no lag of the synchronizing force behind the position displacement, the synchronizing force, that is, the force which tends to bring the rotor back from a position behind or ahead of the position corresponding to the load, would be—or may approximately be assumed as—proportional to the position displacement, p, but with reverse sign, positive for acceleration when p is negative or behind the normal position, negative or retarding when p is ahead. The synchronizing power, that is, the power exerted by the machine to return to the normal position, then is derived by multiplying $-p$ with v, and is shown dotted as w_2 in Fig. 104. As seen, it has a double-frequency alternation with zero as average.

The total resultant power or the resulting damping effect which restores stability, then, is the sum of the synchronizing power w_2 and the damping power w_1, and is shown by the dotted

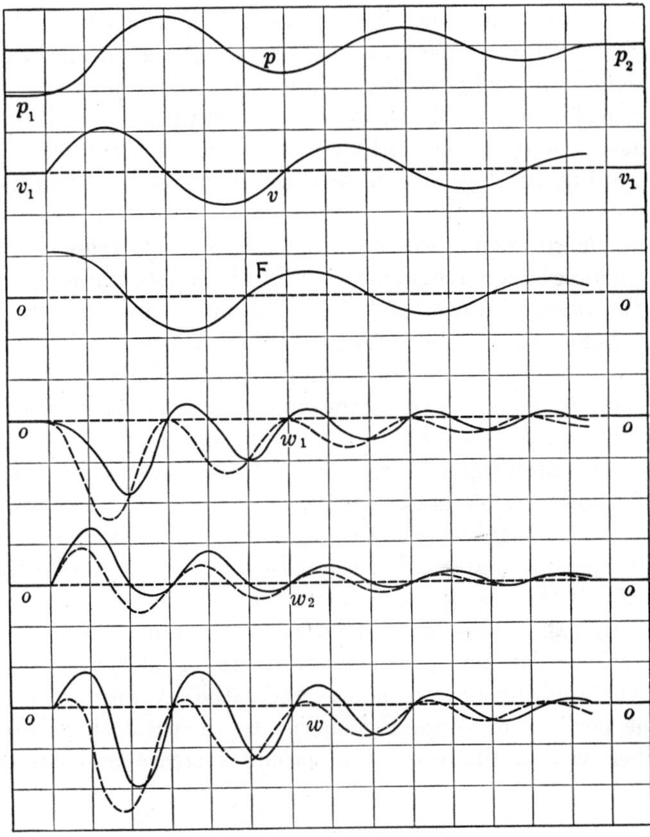

FIG. 104.

curve w. As seen, under the assumption or Fig. 104, in this case a rapid damping occurs.

If the damping winding, which consumes a part of all the power, w_1, is inductive—and to a slight extent it always is—the current in the damping winding lags behind the e.m.f. induced in it by the oscillation, that is, lags behind the speed, v. The power, w_1, or that part of it which is current times voltage, then ceases to be continuously negative or damping, but contains a positive period, and its average is greatly reduced, as shown by the drawn curve, w_1, in Fig. 104, that is, inductivity of the damper winding is very harmful, and it is essential to design the damper winding as non-inductive as possible to give efficient damping.

With the change of position, p, the current, and thus the armature reaction, and with it the magnetic flux of the machine, changes. A flux change can not be brought about instantly, as it represents energy stored, and as a result the magnetic flux of the machine does not exactly correspond with the position, p, but lags behind it, and with it the synchronizing force, F, as shown in Fig. 104, lags more or less, depending on the design of the machine.

The synchronizing power of the machine, Fv, in the case of a lagging synchronizing force, F, is shown by the drawn curve, w_2. As seen, the positive ranges of the oscillation are greater than the negative ones, that is, the average of the oscillating synchronizing power is positive or supplying energy to the oscillating system, which energy tends to increase the amplitude of the oscillation—in other words, tends to produce cumulative hunting.

The total resulting power, $w = w_1 + w_2$, under these conditions is shown by the drawn curve, w, in Fig. 104. As seen, its average is still negative or energy-consuming, that is, the oscillation still dies out, and stability is finally reached, but the average value of w in this case is so much less than in the case above discussed, that the dying out of the oscillation is much slower.

If now, the damping power, w_1, were still smaller, or the average synchronizing power, w_2, greater, the average w would become positive or supplying energy to the oscillating system. In other words, the oscillation would increase and hunting result.

That is:

If the average synchronizing power resulting from the lag of the synchronizing force behind the position exceeds the average damping power, hunting results. The condition of stability of the synchronous machine is, that the average damping power exceeds the average synchronizing power, and the more this is the case, the more stable is the machine, that is, the more rapidly the transient oscillation of readjustment to changed circuit conditions dies out.

Or, if

a = attenuation constant of the oscillating system,

$a<0$ gives cumulative oscillation or hunting.

$a>0$ gives stability.

108. Counting the time, t, from the moment of maximum backward position of the rotor, that is, the moment at which the load on the machine is decreased, and assuming sinusoidal variation, and denoting

$$\phi = 2\pi f t = \omega t \tag{1}$$

where

$$f = \text{frequency of the oscillation} \tag{2}$$

the relative position of the rotor then may be represented by

$$p = -p_0 \epsilon^{a\phi} \cos \phi,$$

where

$p_0 = p_2 - p_1$ = position difference of rotor resulting from change of load, (3)

$$a = \text{attenuation constant of oscillation.} \tag{4}$$

The velocity difference from that of uniform rotation then is

$$v = \frac{dp}{dt} = \omega \frac{dp}{d\phi} = \omega p_0 \epsilon^{-a\phi} (\sin \phi + a \cos \phi). \tag{5}$$

Let

$$a = \tan \alpha; \quad 1 + a^2 = A^2 \tag{6}$$

hence,

$$\sin \alpha = \frac{a}{A}; \quad \cos \alpha = \frac{1}{A} \tag{7}$$

it is

$$v = \omega p_0 A \epsilon^{-a\phi} \sin (\phi + \alpha). \tag{8}$$

Let

γ = lag of damping currents behind e.m.f. induced in damper windings (9)

the damping power is

$$\begin{aligned} w_1 &= -c v v_\gamma \\ &= -c\omega^2 p_0^2 A^2 \epsilon^{-2a\phi} \sin(\phi + \alpha) \sin(\phi + \alpha - \gamma) \end{aligned} \tag{10}$$

where

$c = \frac{w}{v^2}$ = damping power per unit velocity and v_γ is v, lagged by angle γ. (11)

Let

β = lag of synchronizing force behind position displacement p (12)

and

$$\beta = \omega t_0 \tag{13}$$

where

t_0 = time lag of synchronizing force. (14)

The synchronizing force then is

$$F = bp_0\epsilon^{-a\phi} \cos (\phi - \beta) \tag{15}$$

where

$b = \dfrac{F_0}{p_0}$ = ratio of synchronizing force to position displacement, or specific synchronizing force. (16)

The synchronizing power then is

$$w_2 = Fv = b\omega p_0 A\epsilon^{-2a\phi} \sin (\phi + \alpha) \cos (\phi - \beta). \tag{17}$$

The oscillating mechanical power is

$$w = \frac{d}{dt}\frac{mv^2}{e} \doteq m\omega v \frac{dv}{d\phi}$$

$$= m\omega\beta p_0{}^2 A^2 \epsilon^{-2\,a\phi} \sin (\phi + \alpha) \{\cos (\phi + \alpha) - \alpha \sin (\phi + \alpha)\} \tag{18}$$

where

m = moving mass reduced to the radius, on which p is measured. (19)

It is, however,

$$w_1 + w_2 - w = 0 \tag{20}$$

hence, substituting (10), (17), (18) into (20) and canceling,

$$b \cos (\phi - \beta) - c\omega A \sin (\phi + \alpha - \gamma) - m\omega^2 A \cos (\phi + \alpha) + m\omega^2 A a \sin (\phi + a) = 0. \tag{21}$$

This gives, as the coefficients of $\cos \phi$ and $\sin \phi$ the equations

$$\left.\begin{array}{l} b \cos \beta - c\omega A \sin (\alpha - \gamma) - ma^2 A \cos \alpha + m\omega^2 A a \sin \alpha = 0 \\ b \sin \beta - c\omega A \cos (\alpha - \gamma) + m\omega^2 A \sin a + m\omega^2 A \cos \alpha = 0 \end{array}\right\} \tag{22}$$

Substituting (6) and (7) and approximating from (13), for β as a small quantity,

$$\cos \beta = 1; \quad \sin \beta = \omega t_0 \tag{23}$$

gives

$$\left.\begin{array}{r} b - c\omega\,(a \cos \gamma - \sin \gamma) - m\omega^2\,(1 - a^2) = 0 \\ bt_0 - c\,(\cos \gamma + a \sin \gamma) + 2m\omega a = 0 \end{array}\right\} \tag{24}$$

This gives the values, neglecting smaller quantities

$$a = \frac{c \cos \gamma - bt_0}{\sqrt{4\, mb - c^2 \cos^2 \gamma + b^2 t_0^2}} \tag{25}$$

$$\omega = \frac{1}{2m}\left\{\sqrt{4\, mb - c^2 \cos 2\,\gamma + b^2 t_0^2} + c \sin \gamma\right\} \tag{26}$$

$$f = \frac{\omega}{2\,\phi} \tag{27}$$

These equations (25) and (26) apply only for small values of a, but become inaccurate for larger values of a, that is, very rapid damping. However, the latter case is of lesser importance.

$$a = 0$$

gives

$$bt_0 = c \cos \gamma,$$

hence,

$$\left.\begin{aligned} c &> \frac{bt_0}{\cos \gamma} \\ &\text{or,} \\ t_0 &< \frac{c \cos \gamma}{b} \end{aligned}\right\} \tag{28}$$

are the conditions of stability of the synchronous machine.

If

$$t_0 = 0$$

$$\gamma = 0$$

it is

$$a = \frac{c}{\sqrt{4\, mb - c^2}},$$

$$\omega = \frac{\sqrt{4\, mb - c^2}}{2\, m},$$

and, if also,

$$c = 0:$$

it is

$$\omega = \sqrt{\frac{b}{m}}.$$

CHAPTER XII

REACTANCE OF TRANSFORMERS

109. An electric current passing through a conductor is accompanied by a magnetic field surrounding this conductor, and this magnetic field is as integral a part of the phenomenon, as is the energy dissipation by the resistance of the conductor. It is represented by the inductance, L, of the conductor, or the number of magnetic interlinkages with unit current in the conductor. Every circuit thus has a resistance, and an inductance, however small the latter may be in the so-called "non-inductive" circuit. With continuous current in stationary conditions, the inductance, L, has no effect on the energy flow; with alternating current of frequency, f, the inductance, L, consumes a voltage $2\pi fLi$, and is, therefore, represented by the reactance, $x = 2\pi fL$, which is measured in ohms, and differs from the ohmic resistance, r, merely by being wattless or reactive, that is, representing not dissipation of energy, but surging of energy.

Every alternating-current circuit thus has a resistance and a reactance, the latter representing the effect of the magnetic field of the current in the conductor.

When dealing with alternating-current apparatus, especially those having several circuits, it must be realized, however, that the magnetic field of the circuit may have no independent existence, but may merge into and combine with other magnetic fields, so that it may become difficult what part of the magnetic field is to be assigned to each electric circuit, and circuits may exist which apparently have no reactance. In short, in such cases, the magnetic fields of the reactance of the electric circuit may be merely a more or less fictitious component of the resultant magnetic field.

The industrial importance hereof is that many phenomena, such as the loss of power by magnetic hysteresis, the m.m.f. required for field excitation, etc., are related to the resultant magnetic field, thus not equal to the sum of the corresponding effects of the components.

As the transformer is the simplest alternating-current apparatus, the relations are best shown thereon.

Leakage Flux of Alternating-current Transformer

110. The alternating-current transformer consists of a magnetic circuit, interlinked with two electric circuits, the primary circuit, which receives power from its impressed voltage, and the secondary circuit, which supplies power to its external circuit.

For convenience, we may assune the secondary circuit as reduced to the primary circuit by the ratio of turns, that is, assume ratio of turns $1 \div 1$.

Let

$Y_0 = g - jb$ = primary exciting admittance;

$Z_0 = r_0 + jx_0$ = primary self-inductive impedance;

$Z_1 = r_1 + jx_1$ = secondary self-inductive impedance (reduced to the primary).

The transformer thus comprises three magnetic fluxes: the mutual magnetic flux, Φ, which, being interlinked with primary and secondary, transforms the power from primary to secondary, and is due to the resultant m.m.f of primary and secondary circuit; the primary leakage flux, Φ'_0, due to the m.m.f. of the primary circuit, F_0, and interlinked with the primary circuit only, which is represented by the self-inductive or leakage reactance, x_0; and the secondary leakage flux, Φ'_1, due to the m.m.f. of the secondary circuit, F_1, and interlinked with the secondary circuit only which is represented by the secondary reactance, x_1.

As seen in Fig. 105*o*, the mutual flux, Φ—usually—has a closed iron circuit of low reluctance, ρ, thus low m.m.f., F, and high intensity; the self-inductive flux or leakage reactance flux, Φ'_0 and Φ'_1, close through the air circuit between the primary and secondary electric circuits, thus meet with a high reluctance, ρ_0, respectively ρ_1, usually many hundred times higher than ρ. Their m.m.fs., F_0 and F_1, however, are usually many times greater than F; the latter is the m.m.f. of the exciting current, the former that of full primary or secondary current.

For instance, if the exciting current is 5 per cent. of full-load current, the reactance of the transformer 4 per cent., or 2 per cent. primary and 2 per cent. secondary, then the m.m.f. of the leakage flux is 20 times that of the mutual flux, and the mutual flux 50 times the leakage flux, hence the reluctance of leakage flux $50 \times 20 = 1000$ times that of the mutual or main flux: $\rho_1 = 1000\ \rho$.

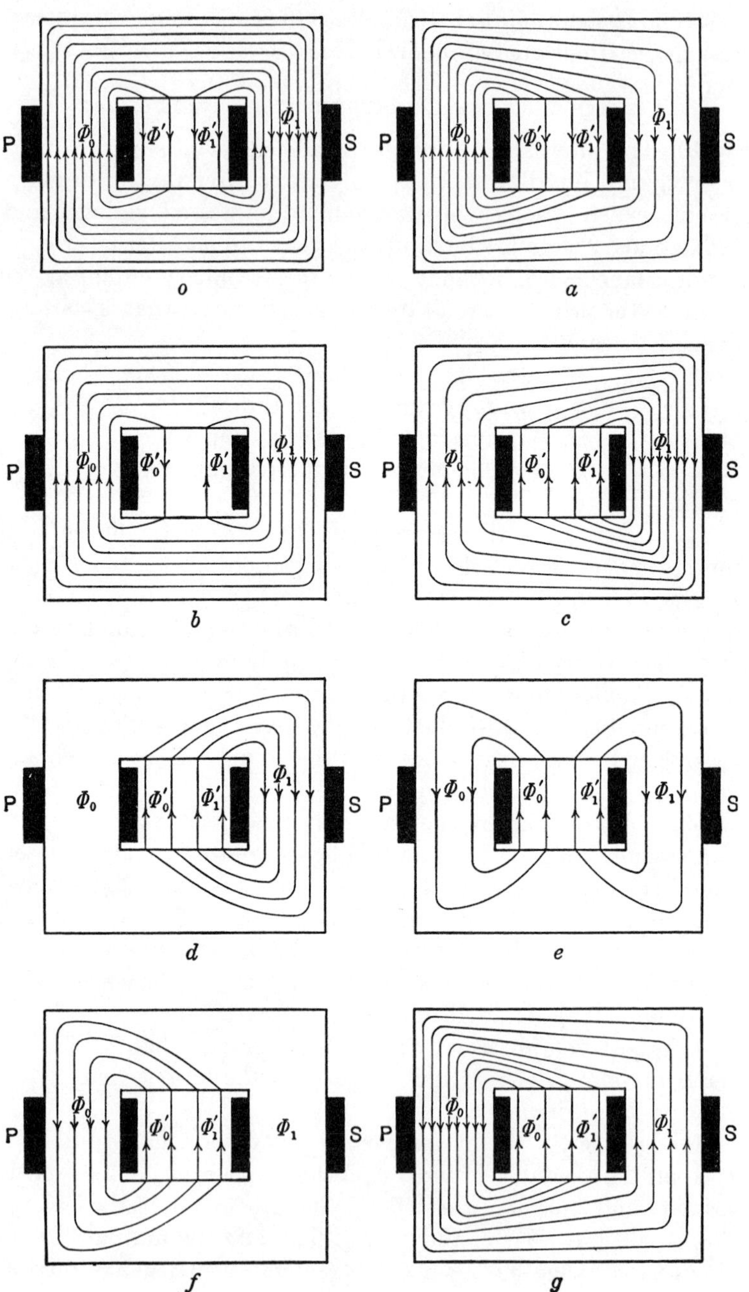

FIG. 105.

111. Usually, as stated, the leakage fluxes are not considered as such, but represented by their reactances, in the transformer diagram. Thus, at non-inductive load, it is, Fig. 106,

$O\Phi$ = mutual, or main magnetic flux, chosen as negative vertical.

OF = m.m.f. required to produce flux, $O\Phi$, and leading it by the angle of hysteretic advance of phase, $FO\Phi$.

OE'_1 = e.m.f. induced in the secondary circuit by the mutual flux, and 90° behind it.

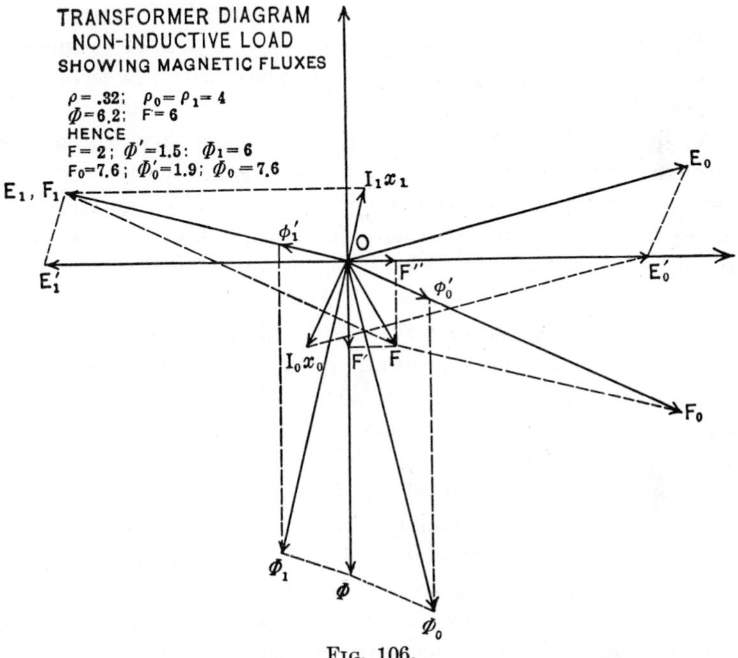

FIG. 106.

I_1x_1 = secondary reactance voltage, 90° behind the secondary current, and combining with OE'_1 to

OE_1 = true secondary induced voltage. From this subtracts the secondary resistance voltage, I_1r_1, leaving the secondary terminal voltage, and, in phase with it at non-inductive load, the secondary current and secondary m.m.f., OF_1.

From component, OF_1, and resultant, OF, follows the other component,

OF_0 = primary m.m.f. and in phase with it the primary current.

OE'_0 = primary voltage consumed by mutual flux, equal and opposite to OE'_1.

I_0x_0 = primary reactance voltage, 90° ahead of the primary current OF_0.

From I_0x_0 as component and E'_0 as resultant follows the other component, $\overline{OE_0}$, and adding thereto the primary resistance voltage, I_0r_0, gives primary supply voltage.

In this diagram, Fig. 106, the primary leakage flux is represented by $O\Phi'_0$, in phase with the primary current, $\overline{OF_0}$, and the secondary leakage flux is represented by $\overline{O\Phi'_1}$, in phase with the secondary current, OF_1.

As shown in Fig. 105*o*, the primary leakage flux, Φ'_0, passes through the iron core inside of the primary coil, together with the resultant flux, Φ, and the secondary leakage flux, Φ'_1, passes through the secondary core, together with the mutual flux, Φ. However, at the moment shown in Fig. 105*o*, Φ'_1 and Φ in the secondary core are opposite in direction. This obviously is not possible, and the flux in the secondary core in this moment is $\Phi - \Phi'_1$, that is, the magnetic disposition shown in Fig. 105*o* is merely nominal, but the actual magnetic distribution is as shown in Fig. 105*a*; the flux in the primary core, $\Phi_0 = \Phi + \Phi'_0$, the flux in the secondary core, $\Phi_1 = \Phi - \Phi'_1$.

As seen, at the moment shown in Fig. 105*o* and 105*a*, all the leakage flux comes from and interlinks with the primary winding, none with the secondary winding, and it thus would appear, that all the self-inductive reactance is in the primary circuit, none in the secondary circuit, or, in other words, that the secondary circuit of the transformer has no reactance.

However, at a later moment of the cycle, shown in Fig. 105*c*, all the leakage flux comes from and interlinks with the secondary, and this figure thus would give the impression, that all the leakage reactance of the transformer is in the secondary, none in the primary winding.

In other words, the leakage fluxes of the transformer and the mutual or main flux are not independent fluxes, but partly traverse the same magnetic circuit, so that each of them during a part of the cycle is a part of any other of the fluxes. Thus, the reactance voltage and the mutual inductive voltage of the transformer are not separate e.m.fs., but merely mathematical fictions, components of the resultant induced voltage, OE_1 and OE_0, induced by the resultant fluxes, $\overline{O\Phi_0}$ in the primary, and $\overline{O\Phi_1}$ in the secondary core.

112. In Fig. 107 are plotted, in rectangular coördinates, the magnetic fluxes:

The mutual or main magnetic flux, Φ;

The primary leakage flux, Φ'_0;

The resultant primary flux, $\Phi_0 = \Phi + \Phi'_0$;

The secondary leakage flux, Φ'_1;

The resultant secondary flux, $\Phi_1 = \Phi - \Phi'_1$;

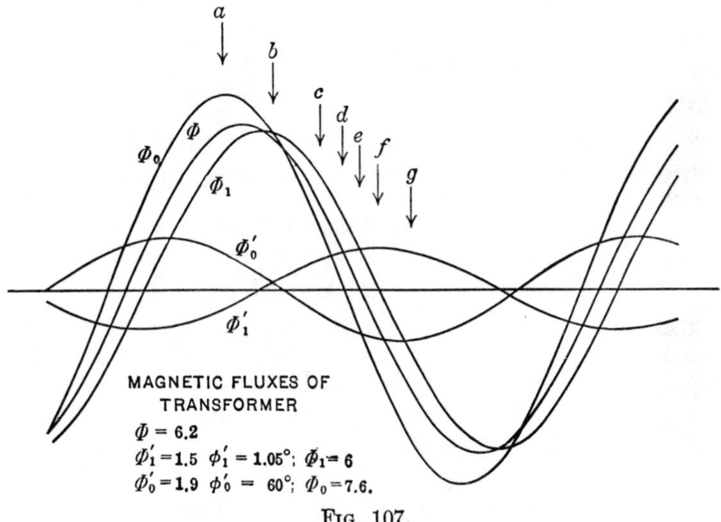

FIG. 107.

and the magnetic distribution in the transformer, during the moments marked as *a*, *b*, *c*, *d*, *e*, *f*, *g*, in Fig. 107, is shown in Fig. 105.

In Fig. 105*a*, the primary flux is larger than the secondary, and all leakage fluxes (x_0 and x_1) come from the primary flux, that is, there is no secondary leakage flux.

In Fig. 105*b*, primary and secondary flux equal, and primary and secondary leakage flux equal and opposite, though small.

In Fig. 105*c*, the secondary flux is larger, all leakage flux (x_0 and x_1) comes from the secondary flux, that is, there is no primary leakage flux.

In Fig. 105*d*, there is no primary flux, and all the secondary flux is leakage flux.

In Fig. 105*e*, there is no mutual flux, all primary flux is primary leakage flux, and all secondary flux is secondary leakage flux.

In Fig. 105*f*, there is no secondary flux, and all primary flux is leakage flux.

In Fig. 105*g*, the primary flux is larger than the secondary, and all leakage flux comes from the primary, the same as in 105*a*.

Figs. 105*a* to 105*f*, thus show the complete cycle, corresponding to diagrams, Figs. 106 and 107.

These figures are drawn with the proportions,

$$\begin{aligned} \rho \div \rho_0 \div \rho_1 &= 1 \div 12.5 \div 12.5 \\ F \div F_0 \div F_1 &= 1 \div 3.8 \div 3 \\ \Phi \div \Phi'_0 \div \Phi'_1 &= 1 \div 0.317 \div 0.25. \end{aligned}$$

thus are greatly exaggerated, to show the effect more plainly. Actually, the relations are usually of the magnitude,

$$\begin{aligned} \rho \div \rho_0 \div \rho_1 &= 1 \div 1000 \div 1000 \\ F \div F_0 \div F_1 &= 1 \div 20.6 \div 20 \\ \Phi \div \Phi'_0 \div \Phi'_1 &= 1 \div 0.02 \div 0.02 \end{aligned}$$

113. In symbolic representation, denoting,

$\dot{\Phi}$ = mutual magnetic flux.
$\dot{E}$ = mutual induced voltage.
$\dot{\Phi}_0$ = resultant primary flux.
$\dot{\Phi}'_0$ = primary leakage flux.
$\dot{E}_0$ = primary terminal voltage.
$\dot{I}_0$ = primary current.
$Z_0 = r_0 + jx_0$ = primary self-inductive impedance.
$\dot{\Phi}_1$ = resultant secondary flux.
$\dot{\Phi}'_1$ = secondary leakage flux.
$\dot{E}_1$ = secondary terminal voltage.
$\dot{I}_1$ = secondary current.
$Z_1 = r_1 + jx_1$ = secondary self-inductive impedance.

and

$$c = 2\pi f n$$

where n = number of turns.

It then is

$$
\begin{aligned}
c\Phi'_1 &= jx_0 I_0 \\
c\Phi'_1 &= jx_1 I_1 \\
c\Phi &= E = E_0 - Z_0 I_0 = E_1 + Z_1 I_1 \\
c\Phi_0 &= E_0 - r_0 I_0 = E + jx_0 I_0 \\
c\Phi_1 &= E_1 + r_1 I_1 = E - jx_1 I_1 \\
\Phi'_0 &= \Phi_0 - \Phi \\
\Phi'_1 &= \Phi + \Phi_1,
\end{aligned}
$$

thus, the total leakage flux

$$\Phi' = \Phi'_0 + \Phi'_1 = \Phi_0 - \Phi_1.$$

114. One of the important conclusions from the study of the actual flux distribution of the transformer is that the distinction between primary and secondary leakage flux, Φ'_0 and Φ'_1, is really an arbitrary one. There is no distinct primary and secondary leakage flux, but merely one leakage flux, Φ', which is the flux passing between primary and secondary circuit, and which during a part of the cycle interlinks with the primary, during another part of the cycle interlinks with the secondary circuit Thus the corresponding electrical quantities, the reactances, x_0 and x_1, are not independent quantities, that is, it can not be stated that there is a definite primary reactance, x_0, and a definite secondary reactance, x_1, but merely that the transformer has a definite reactance, x, which is more or less arbitrarily divided into two parts; $x = x_0 + x_1$, and the one assigned to the primary, the other to the secondary circuit.

As the result hereof, "mutual magnetic flux" Φ, and the mutual induced voltage, E, are not actual quantities, but rather mathematical fictions, and not definite but dependent upon the distribution of the total reactance between the primary and the secondary circuit.

This explains why all methods of determining the transformer reactance give the total reactance $x_0 + x_1$.

However, the subdivision of the total transformer reactance into a primary and a secondary reactance is not entirely arbitrary. Assuming we assign all the reactance to the primary, and consider the secondary as having no reactance. Then the mutual magnetic flux and mutual induced voltage would be

$$c\Phi = E = E_0 - [r_0 + j(x_0 + x_1)] I_0$$

and the hysteresis loss in the transformer would correspond hereto, by the usual assumption in transformer calculations.

Assigning, however, all the reactance to the secondary circuit, and assuming the primary as non-inductive, the mutual flux and mutual induced voltage would be $c\dot{\Phi} = \dot{E} = \dot{E}_0 - r_0\dot{I}_0$, hence larger, and the hysteresis loss calculated therefrom larger than under the previous assumption. The first assumption would give too low, and the last too high a calculated hysteresis loss, in most cases.

By the usual transformer theory, the hysteresis loss under load is calculated as that corresponding to the mutual induced voltage, E. The proper subdivision of the total transformer reactance, x, into primary reactance, x_0, and secondary reactance, x_1, would then be that, which gives for a uniform magnetic flux, Φ, corresponding to the mutual induced voltage, E, the same hysteresis loss, as exists with the actual magnetic distribution of $\Phi_0 = \Phi + \Phi'_0$ in the primary, and $\Phi_1 = \Phi - \Phi'_1$ in the secondary core. Thus, if V_0 is the volume of iron carrying the primary flux, Φ_0, at flux density, B_0, V_1 the volume of iron carrying the secondary flux, Φ_1, at flux density, B_1, the flux density of the theoretical mutual magnetic flux would be given by

$$B^{1.6} = \frac{V_0 B_0{}^{1.6} + V_1 B_1{}^{1.6}}{V_0 + V_1}$$

from B then follows Φ, E, and thus x_0 and x_1.

This does not include consideration of eddy-current losses. For these, an approximate allowance may be made by using 1.7 as exponent, instead of 1.6.

Where the magnetic stray field under load causes additional losses by eddy currents, these are not included in the loss assigned to the mutual magnetic flux, but appear as an energy component of the leakage reactances, that is, as an increase of the ohmic resistances of the electric circuits, by an effective resistance.

115. Usually, the subdivision of x into x_0 and x_1, by this assumption of assigning the entire core loss to the mutual flux, is sufficiently close to equality, to permit this assumption. That is, the total transformer reactance is equally divided between primary and secondary circuit.

This, however, is not always justified, and in some cases, the one circuit may have a higher reactance than the other. Such, for instance, is the case in some very high voltage transformers, and usually is the case in induction motors and similar apparatus.

It is more commonly the case, where true self-inductive fluxes exist, that is, magnetic fluxes produced by the current in one circuit, and interlinked with this circuit, closing upon themselves in a path which is entirely distinct from that of the mutual magnetic flux, that is, has no part in common with it. Such, for instance, frequently is the self-inductive flux of the end connections of coils in motors, transformers, etc. To illustrate: in the high-voltage shell-type transformer, shown diagrammatically in Fig. 108, with primary coil 1, closely adjacent to the core, and high-voltage secondary coil 2 at considerable distance:

The primary leakage flux consists of the flux in spaces, *a*, between the yokes of the transformer, closing through the iron core, *C*, and the flux through the spaces, *b*, outside of the transformer, which enters the faces, *F*, of the yokes and closes through the central core, *C*.

The secondary leakage flux contains the same two components: the flux through the spaces, *a*, between the yokes closing, however, through the outside shells, *S*, and the flux through the spaces, *b*, outside of the transformer, and entering the faces, *F*, but in this case closing through the shells, *S*. In addition to these two components, the secondary leakage flux contains a third component, passing through the spaces, *b*, between the coils, but closing, through outside space, *c*, in a complete air circuit. This flux has no corresponding component in the primary, and the total secondary leakage reactance in this case thus is larger than the total primary reactance.

Similar conditions apply to magnetic structures as in the induction motor, alternator, etc.

In such a case as represented by Fig. 108, the total reactance of the transformer, with (2) as primary and (1) as secondary, would be greater than with (1) as primary and (2) as secondary.

In this case, when subdividing the total reactance into primary reactance and secondary reactance, it would appear legitimate to divide it in proportion of the total reactances with (1) and (2) as primary, respectively. That is,

if x = total reactance, with coil (1) as primary, and (2) as secondary, and

x' = total reactance, with coil (2) as primary, and (1) as secondary, then it is:

With coil (1) as primary and (2) as secondary,

Primary reactance,

$$x_0 = \frac{x}{x + x'}x = \frac{x^2}{x + x'}.$$

Secondary reactance,

$$x_1 = \frac{x'}{x + x'}x = \frac{xx'}{x + x'}.$$

With coil (2) as primary and (1) as secondary,

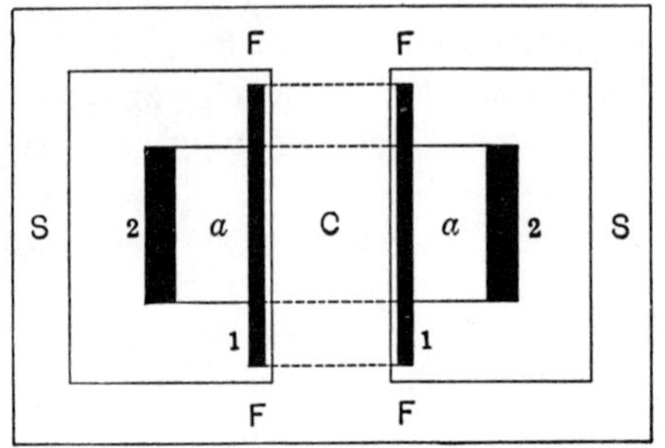

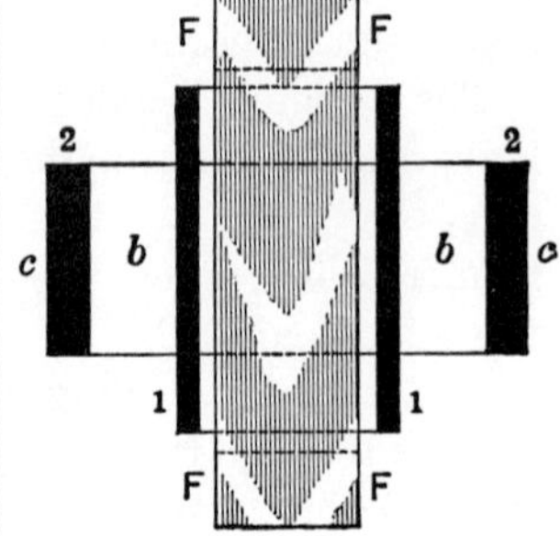

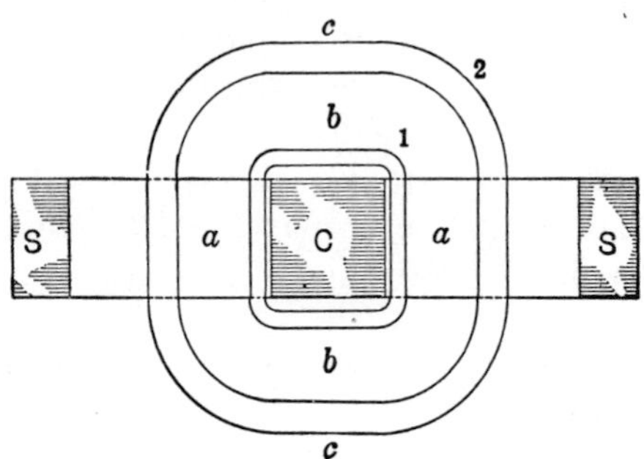

Fig. 108.

Primary reactance,

$$x_0 = \frac{x'}{x + x'}x' = \frac{x'^2}{x + x'}.$$

Secondary reactance,

$$x_1 = \frac{x}{x + x'}x' = \frac{xx'}{x + x'}.$$

116. By test, the two total reactances, x and x', can be derived by considering, that in Fig. 107 at the moments, f and d, the total flux is leakage flux, as more fully shown in Fig. 105f and 105d, and the flux measured from f, gives the reactance, x, measured from d, gives the reactance, d.

Assuming we connect primary coil and secondary coil in series with each other, but in opposition, into an alternating-current circuit, as shown in Fig. 109, and vary the number of primary and secondary turns, until the voltage, e_1, across the secondary coil, s, becomes equal to $r_1 i$. Then no flux passes through the secondary coil, that is, the condition, Fig. 107f, exists, and the voltage, e_0, across the primary coil, p, gives the total reactance, x, for p as primary,

$$e_0^2 = i^2 (r_0^2 + x^2).$$

Varying now the number of turns so that the voltage across the primary coil equals its resistance drop, $e_0 = r_0 i$, then the

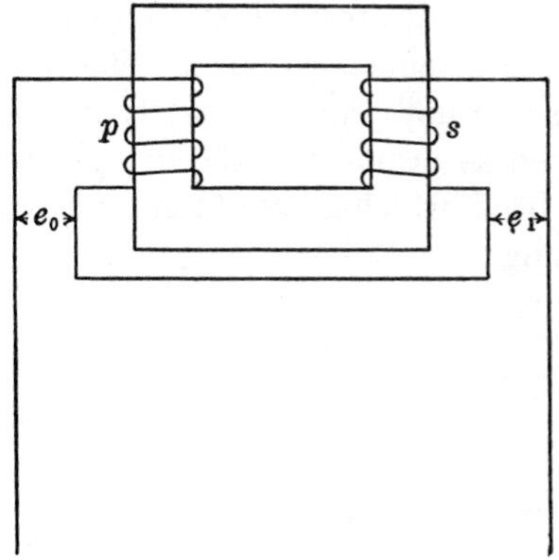

FIG. 109.

voltage across the secondary coil, s, gives the total reactance, x', for s as primary,

$$e_1^2 = i^2 (r_1^2 + x'^2).$$

It would rarely be possible to vary the turns of the two coils, p and s. However, if we short-circuit s and pass an alternating current through p, then at the very low resultant magnetic flux and thus resultant m.m.f., primary and secondary current are practically in opposition and of the same m.m.f., and the magnetic flux in the secondary coil is that giving the resistance drop $r_1 i_1$, that is, $e'_1 = r_1 i_1$ is the true primary voltage in the secondary, and the voltage across the primary terminals thus is that giving primary resistance drop, $r_0 i_0$, total self-inductive reactance, $x i_0$, and the secondary induced voltage, $r_1 i_1$. Thus,

$$e_0^2 = (r_0 i_0 + r_1 i_1)^2 + x^2 i_0^2,$$

or, since i_1 practically equals i_0,

$$e_0^2 = i_0^2 [(r_0 + r_1)^2 + x^2],$$

and inversely, impressing a voltage upon coil, s, and short-circuiting the coil p, gives the leakage reactance, x', for s as primary,

$$e_1^2 = i_1^2 [(r_0 + r_1)^2 + x'^2)].$$

Thus, the so-called "impedance test" of the transformer gives the total leakage reactance $x_0 + x_1$, for that coil as primary, which is used as such in the impedance test.

Where an appreciable difference of the total leakage flux is expected when using the one coil as primary, as when using the other coil, the impedance tests should be made with that coil as primary, which is intended as such. Since, however, the two leakage fluxes are usually approximately equal, it is immaterial which coil is used as primary in the impedance test, and generally that coil is used, which gives a more convenient voltage and current for testing.

INDEX FOR VOLUME I

SOME DOVER SCIENCE BOOKS

SOME DOVER SCIENCE BOOKS

WHAT IS SCIENCE?,
Norman Campbell
This excellent introduction explains scientific method, role of mathematics, types of scientific laws. Contents: 2 aspects of science, science & nature, laws of science, discovery of laws, explanation of laws, measurement & numerical laws, applications of science. 192pp. 5⅜ x 8. 60043-2 Paperbound $1.25

FADS AND FALLACIES IN THE NAME OF SCIENCE,
Martin Gardner
Examines various cults, quack systems, frauds, delusions which at various times have masqueraded as science. Accounts of hollow-earth fanatics like Symmes; Velikovsky and wandering planets; Hoerbiger; Bellamy and the theory of multiple moons; Charles Fort; dowsing, pseudoscientific methods for finding water, ores, oil. Sections on naturopathy, iridiagnosis, zone therapy, food fads, etc. Analytical accounts of Wilhelm Reich and orgone sex energy; L. Ron Hubbard and Dianetics; A. Korzybski and General Semantics; many others. Brought up to date to include Bridey Murphy, others. Not just a collection of anecdotes, but a fair, reasoned appraisal of eccentric theory. Formerly titled *In the Name of Science*. Preface. Index. x + 384pp. 5⅜ x 8.
20394-8 Paperbound $2.00

PHYSICS, THE PIONEER SCIENCE,
L. W. Taylor
First thorough text to place all important physical phenomena in cultural-historical framework; remains best work of its kind. Exposition of physical laws, theories developed chronologically, with great historical, illustrative experiments diagrammed, described, worked out mathematically. Excellent physics text for self-study as well as class work. Vol. 1: Heat, Sound: motion, acceleration, gravitation, conservation of energy, heat engines, rotation, heat, mechanical energy, etc. 211 illus. 407pp. 5⅜ x 8. Vol. 2: Light, Electricity: images, lenses, prisms, magnetism, Ohm's law, dynamos, telegraph, quantum theory, decline of mechanical view of nature, etc. Bibliography. 13 table appendix. Index. 551 illus. 2 color plates. 508pp. 5⅜ x 8.
60565-5, 60566-3 Two volume set, paperbound $5.50

THE EVOLUTION OF SCIENTIFIC THOUGHT FROM NEWTON TO EINSTEIN,
A. d'Abro
Einstein's special and general theories of relativity, with their historical implications, are analyzed in non-technical terms. Excellent accounts of the contributions of Newton, Riemann, Weyl, Planck, Eddington, Maxwell, Lorentz and others are treated in terms of space and time, equations of electromagnetics, finiteness of the universe, methodology of science. 21 diagrams. 482pp. 5⅜ x 8.
20002-7 Paperbound $2.50

CHANCE, LUCK AND STATISTICS: THE SCIENCE OF CHANCE,
Horace C. Levinson

Theory of probability and science of statistics in simple, non-technical language. Part I deals with theory of probability, covering odd superstitions in regard to "luck," the meaning of betting odds, the law of mathematical expectation, gambling, and applications in poker, roulette, lotteries, dice, bridge, and other games of chance. Part II discusses the misuse of statistics, the concept of statistical probabilities, normal and skew frequency distributions, and statistics applied to various fields—birth rates, stock speculation, insurance rates, advertising, etc. "Presented in an easy humorous style which I consider the best kind of expository writing," Prof. A. C. Cohen, Industry Quality Control. Enlarged revised edition. Formerly titled *The Science of Chance*. Preface and two new appendices by the author. xiv + 365pp. 5⅜ x 8. 21007-3 Paperbound $2.00

BASIC ELECTRONICS,
prepared by the U.S. Navy Training Publications Center

A thorough and comprehensive manual on the fundamentals of electronics. Written clearly, it is equally useful for self-study or course work for those with a knowledge of the principles of basic electricity. Partial contents: Operating Principles of the Electron Tube; Introduction to Transistors; Power Supplies for Electronic Equipment; Tuned Circuits; Electron-Tube Amplifiers; Audio Power Amplifiers; Oscillators; Transmitters; Transmission Lines; Antennas and Propagation; Introduction to Computers; and related topics. Appendix. Index. Hundreds of illustrations and diagrams. vi + 471pp. 6½ x 9¼.
61076-4 Paperbound $2.95

BASIC THEORY AND APPLICATION OF TRANSISTORS,
prepared by the U.S. Department of the Army

An introductory manual prepared for an army training program. One of the finest available surveys of theory and application of transistor design and operation. Minimal knowledge of physics and theory of electron tubes required. Suitable for textbook use, course supplement, or home study. Chapters: Introduction; fundamental theory of transistors; transistor amplifier fundamentals; parameters, equivalent circuits, and characteristic curves; bias stabilization; transistor analysis and comparison using characteristic curves and charts; audio amplifiers; tuned amplifiers; wide-band amplifiers; oscillators; pulse and switching circuits; modulation, mixing, and demodulation; and additional semiconductor devices. Unabridged, corrected edition. 240 schematic drawings, photographs, wiring diagrams, etc. 2 Appendices. Glossary. Index. 263pp. 6½ x 9¼. 60380-6 Paperbound $1.75

GUIDE TO THE LITERATURE OF MATHEMATICS AND PHYSICS,
N. G. Parke III

Over 5000 entries included under approximately 120 major subject headings of selected most important books, monographs, periodicals, articles in English, plus important works in German, French, Italian, Spanish, Russian (many recently available works). Covers every branch of physics, math, related engineering. Includes author, title, edition, publisher, place, date, number of volumes, number of pages. A 40-page introduction on the basic problems of research and study provides useful information on the organization and use of libraries, the psychology of learning, etc. This reference work will save you hours of time. 2nd revised edition. Indices of authors, subjects, 464pp. 5⅜ x 8.
60447-0 Paperbound $2.75

A Source Book in Mathematics,
D. E. Smith
Great discoveries in math, from Renaissance to end of 19th century, in English translation. Read announcements by Dedekind, Gauss, Delamain, Pascal, Fermat, Newton, Abel, Lobachevsky, Bolyai, Riemann, De Moivre, Legendre, Laplace, others of discoveries about imaginary numbers, number congruence, slide rule, equations, symbolism, cubic algebraic equations, non-Euclidean forms of geometry, calculus, function theory, quaternions, etc. Succinct selections from 125 different treatises, articles, most unavailable elsewhere in English. Each article preceded by biographical introduction. Vol. I: Fields of Number, Algebra. Index. 32 illus. 338pp. 5⅜ x 8. Vol. II: Fields of Geometry, Probability, Calculus, Functions, Quaternions. 83 illus. 432pp. 5⅜ x 8.
60552-3, 60553-1 Two volume set, paperbound $5.00

Foundations of Physics,
R. B. Lindsay & H. Margenau
Excellent bridge between semi-popular works & technical treatises. A discussion of methods of physical description, construction of theory; valuable for physicist with elementary calculus who is interested in ideas that give meaning to data, tools of modern physics. Contents include symbolism; mathematical equations; space & time foundations of mechanics; probability; physics & continua; electron theory; special & general relativity; quantum mechanics; causality. "Thorough and yet not overdetailed. Unreservedly recommended," *Nature* (London). Unabridged, corrected edition. List of recommended readings. 35 illustrations. xi + 537pp. 5⅜ x 8. 60377-6 Paperbound $3.50

Fundamental Formulas of Physics,
ed. by D. H. Menzel
High useful, full, inexpensive reference and study text, ranging from simple to highly sophisticated operations. Mathematics integrated into text—each chapter stands as short textbook of field represented. Vol. 1: Statistics, Physical Constants, Special Theory of Relativity, Hydrodynamics, Aerodynamics, Boundary Value Problems in Math, Physics, Viscosity, Electromagnetic Theory, etc. Vol. 2: Sound, Acoustics, Geometrical Optics, Electron Optics, High-Energy Phenomena, Magnetism, Biophysics, much more. Index. Total of 800pp. 5⅜ x 8.
60595-7, 60596-5 Two volume set, paperbound $4.75

Theoretical Physics,
A. S. Kompaneyets
One of the very few thorough studies of the subject in this price range. Provides advanced students with a comprehensive theoretical background. Especially strong on recent experimentation and developments in quantum theory. Contents: Mechanics (Generalized Coordinates, Lagrange's Equation, Collision of Particles, etc.), Electrodynamics (Vector Analysis, Maxwell's equations, Transmission of Signals, Theory of Relativity, etc.), Quantum Mechanics (the Inadequacy of Classical Mechanics, the Wave Equation, Motion in a Central Field, Quantum Theory of Radiation, Quantum Theories of Dispersion and Scattering, etc.), and Statistical Physics (Equilibrium Distribution of Molecules in an Ideal Gas, Boltzmann Statistics, Bose and Fermi Distribution. Thermodynamic Quantities, etc.). Revised to 1961. Translated by George Yankovsky, authorized by Kompaneyets. 137 exercises. 56 figures. 529pp. 5⅜ x 8½.
60972-3 Paperbound $3.50

College Algebra, *H. B. Fine*
Standard college text that gives a systematic and deductive structure to algebra; comprehensive, connected, with emphasis on theory. Discusses the commutative, associative, and distributive laws of number in unusual detail, and goes on with undetermined coefficients, quadratic equations, progressions, logarithms, permutations, probability, power series, and much more. Still most valuable elementary-intermediate text on the science and structure of algebra. Index. 1560 problems, all with answers. x + 631pp. 5⅜ x 8. 60211-7 Paperbound $2.75

Higher Mathematics for Students of Chemistry and Physics, *J. W. Mellor*
Not abstract, but practical, building its problems out of familiar laboratory material, this covers differential calculus, coordinate, analytical geometry, functions, integral calculus, infinite series, numerical equations, differential equations, Fourier's theorem, probability, theory of errors, calculus of variations, determinants. "If the reader is not familiar with this book, it will repay him to examine it," *Chem. & Engineering News.* 800 problems. 189 figures. Bibliography. xxi + 641pp. 5⅜ x 8. 60193-5 Paperbound $3.50

Trigonometry Refresher for Technical Men, *A. A. Klaf*
A modern question and answer text on plane and spherical trigonometry. Part I covers plane trigonometry: angles, quadrants, trigonometrical functions, graphical representation, interpolation, equations, logarithms, solution of triangles, slide rules, etc. Part II discusses applications to navigation, surveying, elasticity, architecture, and engineering. Small angles, periodic functions, vectors, polar coordinates, De Moivre's theorem, fully covered. Part III is devoted to spherical trigonometry and the solution of spherical triangles, with applications to terrestrial and astronomical problems. Special time-savers for numerical calculation. 913 questions answered for you! 1738 problems; answers to odd numbers. 494 figures. 14 pages of functions, formulae. Index. x + 629pp. 5⅜ x 8. 20371-9 Paperbound $3.00

Calculus Refresher for Technical Men, *A. A. Klaf*
Not an ordinary textbook but a unique refresher for engineers, technicians, and students. An examination of the most important aspects of differential and integral calculus by means of 756 key questions. Part I covers simple differential calculus: constants, variables, functions, increments, derivatives, logarithms, curvature, etc. Part II treats fundamental concepts of integration: inspection, substitution, transformation, reduction, areas and volumes, mean value, successive and partial integration, double and triple integration. Stresses practical aspects! A 50 page section gives applications to civil and nautical engineering, electricity, stress and strain, elasticity, industrial engineering, and similar fields. 756 questions answered. 556 problems; solutions to odd numbers. 36 pages of constants, formulae. Index. v + 431pp. 5⅜ x 8. 20370-0 Paperbound $2.25

Introduction to the Theory of Groups of Finite Order, *R. Carmichael*
Examines fundamental theorems and their application. Beginning with sets, systems, permutations, etc., it progresses in easy stages through important types of groups: Abelian, prime power, permutation, etc. Except 1 chapter where matrices are desirable, no higher math needed. 783 exercises, problems. Index. xvi + 447pp. 5⅜ x 8. 60300-8 Paperbound $3.00

FIVE VOLUME "THEORY OF FUNCTIONS" SET BY KONRAD KNOPP

This five-volume set, prepared by Konrad Knopp, provides a complete and readily followed account of theory of functions. Proofs are given concisely, yet without sacrifice of completeness or rigor. These volumes are used as texts by such universities as M.I.T., University of Chicago, N. Y. City College, and many others. "Excellent introduction . . . remarkably readable, concise, clear, rigorous," *Journal of the American Statistical Association.*

ELEMENTS OF THE THEORY OF FUNCTIONS,
Konrad Knopp
This book provides the student with background for further volumes in this set, or texts on a similar level. Partial contents: foundations, system of complex numbers and the Gaussian plane of numbers, Riemann sphere of numbers, mapping by linear functions, normal forms, the logarithm, the cyclometric functions and binomial series. "Not only for the young student, but also for the student who knows all about what is in it," *Mathematical Journal.* Bibliography. Index. 140pp. 5⅜ x 8. 60154-4 Paperbound $1.50

THEORY OF FUNCTIONS, PART I,
Konrad Knopp
With volume II, this book provides coverage of basic concepts and theorems. Partial contents: numbers and points, functions of a complex variable, integral of a continuous function, Cauchy's integral theorem, Cauchy's integral formulae, series with variable terms, expansion of analytic functions in power series, analytic continuation and complete definition of analytic functions, entire transcendental functions, Laurent expansion, types of singularities. Bibliography. Index. vii + 146pp. 5⅜ x 8. 60156-0 Paperbound $1.50

THEORY OF FUNCTIONS, PART II,
Konrad Knopp
Application and further development of general theory, special topics. Single valued functions. Entire, Weierstrass, Meromorphic functions. Riemann surfaces. Algebraic functions. Analytical configuration, Riemann surface. Bibliography. Index. x + 150pp. 5⅜ x 8. 60157-9 Paperbound $1.50

PROBLEM BOOK IN THE THEORY OF FUNCTIONS, VOLUME 1.
Konrad Knopp
Problems in elementary theory, for use with Knopp's *Theory of Functions,* or any other text, arranged according to increasing difficulty. Fundamental concepts, sequences of numbers and infinite series, complex variable, integral theorems, development in series, conformal mapping. 182 problems. Answers. viii + 126pp. 5⅜ x 8. 60158-7 Paperbound $1.50

PROBLEM BOOK IN THE THEORY OF FUNCTIONS, VOLUME 2,
Konrad Knopp
Advanced theory of functions, to be used either with Knopp's *Theory of Functions,* or any other comparable text. Singularities, entire & meromorphic functions, periodic, analytic, continuation, multiple-valued functions, Riemann surfaces, conformal mapping. Includes a section of additional elementary problems. "The difficult task of selecting from the immense material of the modern theory of functions the problems just within the reach of the beginner is here masterfully accomplished," *Am. Math. Soc.* Answers. 138pp. 5⅜ x 8.
60159-5 Paperbound $1.50

NUMERICAL SOLUTIONS OF DIFFERENTIAL EQUATIONS,
H. Levy & E. A. Baggott

Comprehensive collection of methods for solving ordinary differential equations of first and higher order. All must pass 2 requirements: easy to grasp and practical, more rapid than school methods. Partial contents: graphical integration of differential equations, graphical methods for detailed solution. Numerical solution. Simultaneous equations and equations of 2nd and higher orders. "Should be in the hands of all in research in applied mathematics, teaching," *Nature*. 21 figures. viii + 238pp. 5⅜ x 8. 60168-4 Paperbound $1.85

ELEMENTARY STATISTICS, WITH APPLICATIONS IN MEDICINE AND THE BIOLOGICAL SCIENCES, *F. E. Croxton*

A sound introduction to statistics for anyone in the physical sciences, assuming no prior acquaintance and requiring only a modest knowledge of math. All basic formulas carefully explained and illustrated; all necessary reference tables included. From basic terms and concepts, the study proceeds to frequency distribution, linear, non-linear, and multiple correlation, skewness, kurtosis, etc. A large section deals with reliability and significance of statistical methods. Containing concrete examples from medicine and biology, this book will prove unusually helpful to workers in those fields who increasingly must evaluate, check, and interpret statistics. Formerly titled "Elementary Statistics with Applications in Medicine." 101 charts. 57 tables. 14 appendices. Index. vi + 376pp. 5⅜ x 8. 60506-X Paperbound $2.25

INTRODUCTION TO SYMBOLIC LOGIC,
S. Langer

No special knowledge of math required—probably the clearest book ever written on symbolic logic, suitable for the layman, general scientist, and philosopher. You start with simple symbols and advance to a knowledge of the Boole-Schroeder and Russell-Whitehead systems. Forms, logical structure, classes, the calculus of propositions, logic of the syllogism, etc. are all covered. "One of the clearest and simplest introductions," *Mathematics Gazette*. Second enlarged, revised edition. 368pp. 5⅜ x 8. 60164-1 Paperbound $2.25

A SHORT ACCOUNT OF THE HISTORY OF MATHEMATICS,
W. W. R. Ball

Most readable non-technical history of mathematics treats lives, discoveries of every important figure from Egyptian, Phoenician, mathematicians to late 19th century. Discusses schools of Ionia, Pythagoras, Athens, Cyzicus, Alexandria, Byzantium, systems of numeration; primitive arithmetic; Middle Ages, Renaissance, including Arabs, Bacon, Regiomontanus, Tartaglia, Cardan, Stevinus, Galileo, Kepler; modern mathematics of Descartes, Pascal, Wallis, Huygens, Newton, Leibnitz, d'Alembert, Euler, Lambert, Laplace, Legendre, Gauss, Hermite, Weierstrass, scores more. Index. 25 figures. 546pp. 5⅜ x 8. 20630-0 Paperbound $2.75

INTRODUCTION TO NONLINEAR DIFFERENTIAL AND INTEGRAL EQUATIONS,
Harold T. Davis

Aspects of the problem of nonlinear equations, transformations that lead to equations solvable by classical means, results in special cases, and useful generalizations. Thorough, but easily followed by mathematically sophisticated reader who knows little about non-linear equations. 137 problems for student to solve. xv + 566pp. 5⅜ x 8½. 60971-5 Paperbound $2.75

AN INTRODUCTION TO THE GEOMETRY OF N DIMENSIONS, *D. H. Y. Sommerville*
An introduction presupposing no prior knowledge of the field, the only book in English devoted exclusively to higher dimensional geometry. Discusses fundamental ideas of incidence, parallelism, perpendicularity, angles between linear space; enumerative geometry; analytical geometry from projective and metric points of view; polytopes; elementary ideas in analysis situs; content of hyper-spacial figures. Bibliography. Index. 60 diagrams. 196pp. 5⅜ x 8.
60494-2 Paperbound $1.50

ELEMENTARY CONCEPTS OF TOPOLOGY, *P. Alexandroff*
First English translation of the famous brief introduction to topology for the beginner or for the mathematician not undertaking extensive study. This unusually useful intuitive approach deals primarily with the concepts of complex, cycle, and homology, and is wholly consistent with current investigations. Ranges from basic concepts of set-theoretic topology to the concept of Betti groups. "Glowing example of harmony between intuition and thought," David Hilbert. Translated by A. E. Farley. Introduction by D. Hilbert. Index. 25 figures. 73pp. 5⅜ x 8. 60747-X Paperbound $1.25

ELEMENTS OF NON-EUCLIDEAN GEOMETRY, *D. M. Y. Sommerville*
Unique in proceeding step-by-step, in the manner of traditional geometry. Enables the student with only a good knowledge of high school algebra and geometry to grasp elementary hyperbolic, elliptic, analytic non-Euclidean geometries; space curvature and its philosophical implications; theory of radical axes; homothetic centres and systems of circles; parataxy and parallelism; absolute measure; Gauss' proof of the defect area theorem; geodesic representation; much more, all with exceptional clarity. 126 problems at chapter endings provide progressive practice and familiarity. 133 figures. Index. xvi + 274pp. 5⅜ x 8. 60460-8 Paperbound $2.00

INTRODUCTION TO THE THEORY OF NUMBERS, *L. E. Dickson*
Thorough, comprehensive approach with adequate coverage of classical literature, an introductory volume beginners can follow. Chapters on divisibility, congruences, quadratic residues & reciprocity. Diophantine equations, etc. Full treatment of binary quadratic forms without usual restriction to integral coefficients. Covers infinitude of primes, least residues. Fermat's theorem. Euler's phi function, Legendre's symbol, Gauss's lemma, automorphs, reduced forms, recent theorems of Thue & Siegel, many more. Much material not readily available elsewhere. 239 problems. Index. I figure. viii + 183pp. 5⅜ x 8.
60342-3 Paperbound $1.75

MATHEMATICAL TABLES AND FORMULAS, *compiled by Robert D. Carmichael and Edwin R. Smith*
Valuable collection for students, etc. Contains all tables necessary in college algebra and trigonometry, such as five-place common logarithms, logarithmic sines and tangents of small angles, logarithmic trigonometric functions, natural trigonometric functions, four-place antilogarithms, tables for changing from sexagesimal to circular and from circular to sexagesimal measure of angles, etc. Also many tables and formulas not ordinarily accessible, including powers, roots, and reciprocals, exponential and hyperbolic functions, ten-place logarithms of prime numbers, and formulas and theorems from analytical and elementary geometry and from calculus. Explanatory introduction. viii + 269pp. 5⅜ x 8½. 60111-0 Paperbound $1.50

MATHEMATICAL PHYSICS, *D. H. Menzel*
Thorough one-volume treatment of the mathematical techniques vital for classical mechanics, electromagnetic theory, quantum theory, and relativity. Written by the Harvard Professor of Astrophysics for junior, senior, and graduate courses, it gives clear explanations of all those aspects of function theory, vectors, matrices, dyadics, tensors, partial differential equations, etc., necessary for the understanding of the various physical theories. Electron theory, relativity, and other topics seldom presented appear here in considerable detail. Scores of definition, conversion factors, dimensional constants, etc. "More detailed than normal for an advanced text . . . excellent set of sections on Dyadics, Matrices, and Tensors," *Journal of the Franklin Institute.* Index. 193 problems, with answers. x + 412pp. 5⅜ x 8. 60056-4 Paperbound $2.50

THE THEORY OF SOUND, *Lord Rayleigh*
Most vibrating systems likely to be encountered in practice can be tackled successfully by the methods set forth by the great Nobel laureate, Lord Rayleigh. Complete coverage of experimental, mathematical aspects of sound theory. Partial contents: Harmonic motions, vibrating systems in general, lateral vibrations of bars, curved plates or shells, applications of Laplace's functions to acoustical problems, fluid friction, plane vortex-sheet, vibrations of solid bodies, etc. This is the first inexpensive edition of this great reference and study work. Bibliography, Historical introduction by R. B. Lindsay. Total of 1040pp. 97 figures. 5⅜ x 8. 60292-3, 60293-1 Two volume set, paperbound $6.00

HYDRODYNAMICS, *Horace Lamb*
Internationally famous complete coverage of standard reference work on dynamics of liquids & gases. Fundamental theorems, equations, methods, solutions, background, for classical hydrodynamics. Chapters include Equations of Motion, Integration of Equations in Special Gases, Irrotational Motion, Motion of Liquid in 2 Dimensions, Motion of Solids through Liquid-Dynamical Theory, Vortex Motion, Tidal Waves, Surface Waves, Waves of Expansion, Viscosity, Rotating Masses of Liquids. Excellently planned, arranged; clear, lucid presentation. 6th enlarged, revised edition. Index. Over 900 footnotes, mostly bibliographical. 119 figures. xv + 738pp. 6⅛ x 9¼. 60256-7 Paperbound $4.00

DYNAMICAL THEORY OF GASES, *James Jeans*
Divided into mathematical and physical chapters for the convenience of those not expert in mathematics, this volume discusses the mathematical theory of gas in a steady state, thermodynamics, Boltzmann and Maxwell, kinetic theory, quantum theory, exponentials, etc. 4th enlarged edition, with new material on quantum theory, quantum dynamics, etc. Indexes. 28 figures. 444pp. 6⅛ x 9¼. 60136-6 Paperbound $2.75

THERMODYNAMICS, *Enrico Fermi*
Unabridged reproduction of 1937 edition. Elementary in treatment; remarkable for clarity, organization. Requires no knowledge of advanced math beyond calculus, only familiarity with fundamentals of thermometry, calorimetry. Partial Contents: Thermodynamic systems; First & Second laws of thermodynamics; Entropy; Thermodynamic potentials: phase rule, reversible electric cell; Gaseous reactions: van't Hoff reaction box, principle of LeChatelier; Thermodynamics of dilute solutions: osmotic & vapor pressures, boiling & freezing points; Entropy constant. Index. 25 problems. 24 illustrations. x + 160pp. 5⅜ x 8. 60361-X Paperbound $2.00

CELESTIAL OBJECTS FOR COMMON TELESCOPES,
Rev. T. W. Webb
Classic handbook for the use and pleasure of the amateur astronomer. Of inestimable aid in locating and identifying thousands of celestial objects. Vol I, The Solar System: discussions of the principle and operation of the telescope, procedures of observations and telescope-photography, spectroscopy, etc., precise location information of sun, moon, planets, meteors. Vol. II, The Stars: alphabetical listing of constellations, information on double stars, clusters, stars with unusual spectra, variables, and nebulae, etc. Nearly 4,000 objects noted. Edited and extensively revised by Margaret W. Mayall, director of the American Assn. of Variable Star Observers. New Index by Mrs. Mayall giving the location of all objects mentioned in the text for Epoch 2000. New Precession Table added. New appendices on the planetary satellites, constellation names and abbreviations, and solar system data. Total of 46 illustrations. Total of xxxix + 606pp. 5⅜ x 8. 20917-2, 20918-0 Two volume set, paperbound $5.00

PLANETARY THEORY,
E. W. Brown and C. A. Shook
Provides a clear presentation of basic methods for calculating planetary orbits for today's astronomer. Begins with a careful exposition of specialized mathematical topics essential for handling perturbation theory and then goes on to indicate how most of the previous methods reduce ultimately to two general calculation methods: obtaining expressions either for the coordinates of planetary positions or for the elements which determine the perturbed paths. An example of each is given and worked in detail. Corrected edition. Preface. Appendix. Index. xii + 302pp. 5⅜ x 8½. 61133-7 Paperbound $2.25

STAR NAMES AND THEIR MEANINGS,
Richard Hinckley Allen
An unusual book documenting the various attributions of names to the individual stars over the centuries. Here is a treasure-house of information on a topic not normally delved into even by professional astronomers; provides a fascinating background to the stars in folk-lore, literary references, ancient writings, star catalogs and maps over the centuries. Constellation-by-constellation analysis covers hundreds of stars and other asterisms, including the Pleiades, Hyades, Andromedan Nebula, etc. Introduction. Indices. List of authors and authorities. xx + 563pp. 5⅜ x 8½. 21079-0 Paperbound $3.00

A SHORT HISTORY OF ASTRONOMY, *A. Berry*
Popular standard work for over 50 years, this thorough and accurate volume covers the science from primitive times to the end of the 19th century. After the Greeks and the Middle Ages, individual chapters analyze Copernicus, Brahe, Galileo, Kepler, and Newton, and the mixed reception of their discoveries. Post-Newtonian achievements are then discussed in unusual detail: Halley, Bradley, Lagrange, Laplace, Herschel, Bessel, etc. 2 Indexes. 104 illustrations, 9 portraits. xxxi + 440pp. 5⅜ x 8. 20210-0 Paperbound $2.75

SOME THEORY OF SAMPLING, *W. E. Deming*
The purpose of this book is to make sampling techniques understandable to and useable by social scientists, industrial managers, and natural scientists who are finding statistics increasingly part of their work. Over 200 exercises, plus dozens of actual applications. 61 tables. 90 figs. xix + 602pp. 5⅜ x 8½.
61755-6 Paperbound $3.50

Applied Optics and Optical Design,
A. E. Conrady
With publication of vol. 2, standard work for designers in optics is now complete for first time. Only work of its kind in English; only detailed work for practical designer and self-taught. Requires, for bulk of work, no math above trig. Step-by-step exposition, from fundamental concepts of geometrical, physical optics, to systematic study, design, of almost all types of optical systems. Vol. 1: all ordinary ray-tracing methods; primary aberrations; necessary higher aberration for design of telescopes, low-power microscopes, photographic equipment. Vol. 2: (Completed from author's notes by R. Kingslake, Dir. Optical Design, Eastman Kodak.) Special attention to high-power microscope, anastigmatic photographic objectives. "An indispensable work," *J., Optical Soc. of Amer.* Index. Bibliography. 193 diagrams. 852pp. 6⅛ x 9¼.
60611-2, 60612-0 Two volume set, paperbound $8.00

Mechanics of the Gyroscope, the Dynamics of Rotation,
R. F. Deimel, Professor of Mechanical Engineering at Stevens Institute of Technology
Elementary general treatment of dynamics of rotation, with special application of gyroscopic phenomena. No knowledge of vectors needed. Velocity of a moving curve, acceleration to a point, general equations of motion, gyroscopic horizon, free gyro, motion of discs, the damped gyro, 103 similar topics. Exercises. 75 figures. 208pp. 5⅜ x 8. 60066-1 Paperbound $1.75

Strength of Materials,
J. P. Den Hartog
Full, clear treatment of elementary material (tension, torsion, bending, compound stresses, deflection of beams, etc.), plus much advanced material on engineering methods of great practical value: full treatment of the Mohr circle, lucid elementary discussions of the theory of the center of shear and the "Myosotis" method of calculating beam deflections, reinforced concrete, plastic deformations, photoelasticity, etc. In all sections, both general principles and concrete applications are given. Index. 186 figures (160 others in problem section). 350 problems, all with answers. List of formulas. viii + 323pp. 5⅜ x 8.
60755-0 Paperbound $2.50

Hydraulic Transients,
G. R. Rich
The best text in hydraulics ever printed in English . . . by former Chief Design Engineer for T.V.A. Provides a transition from the basic differential equations of hydraulic transient theory to the arithmetic integration computation required by practicing engineers. Sections cover Water Hammer, Turbine Speed Regulation, Stability of Governing, Water-Hammer Pressures in Pump Discharge Lines, The Differential and Restricted Orifice Surge Tanks, The Normalized Surge Tank Charts of Calame and Gaden, Navigation Locks, Surges in Power Canals—Tidal Harmonics, etc. Revised and enlarged. Author's prefaces. Index. xiv + 409pp. 5⅜ x 8½. 60116-1 Paperbound $2.50

Prices subject to change without notice.

Available at your book dealer or write for free catalogue to Dept. Adsci, Dover Publications, Inc., 180 Varick St., N.Y., N.Y. 10014. Dover publishes more than 150 books each year on science, elementary and advanced mathematics, biology, music, art, literary history, social sciences and other areas.